Springer Advanced Texts in Chemistry

Charles R. Cantor, Editor

Hermann Dugas
Christopher Penney

Bioorganic Chemistry

A Chemical Approach to Enzyme Action

With 82 Figures

Springer-Verlag
New York   Heidelberg   Berlin

Dr. Hermann Dugas
Département de Chimie
Université de Montréal
Montréal, Québec
Canada H3C 3V1

Dr. Christopher Penney
Connaught Research Institute
Willowdale, Ontario
Canada M2N 5T8

Series Editor:
Prof. Charles R. Cantor
Columbia University
Box 608 Havemeyer Hall
New York, New York 10027 USA

Cover: The green illustration represents the hypothetical mode of binding of a rigid structural analogue of *N*-benzoyl-ʟ-phenylalanine methyl ester at the active site of α-chymotrypsin. The illustration emphasizes the equilibration toward the favored configuration (see text page 224). The background design is taken from a diagrammatic representation of the primary structure of α-chymotrypsin. After *Nature* with permission [B.W. Matthews, P.B. Sigler, R. Henderson, and D.M. Blow (1967), *Nature* **214**, 652−656].

Library of Congress Cataloging in Publication Data

Dugas, Hermann, 1942−
  Bioorganic chemistry.

  (Springer advanced texts in chemistry.)
  Bibliography: p.
  Includes index.
  1. Enzymes. 2. Biological chemistry.
3. Chemistry, Organic. I. Penney, Christopher, 1950−
joint author. II. Title. III. Series.
[DNLM: 1. Biochemistry. 2. Enzymes—Metabolism.
QU135 D866b]
QP601.D78   574.19′25   80-16222

© 1981 by Springer-Verlag New York Inc. 107667

Printed in the United States of America

9 8 7 6 5 4 3 2 1

ISBN 0-387-90491-3 Springer-Verlag New York Heidelberg Berlin
ISBN 3-540-90491-3 Springer-Verlag Berlin Heidelberg New York

Series Preface
Springer Advanced Texts in Chemistry

New textbooks at all levels of chemistry appear with great regularity. Some fields like basic biochemistry, organic reaction mechanisms, and chemical thermodynamics are well represented by many excellent texts, and new or revised editions are published sufficiently often to keep up with progress in research. However, some areas of chemistry, especially many of those taught at the graduate level, suffer from a real lack of up-to-date textbooks. The most serious needs occur in fields that are rapidly changing. Textbooks in these subjects usually have to be written by scientists actually involved in the research which is advancing the field. It is not often easy to persuade such individuals to set time aside to help spread the knowledge they have accumulated. Our goal, in this series, is to pinpoint areas of chemistry where recent progress has outpaced what is covered in any available textbooks, and then seek out and persuade experts in these fields to produce relatively concise but instructive introductions to their fields. These should serve the needs of one semester or one quarter graduate courses in chemistry and biochemistry. In some cases the availability of texts in active research areas should help stimulate the creation of new courses.

New York, New York   CHARLES R. CANTOR

Foreword

In the early 1960s, while at the University of Ottawa, my colleagues of the Chemistry Department agreed that the long-term future of organic chemistry lay in its applications to biochemical problems, apart from its eventual rationalization through theoretical modeling. Accordingly, I proceeded with the preparation of an undergraduate general biochemistry course specifically designed for the benefit of graduating chemistry students lacking any background in classical, descriptive biochemistry. The pedagogical approach centered chiefly on those organic chemical reactions which best illustrated at the fundamental level their biochemical counterparts. Effective chemical modeling of enzymatic reactions was still in an embryonic state, and over the last fifteen years or so much progress has been made in the development of biomimetic systems. It came as a surprise, as word spread around and as years went by, to witness the massive invasion of my classes by undergraduates majoring in biochemistry and biology, so that often enough the chemistry students were clearly outnumbered. As it turned out, the students had discovered that what they thought they really knew through the process of memorization had left them without any appreciation of the fundamental and universal principles at work and which can be so much more readily perceived through the appropriate use of models. By the time I moved to McGill in 1971, the nature of the course had been gradually transformed into what is now defined as bioorganic chemistry, a self-contained course which has been offered at the undergraduate B.Sc. level for the past ten years. The success of the course is proof that it fills a real need. Over these years, I never found the time to use my numerous scattered notes and references as a basis to produce a textbook (the absence of which is still a source of chronic complaint on the part of the students). Fortunately, the present authors (H.D. and C.P.) had first-hand experience at teaching the course (when I was on leave of absence) and as a result felt encouraged to undertake the heroic task of organizing my telegraphic notes into a framework for a textbook which they are now offering. There is little doubt that what they have accomplished will serve most satisfyingly to fill a very serious need in the modern curricula of undergraduate chemists, biochemists, biologists, and all those contemplating a career in medicinal chemistry and medical research. The field is moving so rapidly, however, that revised editions will have to be produced at relatively short intervals. Nevertheless, the substance and the conceptual approach can only have, it is hoped, lasting value.

Montreal   BERNARD BELLEAU
February 1981   MCGILL UNIVERSITY

Preface

Bioorganic chemistry is the application of the principles and the tools of organic chemistry to the understanding of biological processes. The remarkable expansion of this new discipline in organic chemistry during the last ten years has created a new challenge for the teacher, particularly with respect to undergraduate courses. Indeed, the introduction of many new and valuable bioorganic chemical principles is not a simple task. This book will expound the fundamental principles for the construction of bioorganic molecular models of biochemical processes using the tools of organic and physical chemistry.

This textbook is meant to serve as a *teaching* book. It is not the authors' intention to cover all aspects of bioorganic chemistry. Rather, a blend of general and selected topics are presented to stress important aspects underlying the concepts of organic molecular model building. Most of the presentation is accessible to advanced undergraduate students without the need to go back to an elementary textbook of biochemistry; of course, a working knowledge of organic chemistry is mandatory. Consequently, this textbook is addressed first to final-year undergraduate students in chemistry, biochemistry, biology, and pharmacology. In addition, the text has much to offer in modern material that graduate students are expected to, but seldom actually, know.

Often the material presented in elementary biochemistry courses is overwhelming and seen by many students as mainly a matter of memorization. We hope to overcome this situation. Therefore, the chemical organic presentation throughout the book should help to stimulate students to make the "quantum jump" necessary to go from a level of pure memorization of biochemical transformations to a level of adequate comprehension of biochemical principles based on a firm chemical understanding of bioorganic concepts. For this, most chapters start by asking some of the pertinent questions developed within the chapter. In brief, we hope that this approach will stimulate curiosity.

Professor B. Belleau from McGill University acted as a "catalyst" in promoting the idea to write this book. Most of the material was originally inspired from his notes. The authors would like to express their most sincere appreciation for giving us the opportunity of teaching, transforming, and expanding *his* course into a book. It is Dr. Belleau's influence and remarkable dynamism that gave us constant inspiration and strength throughout the writing.

The references are by no means exhaustive, but are, like the topics chosen, selective. The reader can easily find additional references since many of the citations are of books and review articles. The instructor should have a good knowledge of individual references and be able to offer to the students the possibility of discussing a particular subject in more detail. Often we give the name of the main author concerning the subject presented and the year the work was done. This way the students have the opportunity to know the leader in that particular field and can more readily find appropriate references. However, we apologize to all those who have not been mentioned because of space limitation.

The book includes more material than can be handled in a single course of three hours a week in one semester. However, in every chapter, sections of material may be omitted without loss of continuity. This flexibility allows the instructor to emphasize certain aspects of the book, depending if the course is presented to an audience of chemists or biochemists.

We are indebted to the following friends and colleagues for providing us with expert suggestions and comments regarding the presentation of certain parts of the book: P. Brownbridge, P. Deslongchamps, P. Guthrie, J. B. Jones, R. Kluger, and C. Lipsey. And many thanks to Miss C. Potvin, from the Université de Montréal, for her excellent typing assistance throughout the preparation of this manuscript.

Finally, criticisms and suggestions toward improvement of the content of the text are welcome.

Montreal, Canada   HERMANN DUGAS
January 1981   CHRISTOPHER PENNEY

Contents

Chapter 1

Introduction to Bioorganic Chemistry

"It might be helpful to remind ourselves regularly of the sizeable incompleteness of our understanding, not only of ourselves as individuals and as a group, but also of Nature and the world around us."

N. Hackerman
Science **183**, 907 (1974)

1.1 Basic Considerations

Bioorganic chemistry is a new discipline which is essentially concerned with the application of the tools of chemistry to the understanding of biochemical processes. Such an understanding is often achieved with the aid of *molecular models* chemically synthesized in the laboratory. This allows a "sorting out" of the many variable parameters simultaneously operative within the biological system.

For example, how does a biological membrane work? One builds a simple model of known compositions and studies a single behavior, such as an ion transport property. How does the brain work? This is by far a more complicated system than the previous example. Again one studies single synapses and single synaptic constituents and then uses the observations to construct a model.

Organic chemists develop synthetic methodology to better understand organic mechanisms and create new compounds. On the other hand, biochemists study life processes by means of biochemical methodology (enzyme purification and assay, radioisotopic tracer studies in *in vivo* systems). The former possess the methodology to synthesize biological analogues but often fail to appreciate which synthesis would be relevant. The latter possess an appreciation of what would be useful to synthesize in the laboratory, but not the expertise to pursue the problem. The need for the multidisciplinary approach becomes obvious, and the bioorganic chemist will often have two laboratories: one for synthesis and another for biological study. A new dimension results from this combination of chemical and biological sciences; that is the concept of *model building* to study and sort out the various parameters of a complex biological process. By means of simple organic models, many biological reactions as well as the specificity and efficiency of the enzymes involved have been reproduced in the test tube. The success of many of these models indicates the progress that has been made in understanding the chemistry operative in biological systems. Extrapolation of this multidisciplinary science to the pathological state is a major theme of the pharmaceutical industry; organic chemists and pharmacologists working "side-by-side," so that bioorganic chemistry is to biochemistry as medicinal chemistry is to pharmacology.

What are the tools needed for bioorganic model studies? Organic and physical organic chemical principles will provide, by their very nature, the best opportunities for model building—modeling molecular events which form the basis of life. A large portion of organic chemistry has been classically devoted to natural products. Many of those results have turned out to be wonderful tools for the discovery and characterization of specific molecular events in living systems. Think for instance of the development of antibiotics, certain alkaloids, and the design of new drugs for the medicine of today and tomorrow.

All living processes require energy, which is obtained by performing chemical reactions inside cells. These biochemical processes are based on chemical dynamics and involve reductions and oxidations. Biological oxidations are thus the main source of energy to drive a number of endergonic biological transformations.

Many of the reactions involve combustion of foods such as sugars and lipids to produce energy that is used for a variety of essential functions such as growth, replication, maintenance, muscular work, and heat production. These transformations are also related to oxygen uptake; breathing is a biochemical process by which molecular oxygen is reduced to water. Throughout these pathways, energy is stored in the form of adenosine triphosphate (ATP), an energy-rich compound known as the universal product of energetic transactions.

Part of the energy from the combustion engine in the cell is used to perpetuate the machine. The machine is composed of structural components which must be replicated. Ordinary combustion gives only heat plus some visible light and waste. Biological combustions, however, give some heat but a large portion of the energy is used to drive a "molecular engine" which synthesizes copies of itself and which does mechanical work as well. Since these transformations occur at low temperature (body temp., 37°C) and in aqueous media, catalysts are essential for smooth or rapid energy release and transfer. Hence, apart from structural components, *molecular catalysts* are required.

These catalysts have to be highly efficient (a minimum of waste) and highly specific if precise patterns are to be produced. Structural components have a static role; we are interested here in the dynamics. If bond-breaking and bond-forming reactions are to be performed on a specific starting material, then a suitable specific catalyst capable of recognizing the substrate must be "constructed" around that substrate.

In other words, and this is the fundamental question posed by all biochemical phenomena, a substrate molecule and the specific reaction it must undergo must be translated into another structure of much higher order, whose information content perfectly matches the specifically planned chemical transformation. Only large macromolecules can carry enough *molecular information* both from the point of view of substrate recognition and thermodynamic efficiency of the transformation. *These macromolecules are proteins.* They must be extremely versatile in the physicochemical sense since innumerable substrates of widely divergent chemical and physical properties must all be handled by proteins.

Hence, protein composition must of necessity be amenable to wide variations in order that different substrates may be recognized and handled. Some proteins will even need adjuncts (nonprotein parts) to assist in recognition and transformation. These cofactors are called coenzymes. One can therefore predict that protein catalysts or *enzymes* must have a high degree of order and organization. Further, a minimum size will be essential for all the information to be contained.

These ordered biopolymers, which allow the combustion engine to work and to replicate itself, must also be replicated exactly once a perfect translation of substrate structure into a specific function has been established. Hence the molecular information in the proteins (enzymes) must be safely stored into stable, relatively static language. This is where the nucleic acids enter into the picture. Consequently another translation phenomenon involves protein information content written into a linear molecular language which can be copied and distributed to other cells.

The best way to vary at will the information content of a macromolecule is to use some sort of backbone and to peg on it various arrays of side chains. Each side chain may carry well-defined information regarding interactions between themselves or with a specific substrate in order to perform specific bond-making or -breaking functions. Nucleic acid–protein interactions should also be mentioned because of their fundamental importance in the evolution of the genetic code.

The backbone just mentioned is a polyamide and the pegs are the amino acid side chains. *Why polyamide?* Because it has the capacity of "freezing" the biopolymer backbone into precise three-dimensional patterns. Flexibility is also achieved and is of considerable importance for conformational "breathing" effects to occur. A substrate can therefore be transformed in terms of protein conformation imprints and finally, mechanical energy can also be translocated.

The large variety of organic structures known offer an infinite number of structural and functional properties to a protein. Using water as the translating medium, one can go from nonpolar (structured or nonstructured) to polar (hydrogen bonded) to ionic (solvated) amino acids; from aromatic to aliphatics; from reducible to oxidizable groups. Thus, almost the entire encyclopedia of chemical organic reactions can be coded on a polypeptide backbone and tertiary structure. Finally, since all amino acid present are of L (or S) configuration, we realize that *chirality* is essential for order to exist.

1.2 Proximity Effects in Organic Chemistry

Proximity of reactive groups in a chemical transformation allows bond polarization, resulting generally in an acceleration in the rate of the reaction. In nature this is normally achieved by a well-defined alignment of specific amino acid side chains at the active site of an enzyme.

Study of organic reactions helps to construct proper biomodels of enzymatic reactions and open a field of intensive research: medicinal chemistry through rational drug design. Since a meaningful presentation of all applications of organic reactions would be a prodigious task, we limit the present discussion in this chapter to a few representative examples. These illustrate some of the advantages and problems encountered in conceptualizing bioorganic models for the study of enzyme mechanism. Chapter 4 will give a more complete presentation of the proximity effect in relation to intramolecular catalysis.

The first example is the hydrolysis of a glucoside bond. *o*-Carboxyphenyl β-D-glucoside (**1-1**) is hydrolyzed at a rate 10^4 faster than the corresponding

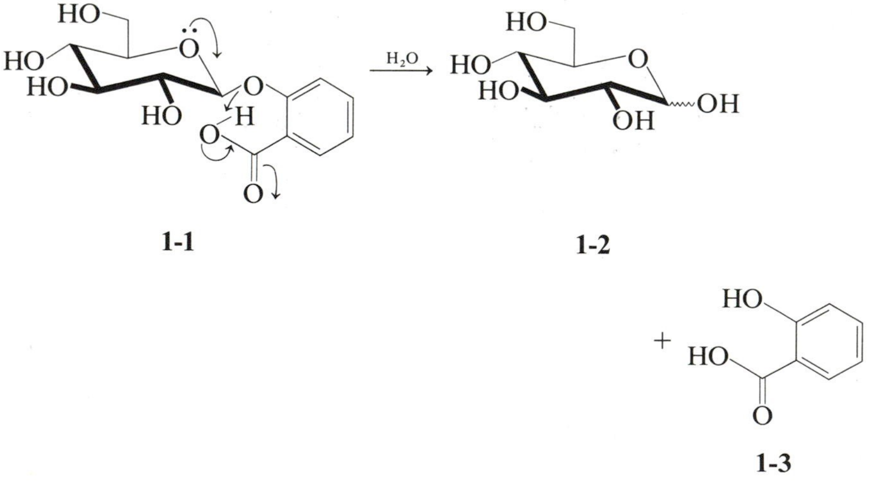

1-1   1-2

1-3

p-carboxyphenyl analogue. Therefore, the carboxylate group in the *ortho* position must "participate" or be involved in the hydrolysis.

This illustrates the fact that the proper positioning of a group (electrophilic or nucleophilic) may accelerate the rate of a reaction. There is thus an analogy to be made with the active site of an enzyme such as lysozyme. Of course the nature of the leaving group is also important in describing the properties. Furthermore, solvation effects can be of paramount importance for the course of the transformation especially in the transition state. Reactions of this type are called *assisted hydrolysis* and occur by an intramolecular displacement mechanism; steric factors may retard the reactions.

Let us look at another example: 2,2'-tolancarboxylic acid (**1-4**) in ethanol is converted to 3-(2-carboxybenzilidene) phthalide (**1-5**). The rate of the reaction is 10^4 faster than with the corresponding 2-tolancarboxylic or 2,4'-tolancarboxylic acid. Consequently, one carboxyl group acts as a general acid catalyst (see Chapter 4) by a mechanism known as *complementary bifunctional catalysis*.

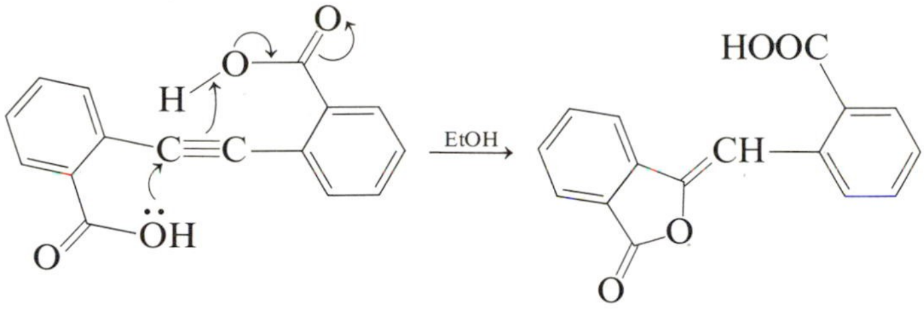

1-4   **1-5**

The ester function of 4-(4'-imidazolyl) butanoic phenyl ester (**1-6**) is hydrolyzed much faster than the corresponding *n*-butanoic phenyl ester. If a *p*-nitro group is present on the aryl residue, the rate of hydrolysis is even faster at neutral pH. As expected, the presence of a better leaving group further accelerates the rate of reaction. This hydrolysis involves the formulation of a tetrahedral intermediate (**1-7**). A detailed discussion of such intermediates

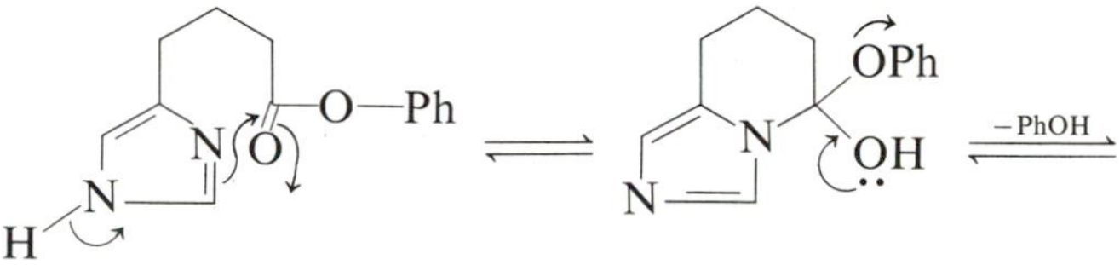

1-6   **1-7**

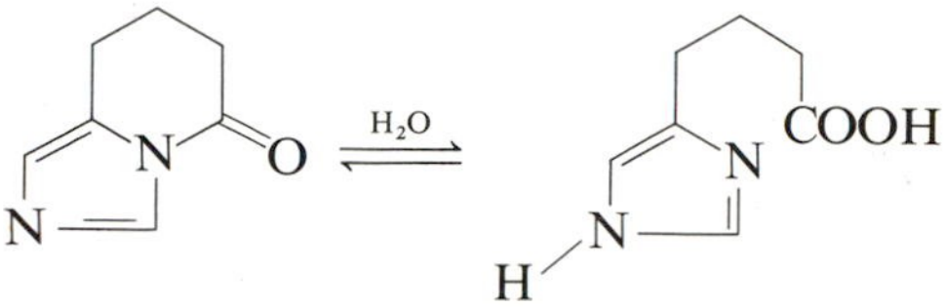

1-8   **1-9**

will be the subject of Chapter 4. The imidazole group acts as a nucleophilic catalyst in this two-step conversion and its proximity to the ester function and the formation of a cyclic intermediate are the factors responsible for the rate enhancement observed. The participation of an imidazole group in the hydrolysis of an ester may represent the simplest model of hydrolytic enzymes.

In a different domain, amide bond hydrolyses can also be accelerated. An example is the following where the reaction is catalyzed by a pyridine ring.

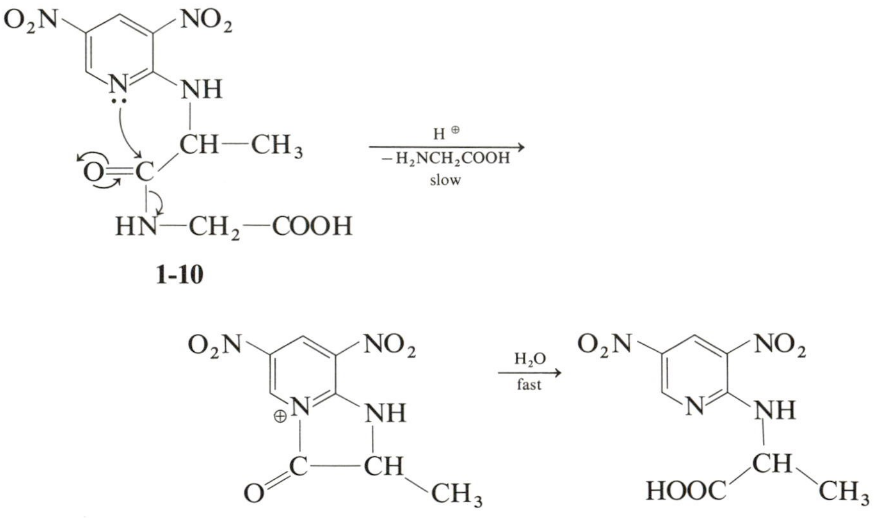

1-10

1-11   **1-12**

The first step is the rate limiting step of the reaction (slow reaction) leading to an acyl pyridinium intermediate (**1-11**), reminiscent of a covalent acyl-enzyme intermediate found in many enzymatic mechanisms. This intermediate is then rapidly trapped by water.

The last example is taken from the steroid field and illustrates the importance of a rigid framework. The solvolysis of acetates (**1-13**) and (**1-14**) in CH_3OH/Et_3N showed a marked preference for the molecule having a β-OH group at carbon 5 where the rate of hydrolysis is 300 times faster.

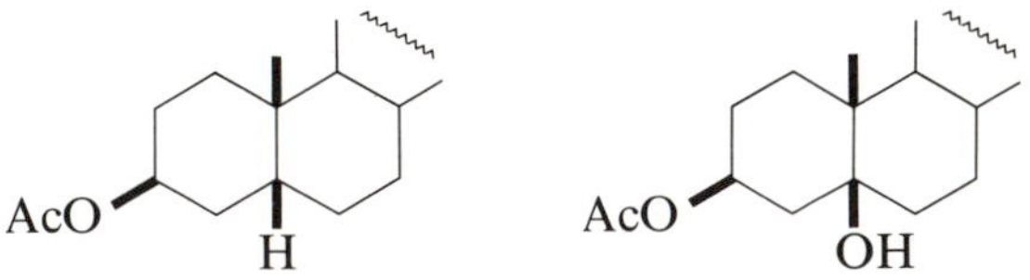

cis junction

1-13   **1-14**

The reason for such a behavior becomes apparent when the molecule is drawn in three-dimensions (**1-15**). The rigidity of the steroid skeleton thus helps in bringing the two functions in proper orientation where catalysis combining one intramolecular and one intermolecular catalyst takes place.

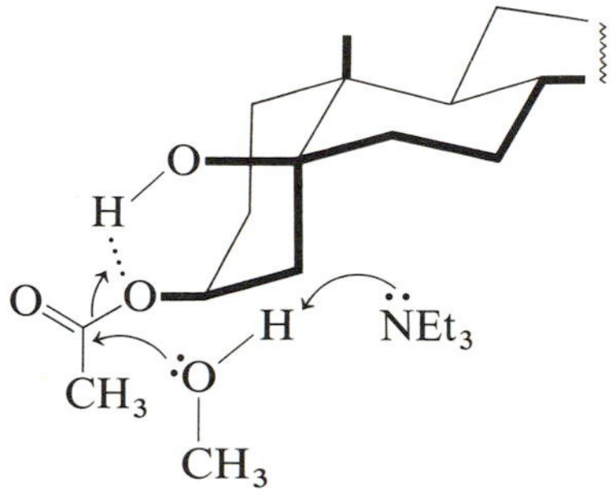

1-15

The proximal hydroxyl group can cooperate in the hydrolysis by hydrogen bonding and the carbonyl function of the ester becomes a better electrophilic center for the solvent molecules. In this mechanism one can perceive a general acid–base catalysis of ester solvolysis (Chapter 4).

These simple examples illustrate that many of the basic active site chemistry of enzymes can be reproduced with simple organic models in the absence of proteins. The role of the latter is of substrate recognition and orientation and the chemistry is often carried out by cofactors (coenzymes) which also have to be specifically recognized by the protein or enzyme. The last chapter of this book is devoted to the chemistry of coenzyme function and design.

1.3 Molecular Adaptation

Other factors besides proximity effects are important and should be considered in the design of biomodels. For instance in 1950, at the *First Symposium on Chemical-Biological Correlation*, H. L. Friedman introduced the concept of *bioisosteric groups* (1). In its broadest sense, the term refers to chemical groups that bear some resemblance in molecular size and shape and as a consequence can compete for the same biological target. This concept has important application in molecular pharmacology, especially in the design of new drugs through the method of variation, or molecular modification (2).

Some pharmacological examples will illustrate the principle. The two neurotransmitters, acetylcholine (**1-16**) and carbachol (**1-17**), have similar muscarinic action.

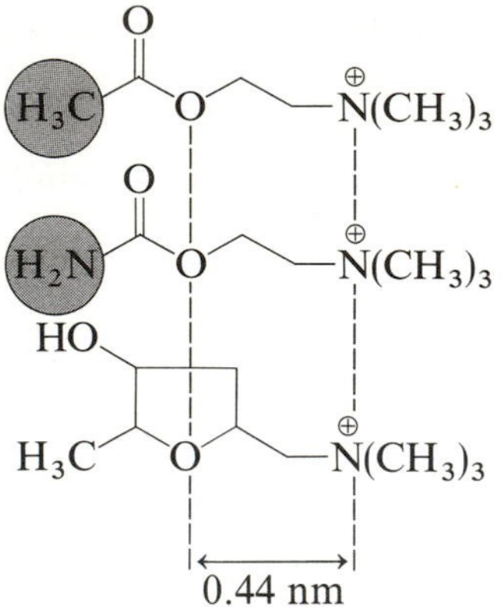

1-16 acetylcholine

1-17 carbachol

1-18 muscarine

0.44 nm

The shaded area represents the bioisosteric equivalence. Muscarine (**1-18**) is an alkaloid which inhibits the action of acetylcholine. It is found for instance in *Amanita muscaria* (Fly Agaric) and other poisonous mushrooms. Its structure infers that, in order to block the action of acetylcholine on receptors of smooth muscles and glandular cells, it must bind in a similar fashion.

5-Fluorocytosine (**1-19**) is an analogue of cytosine (**1-20**) which is commonly used as an antibiotic against bacterial infections. One serious problem in drug design is to develop a therapy that will not harm the patient's tissues but will destroy the infecting cells or bacteria. A novel approach is to "disguise" the drug so that it is chemically modified to gain entry and kill invading microorganisms without affecting normal tissues. The approach involves exploiting a feature that is common to many microorganisms: peptide transport. Hence, the amino function of compound (**1-19**) is chemically joined to a small peptide. This peptide contains D-amino acids and therefore avoids hydrolysis by common human enzymes and entry into human tissues. However, the drug-bearing peptide can sneak into the bacterial cell. It is then metabolized to liberate the active antifungal drug which kills only the invading cell. This is the type of research that the group of A. Steinfeld is undertaking at City University of New York. This principle of using peptides to carry drugs is applicable to many different disease-causing organisms.

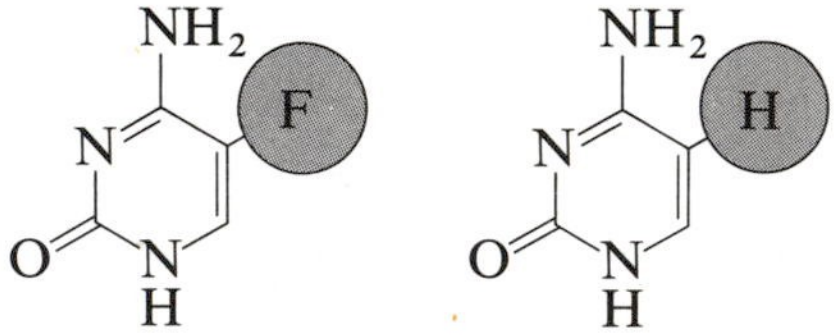

1-19   **1-20**

Similarly, 1-β-D-2'-deoxyribofuranosyl-5-iodo-uracil (**1-21**) is an *antagonist* of 1-β-D-2'-deoxyribofuranosyl thymine, or thymidine (**1-22**). That is, it is able to antagonize or prevent the action of the latter in biological systems, though it may not carry out the same function. Such an altered metabolite is also called an *antimetabolite*.

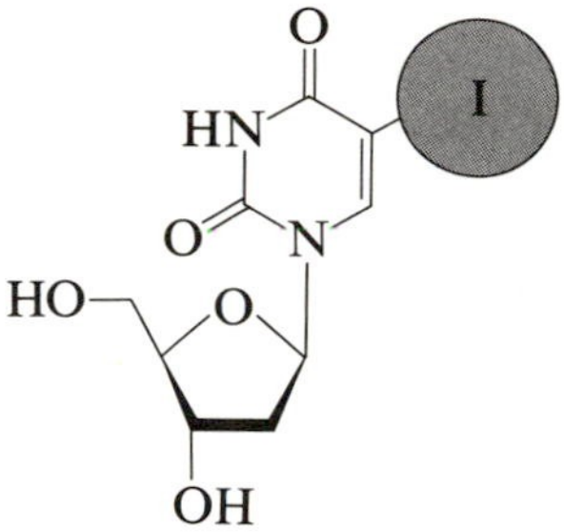

1-21 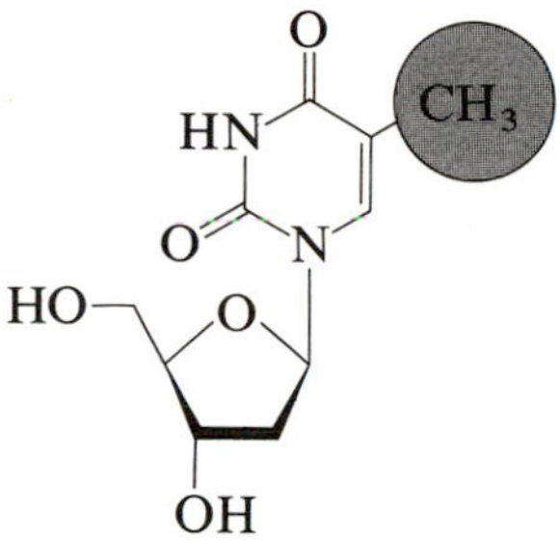   **1-22**

Another example of molecular modification is the synthetic nucleoside adenosine arabinoside (**1-23**). This compound has a pronounced antiviral activity against herpes virus and is therefore widely used in modern chemotherapy.

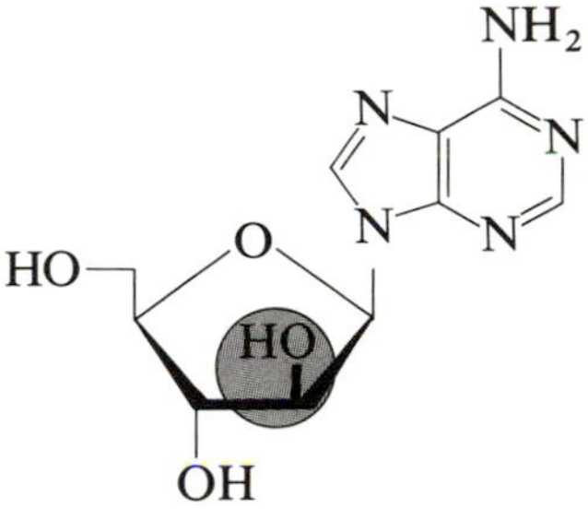

1-23 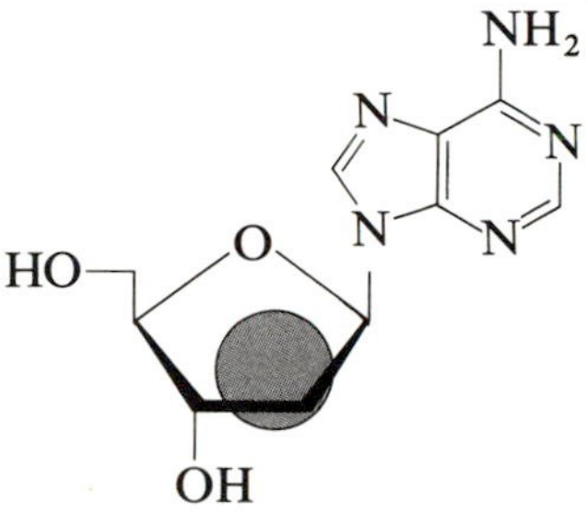   **1-24**

The analogy with deoxyadenosine (**1-24**), a normal component of DNA (see Chapter 3), is striking. Except for the presence of a hydroxyl group at the 2′-sugar position the two molecules are identical. Compared to the ribose ring found in the RNA molecule, it has the inverse or epimeric configuration and hence belongs to the arabinose series. Interestingly, a simple inversion of configuration at C-2′ confers antiviral properties. Its mechanism of action has been well studied and it does act, after phosphorylation, as a potent inhibitor of DNA synthesis (see Chapter 3 for details). Similarly, cytosine arabinoside is the most effective drug for acute myeloblastic leukemia (see Section 3.5).

Most interesting was the finding that this antiviral antibiotic (**1-23**) is in fact produced by a bacterium called *Streptomyces antibioticus*. This allows the production by fermentation of large quantities of this active principle.

A number of organophosphonates have been synthesized as bioisosteric analogues for biochemically important non-nucleoside and nucleoside phosphates (3). For example, the S-enantiomer of 3,4-dihydroxybutyl-1-phosphonic acid (DHBP) has been synthesized as the isosteric analogue of

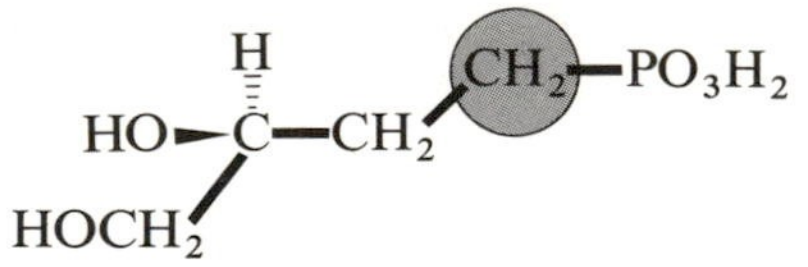

S-DHBP

1-25

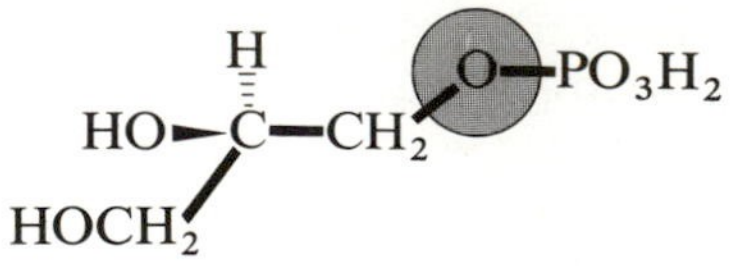

sn-glycerol 3-phosphate

1-26

sn-glycerol 3-phosphate (4). The former material is bacteriostatic at low concentrations to certain strain of *E. coli* and *B. subtilus*. As sn-glycerol 3-phosphate is the backbone of phospholipids (an important cell membrane constituent) and is able to enter into lipid metabolism and the glycolytic pathway, it is sensitive to a number of enzyme mediated processes. The phosphonic acid can participate, but only up to a point, in these cellular reactions. For example, it cannot be hydrolyzed to release glycerol and inorganic phosphate. Of course, the *R*-enantiomer is devoid of biological activity.

The presence of a halogen atom on a molecule sometimes results in interesting properties. For example, substitution of the 9α position by a halogen in cortisone (**1-27**) increases the activity of the hormone by prolonging the half-life of the drug. The activity increases in the following order: X = I > Br > Cl > F > H. These cortisone analogues are employed in the diagnosis and treatment of a variety of disorders of adrenal function and as antiinflammatory agents (2).

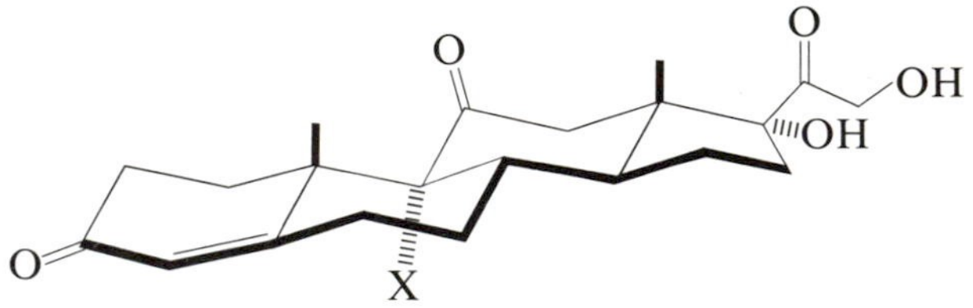

1-27

As another example, the normal thyroid gland is responsible for the synthesis and release of an unusual amino acid called thyroxine (**1-28**). This hormone regulates the rate of cellular oxidative processes (2).

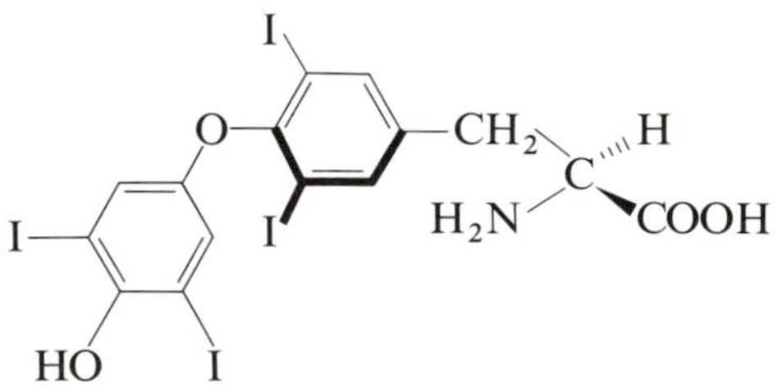

thyroxine

1-28

The presence of the bulky atoms of iodine prevents free rotation around the ether bond and forces the planes of aromatic rings to remain perpendicular to each other. Consequently, it can be inferred that this conformation must be important for its mode of action and it has been suggested that the phenylalanine ring with the two iodines is concerned with binding to the receptor site.

The presence of alkyl groups or chains can also influence the biological activity of a substrate or a drug. An interesting case is the antimalarial compounds derived from 6-methoxy-8-aminoquinoline (**1-29**) (primaquine

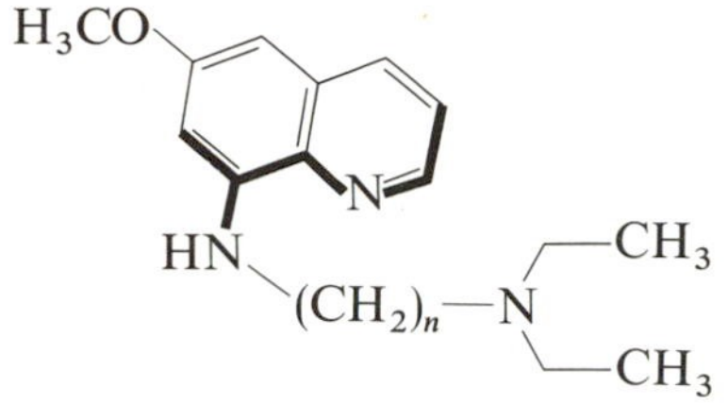

primaquine drug

1-29

analogue). The activity is greater in compounds in which n is an even number in the range of $n = 2$ to 7. So the proper fit of the side chain on a receptor site* or protein is somehow governed by the size and shape of the side chain.

Finally, mention should be made of molecular adaptation at the conformational level. Indeed, many examples can be found among which is the street drug phencyclidine (**1-30**), known as hog (angel dust) by users.

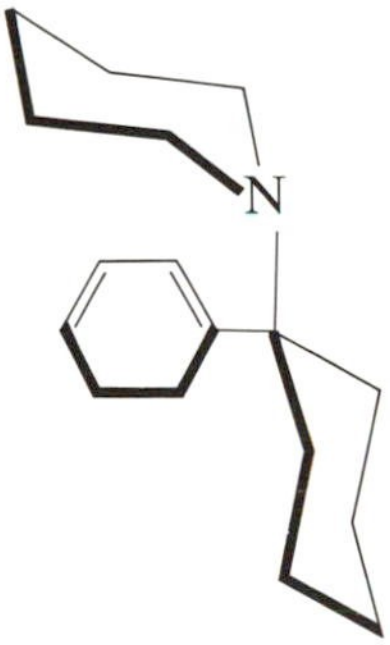

phencyclidine

1-30

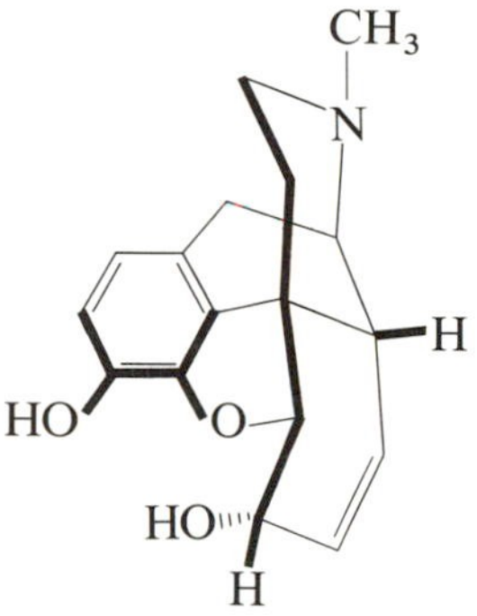

morphine

1-31

---

* A discussion of receptor theory is a topic more appropriate for a text in medicinal chemistry. A general definition is that a receptor molecule is a complex of proteins and lipids which upon binding of a specific organic molecule (effector, neurotransmitter) undergoes a physical or conformational change that usually triggers a series of events which results in a physiological response. In a way, an analogy could be made between receptors and enzymes.

It has strong hallucinogenic properties as well as being a potent analgesic. This is understandable since the corresponding spacial distance between the nitrogen atom and the phenyl ring makes it an attractive mimic of morphine (**1-31**) at the receptor level. This stresses the point that proper conformation can give (sometimes unexpectedly) a compound very unusual thereapeutic properties where analogues can be exploited.

In addition to the steric and external shell factors just mentioned, inductive and resonance contributions can also be important. All these factors must be taken into consideration in the planning of any molecular biomodel system that will hopefully possess the anticipated property. Hence, small but subtle changes on a biomolecule can confer to the new product large and important new properties.

It is in this context, that many of the fundamental principles of bioorganic chemistry are presented in the following chapters.

Chapter 2

Bioorganic Chemistry of the Amino Acids

"L' imagination est plus importante que le savoir."

A. Einstein

Bioorganic chemistry provides a link between the work of the organic chemist and biochemist, and this chapter is intended to serve as a link between organic chemistry, biochemistry, and protein and medicinal chemistry or pharmacology. The emphasis is chemical and one is continually reminded to compare and contrast biochemical reactions with mechanistic and synthetic counterparts. The organic synthesis and biosynthesis of the peptide bond and the phosphate ester linkage (see Chapter 3) are presented "side-by-side"; this way, a surprising number of similarities are readily seen. Each amino acid is viewed separately as an organic entity with a unique chemistry. Dissociation behavior is related in terms of other organic acids and bases, and the basic principles are reviewed so that one is not left with the impression of the amino acid as being a peculiar species. The chemistry of the amino acids is presented as if part of an organic chemistry text, (alkylations, acylations, etc.), and biochemical topics are then discussed in a chemical light.

2.1 General Properties

If we were to consider the protein constituents of ourselves (hair, nails, muscles, connective tissues, etc.), we might suspect that the molecules which constitute a complex organism must be of a complex nature. As such, one might investigate the nature of these "life molecules." Upon treating a protein sample with aqueous acid or base, one would no longer observe the intact protein molecule, but instead a solution containing many simpler, much smaller molecules: the amino acids. The protein molecule is a polymer or biopolymer, whose monomeric units are these amino acids. These monomeric units contain an amino group, a carboxyl group, and an atom of hydrogen all linked to the same atom of carbon. However, the atom or atoms that provide the fourth linkage to this central carbon atom vary from one amino acid to the other. As such, the monomeric units that make up the protein molecule are not the same, and the protein is a complex copolymer. Remember that most man-made polymers are composed of only one monomeric unit. In nature there are about twenty amino acids which make up all protein macromolecules. Two of these do not possess a primary amino function and are thus α-imino acids. These amino acids, proline and hydroxyproline, instead contain a secondary amino group.

$$
\underset{H_2N}{}\overset{R}{\underset{}{\overset{|}{\underset{}{C_{\text{\tiny{III}}}H}}}}\overset{}{COOH}
$$

General form of the α-amino acids. Note that the fourth (R) substituent about the tetrahedral carbon atom provides the variability of these monomeric units.

With such variability in the "R" substituents or side chains of these amino acids, it is possible to divide them into three groups based on their polarity.

(1) *Acidic Amino Acids*

Acidic amino acids are recognized, for example, by their ability to form insoluble calcium or barium salts in alcohol. The side chains of these amino acids possess a carboxyl group, giving rise to their acidity. The two acidic amino acids are:

(a) Aspartic acid (abbreviation: Asp)

$$R = —CH_2COOH; \; pK_a(\beta\text{-}CO_2H) = 3.86$$

(b) Glutamic acid (abbreviation: Glu)

$$R = —CH_2CH_2COOH; \; pK_a(\gamma\text{-}CO_2H) = 4.25$$

(2) *Basic Amino Acids*

Basic amino acids are recognized, for example, by their ability to form precipitates with certain acids. Members of this group include

(a) Lysine (abbreviation: Lys)

$$R = —(CH_2)_4—NH_2; \; pK_a(\varepsilon\text{-}NH_2) = 10.53$$

The four methylene groups are expected to give a flexible amino function to protein molecules.

(b) Hydroxylysine (abbreviation: Hylys)

$$R = -(CH_2)_2-CHOHCH_2NH_2 \; ; \; pK_a(\varepsilon\text{-}NH_2) = 9.67$$

This amino acid is found only in the structural protein of connective tissues, collagen.

(c) Arginine (abbreviation: Arg)

$$R = -(CH_2)_3-NHCNH_2 \; ; \; pK_a = 12.48$$

with NH double-bonded above.

This amino acid is characterized by the guanidine function, which gives rise to its high basicity. Indeed, guanidine is one of the strongest organic bases known, being comparable in strength with sodium hydroxide. Hence, at physiological pH (7.35) this group is always ionized. Most likely this arrangement has been selected because of the special ability of this function to form specific interactions with phosphate groups.

The strong basicity of the guanidine function (guan) may be understood by noting that protonation of the imine function ($>C=NH$) would form a more stable cation than is possible by protonation of a primary function, as can be seen by the following.

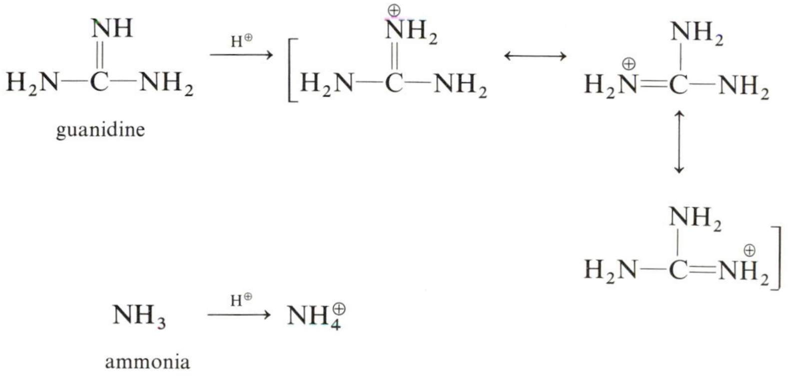

guanidine

ammonia

$$NH_3 \xrightarrow{H^\oplus} NH_4^\oplus$$

That is, guanidine will be more easily protonated.

(d) Histidine (abbreviation: His)

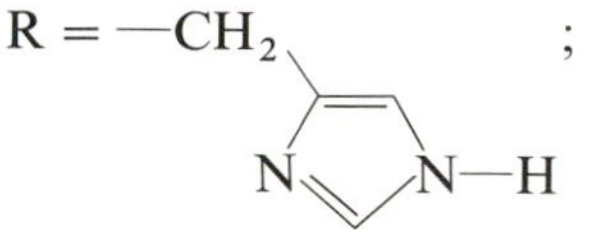

$$R = -CH_2 \qquad ; \qquad pK_a = 6.00$$

with imidazole ring N and N—H.

This amino acid contains the heterocyclic imidazole ring and possesses a unique chemistry. It is both a weak acid and weak base as well as an excellent nucleophile, and the only amino acid that has a pK_a which approximates physiological pH (7.35). As such, it can both pick up and dissociate protons within the biological milieu. Further, it may so function simultaneously by picking up a proton on one side of the ring, and donating it to the other. It has the potential of acting as a *proton-relay system* (detailed in Section 4.4.1).

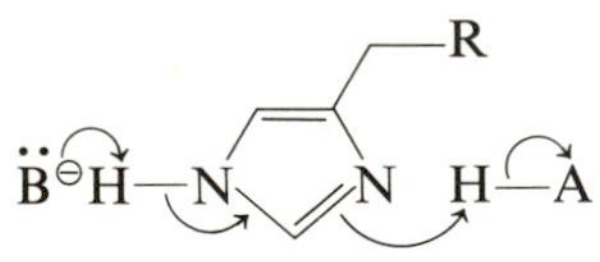

$$\ddot{B}^{\ominus}\!H\!-\!N \diagdown\!\!\!\diagup N \quad H\!-\!A$$

with R substituent at top

(3) *Neutral Amino Acids*

Neutral amino acids contain organic side chains which can neither donate nor accept protons. The simplest (and the only optically inactive amino acid) is:

(a) Glycine (abbreviation: Gly)

$$R = -H$$

Obviously, little chemistry is associated with this amino acid, and its biological role is that of a structural component where limited space (compactness) is important. A number of structural proteins (collagen, silk, wool) contain significant amounts of glycine.

There are a number of amino acids which are *hydrophobic* by virtue of hydrocarbon side chains. These include:

(b) Alanine (abbreviation: Ala)

$$R = -CH_3$$

(c) Valine (abbreviation: Val)

$$R = -CH(CH_3)_2$$

(d) Leucine (abbreviation: Leu)

$$R = -CH_2CH(CH_3)_2$$

(e) Isoleucine (abbreviation: Ile)

$$R = -\overset{\displaystyle CH_3}{\underset{\displaystyle |}{C}}HCH_2CH_3$$

These side chains produce a specific effect on water structure: subtle differences being expected with the different side chains.

Amino acids exist which contain aromatic hydrocarbon side chains.

(f) Phenylalanine (abbreviation: Phe)

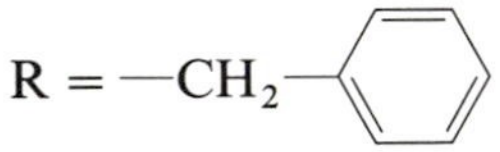

$$R = -CH_2 - \bigcirc \qquad \lambda_{max} = 259 \text{ nm}$$

Here we note the presence of a polarizable π electron cloud. Hydroxylation introduces a functional group, and gives rise to the amino acid:

(g) Tyrosine (abbreviation: Tyr)

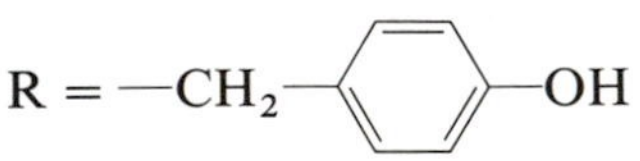

$$R = -CH_2 - \bigcirc - OH \qquad \lambda_{max} = 288 \text{ nm}$$

This amino acid possesses a dissociable (phenolic) hydroxyl of $pK_a = 10.07$. The similarity between phenylalanine and tyrosine allows the former to be converted to the latter in the human. As such, it is phenylalanine, and not tyrosine, which is an essential amino acid to the diet. These amino acids are the precursors for the synthesis of the hormone adrenaline.

Other neutral hydroxylic amino acids include:

(h) Serine (abbreviation: Ser)

$$R = -CH_2OH$$

The hydroxymethyl function ($pK_a \sim 15$) is not dissociable under typical physiological conditions. However, serine does serve an important function in a number of biochemical reactions, because of the ability of the primary hydroxyl to act as a nucleophile under appropriate conditions.

(i) Threonine (abbreviation: Thr)

$$R = -CHOHCH_3$$

The secondary hydroxyl is not known to participate in any biochemical reactions.

Replacement of the serine oxygen with sulfur gives rise to a dissociable proton and the amino acid:

(j) Cysteine (abbreviation: Cys)

$$R = -CH_2SH; \qquad pK_a = 8.33$$

The sulfur atom (with its polarizable or elastic electron cloud) is one of the best nucleophiles known and cysteine, like serine, can participate in a number of biochemical reations. Also, the sulfhydryl of cysteine is quite oxidizable to give rise to the disulfide, cystine. Another sulfur-containing amino acid is:

(k) Methionine (abbreviation: Met)

$$R = -(CH_2)_2-SCH_3$$

This amino acid contains a center of high polarizability in the otherwise inert hydrocarbon side chain. Nucleophilic attack of this sulfur atom on the biological energy store adenosine triphosphate (ATP) gives rise to the cationic biochemically important methyl group donor: S-adenosylmethionine.

Another amino acid, based on the indole ring system is:

(l) Tryptophan (abbreviation: Trp)

$$R = \quad \text{(indole ring with } -CH_2- \text{ substituent)} \qquad \lambda_{max} = 279 \text{ nm}$$

Indeed, a common bacteriological test consists of measuring the ability of some bacteria to form indole from tryptophan. The indole ring is an excellent

π electron donor (electron source). In the presence of an "electron sink," it may give rise to a *charge-transfer* complex, or an electron overlap (a very weak bond) between this source and the sink.

Many examples of charge-transfer complexes are known in organic chemistry. A simple example is the ability of hydrogen chloride, when dissolved in benzene, to form a one to one complex with the latter: a so-called *π-complex*.

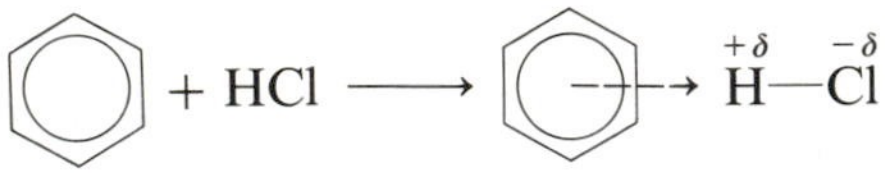

Here the proton of HCl has a weak interaction with the π electrons of the benzene ring.

Note that it is the amino acids phenylalanine, tyrosine, and tryptophan which give proteins an absorption spectra in the ultraviolet. For proteins, this absorption maximum is commonly considered to be 280 nm.

Two neutral amino acids arise from the formation of the primary amides of aspartic and glutamic acid. These are:

(m) Asparagine (abbreviation: Asn)

$$R = -CH_2CONH_2$$

(n) Glutamine (abbreviation: Gln)

$$R = -CH_2CH_2CONH_2$$

Transformation of the carboxylate to amide functions gives rise to a species which can participate in hydrogen bonding important to biological function.

The two α-imino acids are:

(o) Proline (abbreviation: Pro)

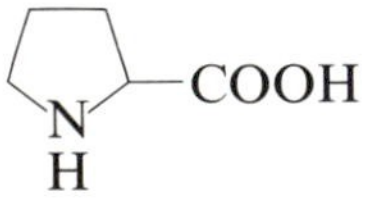

(p) Hydroxyproline (abbreviation: Hypro)

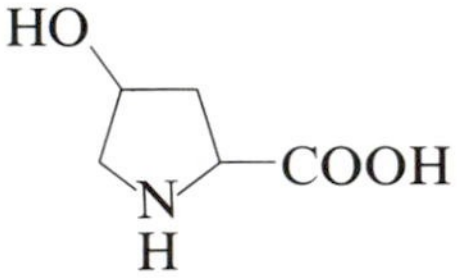

It is this secondary amino function which gives rigidity, and a change in direction, to the peptide backbone of which proteins are built. Thus, the direction of the collagen helix (collagen is a "triple helix" with three separate polypeptide chains wrapped around one another) is continually changing as a result of its proline and hydroxyproline content. Collagen is the only protein in which hydroxyproline is found.

In addition to these common amino acids which make up the protein molecule, a number of other amino acids exists which are not present in the protein but are biochemically important. These may be α-, β-, γ-, or δ-substituted. Two important examples are the neurotransmitter γ-aminobutyric acid (GABA) and the thyroid hormone (**1–28**, Chap. 1) precursor 2,5-diiodotyrosine. Other examples include β-alanine (precursor of the vitamin pantothenic acid), β-cyanoalanine (a plant amino acid), and penicillamine (a clinically useful metal chelating agent).

$$H_2N-(CH_2)_3-COOH$$

GABA

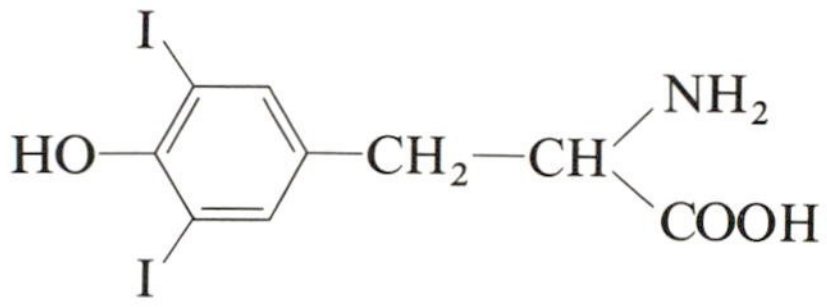

2,5-diiodotyrosine

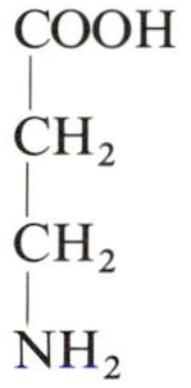

β-alanine

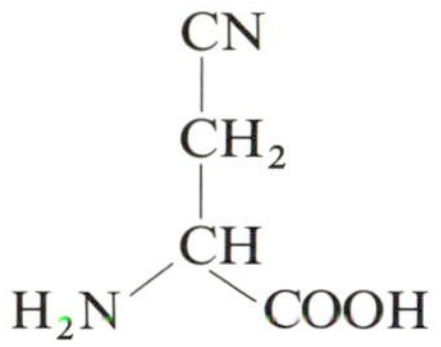

β-cyanoalanine

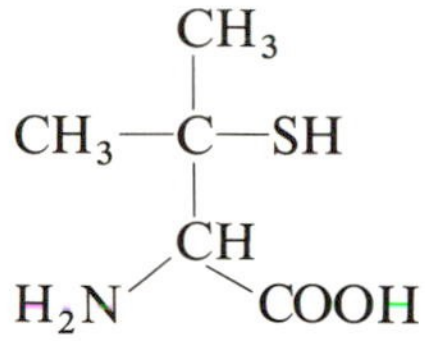

penicillamine

While all amino acids present in proteins exist in the L-configuration, a few nonprotein amino acids do exist in the D-configuration. The importance of the absolute configuration with regard to protein structure and function will become obvious with a deeper consideration of bioorganic processes.

2.2 Dissociation Behavior

Amino acids are crystalline solids which usually decompose or melt in the range of 200°–350°C, and are poorly soluble in organic solvents. These properties suggest that they are organic salts and the evidence is that they exist in the crystal lattice as a dipolar ion or *zwitterion*. That is, the acidic proton from the carboxyl function protonates the amino function on the same molecule. This is not peculiar to amino acids but can instead be representative of any organic salt (i.e., nucleotides; organic molecules which can contain cationic nitrogen and anionic phosphate within the same molecule).

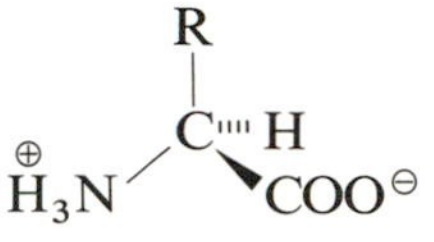

zwitterionic form of the L-amino acids (inner salt)

In solution two possible dissociation pathways exist for all amino acids.

(1)
$$R-CH\begin{array}{c}COOH\\NH_3^{\oplus}\end{array} \rightleftharpoons R-CH\begin{array}{c}COO^{\ominus}\\NH_3^{\oplus}\end{array} + H^{\oplus}$$

$pK_a = 2.1 \pm 0.3$ for most amino acids.

(2)
$$R-CH\begin{array}{c}COO^{\ominus}\\NH_3^{\oplus}\end{array} \rightleftharpoons R-CH\begin{array}{c}COO^{\ominus}\\NH_2\end{array} + H^{\oplus}$$

$pK_a = 9.8 \pm 0.7$ for most amino acids.

This suggests that in solution there will be an acidity (pH) at which the amino acids will exist in the zwitterionic form, or have no net charge. The pH at which this occurs *as a result of* the proton condition is referred to as the *isoionic point* (pI_i). Similarly, when it is observed that there is no net charge on the molecule within the system as judged by experimental conditions (i.e., no mobility during an electrophoresis experiment) the pH at which this occurs is referred to as the *isoelectric point* (pI_e). For an aqueous solution of amino acids:

$$(pI_i) \cong (pI_e) \tag{2-1}$$

However, for proteins this is not necessarily the case, since they may be binding ions other than protons which contribute to an overall charge balance (no net charge). It might be expected and is observed that proteins at their respective isoelectric points will be less soluble than at pH values above or below this point. As they will have no net charge, they will more readily aggregate and precipitate. Further, since different proteins will have different amino acid compositions, they will possess characteristic pI_e values. Such is the basis for protein purification by *isoelectric precipitation*. The protein mixture is adjusted to a pH that is equivalent to the pI_e of the desired protein, allowing the latter to precipitate out of the mixture. The pI_e for amino acids with neutral side chains is 5.6 ± 0.5; it is lower for those amino acids with acidic side chains, and higher for those amino acids with basic side chains. On the other hand, for proteins it can range anywhere from pH equal 0 to 11. Derivation of formulas for the calculation of the pI_i for amino acids is described in most biochemistry texts.

Noting that the amino acids do have acidic properties, it is of interest to compare these with typical organic acids and bases. Remembering that the pK_a of a dissociable function is the pH at which it is half-ionized (see any biochemistry text for the mathematical expression that relates pH and pK_a, the Henderson–Hasselbalch equation), the pK_a values of any compound may serve as index of that compound's acidity.

Comparison of glycine with acetic acid reveals the former to be more than one hundred times stronger than the latter (see Table 2.1), yet both are structurally similar. The amino group then must exert a profound effect on

Table 2.1. The pK_a Values of Various Amino Acids, Organic Acids, and Bases[a]

Compound	pK_a
(A) Amino acids	
(1) $NH_3{}^{\oplus}$—CH_2—COOH Glycine	(COOH) = 2.34 (NH_2) = 9.60
(2) 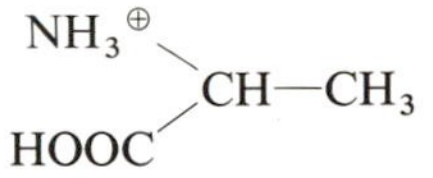 $NH_3{}^{\oplus}$\CH—CH_3 / HOOC Alanine	(COOH) = 2.35 (NH_2) = 9.69
(3) 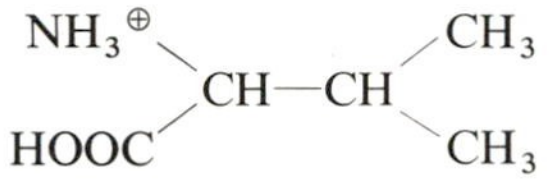 $NH_3{}^{\oplus}$\CH—CH / CH_3 \CH_3 HOOC Valine	(COOH) = 2.32 (NH_2) = 9.62
(4) $NH_3{}^{\oplus}$—CH_2—CH_2—COOH β-Alanine	(COOH) = 3.60 (NH_2) = 10.20
(5) $NH_3{}^{\oplus}$—CH_2—CH_2—CH_2—COOH GABA	(COOH) = 4.23 (NH_2) = 10.43
(6) $NH_3{}^{\oplus}$—CH_2—CH_2—CH_2—CH_2—COOH δ-Aminovaleric acid	(COOH) = 4.27 (NH_2) = 10.79
(7) $NH_3{}^{\oplus}$—CH_2—CH_2—CH_2—CH_2—CH_2—COOH ε-Aminocaproic acid	(COOH) = 4.43 (NH_2) = 10.79
(8) 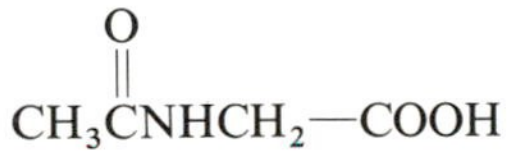 $CH_3\overset{\displaystyle O}{\overset{\displaystyle \|}{C}}NHCH_2$—COOH N-Acetylglycine	(COOH) = 3.60
(9) $NH_3{}^{\oplus}$—CH_2—$CONH_2$ Glycylamide	(NH_2) = 8.0
(10) 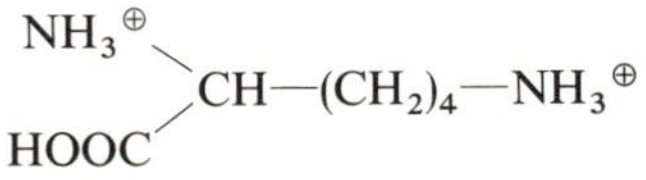 $NH_3{}^{\oplus}$\CH—$(CH_2)_4$—$NH_3{}^{\oplus}$ / HOOC Lysine	(COOH) = 2.18 (α-NH_2) = 8.95 (ε-NH_2) = 10.53
(11) 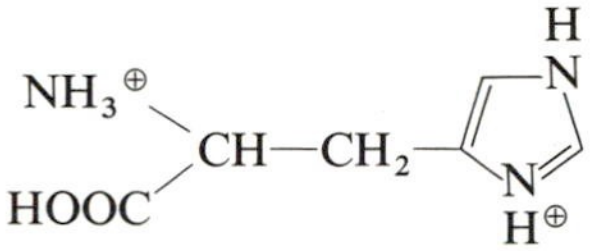 $NH_3{}^{\oplus}$\CH—CH_2— (imidazole) / HOOC Histidine	(COOH) = 1.82 (Im) = 6.0 (NH_2) = 9.17

Table 2.1. (*Continued*)

Compound	pK_a
(12) 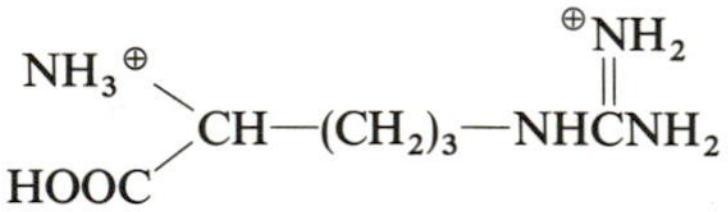 $NH_3^\oplus$ —CH—$(CH_2)_3$—NHCNH$_2$ ($\overset{\oplus NH_2}{\overset{\|\|}{}}$), HOOC Arginine	(COOH) = 2.17 (NH$_2$) = 9.04 (guan) = 12.48
(13) 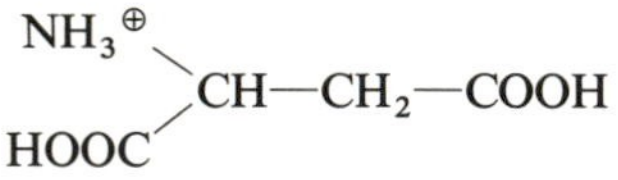 $NH_3^\oplus$ —CH—CH_2—COOH, HOOC Aspartic acid	(α-COOH) = 2.09 (β-COOH) = 3.86 (NH$_2$) = 9.82
(14) 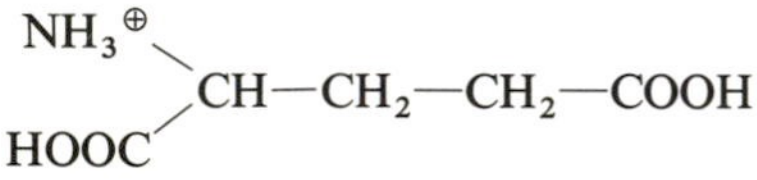 $NH_3^\oplus$ —CH—CH_2—CH_2—COOH, HOOC Glutamic acid	(α-COOH) = 2.19 (γ-COOH) = 4.25 (NH$_2$) = 9.67
(15) 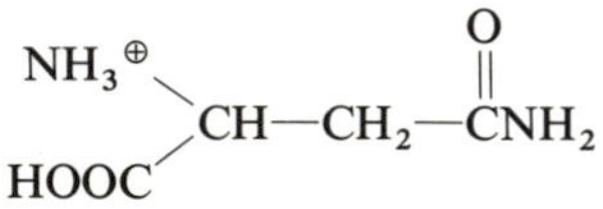 $NH_3^\oplus$ —CH—CH_2—$\overset{O}{\overset{\|\|}{C}}NH_2$, HOOC Asparagine	COOH) = 2.02 (NH$_2$) = 8.8
(16) 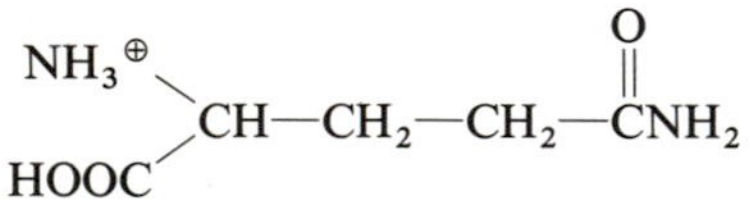 $NH_3^\oplus$ —CH—CH_2—CH_2—$\overset{O}{\overset{\|\|}{C}}NH_2$, HOOC Glutamine	(COOH) = 2.17 (NH$_2$) = 9.13
(17) 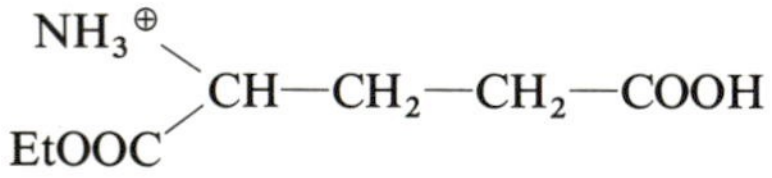 $NH_3^\oplus$ —CH—CH_2—CH_2—COOH, EtOOC α-Ethyl glutamate	(COOH) = 3.84 (NH$_2$) = 7.84
(18) $NH_3^\oplus$ —CH—CH_2—CH_2—$\overset{O}{\overset{\|\|}{C}}$OEt, HOOC γ-Ethyl glutamate	(COOH) = 2.15 (NH$_2$) = 9.19
(19) 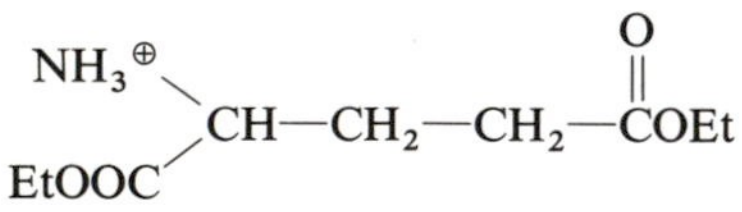 $NH_3^\oplus$ —CH—CH_2—CH_2—$\overset{O}{\overset{\|\|}{C}}$OEt, EtOOC Diethyl glutamate	(NH$_2$) = 7.03

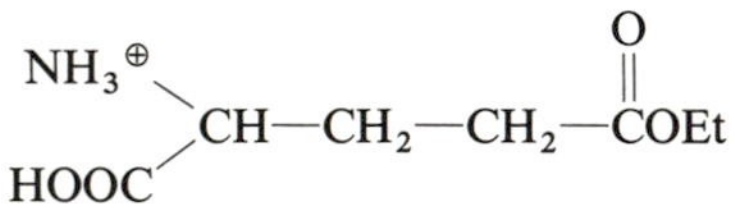

Table 2.1. (*Continued*)

Compound	pK_a
(20) 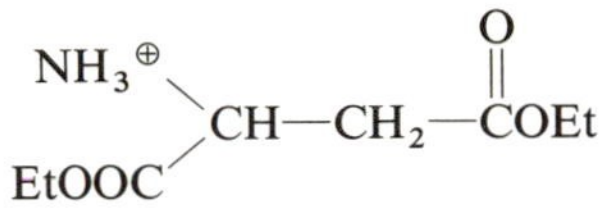 Diethyl aspartate	$(NH_2) = 6.50$
(21) 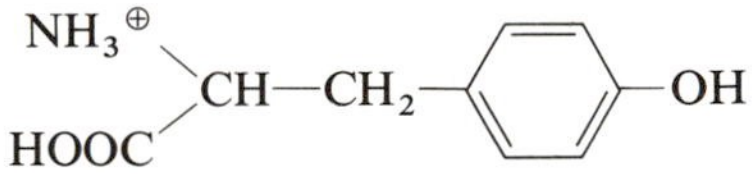 Tyrosine	$(COOH) = 2.20$ $(NH_2) = 9.11$ $(OH) = 10.07$
(22) 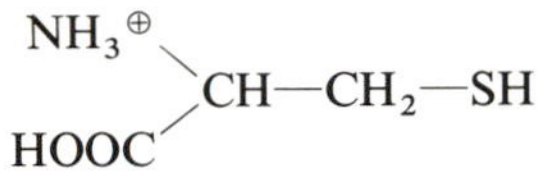 Cysteine	$(COOH) = 1.71$ $(SH) = 8.33$ $(NH_2) = 10.78$
(23) 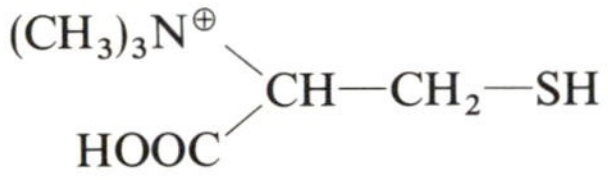 Cysteine betaine	$(SH) = 8.65$
(24) 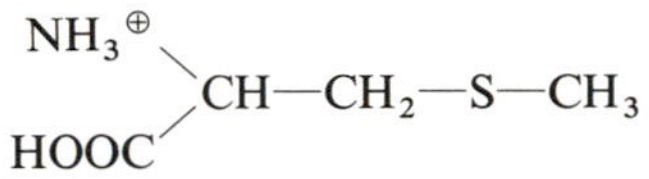 S-Methyl cysteine	$(NH_2) = 8.75$
(25) 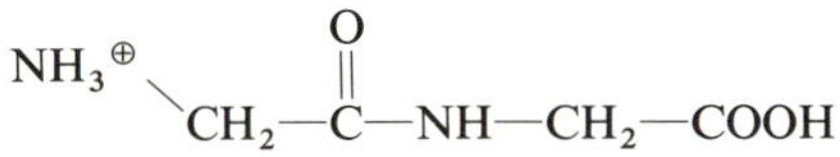 Glycylglycine	$(COOH) = 3.06$ $(NH_2) = 8.13$

(20)

$$\begin{array}{c} NH_3^{\oplus} \\ \diagdown \\ \qquad CH{-}CH_2{-}\overset{\displaystyle O}{\overset{\|}{C}}OEt \\ \diagup \\ EtOOC \end{array}$$

Diethyl aspartate

(21)

$$\begin{array}{c} NH_3^{\oplus} \\ \diagdown \\ \qquad CH{-}CH_2{-}\!\!\!\bigcirc\!\!\!-OH \\ \diagup \\ HOOC \end{array}$$

Tyrosine

(22)

$$\begin{array}{c} NH_3^{\oplus} \\ \diagdown \\ \qquad CH{-}CH_2{-}SH \\ \diagup \\ HOOC \end{array}$$

Cysteine

(23)

$$\begin{array}{c} (CH_3)_3N^{\oplus} \\ \diagdown \\ \qquad CH{-}CH_2{-}SH \\ \diagup \\ HOOC \end{array}$$

Cysteine betaine

(24)

$$\begin{array}{c} NH_3^{\oplus} \\ \diagdown \\ \qquad CH{-}CH_2{-}S{-}CH_3 \\ \diagup \\ HOOC \end{array}$$

S-Methyl cysteine

(25)

$$\begin{array}{c} NH_3^{\oplus} \\ \diagdown \\ \qquad CH_2{-}\overset{\displaystyle O}{\overset{\|}{C}}{-}NH{-}CH_2{-}COOH \end{array}$$

Glycylglycine

(B) Organic compounds

(1) CH_3COOH Acetic acid	4.76
(2) $HCOOH$ Formic acid	3.77
(3) $ClCH_2COOH$ Chloroacetic acid	2.85
(4) $Cl_2CHCOOH$ Dichloroacetic acid	1.30
(5) Cl_3CCOOH Trichloroacetic acid	1.00

Table 2.1. (*Continued*)

Compound	pK_a
(6) CF_3COOH Trifluoroacetic acid	−0.25
(7) $CH_3NH_3^{\oplus}$ Methylamine	10.64
(8) 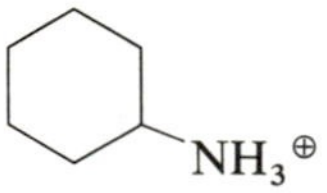 Cyclohexylamine	10.60
(9) 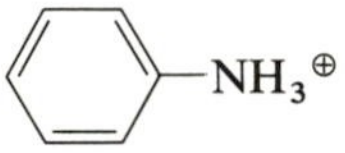 Aniline	4.60
(10)  Pyridine	5.17
(11)  Piperidine	11.13
(12)  Imidazole	7.08
(13) 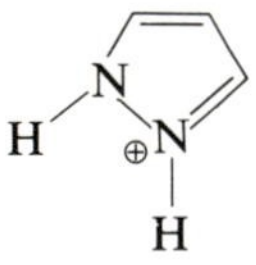 Pyrazole	2.50
(14) CH_3—SH Methyl mercaptan	10.7
(15) 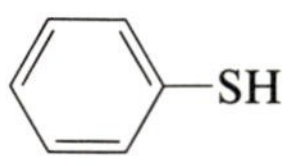 Thiophenol	7.2
(16) 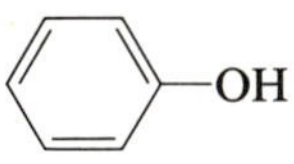 Phenol	9.8

Table 2.1. (*Continued*)

Compound	pK_a
(17) 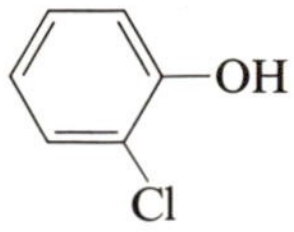 *o*-Chlorophenol	8.5
(18) 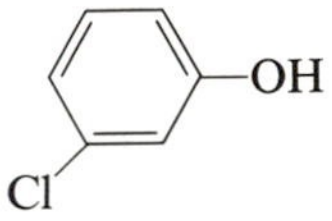 *m*-Chlorophenol	8.2
(19)  *p*-Chlorophenol	9.2
(20) 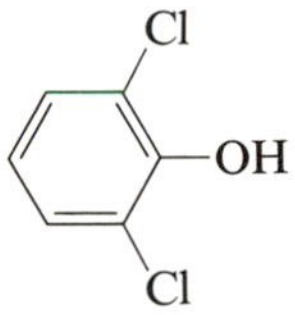 2,6-Dichlorophenol	6.8
(21) 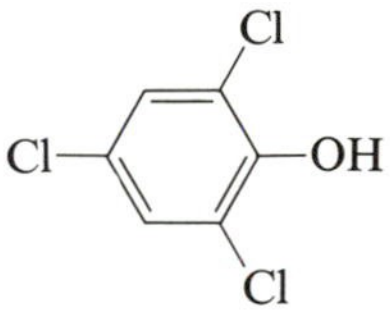 2,4,6-Trichlorophenol	6.4
(22) 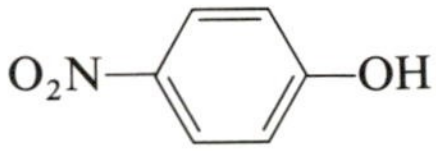 O_2N——OH *p*-Nitrophenol	7.1
(23) CH_3—OH Methanol	16
(24) 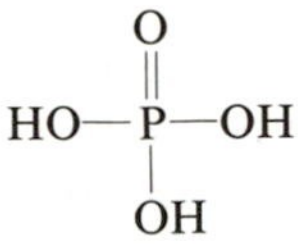 Phosphoric acid	2.12 7.21 12.32

Table 2.1. (*Continued*)

Compound	pK$_a$
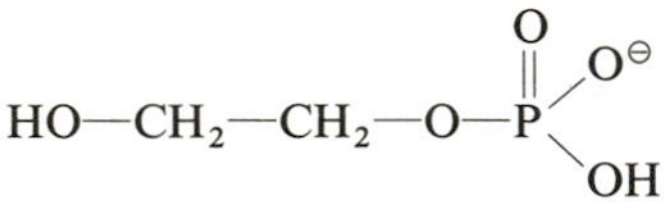 (25) HO—CH$_2$—CH$_2$—O—P (O)(O$^\ominus$)(OH) 2-Hydroxyethyl phosphate	6.4
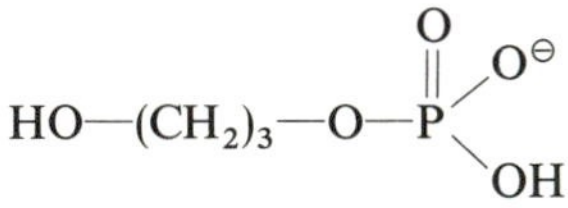 (26) HO—(CH$_2$)$_3$—O—P (O)(O$^\ominus$)(OH) 3-Hydroxypropyl phosphate	6.5
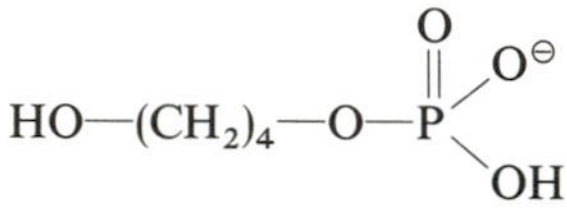 (27) HO—(CH$_2$)$_4$—O—P (O)(O$^\ominus$)(OH) 4-Hydroxybutyl phosphate	6.7
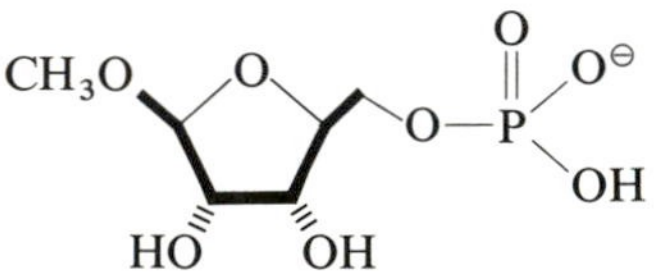 (28) CH$_3$O ... O—P (O)(O$^\ominus$)(OH) HO OH Methyl β-D-ribofuranoside 5-phosphate	6.4

^{a} Values refer to proton dissociation from the indicated structures.

the carboxyl function. Acetylation of the amino group (eliminating the positive charge) reduces the acidity of glycine roughly by a factor of ten, but still the carboxyl of the acetylated glycine is significantly more acidic than that of acetic acid. Two effects are operative which account for the greater acidity of glycine relative to acetic acid. *Inductive pull* (I$\ominus$) of the positive ammonium and acetylated amine decreases electron density at the carboxyl so that the latter more readily gives up its dissociable proton. This is further enforced in the case of the ammonium ion by the presence of a cationic charge in close proximity of the carboxyl: a so-called *field effect*.

Both effects are routinely observed in the dissociation behavior of most organic acids and bases (Table 2.1) and amino acids are not peculiar in this regard. For example, halogenation of acetic acid produces an even stronger acid so that acid strength increases in the order acetic acid < monochloro- < dichloro- < trichloroacetic acid. This reflects the inductive (I$\ominus$) pull, or withdrawal of electron density from the carboxyl, by the electronegative chlorine atoms through the covalent linkages of the acid molecule. A perturbation or change in electron density through the σ bonds may be imagined.

Hence, replacement of three chlorine atoms with three more electronegative fluorine atoms results in an even stronger acid: trifluoroacetic acid.

On the other hand, an acid enhancing effect through space, independent of the covalent bond linkages, may also contribute to the increased acidity of a carboxylate function. Again, electrostatic effects through space are referred to as field effects. An illustration of these is seen in the consideration of the dissociation behavior of malonic and diethyl malonic acid. The ratio of the two dissociation constants (K_1 and K_2) in aqueous solution is in the former case 700, in the latter case, 120,000.

malonic acid
$K_1/K_2 = 700$

diethyl malonic acid
$K_1/K_2 = 120,000$

It may be noticed that in the case of malonic acid, much of the space between the two carboxyl groups is occupied by water, but in the case of the diethyl malonic acid, this space is occupied by two alkyl groups. The two ethyl groups provide a medium of low dielectric constant, so that development of a full negative charge by the first carboxyl function is strongly felt by the second carboxyl. Subsequently, a strong electrostatic repulsion will be felt upon loss of the second proton, and hence the high ratio of the two dissociation constants. On the other hand, water molecules will more readily shield the negative charge upon dissociation of the first proton from malonic acid so that dissociation of the second proton will result in less electrostatic repulsion. Note that a positive ammonium cation would stabilize a developing negative charge on a carboxylate function.

With these considerations in mind, the dissociation behavior of amino acids can then be understood. For example, substitution of the α-proton of glycine for a methyl group to give alanine should not markedly alter the pK_a of the carboxyl function. This is observed (Table 2.1), as it is for other amino acids with neutral side chains. However, in the case of β-alanine, where the amino group is now two carbon atoms away from the carboxyl function, the two interact less closely and the pK_a value is approximately midway between glycine and acetic acid. In the case of the neurotransmitter GABA where the amino and carboxyl groups are separated by three carbon atoms, the carboxyl group has a pK_a close to acetic acid. A similar situation is observed for the dicarboxylic amino acids. Aspartic acid possesses a side chain (β) carboxyl which, like the α-carboxyl but to a lesser extent, feels the presence of the amino function. The pK_a is 3.86 (Table 2.1). On the other hand, glutamic acid possesses a side-chain carboxyl (γ) which is situated further from the amino function, and barely feels the latter. The pK_a is 4.25. Hence, the difference between aspartic and glutamic acid reflects, in addition to structural parameters, acid strengths which can be expected to be important in biological processes.

In turn, consideration of the expected basicity of the amino function of glycine might lead to the conclusion that it is a stronger base (higher pK_a) than a typical organic amine. It might be expected that the full negative charge on the carboxylate function would donate electron density to the amino function and that electrostatic attraction (field effect) between the cationic amine and the anionic carboxylate would make it more difficult to lose a proton from the amino group. Indeed this is correct, and both effects are operative to a significant extent. Yet, the pK_a of the amino function of glycine is 9.60 while methylamine is 10.64 (Table 2.1). This is because the most important or net effect is electron withdrawal ($I\ominus$) by the carboxylate (carbonyl) function. This is illustrated by neutralization of the full charge on the carboxylate residue by conversion to an amide. The pK_a of the amino function of glycyl amide is 8.0 and glycyl glycine 8.13. Here, no donation of electron density by an anionic carboxylate or field (electrostatic) effect is possible and the only effect operative is electron withdrawal by the amide (carbonyl) function. Note that esterification of aspartic and glutamic acids reflects such considerations (Table 2.1). In the case of the diethyl esters, the amino functions are quite acidic.

In addition to inductive and field effects, *resonance effects* can also play an important role in determining the strength of organic acids and bases. For example, the pK_a of a simple alkyl alcohol is ~15, but that of the hydroxyl of tyrosine is 9.11. By analogy, the pK_a of the hydroxyl of phenol is 9.8 (see Table 2.1). Such may be understood by the realization that once the ionization has occured, the phenoxide anion may be stabilized by electron resonance:

phenol

Extending this argument accounts for the even lower pK_a of p-nitrophenol (7.1), which has an additional more important resonance form.

p-nitrophenol
(colorless)

orange
($\lambda_{max} = 400$ nm)

This should not be confused with the lower pK_a value that results upon halogenation of phenol (see Table 2.1), for this again represents inductive pull. Nonetheless, resonance effects can extend themselves to other important examples as is noted by a comparison of the basicity of cyclohexylamine and aniline. The former has a pK_a typical of an organic amine (Table 2.1), but the latter is considerably more acidic. That is because at any given time the electron density of the amino function of aniline is much less than that of cyclohexylamine. The nonbonding electrons are smeared into the aromatic ring via resonance:

aniline

While an amino acid analogue of aniline does not exist, examples are observed in biological systems where an exocyclic amino function is adjacent to an aromatic heterocycle. Most notable are the purines adenine and guanine and the pyrimidine cytosine. These will be discussed in Chapter 3.

Three important parameters, inductive, field, and resonance effects, can then greatly influence the behavior of organic acids and bases, including the biologically important α-amino acids. In aqueous solution, the medium in which biological reactions occur, such effects will allow for a variety of behavior so that dissociation processes are not limited, but instead may reflect the entire range of the pH scale. This is important when it is considered that proteins are built of amino acids, and so they may participate in acid–base chemistry in a manner which reflects their amino acid constituents. Indeed, our examination of amino acid dissociation may, in a simplistic fashion, be regarded as that of a miniature protein model. Proteins play a functional role in biochemical reactions, so that such an analogy may provide a basis for understanding proton transfer processes. However, such a model may be too simple. It does not allow for a consideration of cooperative interactions. For example, how is the dissociation behavior of a lysine residue affected by a linear array of cationic amino acid residues covalently connected by the protein? Further, what sort of dissociation behavior can be expected for a chemical process that might occur close to the hydrophobic (lower dielectric constant) interior of a protein molecule? That significant changes would be expected is indicated by examining the dissociation of glycine in a medium of lower dielectric constant: 95% ethanol. Here the pK_a of the carboxyl function is 3.8, and the amino function 10.0. It may be thought that the acidity now approaches acetic acid, but such is not the case, for the latter has a pK_a of 7.1.

That dissociation pathways on a protein surface represent a complex interplay among the monomeric amino acid residues may be illustrated by examining the ionization of the amino acid cysteine as a simple model. The ionization scheme may be described as follows:

$$
\begin{array}{ccc}
\text{CH}_2\text{SH} & \text{CH}_2\text{SH} & \text{CH}_2\text{SH} \\
| & | & | \\
\text{H}-\overset{}{\text{C}}-\text{COOH} \underset{}{\overset{K_1}{\rightleftharpoons}} \text{H}-\overset{}{\text{C}}-\text{COO}^{\ominus} \underset{}{\overset{K_2}{\rightleftharpoons}} \text{H}-\overset{}{\text{C}}-\text{COO}^{\ominus} \\
| & | & | \\
\text{NH}_3{}^{\oplus} & \text{NH}_3{}^{\oplus} & \text{NH}_2
\end{array}
$$

$$
\Big\updownarrow K_3
$$

$$
\begin{array}{c}
\text{CH}_2\text{S}^{\ominus} \\
| \\
\text{H}-\overset{}{\text{C}}-\text{COO}^{\ominus} \\
| \\
\text{NH}_2
\end{array}
$$

where $pK_1 = 1.71$
pK_2 and/or $pK_3 = 8.33$ and/or 10.78 (see Table 2.1)

The carboxyl ionization (pK_1) is low and easily identified. However, the ammonium and thiol groups have similar pK values (compare methylamine with methyl mercaptan, Table 2.1) and so an uncertainty exists as to which group ionizes first. K_1, K_2, and K_3 represent macroscopic ionization constants determined experimentally from a titration curve. K_2 and K_3 are the composite of four microscopic ionization constants. Once the proton is lost from the carboxyl group, one of two ionization pathways may be followed:

$$
\begin{array}{c}
\text{CH}_2\text{S}^{\ominus} \\
| \\
\text{H}-\overset{}{\text{C}}-\text{COO}^{\ominus} \\
| \\
\text{NH}_3{}^{\oplus}
\end{array}
$$

$$
\begin{array}{ccc}
\text{CH}_2\text{SH} & & \text{CH}_2\text{S}^{\ominus} \\
| & & | \\
\text{H}-\overset{}{\text{C}}-\text{COO}^{\ominus} & & \text{H}-\overset{}{\text{C}}-\text{COO}^{\ominus} \\
| & & | \\
\text{NH}_3{}^{\oplus} & & \text{NH}_2
\end{array}
$$

$$
\begin{array}{c}
\text{CH}_2\text{SH} \\
| \\
\text{H}-\overset{}{\text{C}}-\text{COO}^{\ominus} \\
| \\
\text{NH}_2
\end{array}
$$

$pk_1 = 8.5$
$pk_2 = 8.9$
$pk_3 = 10.4$
$pk_4 = 10.0$

As the four microscopic constants cannot be determined by a titration curve, spectrophotometric analysis (UV absorption of $R\text{-}S^-$) was necessary. The pK (8.65) of cysteine betaine (ionization of a thiol in the presence of a positive nitrogen) and the pK (8.75) of S-methyl cysteine (ionization of an amino group in the presence of neutral sulfur) closely mimic the k_1 and k_2 dissociation pathways and suggest that these values should be close to each other

(Table 2.1). Here again, note the importance of inductive and field effects which significantly alter the pK_a values from simple alkyl mercaptans and amines.

Knowing that a wide variety of states of ionization are available for amino acids and proteins at any pH (say physiological pH, 7.35), ionic strength, or even dielectric constant, it may be expected that as a result of potential ionic (electrostatic) and hydrogen bonding interactions, a close association would be expected between these molecules and the aqueous medium. Such is the case to the point that every protein possesses a specific degree of *hydration*, or a certain amount of water that must be associated with it in order to preserve its structural integrity. Water does not only become bound, but ordered as well, acquiring specific geometric orientations about the protein. Various physical probes (apparent molal quantities) of simple model systems (amino acids and neutral analogues) indicate the possible ordering of solvent water that can occur on a protein surface. An apparent molal quantity (Φ_X) is defined by the equation:

$$\Phi_X = \left(\frac{X - n_1 X_1{}^0}{n_2} \right) \tag{2-2}$$

where $X_1{}^0$ = molar property for pure solvent
$\quad n_1, n_2$ = mole fractions of solvent and solute.

Φ_X reflects the contribution to property "X" of solvent by the solute, if the solvent behaves as if pure.

Apparent molal volumes (Φ_v) and heat capacities (Φ_{cp}) of glycine, alanine, β-alanine and their neutral analogues clearly indicate the tighter more ordered packing of water (shell of hydration) about the charged species.

$$NH_3{}^\oplus—CH_2—CO_2{}^\ominus \qquad\qquad HO—CH_2—CONH_2$$

glycine   glycolamide

Φ_v = 43.5 ml/mol Φ_v = 56.0 ml/mol
Φ_{cp} = 8.8 cal Φ_{cp} = 36 cal

$$\begin{array}{c} CH_3 \\ | \\ NH_3{}^\oplus—CH—CO_2{}^\ominus \end{array} \qquad\qquad \begin{array}{c} CH_3 \\ | \\ HO—CH—CONH_2 \end{array}$$

alanine   lactamide

Φ_v = 60.6 ml/mol Φ_v = 73.8 ml/mol
Φ_{cp} = 33 cal Φ_{cp} = 58 cal

$$\overset{\oplus}{N}H_3—CH_2—CH_2—COO^\ominus$$

β-alanine

Φ_v = 58.9 ml/mol
Φ_{cp} = 18 cal

The smaller Φ_v values for the charged species reflect the ability of water molecules to pack about the charge. The methyl group of alanine provides a steric repulsion which interferes with this solvation. On the other hand, this is partially overcome by charge separation, or breaking up of the zwitterionic attraction, as is indicated by the data for β-alanine. The Φ_{cp} values agree with this interpretation; a smaller value reflects a more ordered system or lesser degrees of freedom and hence a lesser ability to absorb heat per degree rise in temperature.

Clearly, ionization processes will be varied and important in the aqueous (biological) reaction medium. However, they do not constitute the only chemical process that can occur within the biological organism. The amino acids, being organic molecules, are subject to reactions that are familiar to the organic chemist. It would then be expected that similar reactions could occur in the biological system, which are familiar to the biochemist. The problem is as noted earlier: one cannot extrapolate conditions of high temperature, anhydrous organic solvent systems and so forth, but instead one must limit oneself to the aqueous milieu, body temperature, and the use of protein catalysts: the enzymes. Nonetheless, it is of interest to the bioorganic chemist to compare reaction pathways *in vitro* or as undertaken by the synthetic chemist, and *in vivo* or as undertaken by the organism. The differences, similarities, advantages, and disadvantages shall be best seen in a "side-by-side" comparison of the two, beginning first with the chemistry of the amino acids and finally ending with a consideration of the organic and biosynthesis of protein molecules.

2.3 Alkylations

Because of the good nucleophilicity of amines, alkylation of amino acids is an important widespread reaction in both organic and biological systems. A simple methylation may proceed as follows:

$$H_3C\text{---}X + :NH_2\text{---}R \rightarrow CH_3\overset{\oplus}{N}H_2\text{---}R + X:^{\ominus}$$

X = halide, sulfate, etc.

$$CH_3\overset{\oplus}{N}H_2\text{---}R + :NH_2\text{---}R \rightleftharpoons CH_3NHR + \overset{\oplus}{N}H_3\text{---}R \cdot X:^{\ominus}$$

The reaction usually does not stop with monoalkylation, but instead continues via nucleophilic attack of the monoalkylated amino acid on another molecule of alkylator. This can ultimately give rise to the quaternary ammonium salt of the amino acid, also referred to as the *betaine* of the amino acid. Hence, monomethylation of glycine produces sarcosine, a molecule that is involved in the metabolism of muscle, while trimethylation produces glycine betaine. Other quaternary amines derived from amino acids of biological importance include acetylcholine (which functions as a neurotransmitter) and carnitine (which transfers acyl compounds through cell membranes via esterification with its hydroxyl group). However, amino acid betaines do not function as biological alkylators or methyl group donors.

$$CH_3-\overset{\overset{\displaystyle CH_3}{|}}{\underset{\underset{\displaystyle CH_3}{|}}{\overset{\oplus}{N}}}-CH_2-\underset{\underset{\displaystyle OH}{|}}{CH}-CH_2-COOH$$

carnitine

$$CH_3-\overset{\overset{\displaystyle CH_3}{|}}{\underset{\underset{\displaystyle CH_3}{|}}{\overset{\oplus}{N}}}-CH_2-CH_2-O-\overset{\overset{\displaystyle O}{||}}{C}-CH_3$$

acetylcholine

Methyl group donation does not readily occur in organic or biological systems.

$$Nu\!: + H_3C-\overset{\oplus}{N}(CH_3)_2R \xrightarrow{\;/\!/\;} \overset{\oplus}{Nu}-CH_3 + (H_3C)_2NR$$

$Nu\!:$ = nucleophile

However, sulfonium salts will function as alkyl donors. Further, due to the excellent nucleophilicity of the sulfur atom, such compounds are readily prepared, as is illustrated in the following model reaction:

$$\underset{\underset{\textstyle + CH_3-I}{}}{\overset{\overset{\textstyle Ph\diagdown\;\diagup CH_3}{S}}{}} \longrightarrow \underset{|}{\overset{\overset{\textstyle \oplus \diagup CH_3}{Ph-S}}{\diagdown CH_3}} \longleftarrow \quad :Nu = \text{nucleophile}$$

$$Ph-S-CH_3 + CH_3-\overset{\oplus}{Nu}$$

In biological systems, the universal methyl group donor is the sulfonium compound S-adenosyl methionine (SAM). In turn this is synthesized from the amino acid methionine and another biologically important compound (a "high-energy" compound or biological energy store) adenosine triphosphate (ATP). As is the case for all chemical reactions that occur in the biological organism, this reaction is catalyzed by a specific enzyme. The reaction is thermodynamically favorable without the presence of the protein catalyst, but the enzyme gives specificity to the reaction. Without the catalyst, other reaction pathways become possible, such as breakdown of the triphosphate chain, but the catalyst serves to bind and orientate the sulfur nucleophile in such a way that only attack on the methylene carbon becomes possible. Much more shall be said later about the importance of such binding and proximity effects, but it should be noted that while the adenosine portion of the ATP molecule does not participate in the chemistry of the reaction, it does serve a recognition function. It is recognized by the enzyme catalyst and subsequently binds to the surface of the ATP molecule. In mammals, the methionine substrate is an essential amino acid in the diet.

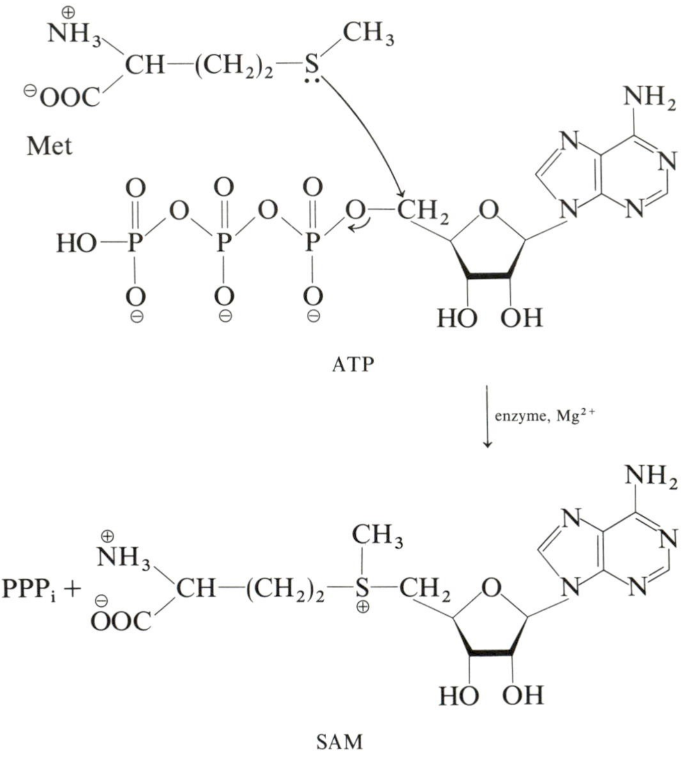

Homocysteine is the amino acid that results after S-adenosyl methionine donates its methyl group, and then the product S-adenosyl homocysteine is hydrolyzed by a molecule of water. In mammals, homocysteine can be converted to cysteine so that the latter is not an essential amino acid, or it can be converted back to methionine by methylation with a compound which serves as a source of one carbon fragments (methyl, formyl, etc.) in biological systems: tetrahydrofolic acid. In some bacteria, homocysteine can be converted back to methionine by methylation with methylcobalamin (the methyl derivative of vitamin B_{12}), in the presence of other required compounds or cofactors. This latter methylation reaction is of interest because it can occur, at a reduced rate, in the absence of any enzymes, and a simpler model system has been developed to mimic this reaction (see details in Chapter 6).

Alkylations thus may proceed under typical organic conditions making use of alkylators such as methyl iodide, dimethyl sulfate, or methyl fluorosulfonate or they may proceed under biological conditions with S-adenosyl methionine and an appropriate enzyme catalyst. However, with the use of potent alkylators in stoichiometric excess, it is possible to alkylate amino acids and proteins under physiological conditions. Such provides the basis for a number of important biochemical probes and pharmacologically active substances.

A potent group of alkylators are the nitrogen or sulfur mustards. These were the active principle of the mustard gases used during World War I. Here the nucleophile and the leaving group are in the same molecule, giving rise to intramolecular (and not intermolecular) nucleophilic attack. This represents a high probability event, or a more favorable system in terms of entropy.

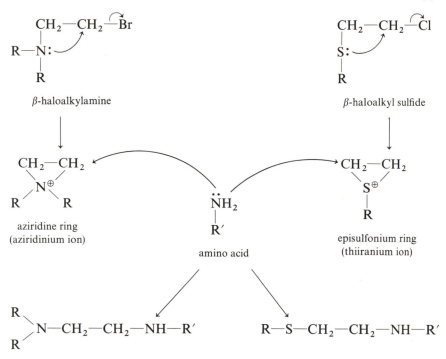

As result of the high probability nucleophilic attack, a strained aziridine or episulfonium ring system is formed which is a potent alkylator. Due to the poorer nucleophilicity of the oxygen atom, an analogous oxonium ring does not readily form, and β-haloethers are not potent alkylators. These compounds are important pharmacologically, and drugs such as phenoxybenzamine have some clinical value. However, such compounds lack specificity in that they will react with any protein with which they make contact, or bind, instead of interacting with just one protein of interest.

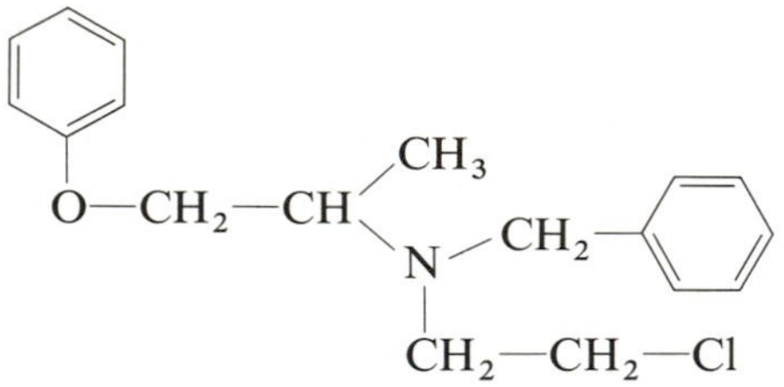

phenoxybenzamine

Some specificity has been achieved in the case of the anti-tumor drug cyclophosphamide. This drug takes advantage of the fact in the cancer patient there will be a greater dephosphorylating (loss of phosphate) activity in the cancer cells than in the normal cells, and hence it will interact to a greater extent with the former.

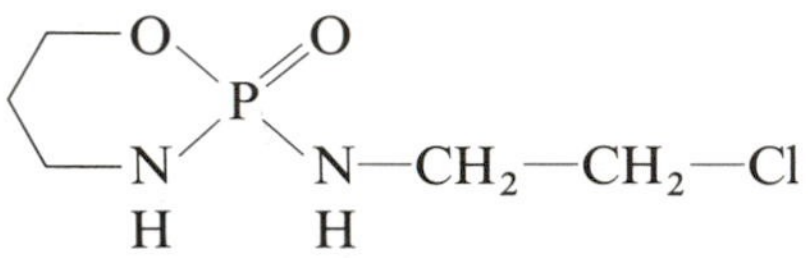

cyclophosphamide

Another potent group of alkylators are the α-haloketones. The classic example is the alkylator TPCK (tosyl-L-phenylalanylchloromethyl ketone) which interacts specifically with one imidazole ring (histidine residue) in the enzyme α-chymotrypsin (see Section 7.2.3). These compounds are significantly more reactive in S_N2 displacements than alkyl halides. For example, nucleophilic attack by iodide will proceed 33,000 times as quickly on chloroacetone as on n-propyl chloride (acetone solvent, $50°C$). Inductive pull of the carbonyl function would be expected to both increase the electrophilicity of the methylene function, and stabilize the approaching anionic nucleophile. α-Haloacids and amides will exhibit a similar effect, and both iodoacetic acid and iodoacetamide have found use as biochemical probes for the alkylation of purified enzymes.

Benzyl halides are also potent alkylators in S_N2 displacement reactions. It will be remembered that the S_N2 reaction proceeds via an sp^2 hybridized transition state, leaving a p orbital bonded to the incoming and departing

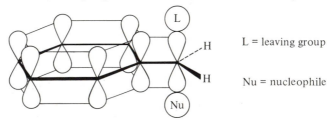

L = leaving group

Nu = nucleophile

Stabilization of the sp^2 transition state by the p orbitals of the benzene ring of a benzyl halide

groups. This p orbital can be conjugated to the π system of the benzene ring giving rise to a more stable transition state.

An alkylator which has found use in analytical work (sequencing of proteins; the determination of the sequence in which the amino acid monomers are arranged in the protein polymer) is 2,4-dinitrofluorobenzene (*Sanger's reagent*).

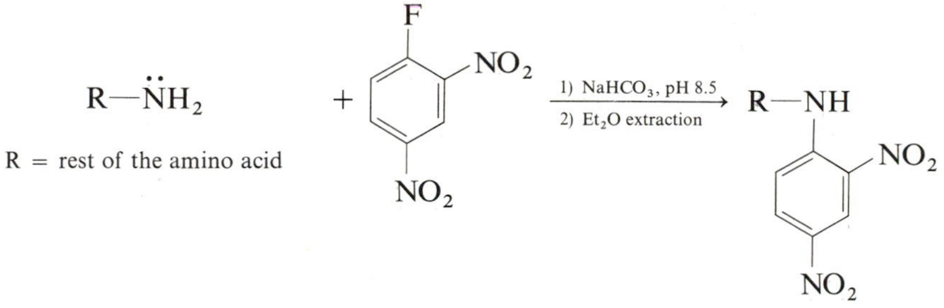

R = rest of the amino acid

The nucleophilic terminal amino function of the protein (i.e., the first amino acid residue or monomer of the protein; the rest of the amino acids have their amino functions involved in the polymer linkage or nonnucleophilic amide linkages) will displace the fluorine atom by means of an addition–elimination mechanism. This occurs with the rate-limiting formation of a negatively charged intermediate:

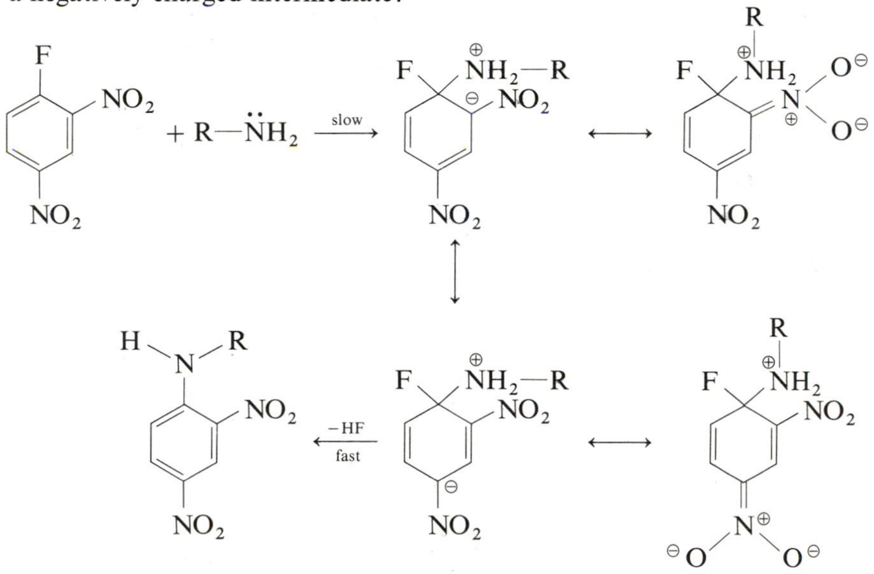

As can be seen, this reaction readily proceeds only when stabilization of the negative charge can occur. Hence the necessity of the nitro function. The fluorine compound is more reactive than the chlorine analogue which might be surprising since the latter is a better leaving group. However, ejection of the leaving group is not the rate-limiting step. Further, the more electronegative fluorine atom will, via inductive pull, offer greater stabilization of the anionic intermediate, as well as increase the electrophilicity of the carbon atom undergoing nucleophilic attack.

Sanger's reagent has unique chromophoric (UV, visible; $\lambda_{max} = 350$ nm) properties which further add to its analytical usefulness. 2,4-Dinitrophenylated amino acids appear as bright orange which allows them to be easily distinguished from other amino acids during a chromatographic separation. Hence, a protein may be reacted with Sanger's reagent, then decomposed by acid hydrolysis (the alkylated terminal amino acid is stable to 6 N HCl; the amino nitrogen is weakly basic or "aniline-like") to the monomeric amino acids, and the terminal amino acid readily separated and identified.

Other analogues of Sanger's reagent have found use as biochemical probes for protein function. This reflects the fact that the terminal amino

| 4-fluoro-3-nitrophenyl azide | 2,4-dinitro-5-fluoroaniline (Bergmann's reagent) | 1,5-difluoro-2,4-dinitrobenzene |

Structure of some Sanger's reagent analogues

acid of the protein is not the only nucleophile present; many of the amino acid side chains also possess nucleophilic functions. As such, the phenolic hydroxyl of tyrosine, the sulfur of cysteine or methionine, ε-amino function of lysine or the imidazole ring of histidine may react with Sanger's reagent or its analogues. 2,4-Dinitro-5-fluoroanaline reacts with amino acids to give derivatives that absorb maximally in the visible ($\lambda_{max} = 400-410$ nm) and which crystallize easily. Further, the amino function may be derivatized and coupled with other dyes to give a more sensitive method of detection of small quantities of amino acids. 4-Fluoro-3-phenyl azide may be used as a *photoaffinity label* for proteins. That is, the protein may be alkylated with the reagent, then subjected to UV light, and the azide will be converted to a reactive nitrene which will also couple to the protein if an appropriate reaction site is available. 1,5-Difluoro-2,4-dinitrobenzene will serve as a bifunctional cross-linking agent. Two nucleophiles in close proximity of each other on a protein surface may displace both fluorine atoms.

From this consideration of the various alkylation reactions of amino acids a number of observations arise. First, while the end product is the same, the methodology for the synthesis of a compound in the laboratory, and with the organism, is markedly different. Nonetheless, both are subject to the same physical laws of the universe: thermodynamic considerations, conservation principles, and so forth. Second, adaptation of some of this synthetic technology to make it compatible with the biological system provides the basis for biochemical probes (i.e., tools for the biologist to use in the study of the life process) and pharmacological action (i.e., chemicals or drugs that effectively act and alter a chemical process from a pathological to a normal state). Not only alkylations, but other reactions are valuable in this regard. For example, sulfonylation of the terminal amino function with dimethyl-aminonaphathalene-5-sulfonyl chloride (dansyl chloride; a highly sensitive fluorescent probe) or with 4-dimethylaminoazobenzene-4′-sulfonyl chloride (dabsyl chloride; a highly sensitive chromophoric probe) followed by hydrolysis of the protein (the sulfonamide linkage is resistant to 6 N HCl) provides an alternative method to Sanger's reagent for the determination of the N-terminal amino acid (free amino) of a protein. Numerous other reagents have been designed which interact specifically with and modify the structure and function of proteins in a manner that facilitates their study. An in-depth examination of these provides the basis for a protein chemistry course and more detailed information is available in protein or organic chemistry texts.

dansyl chloride   dabsyl chloride

One reaction type that merits special consideration is the acylation of amino acids. Other reactions of the amino acids are biologically important. For example, it will be seen later that it is Schiff base formation of the amino function with an aldehyde that provides the basis for all reactions with vitamin B_6 (see Chapter 7). However, it is acylation of the amino function of one amino acid with the carboxyl (activated) of another amino acid that leads to formation of the peptide bond and subsequent protein, or polymer formation. It then becomes of interest for the bioorganic chemist to consider and compare the synthesis of the most complex macromolecule in the test tube and the organism.

2.4 Acylations

A simple example would be the benzoylation of the methyl ester of glycine (in the presence of base):

The result is the formation of a stable amide linkage. Other acylations may be accomplished by classical methodology, such as acetylation with acetic anhydride. However, as was observed with alkylations and other reactions, acylations have been developed which readily occur under mild reaction conditions. For example, amino acids will react with isocyanates (R—N=C=O) and isothiocyanates (R—N=C=S) to form hydantoins and thiohydantoins.

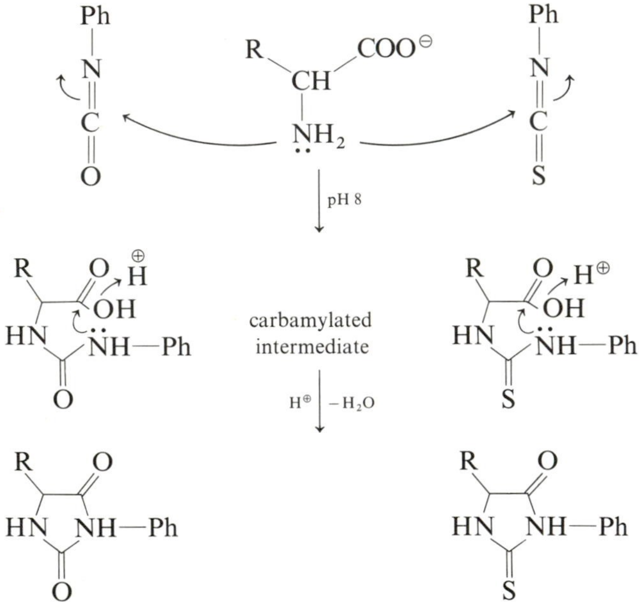

N-phenylhydantoin   N-phenythiohydantoin

It might be noticed that under acidic conditions (i.e., anhydrous trifluoro-acetic acid) the hydroxyl (protonated) becomes a leaving group, and so it is expected an amine (amide linkage) could also leave. Thus, the reaction of a protein with phenylisothiocyanate provides an important method for the determination of the N-terminal amino acid and subsequent determination of the amino acid sequence of the protein. Remembering that the monomeric units of the protein are connected by amide (so-called peptide) linkages, this may be illustrated for the simple dipeptide glycylalanine.

As alanine is the leaving group, glycine must be the N-terminal amino acid. The thiohydantoin and alanine formed are easily separated by chro-matography, the former being detected by UV absorption and fluorescence. Further note that had the sample been a tripeptide, then a dipeptide would have been the leaving group. This dipeptide could then be treated with phenylisothiocyanate so that the second amino acid residue could be identified as its thiohydantoin derivative. The sequence of the tripeptide is then established, and this methodology can be extended to determining the sequential arrangement of the monomeric amino acids in a small protein

$$
\begin{array}{c}
S \\
\| \\
C \\
\| \\
N \\
| \\
Ph
\end{array}
+ H_2\overset{..}{N}-CH_2-\overset{\overset{O}{\|}}{C}-NH-\overset{\overset{CH_3}{|}}{CH}-COO^{\ominus}
$$

glycylalanine

$$
\underset{\underset{\underset{S}{\|}}{\underset{HN \quad \overset{..}{N}H}{\diagdown \diagup}}}{CH_2 - \overset{\overset{O}{\|}}{C}} - NH - \overset{\overset{CH_3}{|}}{CH} - COO^{\ominus}
$$

with $H^{\oplus}$ and Ph labels

$$\downarrow H^{\oplus}$$

$$
\underset{\underset{S}{\|}}{\overset{\overset{O}{\|}}{HN \quad N}}-Ph \quad + \quad H_3\overset{\oplus}{N}-\overset{\overset{CH_3}{|}}{CH}-COOH
$$

starting from the N-terminus. This procedure is referred to as the *Edman degradation*. Other acylation reactions are of importance for the protection of the amino function during protein synthesis. These shall be discussed shortly.

The amide bond resulting from the acylation of amino acids is a planar hybrid structure, with an approximately equal distribution between two resonance forms. Because of the greater predominance of the ($C=N^{\oplus}<$) form, relative to esters, and the ($C=O^{\oplus}$) form, the amide bond is a stronger bond. As has been indicated, the amino function of one amino acid may be acylated by the acid function of a second amino acid. The amide bond so-formed is referred to as a peptide linkage and the product termed a *dipeptide*. Proteins are polymers which have monomeric units connected by amide (peptide) linkages. The amide bond formed between two amino acids is a secondary amide in a *trans* geometry. Free rotation does not readily occur about the C—N bond as this would destroyed the π resonance overlap, with the *trans* geometry being preferable for steric reasons.

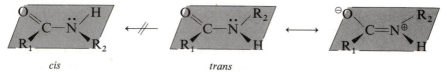

cis   trans

Structure of the planar sp^2 hydridized amide (peptide) bond;
free rotation from a *trans* to a *cis* geometry, with an energy barrier
of ~41.8 kJ/mol (10 kcal/mol), does not readily occur.

The peptide bond is a strong bond, and energy is required for its formation. Mixing an aqueous solution of two amino acids, one with an unprotonated amino function (potentially nucleophilic) and the other with a protonated carboxyl function, at room temperature would only result in salt formation. Chemically, the carboxyl function must be converted to a good leaving group. Energetically, the carboxyl function must be activated to compensate for the work done during peptide bond formation. This is reflected in the *free energy of hydrolysis* (ΔG_{hydro}) of the amide bond which is in the range of -12 kJ/mol (-3 to -4 kcal/mol). On the other hand, the ΔG_{hydro} for an acyl chloride is approximately -29.3 kJ/mol (-7 kcal/mol), and the chlorine atom is a good leaving group. It is therefore possible to convert the carboxyl of an amino acid to an acyl chloride (thionyl chloride, phosphorous pentachloride) and react this with the amino function of a second amino acid to form a peptide bond. This is an oversimplification of the problem of peptide bond (protein) synthesis and as shall be seen shortly a rather elaborate methodology has been developed. However, it may be of interest to first examine how the peptide bond is synthesized within the biological system, with particular attention to the problem of carboxyl function activation. Remember, the same energetics prevail in both the test tube and the organism.

$$R-C\underset{Cl}{\overset{O}{\diagup}} + H_2\ddot{N}-R' \longrightarrow R-\overset{O}{\overset{\|}{C}}-NH-R' + HCl$$

peptide bond formation

Within the biological context, an anhydride structure, as indicated below, is a high-energy structure or potential energy store. The definition of *high-*

$$(-\overset{O}{\overset{\|}{C}}-\ddot{X}-\overset{Z}{\overset{\|}{Y}}-)$$

energy for biological systems and more background into the chemistry of anhydrides is presented in Chapter 3 on bioorganic phosphorus. For the moment, it will suffice to say that any compound with an anhydride structure will exhibit $\Delta G_{hydro} > 29.3$ kJ/mol (7.0 kcal/mol). Examples include acetic anhydride, acetyl phosphate, and acetyl imidazole. The biological energy store ATP also possesses the anhydride structure in its triphosphate side

$$CH_3-\overset{O}{\overset{\|}{C}}-O-\overset{O}{\overset{\|}{C}}-CH_3 \qquad CH_3-\overset{O}{\overset{\|}{C}}-O-\overset{O}{\overset{\|}{P}}\overset{O^{\ominus}}{\underset{OH}{\diagup}} \qquad CH_3-\overset{O}{\overset{\|}{C}}-N\diagdown\overset{N}{\diagup}$$

| acetic anhydride | acetyl phosphate | acetyl imidazole |

chain. The fact that ATP can transfer this energy to activate a carboxyl function is shown in a simple biochemical experiment. Incubation of a liver homogenate (liver of an animal homogenized so as to release the enzymes from inside the cell) with glycine, benzoic acid, and ATP leads to the formation of *N*-benzoyl glycine (hippuric acid). Actually, this represents a mechanism of *detoxication* in the mammal or making a harmful substance (benzoic acid) harmless by conjugation with a freely available substance in the body (glycine) and then eliminating the product in the urine. The first step would be nucleophilic attack of the benzoate on ATP, to give an activated benzoate (carboxylate) and inorganic pyrophosphate. The amino function of glycine may now attack this anhydride intermediate to give the N-benzoylated product.

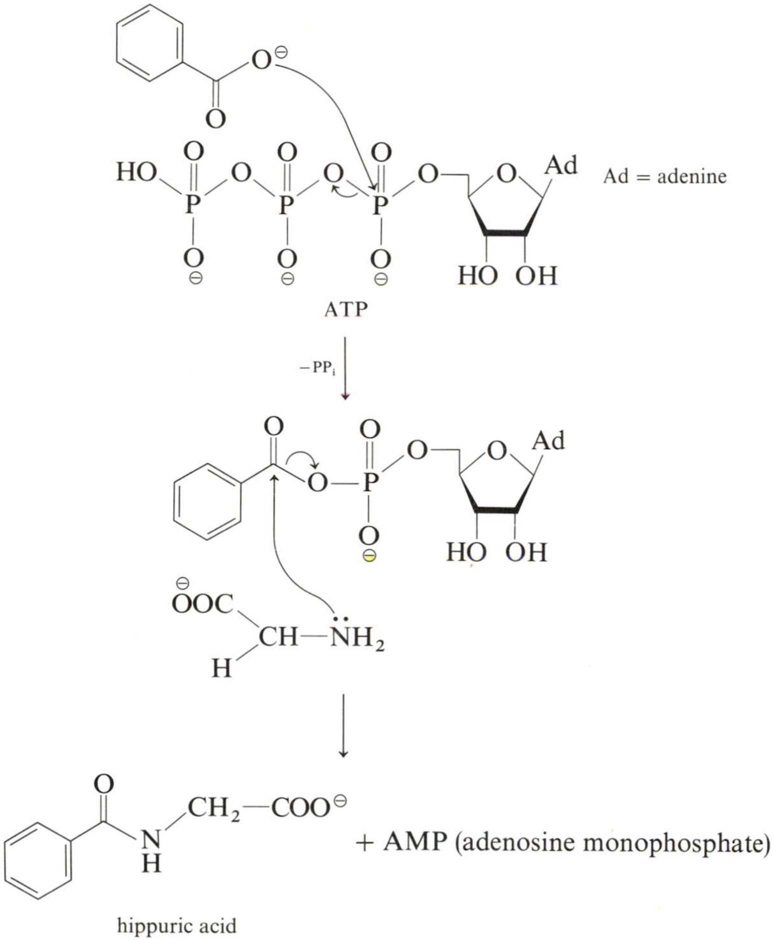

ATP

Ad = adenine

hippuric acid

+ AMP (adenosine monophosphate)

2.5 Biological Synthesis of Proteins

Much the same principle of carboxylate activation is applicable to the *in vivo* synthesis of proteins. Again the carboxylate of an amino acid becomes activated by reaction with ATP to form an anhydride intermediate. The next step does not simply involve attack of a second amino acid of this anhydride since the synthesis of a protein involves the precise sequential coupling of many (up to a few hundred) amino acids. A *template* or "ordered surface" must be available to ensure correct sequencing of the protein molecule. The macromolecule that serves such a function is a polynucleotide, transfer ribonucleic acid (tRNA); the constitution of polynucleotides will be described in the next chapter.

Activation of the amino acid by ATP is only an intermediate step catalyzed by the enzyme aminoacyl-tRNA synthetase. The 3′- or 2′-hydroxyl of the terminal adenylic acid of the tRNA molecule then attacks the anhydride intermediate to give an aminoacyl-tRNA molecule.

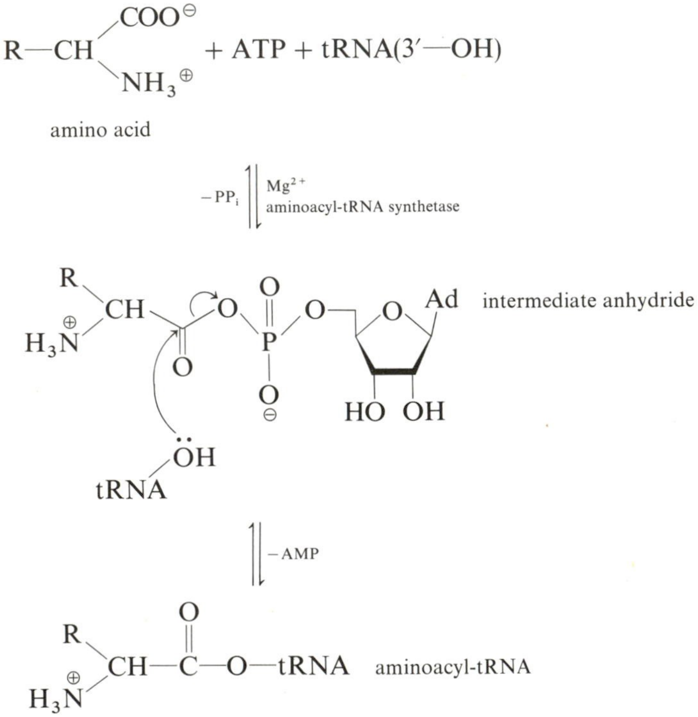

$$R\!-\!CH \begin{cases} COO^{\ominus} \\ NH_3^{\oplus} \end{cases} + ATP + tRNA(3'\!-\!OH)$$

amino acid

$$-PP_i \quad \Big\Updownarrow \quad \begin{array}{l} Mg^{2+} \\ \text{aminoacyl-tRNA synthetase} \end{array}$$

intermediate anhydride

$$-AMP$$

aminoacyl-tRNA

The ester linkage between the hydroxyl of the tRNA is a high energy bond (due to the adjacent 2′-hydroxyl and cationic amino functions) so that the overall enzyme catalyzed reaction has a free energy change close to zero. Each amino acid has a specific tRNA molecule and one specific aminoacyl-tRNA synthetase enzyme. In turn, each aminoacyl-tRNA synthetase enzyme will accept only its particular amino acid as a substrate. However, it has been possible to slightly modify the naturally occuring amino acids and have them serve as substrates for the enzyme. For example, *p*-fluorophenylalanine can substitute to some extent for phenylalanine.

Once synthesis of the aminoacyl-tRNA is complete, the amino acid no longer serves a recognition function. Specificity is dictated by the tRNA portion of the molecule by its interaction with the genetic message (mRNA) and another large surface upon which protein synthesis takes place; a cellular organelle referred to as the ribosome.

This was demonstrated by taking the aminoacyl-tRNA complex specific for cysteine (abbreviation, cysteinyl-tRNA^{Cys}, showing that the cysteine has combined with its specific tRNA) and modifying the amino acid side chain to form alanine by catalytic reduction.

$$tRNA^{Cys}-O-\overset{\overset{\displaystyle O}{\|}}{C}\underset{\underset{\displaystyle CH_2SH}{}}{\overset{\oplus}{\diagup}NH_3} \xrightarrow[\text{Raney Ni}]{H_2} tRNA^{Cys}-O-\overset{\overset{\displaystyle O}{\|}}{C}\underset{\underset{\displaystyle CH_2-H}{}}{\overset{\oplus}{\diagup}NH_3}$$

cysteinyl-tRNA^{Cys} alanyl-tRNA^{Cys}

The modified aminoacyl-tRNA was then incubated *in vivo* and incorporated alanine into the protein at those positions which normally accepted cysteine.

Protein synthesis begins with the N-terminal amino acid and proceeds from this point. In some bacteria, yeast, and higher organisms, this first aminoacyl-tRNA is known to be *N*-formylmethionyl-tRNA^{fMet}. Formylation of the amino function can be considered as a protecting group to prevent participation of the amino function is peptide bond formation. The fMet-tRNA^{fMet} is then the first aminoacyl-tRNA to bind to the ribosome and mRNA. After the protein is synthesized, the formyl group is removed by enzymatic cleavage (formylase).

The ribosome is a large cellular organelle, composed of RNA and a number of different proteins, and built of two dissociable subunits. This provides the ultimate ordered surface for protein synthesis, being able to interact precisely with the large tRNA portions of the various aminoacyl-tRNA's. While some of the proteins of the ribosome presumably have a catalytic function, the rest as well as the ribosomal RNA (rRNA) participate in specific conformational interactions that occur during protein synthesis. Protein synthesis is a dynamic process, but again this process occurs in an ordered fashion which the sequential coupling of amino acids demands. The site to which the fMet-tRNA^{fMet} binds on the ribosome is referred to as the *peptidyl site*. The stage is now set for the synthesis of the peptide linkage.

The second amino acid (aminoacyl-tRNA) also binds on the ribosome (at the so-called *aminoacyl site*) in close proximity to the fMet-tRNA^{fMet}. While no chemical reaction has yet occured, this binding process requires work, the energy of which comes from a molecule of GTP (guanosine triphosphate; like ATP except adenine is replaced with guanine). The amino function of the aminoacyl-tRNA now attacks the f-methionine, at which point the tRNA^{fMet} becomes a leaving group, and the peptide bond is formed. This reaction is enzyme catalyzed, but neither ATP or GTP is required for the peptide bond formation as the energy for this process comes from the cleavage of the high energy tRNA^{fMet} ester. While this completes peptide bond formation, obviously some physical changes must occur in order that another incoming aminoacyl-tRNA may attach to the dipeptide. The di-peptidyl-tRNA at the aminoacyl site is physically shifted to the peptidyl site, simultaneously displacing the tRNA^{fMet}. This, very likely, results from a conformational change on the ribosome which again requires energy at the expense of a molecule of GTP. The aminoacyl site is now empty, and the mRNA has also shifted, (translocation) so that it can now dictate the entrance

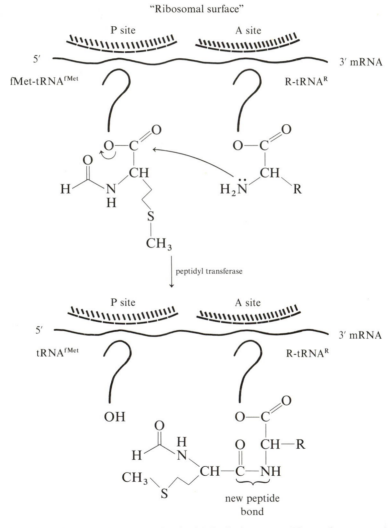

Fig. 2.1. Peptide bond formation in the biological system. The entire process takes place on the ribosome and involves two binding sites, the P (peptidyl) and A (aminoacyl) sites. The reaction is catalyzed by the enzyme peptidyl transferase. It is the message carried by mRNA (which in turn is dictated by the genetic material DNA) which determines by specific interactions as to which aminoacyl-tRNA will bind at the P and A sites.

of a new aminoacyl-tRNA to the aminoacyl site. Once this has occurred, peptide bond formation (to give a tripeptide) can take place, thus repeating the sequence of events as described (Fig. 2.1).

This provides a basic description of *in vivo* protein synthesis, neglecting details such as the involvement of protein elongation factors. Obviously the process is very complex, but underlying this complexity is the basic theme; activation of the carboxyl function followed by sequential coupling of amino acids on an ordered matrix that makes an incorrect sequence and other side reactions a virtual impossibility. The importance of such considerations will be realized when the chemical synthesis of proteins is discussed shortly. Nonetheless, with this knowledge of *in vivo* protein synthesis, it is possible to appreciate some of the pharmacological actions of some drugs or antibiotics that interfere with this process. Such antibiotics tend to be of a potentially toxic nature as they can interfere with the protein synthesis of both the infecting bacteria and the patient, but depending upon the drug, they can be quite therapeutically useful.

The antibiotic puromycin has a structure very similar to the terminal adenylic acid of tRNA (Fig. 2.2). However, the primary 5'-hydroxyl of

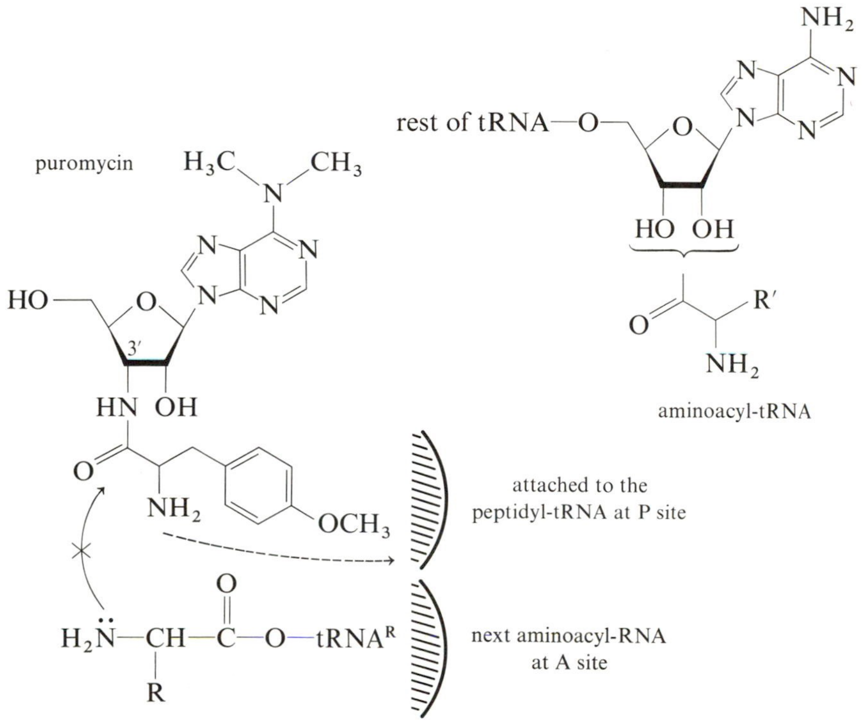

Fig. 2.2. Structure of the antibiotic puromycin. Puromycin interrupts the synthesis of the protein chain by mimicking an aminoacyl-tRNA complex, binding to the ribosome, attacking a peptidyl-tRNA (at the P site) with its amino function, but then not undergoing attack by an aminoacyl-tRNA, hence terminating the synthesis. In the upper right corner is shown the adenylic acid terminus of tRNA.

puromycin is not esterified to a large tRNA molecule, the amino function of the adenine is dimethylated and the 3′-hydroxyl is replaced by an amide linkage to the carboxyl of p-methoxyphenylalanine. During synthesis of the protein chain, a molecule of puromycin can enter the aminoacyl site when, for example, tyrosyl-tRNA^{Tyr} is called for at the ribosome. Once bound, it can via its amino function, attack the chain at the peptidyl site. It cannot, however, be attacked by the amino function of another incoming aminoacyl-tRNA, for the latter cannot break apart the stronger peptide bond at the 3′-position of the antibiotic. Further, after translocation, it is no longer bound to the ribosome by a tRNA molecule, and so can dissociate from the surface, taking with it the partially completed protein molecule. Puromycin will act equally well on protein synthesis in mammalian cells and so is not a practical antibacterial agent.

On the other hand, the antibiotic chloramphenicol binds to the bacterial ribosomes, but not the larger ribosomes of animals, thereby showing greater specificity. This inhibits peptide bond formation. Chloramphenicol is relatively nontoxic, but it still must be used with caution as side effects such as anemia can result. Other antibiotics which inhibit protein synthesis include the tetracyclines, streptomycin, and cycloheximide.

$$NO_2$$

$$CH_2OH$$
$$HO$$
$$HN-C-CH \begin{array}{c} Cl \\ Cl \end{array}$$
$$O$$

chloramphenicol

Some microorganisms are able to use a simpler, more primitive method of peptide bond synthesis. While perfectly effective for the peptide formation, the system lacks the highly ordered apparatus provided by the ribosome and tRNA structures. As such, only small proteins (polypeptides) are synthesized by this means; the antibiotic gramicidin S providing an important example (5). Gramicidin S is an interesting antibiotic for a number of reasons. First, it contains phenylalanine in the D-configuration. The occurrence of D-amino acids in nature is very rare and only L-amino acids are found in proteins. Second, gramicidin S contains the amino acid ornithine, which is not a normal constituent of proteins.

Gramicidin S is a cyclic decapeptide isolated from *Bacillius brevis* whose antibiotic activity is derived from its ability to complex alkali metal ions and transport them across membranes. It acts as a channel for ion leakage which is not difficult to imagine when it is noted that it has a doughnut

2.5 Biological Synthesis of Proteins

D-Phe $\longrightarrow$ L-Pro $\longrightarrow$ L-Val $\longrightarrow$ L-Orn $\longrightarrow$ L-Leu

L-Leu $\longleftarrow$ L-Orn $\longleftarrow$ L-Val $\longleftarrow$ L-Pro $\longleftarrow$ D-Phe

$$\text{Orn = ornithine}; \ H_2N\text{-}(CH_2)_3\text{—}CH \Big\langle {\overset{\oplus}{N}H_3 \atop COO^{\ominus}}$$

Structure of the antibiotic gramicidin S; decapeptide consists
of two complementary pentapeptide strands; molecule has a
twofold axis of symmetry

shape capable of encapsulating the metal ion (see Section 5.1.3). Unfortunately this action is not restricted to bacterial membranes and so gramicidin S is only used topically.*

We may now examine how a dissymmetric macromolecular complex of proteins, found in nature, is responsible for the synthesis of this cyclic polypeptide without the need of ribosomes. This process, which occurs only in simple organisms like bacteria, was deciphered by the biochemist F. Lipmann, from Rockefeller University.

Lipmann found that the mere addition of ATP to crude cell homogenate and extracts prepared from *B. brevis* promotes the synthesis of gramicidin S. This activity was found in particle-free supernatant fractions and was resistant to known inhibitors of protein synthesis and to treatment with ribonuclease, excluding the participation of an RNA molecule. In this biological transformation, ATP fulfills the function of a condensing reagent in a manner analogous, for example, to a carbodiimide group (see Section 2.6.3).

Two enzymes are involved in the synthesis of gramicidin S: a light (MW = 100,000) and a heavy (MW = 280,000). Synthesis begins on the light enzyme, which also functions as a "racemase," converting L-phenyl-alanine to the D-enantiomer. A thiol nucleophile on the light enzyme attacks an activated phenylalanine (ATP and the amino acid reacted to form an anhydride) to give a high energy thiol ester. The ΔG_{hydro} of a thiol ester (base catalyzed) is approximately 38 kJ/mol (-8 kcal/mol). The difference between thiol and acyl esters is accounted for by the fact that the oxygen atom will more readily delocalize its nonbonding electrons into the carbonyl function than will the sulfur atom. Such delocalization will reduce the electrophilicity of the carbonyl function. Furthermore, the thiol is a much better leaving group than the corresponding alcohol. Remember, the pK_a of a mercaptan is approximately 10, while that of an alcohol is approximately 15 (Table 2.1).

---

* For the most recent chemical synthesis of gramicidin S, see reference 364.

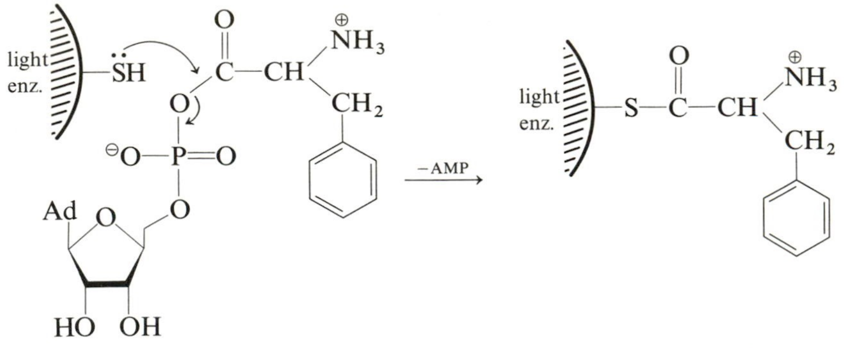

In the next step, proline attached to the heavy enzyme by a thiol ester linkage attacks the light enzyme, thus transferring the dipeptide to the heavy enzyme. The function of the light enzyme is terminated.

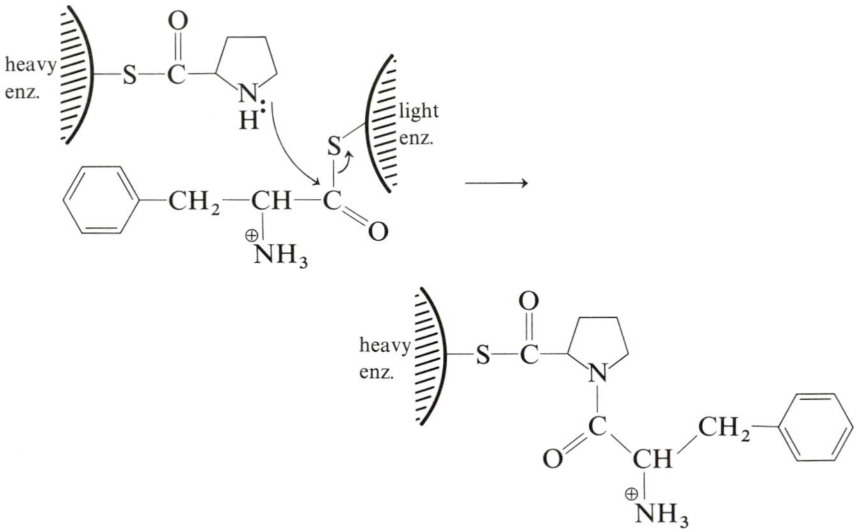

Similarly, on the heavy enzyme, the amino function of valine (attached as the thiol ester) attacks the dipeptide, forming a tripeptide product. Remember, the energy for peptide bond formation is being derived from the aminolysis of the thiol ester.

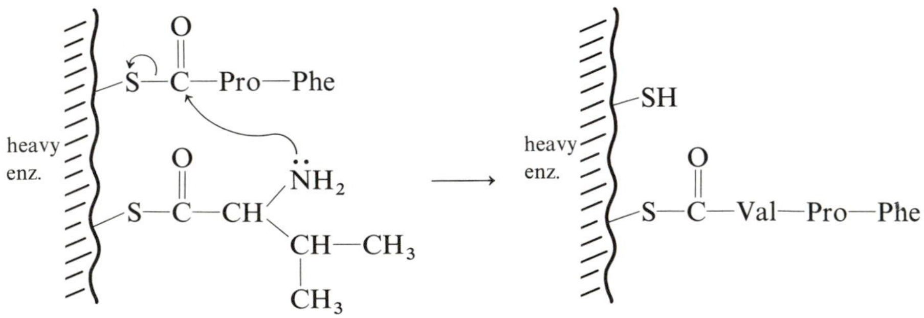

The next step involves an "arm" on the heavy enzyme. This arm "picks up" or attacks the tripeptide via its thiol function. It is then attacked by ornithine to form a tetrapeptide, but picks up the tetrapeptide (via the same thiol) and is attacked by leucine to form a pentapeptide. The arm again retrieves the pentapeptide, but this time the pentapeptide is attacked by a complementary pentapeptide (attached to another arm) to give rise to the cyclic decapeptide.

Structure of the thiol containing "swinging arm" present in "heavy enzyme"; arm consists of β-mercaptoethylamine and pantothenic acid (an important component of coenzyme A) esterified to a serine phosphate of the enzyme

In brief, quaternary interactions between the two enzymes allow the two amino acids to get close enough for peptide bond formation. This process is repeated five times to form the two halves of the molecule. The proteinic complex collapses the two identical chains to give finally the cyclic peptide, schematized in Fig. 2.3.

Figure 2.4 gives a different representation to amplify the fact that only the light chain is responsible for the presence of D-Phe in the sequence. The remaining part of the antibiotic is synthesized by the heavy chain.

In conclusion, these two enzymes have enough tridimensional orientational effect to recognize the right amino acids and to bring them together for the synthesis of a well-defined asymmetric and cyclic peptide molecule. It is the presence of specific protein–protein interactions that enable such an efficient polypeptide synthesis to take place. In other words, all the molecular information needed for the synthesis of gramicidin S is present within this macromolecular enzymatic complex to allow amino acid recognition and proper peptide bond formation. Again, this nonribosomal polypeptide synthesis is a remarkable example of the importance of chirality (see Section 4.2.1). We are still far from being able to produce such an efficient peptide synthesis in the laboratory, even with Merrifield's solid phase method (see page 77). However, the analogy is apparent.

The team of R. L. Letsinger and I. M. Klotz of Northwestern University developed recently a method for peptide synthesis based on a template-directed scheme that parallels that of the natural mechanism using ribosomes, presented earlier in Fig. 2.1. The strategy employed makes use of a resin

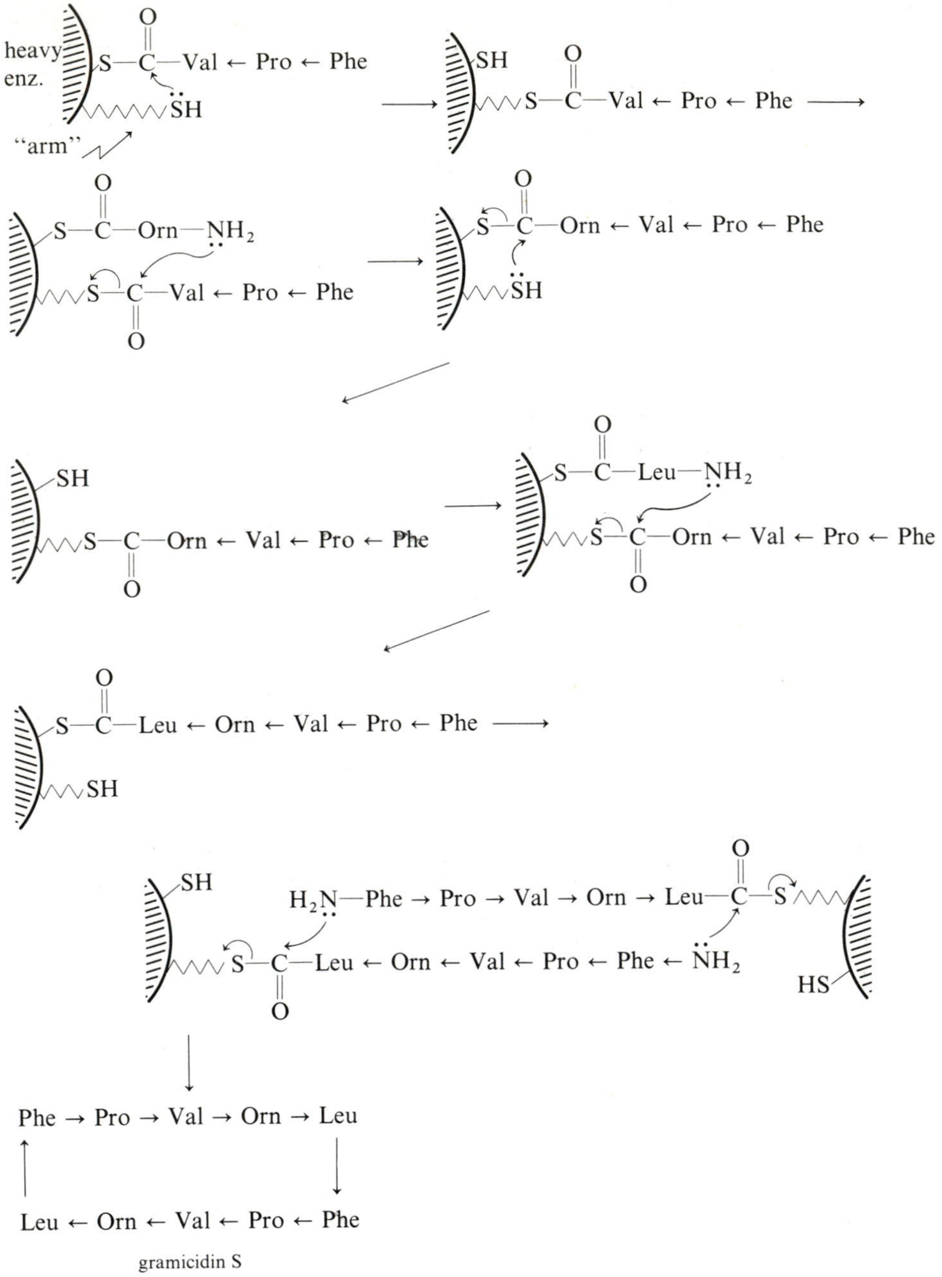

Fig. 2.3. Lipmann's enzyme-controlled synthesis of the antibiotic peptide gramicidin S.

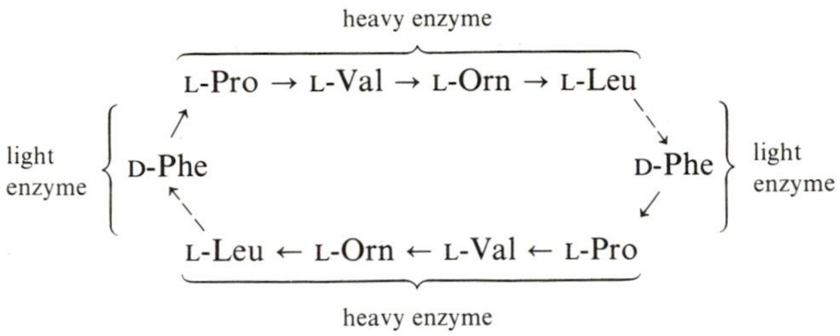

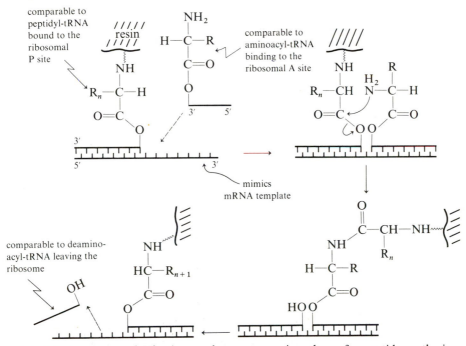

heavy enzyme

L-Pro → L-Val → L-Orn → L-Leu

light enzyme { D-Phe   D-Phe } light enzyme

L-Leu ← L-Orn ← L-Val ← L-Pro

heavy enzyme

Fig. 2.4. Sequence of amino acids in gramicidin S. The brackets indicate the enzymes activating the enclosed amino acids.

support and a polynucleotide template but does not require (like the natural system) temporary amino acid protecting groups in order to achieve fidelity of coupling. The approach as been called the *complementary carrier method for peptide synthesis* and is summarized in Fig. 2.5.

In practice the growing polypeptide chain is attached to an oligonucleotide at the ribose 5′-OH by an ester linkage to the terminal carboxyl group. The incoming new amino acid is also linked as an ester, but to the 3′-OH group of a second oligonucleotide. From a variety of studies on helix stability

Fig. 2.5. Letsinger and Klotz's complementary carrier scheme for peptide synthesis which has a great similarity with the natural ribosomal mechanism (6).

it appears that carriers of eight nucleotides in length are sufficient to bind properly to a longer oligonucleotide template by virtue of Watson–Crick complementary base pairings (see page 103). Current chemical methods of peptide bond formation are used to extend the polypeptide chain by one amino acid. To proceed to the next coupling step of the synthesis, the un-bound oligonucleotides are simply washed from the insoluble solid support under conditions that disrupt base pairing, such as temperature or decreased ionic strength. This circular sequence of reactions can be repeated until the desired polypeptide molecule is constructed. Notice that at each step, the growing polypeptide chain is transferred to the complementary carrier oligonucleotide and at the end of the synthesis the final carrier can then be cleaved from the solid support. The remaining attached oligonucleotide can serve as a chromatographic handle during the purification of the poly-peptide molecule and eventually released under mild basic conditions.

The important feature of this methodology is that fidelity of the amide bond formation is achieved by specific and template-directed coupling of the desired acyl transfer. Therefore, proper juxtaposition of peptidyl- and aminoacyl-oligonucleotides on a polynucleotide template control the direction of polypeptide synthesis. This is essentially the way that fidelity of translation in protein biosynthesis is taking place in natural systems.

This scheme (Fig. 2.5) thus shows how oligonucleotides can direct the synthesis of polypeptides in the absence of protein or ribosomal machinery and, as such, is an appealing bioorganic model for the origin of prebiotic protein synthesis (see Section 3.7.2.1). Indeed, it seems most probable that primitive biosystems used a similar concept to carry out primitive protein synthesis where Watson–Crick base pairing provided the intrinsic mechanism for achieving fidelity of replication and direction of protein synthesis. In time, the carrier oligonucleotides could have evolved into more efficient species such as the present-day tRNA molecules.

2.6 Chemical Synthesis of Proteins

Biological synthesis of proteins is a complex process requiring ordered macromolecular surfaces, protein catalysts, energy storage forms, etc. This alone would indicate that the chemical synthesis of a protein would be a difficult undertaking. Therefore, a number of considerations must be made before attempting such a project.

First, specificity or sequential coupling of the amino acids must be achieved. Unwanted side reactions must be avoided. Consider, for example, the synthesis of the dipeptide glycylalanine (Gly-Ala). It is not possible to mix the two amino acids together and let them react as the desired sequence will not be the only product. For example, it is possible that glycine will react with itself to give the dipeptide glycylglycine. In fact, the number of possible dipeptides is four (Gly-Gly, Gly-Ala, Ala-Gly, Ala-Ala), of which only one is the correct sequence. The problem becomes worse with increasing chain length. If random reaction were allowed for the synthesis of half of the gramicidin S molecule (i.e., five amino acids were placed in a reaction vessel and allowed to react), then product pentapeptides of 3125 possible sequences would form. Only one would be the desired sequence, with the other 3124 pentapeptides having to be separated and discarded as by-products.

Clearly, a successful protein synthesis demands ordered sequential coupling of amino acids, with little by-product formation. Such is achieved by the use of protecting groups for amino functions, carboxyl functions, and potentially reactive amino acid side chains. For example, returning to the synthesis of Gly-Ala, if the amino function of glycine is protected (rendered chemically unreactive) then reaction of glycine with itself becomes impossible. Further, if the carboxyl function of alanine is also protected, then the only possible reaction is that between the carboxyl (activated) of glycine and the amino of alanine to give the desired dipeptide product.

Of course, once the peptide bond is synthesized it becomes necessary to remove the protecting groups under conditions that will not affect the product. Consequently, protecting groups must easily be both attached to the reactants and removed from the product under mild conditions and in high yield. With these challenges in mind, perhaps it can now be realized that the chemistry of protecting groups is by itself an important field of organic chemistry. These concepts are not limited to the synthesis of proteins, but are applicable to the synthesis of any difficult organic molecule (for example, as shall be seen in the next chapter for the synthesis of polynucleotides) whose construction involves potentially unwanted side reactions from numerous reactive centers.

Knowing that each amino acid must be added sequentially in a stepwise fashion, the chemical yield of each step in the synthesis of the protein becomes very important. Again considering the synthesis of Gly-Ala, if the peptide bond forming reaction occurs in 90% yield, this might be considered a good synthesis. However, if the same procedure were used to make the gramicidin S decapeptide, the overall yield would be $(0.90^{10} \times 100\%) = 35\%$. This does not include any losses during protection and deprotection procedures. Consequently, in the synthesis of a protein macromolecule, peptide bond formation must be a high yield reaction.

Another factor peculiar to biological systems must be considered: *optical purity*. Proteins are made of L-amino acids. As such, a chemical synthesis must start with L-amino acids and racemization must be minimized during the synthesis. This is especially true in the synthesis of an enzyme, as catalytic activity is dependent upon optical integrity. Amino acids are particularly susceptible to racemization once they have been acylated (i.e., addition of an acyl protecting group to the amino function) via intermediate azlactone formation. This can occur during protection or coupling procedures:

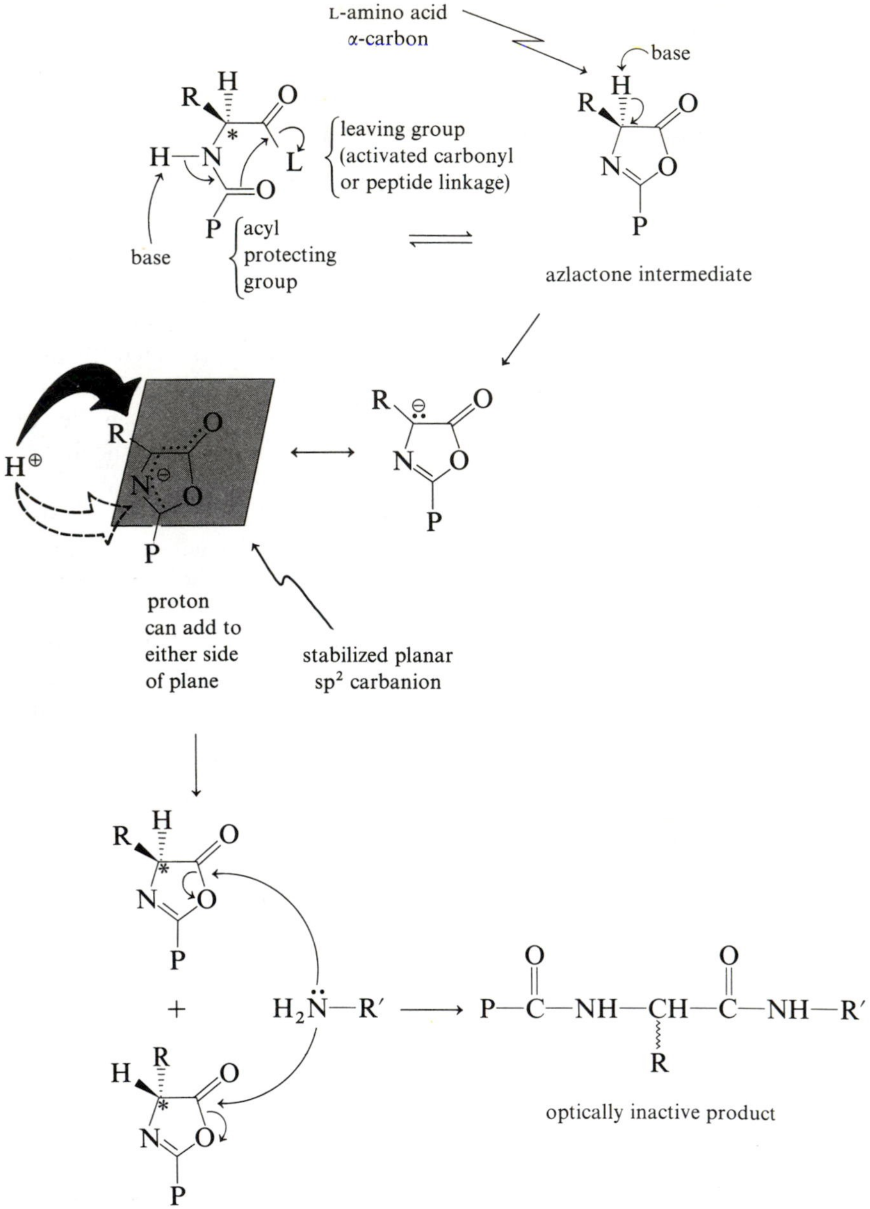

The α-proton of the azlactone is quite base labile (stabilized carbanion) so that a proton may be lost, but can later add to either side of the azlactone plane with subsequent loss of optical purity. Further, the azlactone may participate in peptide bond forming reactions, so that it disappears during the synthesis, but racemic amino acids are introduced into the product protein.

2.6.1 Amino Protecting Groups

(1) tert-Butoxycarbonyl Function (tert-BOC)

tert-Butoxycarbonyl function (*tert*-BOC) is an acyl protecting group for the amino function and is readily removed under mild acidic conditions. Once acylated, the amino function is rendered chemically unreactive or non-nucleophilic as a result of electron delocalization into the amide bond (carbamate). This may be introduced by reaction of the chloride with the amino acid. The chloride is synthesized from *tert*-butanol and phosgene. However, it is too unstable for convenient use and storage:

$$tert\text{-Bu}-\text{OH} + \text{Cl}-\overset{\overset{\text{O}}{\|}}{\text{C}}-\text{Cl} \xrightarrow{-\text{HCl}} \quad tert\text{-Bu}-\text{O}-\overset{\overset{\text{O}}{\|}}{\text{C}}-\text{Cl} \quad (tert\text{-BOC}-\text{Cl})$$

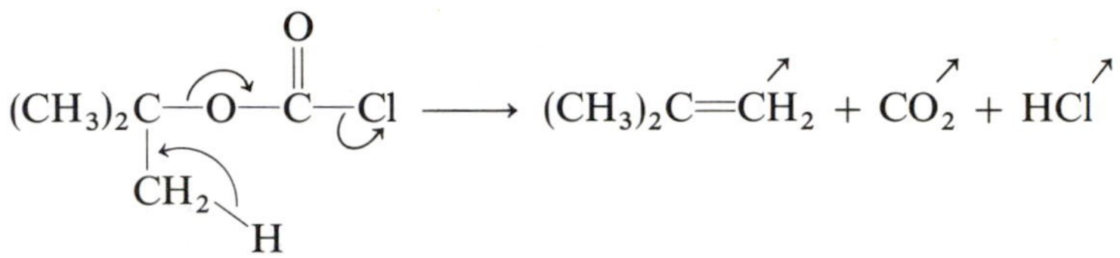

$$(\text{CH}_3)_2\text{C}-\text{O}-\overset{\overset{\text{O}}{\|}}{\text{C}}-\text{Cl} \longrightarrow (\text{CH}_3)_2\text{C}=\text{CH}_2 + \text{CO}_2 + \text{HCl}$$
$$\underset{\quad \text{H}}{|}$$
$$\text{CH}_2$$

This has been replaced by the less reactive *tert*-BOC azide. The azide has been used extensively until recently when it was reported to be dangerous (explosive). Presently, the reagents of choice are "BOC-ON" [2-(*tert*-butoxycarbonyloxyimino)-2-phenylacetonitrile] and di-*tert*-butyl dicarbonate. Both give *tert*-BOC amino acids in high yields.

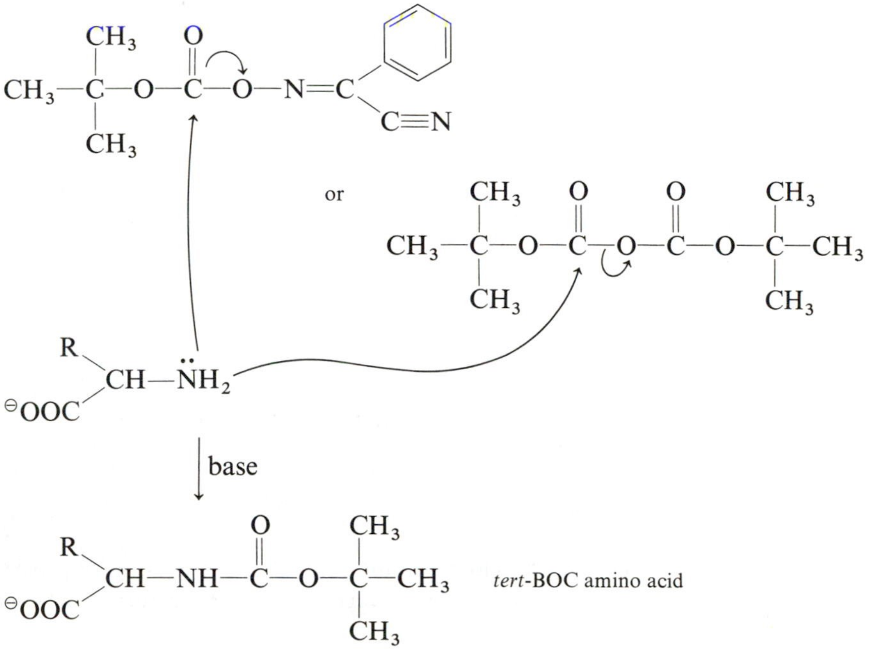

tert-BOC amino acid

The *tert*-BOC function is resistant to conditions that are typically used to remove other protecting groups: hydrogenolysis, sodium in liquid ammonia, and alkaline conditions (more so than the CBz group; see below).

The *tert*-BOC group is readily removed under mildly acidic conditions to give gaseous isobutylene:

$$CH_3-\overset{\displaystyle \underset{|}{CH_3}}{\underset{\displaystyle \underset{|}{CH_2}}{C}}-O-\overset{\displaystyle \overset{O}{\|}}{C}-NH-\overset{\displaystyle \underset{|}{R}}{CH}-COO^{\ominus} \;\; \xrightarrow{\hspace{1cm}}$$

$$\overset{\displaystyle CH_3}{\underset{\displaystyle CH_3}{>}}C{=}CH_2 + \overset{\oplus}{C}O_2 + H_3\overset{\oplus}{N}-CH\overset{\displaystyle COO^{\ominus}}{\underset{\displaystyle R}{<}}$$

isobutylene

and carbon dioxide (further pushing the equilibrium to the right). Acidic conditions used to remove the group include hydrogen bromide in acetic acid, hydrogen chloride in acetic acid, warm acetic acid (80%), anhydrous trifluoroacetic acid (the latter three will not affect the CBz group, see below), and liquid hydrogen fluoride.

(2) Carbobenzoxy Function (CBz)

Carbobenzoxy function (CBz) is also an acyl protecting group for the amino function which is removed by hydrogenolysis or under acidic conditions.

The chloride is prepared by reaction of benzyl alcohol with phosgene:

$$PhCH_2OH + Cl-\overset{\displaystyle \overset{O}{\|}}{C}-Cl \;\longrightarrow\; PhCH_2O-\overset{\displaystyle \overset{O}{\|}}{C}-Cl$$

The chloride may be reacted with amino acids in mildly alkaline, aqueous media:

$$PhCH_2O-\overset{\displaystyle \overset{O}{\|}}{C}-Cl$$

$$+$$

$$H_2\overset{\displaystyle ..}{N}-CH\overset{\displaystyle COO^{\ominus}}{\underset{\displaystyle R}{<}} \;\; \xrightarrow{\text{base}} \;\; PhCH_2O-\overset{\displaystyle \overset{O}{\|}}{C}-NH-CH\overset{\displaystyle COO^{\ominus}}{\underset{\displaystyle R}{<}}$$

CBz amino acid

The CBz function is more acid resistant than the *tert*-BOC group (see above), and it will not be removed by warm acetic acid for the period of time needed to remove the trityl and *p*-methoxytrityl functions (see below). It is base resistant (i.e., dilute NaOH at room temperature), but not to the same extent as the *tert*-BOC function due to the better leaving ability of the benzyl group.

The CBz function may be removed by catalytic hydrogenation:

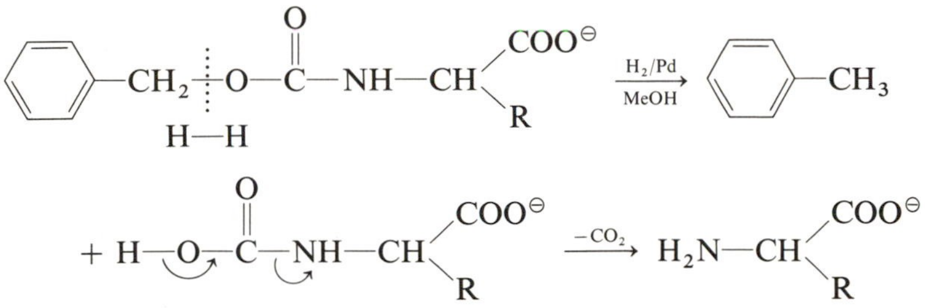

Hydrogenation of the benzene ring would occur to some extent only under high pressure, and in the presence of platinum. However, poisoning of the metal catalyst may result from the presence of sulfur-containing amino acids. In some cases this may be overcome by the use of excess catalyst. Also, benzyl groups (and to a lesser extent trityl) will be removed under typical hydrogenation conditions. The CBz group may also be removed by HBr in anhydrous acetic acid (room temperature). The mechanism involves alkyl fission:

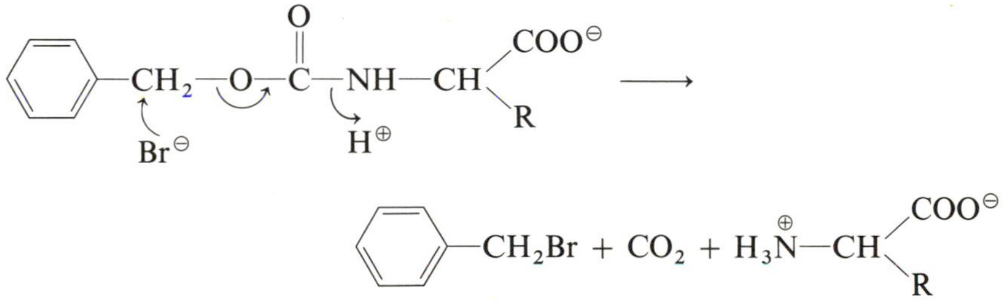

Other acids may also be used, but these usually require elevated temperatures (i.e., boiling trifluoroacetic acid, boiling methanolic HCl). Another means of removal is reduction with sodium in liquid ammonia. However, this method also removes tosyl (see below) and benzyl groups and demethylates any methionine present to homocysteine.

Other derivatives of the CBz group have also found use for amino acid protection. For example, the *p*-methoxycarbobenzoxy group can be removed in anhydrous trifluoroacetic acid. On the other hand, the *p*-nitrocarbobenzoxy group is significantly more resistant to HBr (electron withdrawing nitro group) than the CBz group. The CBz function may be removed in the presence of the *p*-nitro analogue. Also, the *p*-nitro-CBz amino acid derivatives tend to crystallize more readily.

(3) *Phthaloyl Function*

Phthaloyl function is an acyl anhydride which is removed by hydrazine or phenylhydrazine.

The classical method of preparation of *N*-phthaloyl amino acids involves fusing together a mixture of phthalic anhydride and the amino acid:

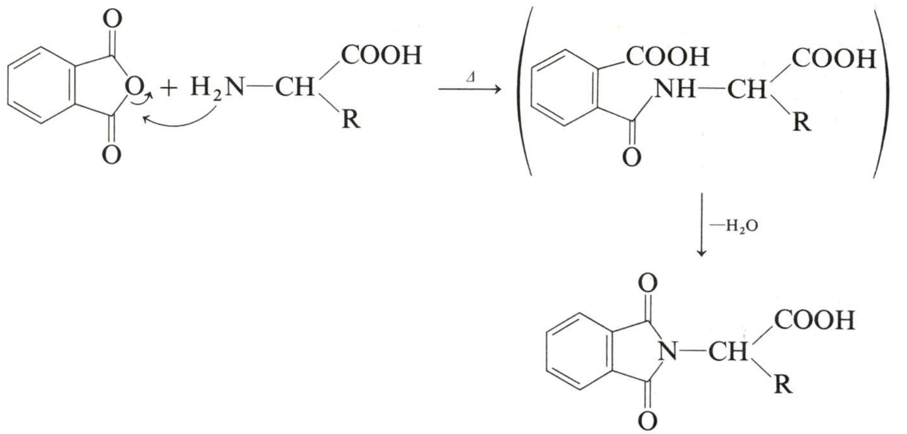

N-phthaloyl amino acid

However, heating can be damaging to sensitive peptides and significant racemization can result. Other methods for introduction of the phthaloyl group have been explored. For example, preparation of the *N*-phthaloyl amino acid has been achieved by reaction of *o*-carboethoxythiobenzoic acid with an amino acid ester, followed by reaction with aqueous HBr in acetic

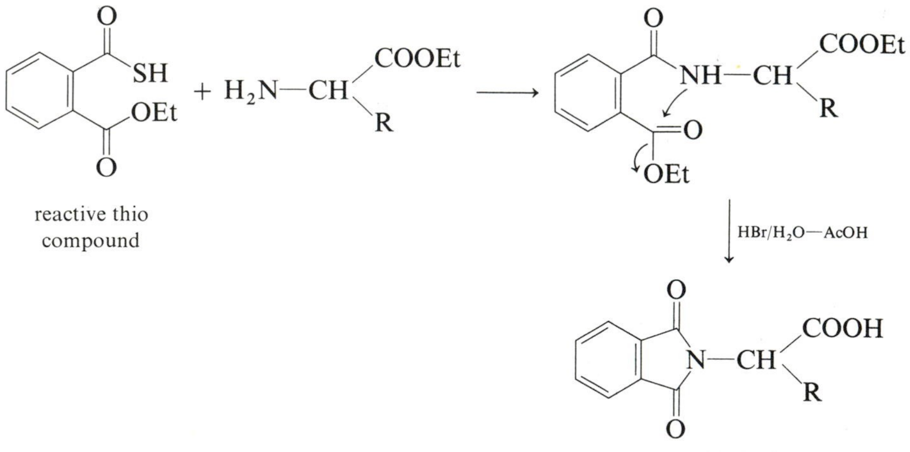

reactive thio
compound

acid. The most general method used is the reaction of an amino acid with *N*-carboethoxyphthalimide. This reagent is prepared by reaction of phthalimide with ethyl chloroformate:

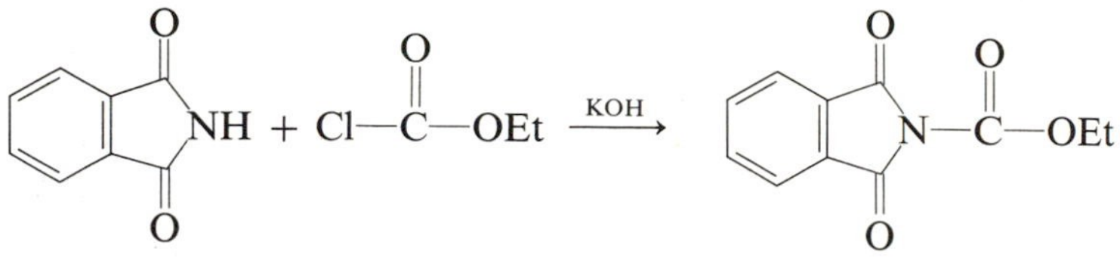

N-carboethoxyphthalimide

Reaction with amino acids takes place in dilute alkaline, aqueous solution, without heat.

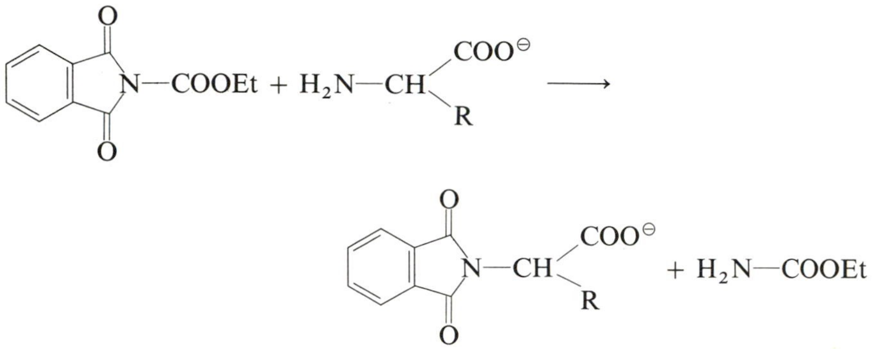

The phthaloyl group is unaffected by catalytic hydrogenation, sodium in liquid ammonia, and is acid resistant (i.e., HCl or HBr in acetic acid water, at room temperature). However, this group is base labile. The product formed upon base treatment is cleaved by slightly acidic conditions to give the free amino acid:

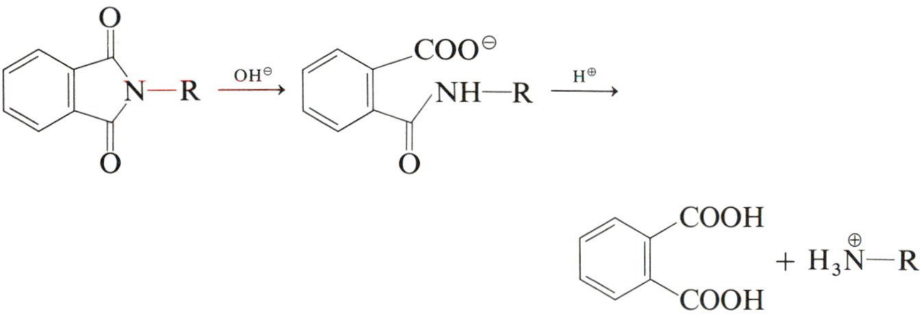

The classical method for removal of the phthaloyl function is treatment with aqueous or ethanolic hydrazine hydrate. The hydrazide by-product separates out by precipitation. Hydrazine is a good nucleophile (so-called α-effect) which will attack the anhydride-like phthaloyl function under conditions that will not affect peptide linkages. However, there is a danger of hydrazinolysis of any ester linkages, if present. Other protecting groups will not be affected by the hydrazinolysis. The use of hydrazine in hydroxylic solvents may be replaced by phenylhydrazine in organic solvents, in which case the hydrazide remains in solution, and the amino acid can be precipitated.

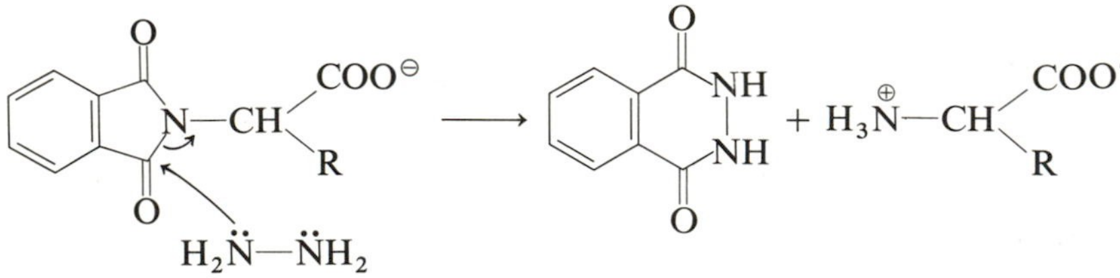

(4) p-Toluenesulfonyl (Tosyl) Function

p-Toluenesulfonyl (tosyl) function allows protection of the amino function by sulfonylation, instead of acylation. Removal is accomplished by sodium in liquid ammonia. Sulfonylation proceeds in aqueous base or pyridine:

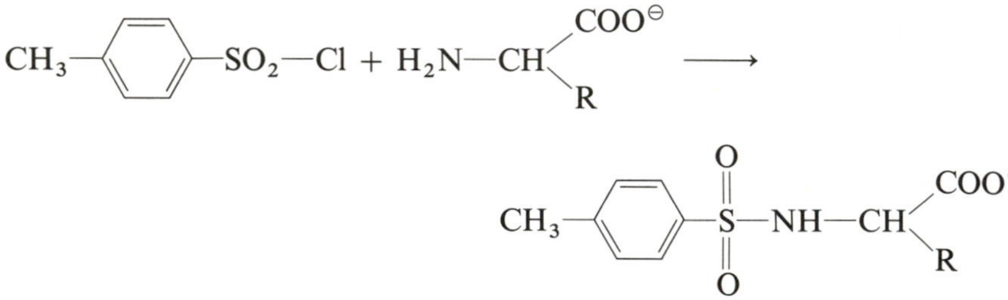

The tosyl function is unaffected by catalytic hydrogenation, and is resistant to most conditions of acidity and alkalinity (i.e., HBr in acetic acid, at room temperature).

Removal of the tosyl function is undertaken by dissolving the protected amino acid/peptide in liquid ammonia and slowly adding sodium. Excess sodium is destroyed at the end of the reaction by the addition of ammonium chloride, iodide, or acetic acid.

(5) Triphenylmethyl (Trityl) Function

Triphenylmethyl (trityl) function may be used to protect the amino function by an S_N1 alkylation. The trityl group is very acid labile. Whereas acyl and sulfonyl groups protect amino functions by rendering the nitrogen non-nucleophilic, the trityl function preserves this nucleophilicity (basicity) and protects the amino group by steric bulk. In fact, this can be a disadvantage, because the bulky trityl group can also interfere with peptide coupling at the (activated) carboxyl.

The trityl function is introduced by reaction of trityl chloride with an amino acid ester in the presence of an organic base and in an organic solvent. The reaction cannot be done in aqueous media. If the amino acid is desired, then the ester must be hydrolyzed with hot alkali, as the bulky trityl group will slow the hydrolysis. Alternatively, the benzyl ester may be used and removed by selected hydrogenolysis, as the O-benzyl group will be removed more quickly.

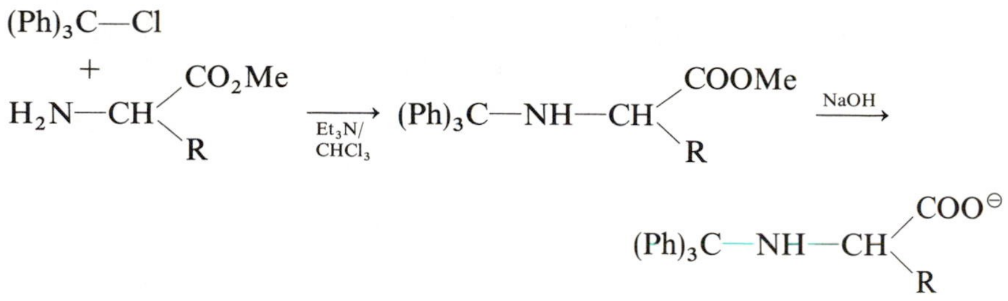

$(Ph)_3C-Cl$

$$+ \quad H_2N-CH \underset{R}{\overset{CO_2Me}{<}} \quad \xrightarrow[CHCl_3]{Et_3N/} \quad (Ph)_3C-NH-CH \underset{R}{\overset{COOMe}{<}} \quad \xrightarrow{NaOH}$$

$$(Ph)_3C-NH-CH \underset{R}{\overset{COO^{\ominus}}{<}}$$

The trityl group is stable to base, may be removed by hydrogenolysis, but is very susceptible to acid. It can be removed under conditions (i.e., 80% acetic acid, $\frac{1}{2}$ hour, 30°C) that will not affect the *tert*-BOC and CBz functions. A variety of acids may be used which include acetic acid, HCl in water, acetic acid, methanol and chloroform, and trifluoroacetic acid. The case of acidolysis is the result of the stabilized carbonium ion:

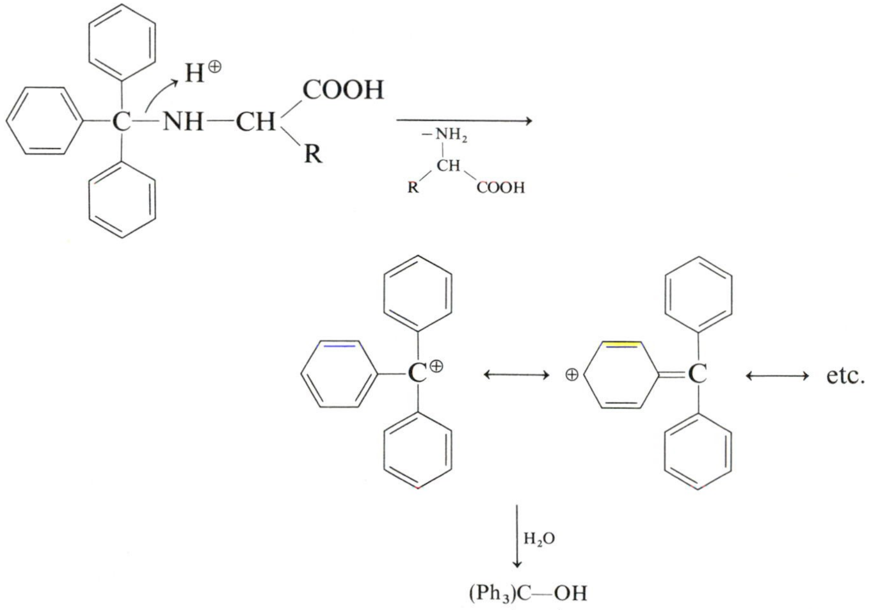

$$\downarrow H_2O$$

$$(Ph_3)C-OH$$
tritanol

An even more acid labile trityl analogue which has found some use as an amino protecting group is the *p*-methoxytrityl function.

(6) *Formyl Function*

Formyl function may be used as a simple acyl protecting group, as was observed for the N-terminal amino acid in the biological synthesis of proteins. It may be removed under mildly acidic conditions that will not affect peptide bonds.

Amino acids may be formylated in formic acid–acetic anhydride (formic anhydride is unstable). The formyl group is not affected by catalytic hydrogenation and sodium in liquid ammonia, but is only resistant to gentle base (i.e., conditions that will saponify ester linkages; the amide would be a poorer leaving group than alkoxide). It is easily removed by acid or hydrazine (and AcOH) in alcohol.

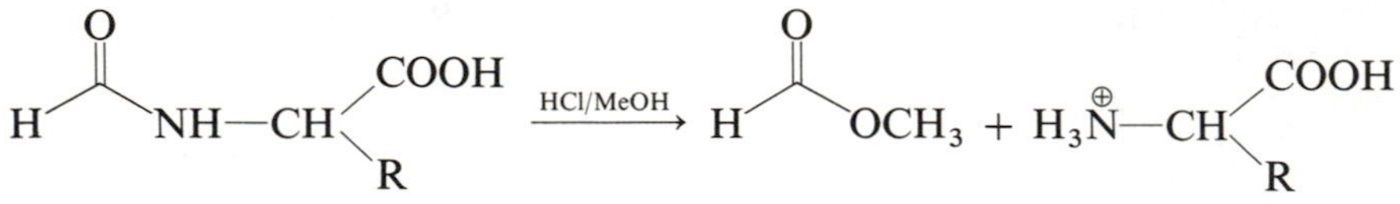

The CBz, phthaloyl, and tosyl groups will not be affected, but the trityl and *tert*-BOC groups will be removed. It can also be removed from the amino acid by NH_2NH_2 in 60% EtOH + HOAc or from both amino acids and peptides by 15% aq H_2O_2, 60°C, 2 hr. However, esters are probably affected by these procedures.

(7) *Trifluoroacetyl Function*
Trifluoroacetyl function may be used as an acyl protecting group which is removed in mild base. It may be introduced upon reaction of the amino acid with trifluoroacetic anhydride or trifluoroacetic acid thioethyl ester:

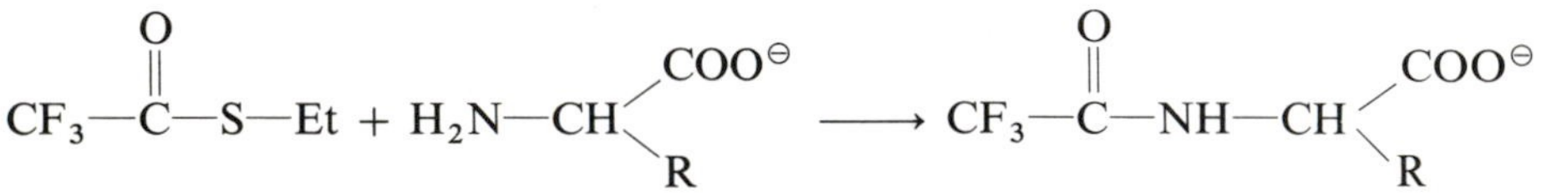

The disadvantage is that trifluoroacetyl amino acids readily racemize. The group may be removed by sodium hydroxide at room temperature; conditions that do not affect most other protecting groups.

(8) *Other Protecting Groups*
Other protecting groups, which find less use, have been developed. Many are unique by virtue of their method of application or removal. For example, *o*-nitrobenzyl derivatives can be used as photosensitive blocking agents for both amino and carboxylate functions. They are removed by irradiation with light of longer wavelength than 320 nm. An organometallic protecting group {pentacarbonyl[methoxy-(phenyl)carbene] chromium(0)} has also been developed for the amino group. It is readily removed with trifluoroacetic acid at room temperature. The amino group may also be protected by sulfenylation with *o*-nitrophenylsulfenyl chloride. Removal is readily accomplished with HCl, but the nitrophenylsulfenyl amino acid must be stored as the carboxylate salt since the free acid is unstable.

2.6.2 Carboxyl Protecting Groups

If two amino acids, neither of which has a protected or activated carboxyl function, are to be coupled together to form a dipeptide, then treatment with a reagent that simultaneously activates the carboxyls and brings about peptide bond formation will result in self-condensation. This may be avoided by protection of the carboxyl group of one amino acid.

Generally, this is accomplished by esterification of the carboxyl group to be protected. The methyl or ethyl ester may be synthesized by reaction of the amino acid with gaseous HCl in methanol or ethanol (Fisher esterification). However, it is desirable to have an ester function which can be removed under mild conditions. While esters may be saponified by base more easily than peptides (the alkoxide is a better leaving group), the alkaline conditions that are required may not be compatible to the polypeptide. Use of benzyl esters allows removal of the protecting group under neutral conditions via catalytic hydrogenation. The benzyl ester may be prepared by reaction of the acid and benzyl alcohol in the presence of acid or thionyl chloride (converts the alcohol to the chlorosulfite which then undergoes displacement by the acid) or by reaction of the cesium salt of the acid with benzyl bromide or transesterification:

$$R-C\underset{OMe}{\overset{O}{\big\langle}} + \underset{}{\text{(C}_6\text{H}_5)\text{CH}_2\text{OH}} \rightleftharpoons R-C\underset{O-CH_2-\text{(C}_6\text{H}_5)}{\overset{O}{\big\langle}} + CH_3OH$$

In the case of acid catalysis or transesterification, it is possible to push the reaction to the right by removal of water (azeotropic distillation) or the volatile methanol. Alternatively, preparation of the *tert*-butyl ester confers resistance to alkali, but allows removal of the protecting group under mildly acidic conditions. The ester, which for steric reasons cannot be prepared from *tert*-butanol, may be synthesized from isobutylene gas in the presence of acid, or by transesterification with *tert*-butyl acetate.

$$R-\overset{O}{\overset{\|}{C}}-OH + \underset{CH_3}{\overset{CH_3}{>}}C{=}CH_2 \xrightarrow{\text{H}^{\oplus}} R-\overset{O}{\overset{\|}{C}}-O-\underset{CH_3}{\overset{CH_3}{\underset{|}{\overset{|}{C}}}}-CH_3$$

$$R-\overset{O}{\overset{\|}{C}}-OH + CH_3-\overset{O}{\overset{\|}{C}}-O-C(CH_3)_3 \xrightarrow{\text{H}^{\oplus}} \rightleftharpoons$$

$$R-\overset{O}{\overset{\|}{C}}-O-C(CH_3)_3 + CH_3-\overset{O}{\overset{\|}{C}}-OH$$

Recently, the 2-trimethylsilylethyl function has been prepared as a carboxyl protecting group. This may be synthesized by the reaction of the acid with 2-trimethylsilylethanol and dicyclohexylcarbodiimide (DCC, a dehydrating or coupling agent, see below):

$$R-C\overset{O}{\underset{OH}{\diagdown}} + (CH_3)_3-Si-CH_2CH_2OH \xrightarrow[\text{pyridine}]{\bigcirc-N=C=N-\bigcirc}$$

$$R-C\overset{O}{\underset{O-CH_2-CH_2-Si-(CH_3)_3}{\diagdown}}$$

Removal of the protecting group may be achieved under neutral conditions by treatment with fluoride. As shall be seen later, fluoride has found much use for the removal of silyl protecting groups on carbohydrates. In anhydrous solvent systems, it is a good nucleophile:

$$R-C\overset{O}{\underset{O-CH_2-CH_2-Si-CH_3}{\diagdown}}\overset{CH_3}{\underset{CH_3}{\diagup}} \longrightarrow$$

$$\ddot{F}^{\ominus}$$

$$R-C\overset{O}{\underset{O^{\ominus}}{\diagdown}} + CH_2=CH_2 + (CH_3)_3Si-F$$

During the course of a peptide synthesis, protection of the amino and carboxyl functions may not be adequate. Under the conditions of peptide bond formation nucleophilic or other chemically reactive amino acid side chains may participate in unwanted side reactions. This may be avoided by protection of these side chains with protecting groups that meet the same general requirements of all protecting groups: introduction and removal under mild conditions, etc.

The means of protection of various amino acid side chains shall not be dealt with here in any detail, as this is adequately described elsewhere (see General Reading). More important, familiarity with the chemistry of the protection methods for amino and carboxyl functions allows comprehension of the methodology developed for side chain protection. Often, similar or the same protecting groups may be used. As an example, consider the protection of the thiol group of cysteine and the amino functions of ornithine and lysine.

The thiol group is a potent nucleophile, and is susceptible to oxidation to the disulfide form. It therefore must be protected during peptide synthesis. Use may be made of the benzyl group, which is introduced by the reaction of the thiol with benzyl chloride (remember, benzyl groups undergo S_N2 displacements very well). Benzyl functions are susceptible to hydrogenolysis (CBz, trityl) and so this group may be removed by reduction with sodium in liquid ammonia. Catalytic hydrogenation is not desirable as the catalyst will be poisoned by sulfur, as noted earlier.

The same protecting group may be used for the protection of α-, δ- (ornithine), and ε- (lysine) amino functions. Discrimination of the side chain amino functions may be achieved by first forming the copper chelate of the amino acid:

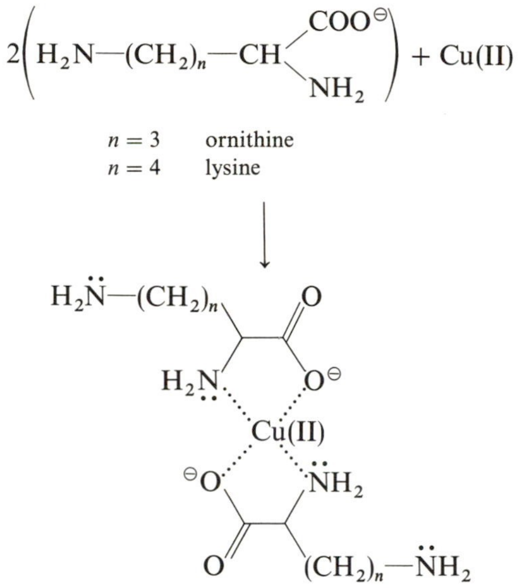

$$2\left(H_2N-(CH_2)_n-CH \underset{NH_2}{\overset{COO^{\ominus}}{\diagdown}} \right) + Cu(II)$$

$n = 3$ ornithine
$n = 4$ lysine

$$H_2\ddot{N}-(CH_2)_n \diagdown \qquad \diagup O$$

Chelation with various metal ions is a general reaction of amino acids. As the nonbonding electrons of the α-amino group are coordinated to the metal, only the side chain amino function is free to react with the acylating agent. The α-amino function may be freed by treatment with hydrogen sulfide.

2.6.3 Peptide Bond Formation

Two general approaches are possible. The first is to convert the amino protected amino acid to an activated form and then react this with the amino function of a second amino acid. As will be recalled, activation is necessary as work must be done during peptide bond formation. Alternatively, it is possible to react two amino acids together (one amino protected, the other carboxyl protected) in the presence of a *coupling agent* which activates the carboxyl function *in situ*. The former approach shall be examined first.

(1) *Acyl Chlorides*
Amino acids may be converted to their acyl chlorides:

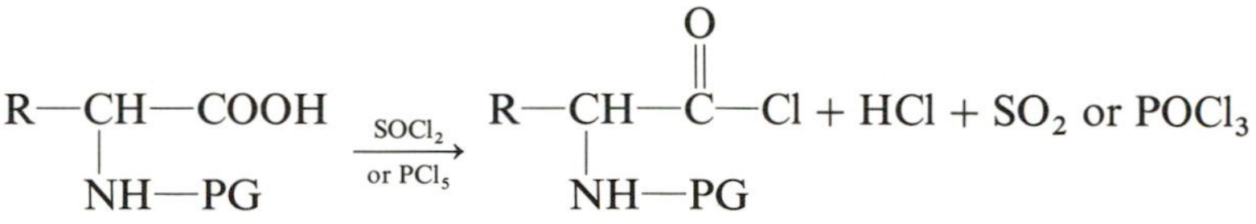

$$\underset{NH-PG}{R-CH-COOH} \xrightarrow[\text{or PCl}_5]{SOCl_2} \underset{NH-PG}{R-CH-\overset{O}{\overset{\|}{C}}-Cl} + HCl + SO_2 \text{ or } POCl_3$$

PG = protecting group

The acyl chloride formed will readily react with the amino function of a second amino acid to give peptide bond formation. However, because the chlorine is such a good leaving group, the acyl chloride possesses a high probability of racemization via intermediate azlactone formation (see page 56).

(2) Anhydrides

Mixed anhydrides tend to be less reactive than acyl chlorides (poorer leaving group) giving less racemization upon their formation. Synthesis may be accomplished by nucleophilic attack of the amino acid carboxylate on an

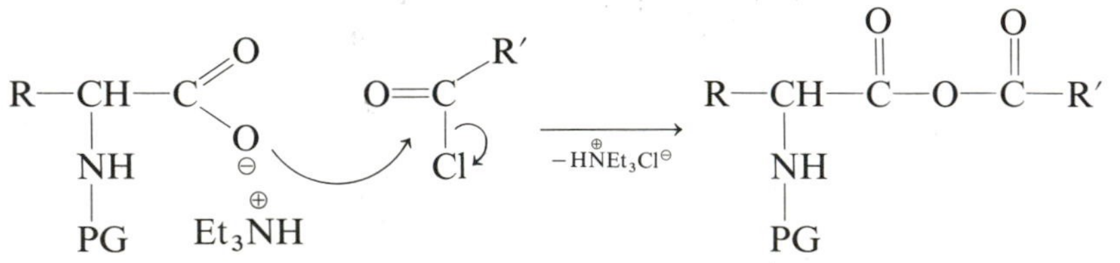

acyl halide. Organic solubility will be increased by use of an organic cation. The anhydride formed will readily react with the amino function of the second amino acid. However, the amino function can react with one of two carbonyls, so that considerable by-product formation can occur. This may be avoided by synthesis of a mixed anhydride in which nucleophilic attack will be preferred at one of the two possible carbonyls. Reaction of an amino acid with ethyl chloroformate forms such an anhydride. Only one carbonyl is attacked, as it is more electrophilic (the other carbonyl is flanked by two oxygen atoms which can delocalize their nonbonding electrons) and it is a poorer leaving group. Racemization can still occur and synthesis of the anhydride must be undertaken at low temperature to avoid decomposition.

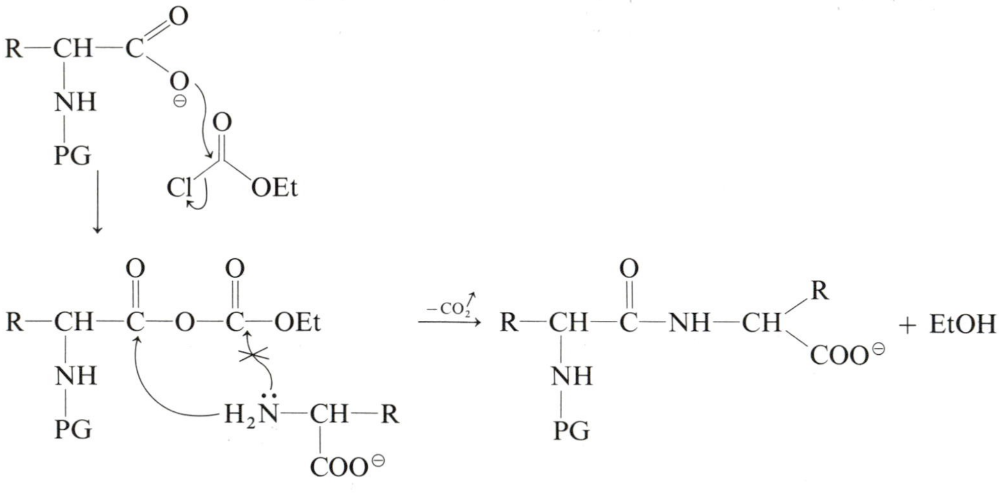

(3) Acyl Azides

Use of acyl azides leads to less racemization (not significant) than acyl chlorides or anhydrides. The azide could be synthesized from the acyl chloride:

$$R-\underset{\underset{PG}{\overset{\displaystyle |}{NH}}}{\overset{\displaystyle |}{CH}}-\overset{\overset{\displaystyle O}{\|}}{C}-Cl + NaN_3 \xrightarrow[-NaCl]{} R-\underset{\underset{PG}{\overset{\displaystyle |}{NH}}}{\overset{\displaystyle |}{CH}}-\overset{\overset{\displaystyle O}{\|}}{C}-N_3$$

However, this would require the conversion of the amino acid to the unfavorable acyl chloride, thus defeating the purpose of the synthesis. Synthesis is instead undertaken with the amino acid ester:

$$R-\underset{\underset{PG}{\overset{|}{NH}}}{\overset{|}{CH}}-\overset{\overset{O}{\|}}{C}-OCH_3 \xrightarrow{H_2N-NH_2} R-\underset{\underset{PG}{\overset{|}{NH}}}{\overset{|}{CH}}-\overset{\overset{O}{\|}}{C}-NH-NH_2 \xrightarrow{HNO_2} R-\underset{\underset{PG}{\overset{|}{NH}}}{\overset{|}{CH}}-\overset{\overset{O}{\|}}{C}-N_3$$

stable
(usually crystalline)

Nitrous acid (a source of dinitrogen trioxide) is a well-known reagent for the synthesis of azides. The mechanism for the conversion of hydrazides is similar to the diazotization of primary amines with nitrous acid:

$$R-\overset{\overset{O}{\|}}{C}-NH-\overset{..}{N}H_2 \quad \underset{\underset{O}{\|}}{N}-O-N \xrightarrow{-HNO_2} R-\overset{\overset{O}{\|}}{C}-NH-\overset{\overset{H}{|}}{N}-N=O$$

$$R-\overset{\overset{O}{\|}}{C}-\overset{\ominus}{\underset{..}{N}}-\overset{\oplus}{N}\equiv N: \qquad \xleftarrow{-H_2O} \quad R-\overset{\overset{O}{\|}}{C}-NH-\overset{..}{N}=N-OH$$

$$R-\overset{\overset{O}{\|}}{C}-\overset{..}{N}=\overset{\oplus}{N}=\overset{..}{N}:^{\ominus}$$

Storage of the azide is difficult, and all handling of the azide must occur at low temperature to prevent conversion to the isocyanate via the Curtius rearrangement:

$$R-C\overset{\overset{\displaystyle O}{\diagup}}{\underset{\diagdown N_3}{}} \xrightarrow{-N_2} \left[R-C\overset{\overset{\displaystyle O}{\diagup}}{\underset{\diagdown \overset{..}{N}}{}} \right] \longrightarrow R-N=C=O$$

reactive nitrene intermediate

It is possible to convert the azide to a stable form that may be conveniently handled by reaction with N-hydroxysuccinimide:

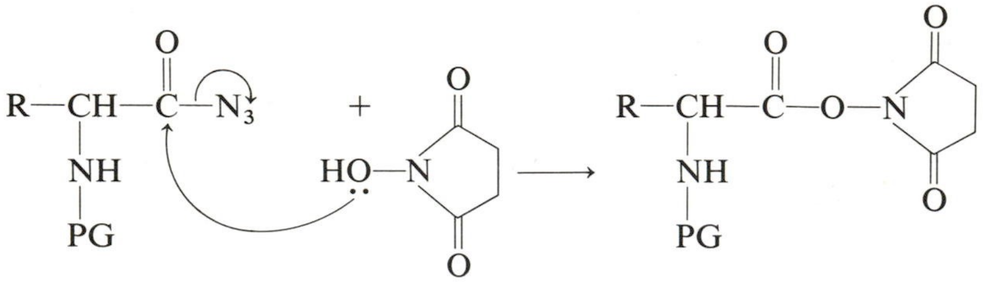

The product is a stable, crystalline "activated ester."

(4) Activated Esters*

The aminolysis of alkyl esters is a slow, almost equilibrium, process. Thermodynamically, the peptide bond is slightly stronger. Chemically, the alkoxide is not a good leaving group. However, it is possible to accelerate peptide bond formation by using an ester that has a better leaving group: an "activated ester." Aminolysis of the activated ester will provide the energy necessary for peptide bond formation. p-Nitrophenol is a much stronger acid (resonance stabilization of the anion, see above) than methanol so that the p-nitrophenyl ester of amino acids is activated. This ester may be prepared by reaction of the acid with p-nitrophenol and the coupling (dehydrating) agent DCC (see below). Pentachlorophenol is also a stronger acid (inductive pull of the chlorine atoms, see above) than methanol, so that it may be used for the preparation of activated esters.

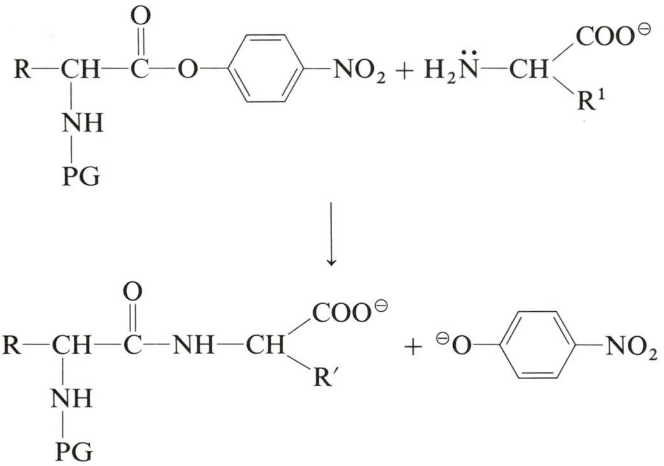

---

* New methods for the synthesis of "activated esters" are a topic of continuous interest. The most recent of these was described during preparation of this text (361).

In situ activation of the carboxyl function, in the presence of a nucleophilic amino function, may be achieved with a variety of coupling agents.

(1) *The Carbodiimides*
The carbodiimides are a family of coupling (dehydrating, condensing) agents of the general structure:

$$R—\overset{..}{N}=C=\overset{..}{N}—R$$
$$\underset{+\delta}{}$$

The central carbon atom, flanked by two nitrogen atoms, is electrophilic and thus subject to nucleophilic attack by a variety of reagents. The most well-known reactions are those between carbodiimides and acids. For example, carbodiimides will undergo addition reactions with a variety of weak acids:

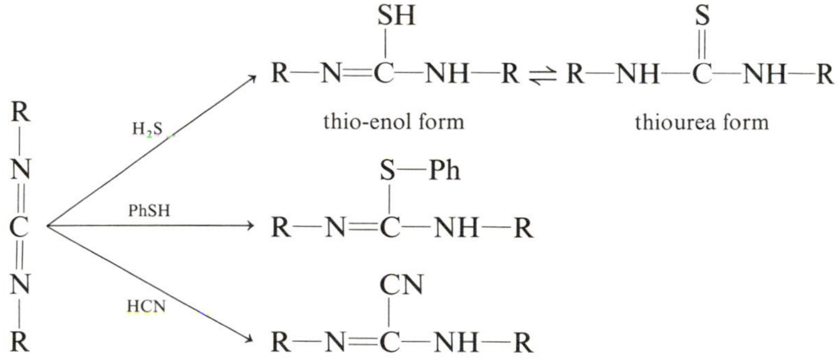

Anhydrous HCl will add exothermically to form the dichloro compound, while mineral acids will catalyze hydration of the carbodiimide. Perhaps

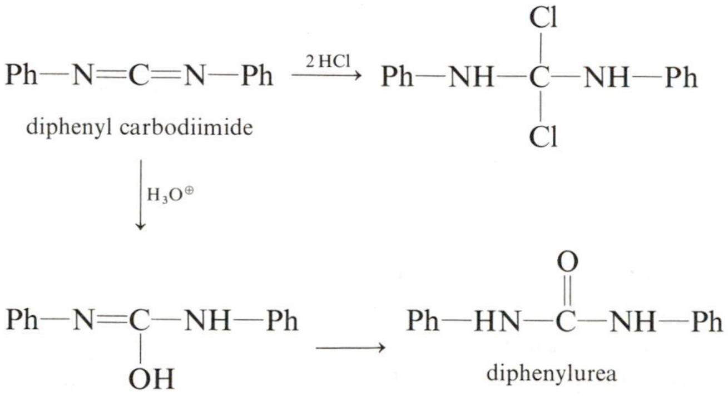

most important, at least for the synthesis of biologically interesting compounds, is the reaction of carboxylic, phosphoric, and sulfonic acids. Under appropriate conditions, these will add to carbodiimides to form anhydride-like intermediates, and eventually acid anhydrides. This is illustrated for a carboxylic acid (in Chapter 3, the reaction of phosphoric acids in regard to nucleotide synthesis will be discussed):

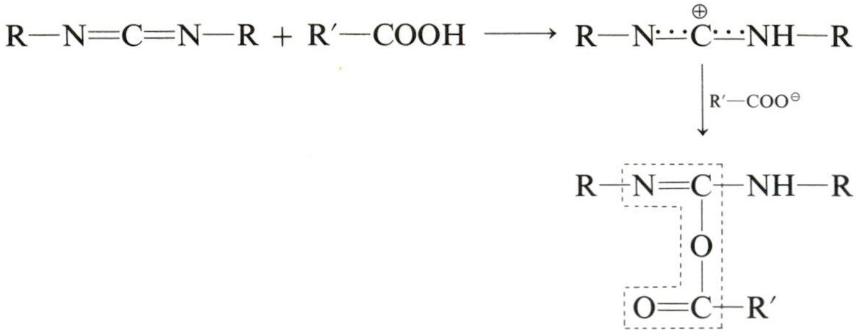

$$R-N{=}C{=}N-R + R'{-}COOH \longrightarrow R-N\cdots\overset{\oplus}{C}\cdots NH-R$$

$$\downarrow {\scriptstyle R'-COO^{\ominus}}$$

$$R{-}N{=}C{-}NH{-}R$$
$$|$$
$$O$$
$$|$$
$$O{=}C{-}R'$$

unstable *O*-acyl urea
(anhydride-like intermediate)

The unstable intermediate formed can either rearrange to form a urea, or a second molecule of acid can attack to give the acid anhydride product.

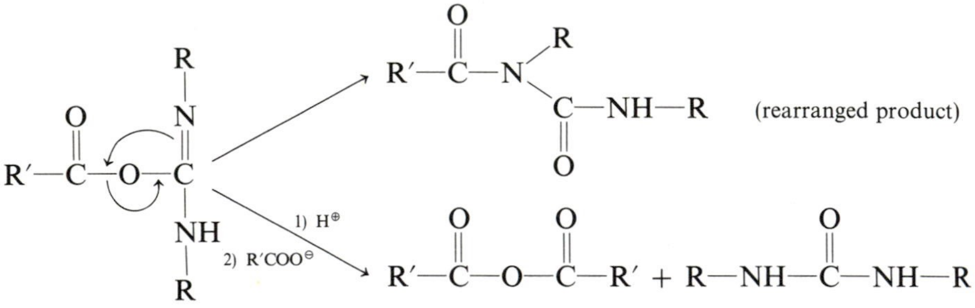

(rearranged product)

Reaction of the carbodiimide with the acid then provides a means for the synthesis of a biologically high-energy acid anhydride. This suggests a potential route for the synthesis of a peptide bond. For example, it has already been pointed out that amines will react with acid anhydrides to form peptide bonds. Further, under the mild conditions (i.e., room temperature) that carbodiimide will react with carboxylic acids, competing reactions will not occur. For example, carbodiimides will only undergo nucleophilic attack by alcohols under refluxing conditions.

In anticipating the use of carbodiimides for amide synthesis, thought must be given to the selection of the "R" group. The stability of aliphatic and aromatic carbodiimides is a function of the substituting groups, so that decomposition and polymerization can occur upon standing. Alkyl chain length has little influence upon stability. Instead, stability increases markedly with the branching of the alkyl substituent, on the two nitrogen atoms. Thus, diethylcarbodiimide polymerizes in a few days, while dicyclohexylcarbodiimide is stable for months. It is this latter reagent that has found the most use in protein synthesis.

Dicyclohexylcarbodiimide (DCC) may then be used for the formation of peptide bonds:

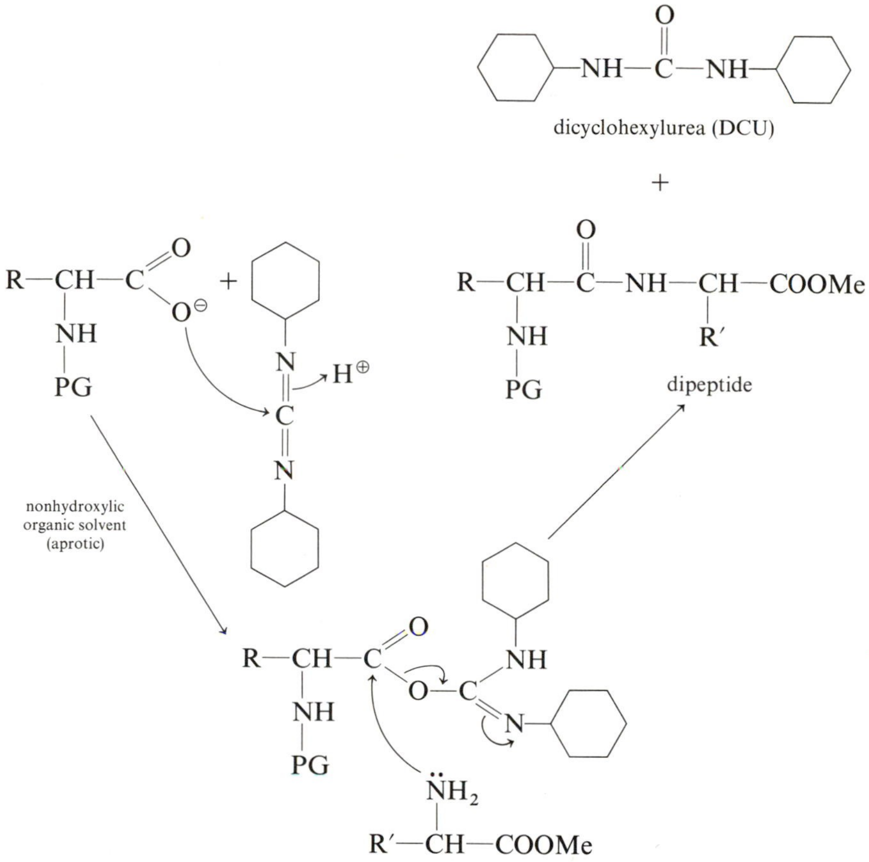

$$
\text{dicyclohexylurea (DCU)}
$$

$$
+
$$

R—CH—C(=O)—O⁻ + [cyclohexyl-N=C(H⁺)=N-cyclohexyl] → R—CH—C(=O)—NH—CH—COOMe

with NH / PG groups on left and NH / PG, R′ groups on right

nonhydroxylic
organic solvent
(aprotic)

dipeptide

R—CH—C(=O)—O—C(NH-cyclohexyl)=N-cyclohexyl
|
NH
|
PG

:NH₂
|
R′—CH—COOMe

second amino acid reacts with anhydride intermediate

Under the mild reaction conditions (0°C), the second amino acid (carboxyl protected) will not react with DCC, but only with the anhydride-like intermediate. The DCU formed is insoluble in most organic solvents so that most often separation from the product merely involves filtrations. However, it occasionally may be difficult to remove the last trace of DCU by filtration. As such, the use of carbodiimides that are soluble in aqueous solvents may be necessary. One example is 1-ethyl-3-(3-dimethylaminopropyl)-carbodiimide hydrochloride. The urea by-product formed is also water soluble.

$$
H_3C-CH_2-N=C=N-(CH_2)_3-N(CH_3)_2 \cdot HCl
$$

So that after reaction in an organic solvent, it is possible to remove any unreacted carbodiimide and the urea simply by washing with water. It should be noted that significant racemization, via azlactone formation of the anhydride-like intermediate, can occur during the synthesis of a large peptide (protein) molecule. However, for a smaller polypeptide, this is minimal.

(2) Woodward's Reagent

This reagent is a dehydrating agent which has been used much less for peptide bond formation than the carbodiimides. Reaction first occurs with a carboxylate function after base catalyzed decomposition of the isoxazolium salt:

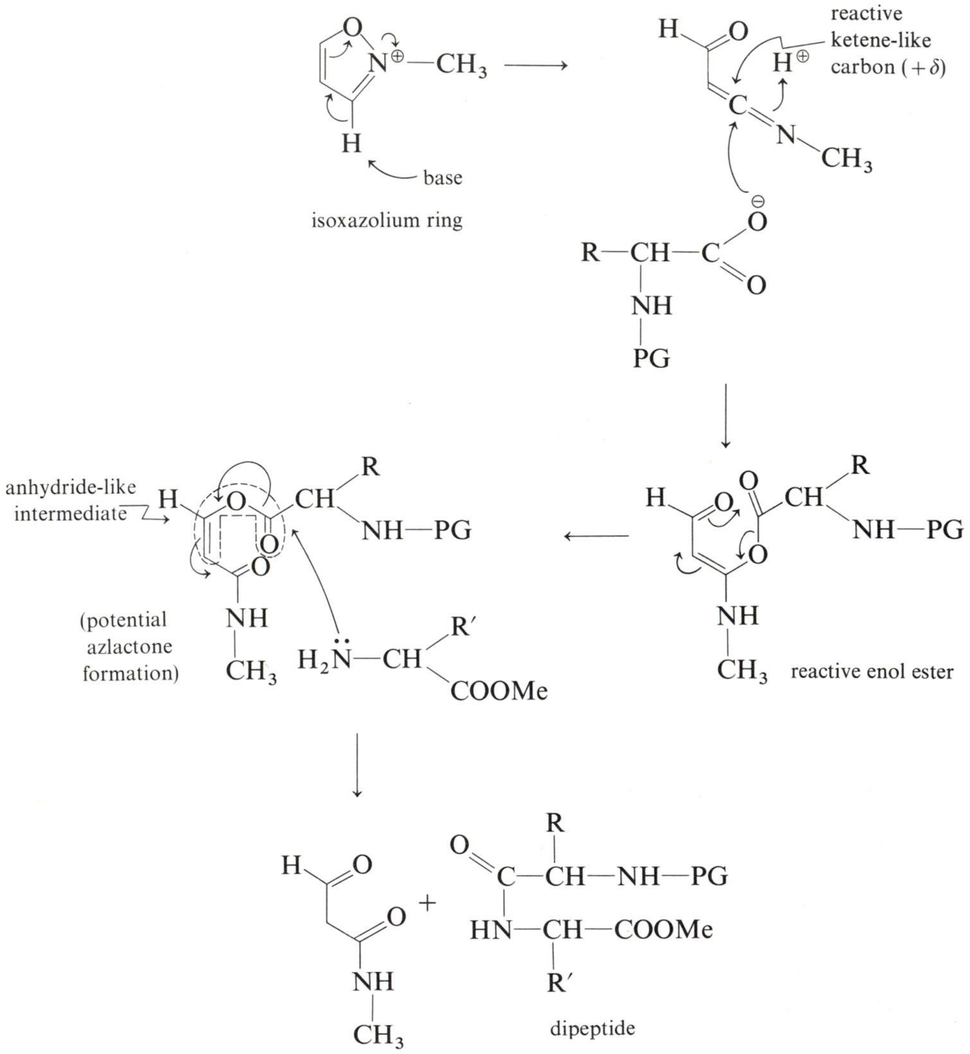

The presence of base in the reaction media is unfavorable, and a variety of unwanted reactions, including significant racemization, can occur.

(3) EEDQ (Belleau's Reagent)

2-Ethoxyethoxycarbonyl-1,2-dihydroquinoline (EEDQ) is also a coupling agent that may be used for peptide bond synthesis (7). EEDQ is readily synthesized from quinoline and ethyl chloroformate:

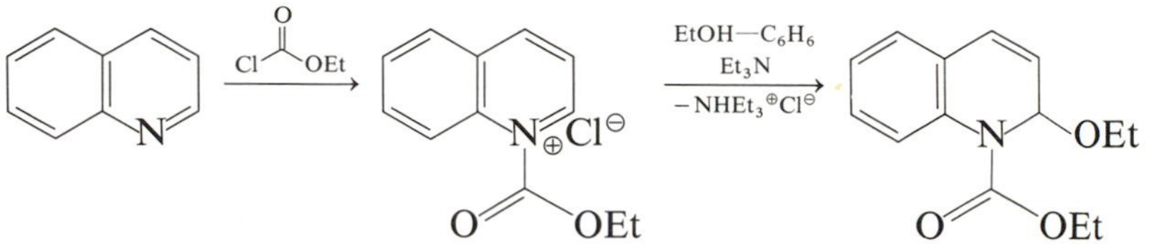

Under mild (room temperature) reaction conditions, EEDQ will not react with amines. In the presence of a hydroxylic solvent, alcoholysis of the carboethoxy function will not occur, but instead the ethoxy function will be exchanged, if a trace of Lewis acid (i.e., BF_3) is present. The ethoxy function is a good leaving group which may be induced to do so upon protonation. The driving force is the opportunity to gain aromaticity:

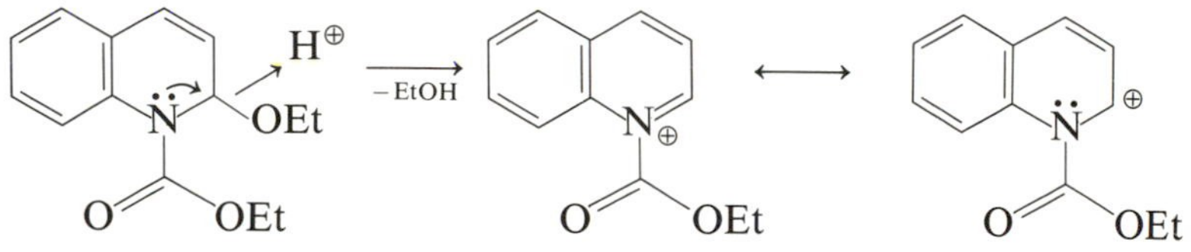

The carboxylate function will readily add to the cation, after which mixed anhydride formation will occur. This will in turn react with the amino function of a second amino acid to give peptide bond formation. Further, the mixed anhydride so formed does not accumulate in solution (its formation is rate limiting) but instead suffers immediate nucleophilic attack by the amine. Azlactone formation does not have a chance to occur, and so no significant racemization is observed during polypeptide synthesis. The mixed anhydride formed will be attacked by the second amino acid only at one of the two carbonyl functions, giving carbon dioxide and ethanol by-products. The reason for this has been discussed earlier (see peptide bond formation via acid anhydrides).

The special reactivity of EEDQ toward carboxylate functions can be understood by noting that displacement of the ethoxy function by any other nucleophile (i.e., alcohol or thiol exchange reaction) leads to an analogue of

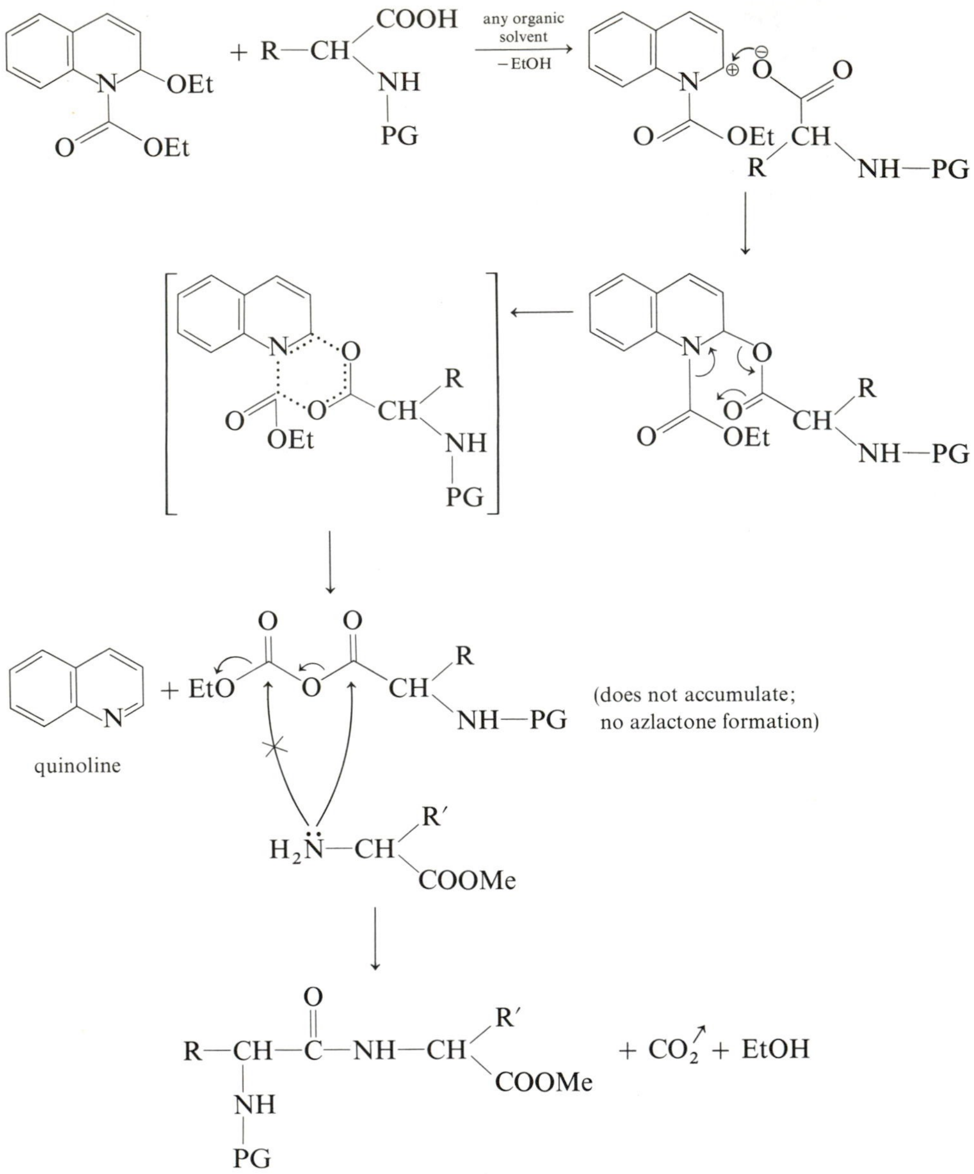

quinoline

(does not accumulate; no azlactone formation)

the starting material. However, displacement by the carboxylate proceeds through a six-membered transition state, and leads to the formation of a stable 10 π-electron aromatic structure.

(4) Leuch's Anhydride

Reaction of an unprotected amino acid with phosgene provides a simple method of peptide bond formation. The reaction begins with N-acylation of the amino acid, followed by cyclization to give a reactive anhydride-like intermediate:

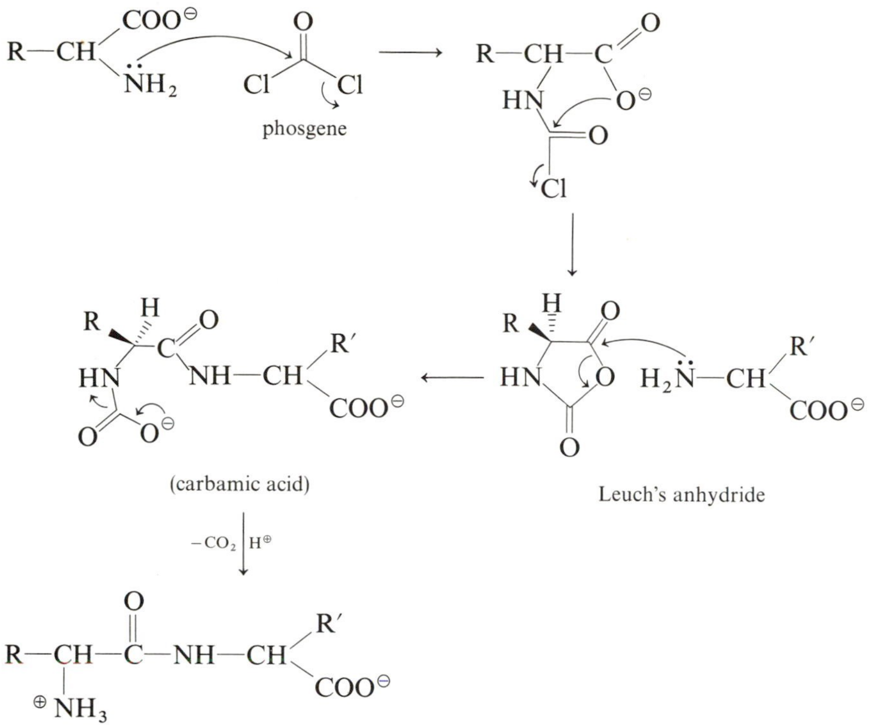

(carbamic acid)

Leuch's anhydride

dipeptide

The anhydride formed will only undergo nucleophilic attack at one of the two carbonyl functions, as the untouched carbonyl is both less electrophilic (flanked by a nitrogen and an oxygen atom) and a better leaving group. The carbamate formed can be readily decomposed to the dipeptide and carbon dioxide. The phosgene acts as a coupling agent since once the anhydride is formed, it undergoes displacement by a second amino acid. The anhydride is not isolated. Further, the carbamate is not readily isolated, and the reaction is difficult to control. It is therefore possible for the dipeptide product to be N-acylated, undergo anhydride formation, and then be attacked by a third amino acid, etc. The Leuch's anhydride methodology is best suited for homopolymer synthesis. However, by careful control of reaction conditions (0°C, pH = 10.2) it is possible to stop the reaction at the stage of the carbamate and sequentially add the next monomeric unit, and then allow the reaction to proceed by freeing the amino function. By so doing, large polypeptide chains of desired sequence have been synthesized.

(5) *Solid Phase Synthesis* (*Merrifield's Method*) (8)
This technique combines the above methodology (amino and carboxyl protecting groups, coupling agents, etc.) but greatly reduces the processing of intermediates or "workup" associated with these techniques. The necessity

to obtain the product of each coupling reaction in pure crystalline form is eliminated, and excess reagents and by-products are simply removed by filtration and washing with appropriate solvent systems. This amounts to a lot of work if the synthesis of a large polypeptide molecule is to be considered. This is accomplished by the use of an insoluble polystyrene matrix.

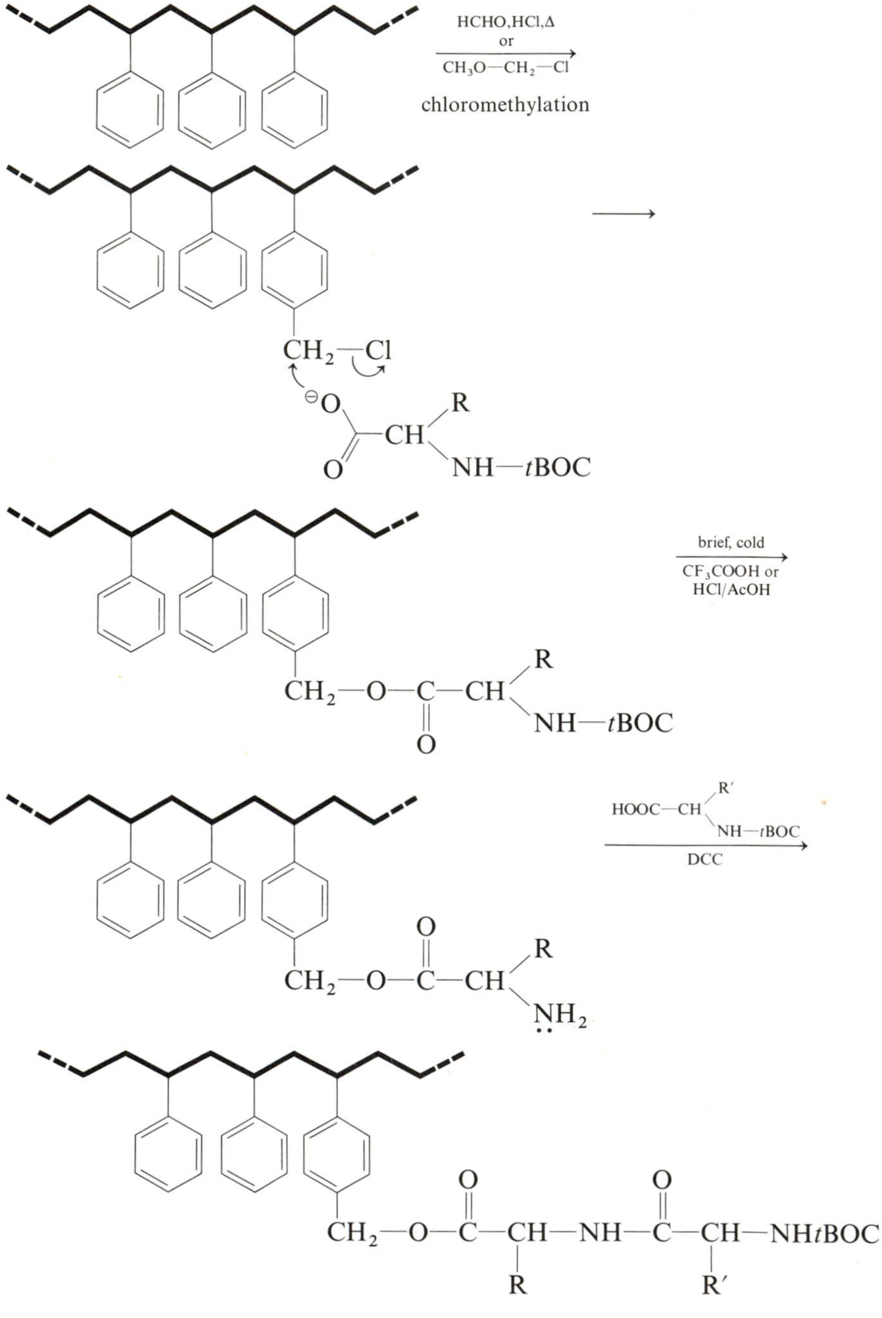

A reactive center is introduced into the polymer support by chloromethyl-ation. The benzyl chloride linkage formed has already been noted to be very susceptible to S_N2 displacements and so is easily replaced by the carboxyl function, which can in turn be removed (by acidolysis) at a later date. Note that had the amino function been used to displace the chlorine, then removal of the amino acid from the resin later would be impossible. Once the amino acid is covalently linked to the resin, the *tert*-BOC function may be removed by treatment with acid which will not affect the benzyl linkage to the support. The freed amino function may now react with a second (*tert*-BOC protected) amino acid to form the peptide linkage, the *tert*-BOC group again removed and the bound dipeptide may react with a third amino acid, etc. Again, workup after each reaction consists of simply filtering the bound product and washing away unreacted reagents and unwanted side products. After the peptide of specified sequence has been synthesized, it is removed from the resin by treatment with HBr or HF in trifluoroacetic acid.

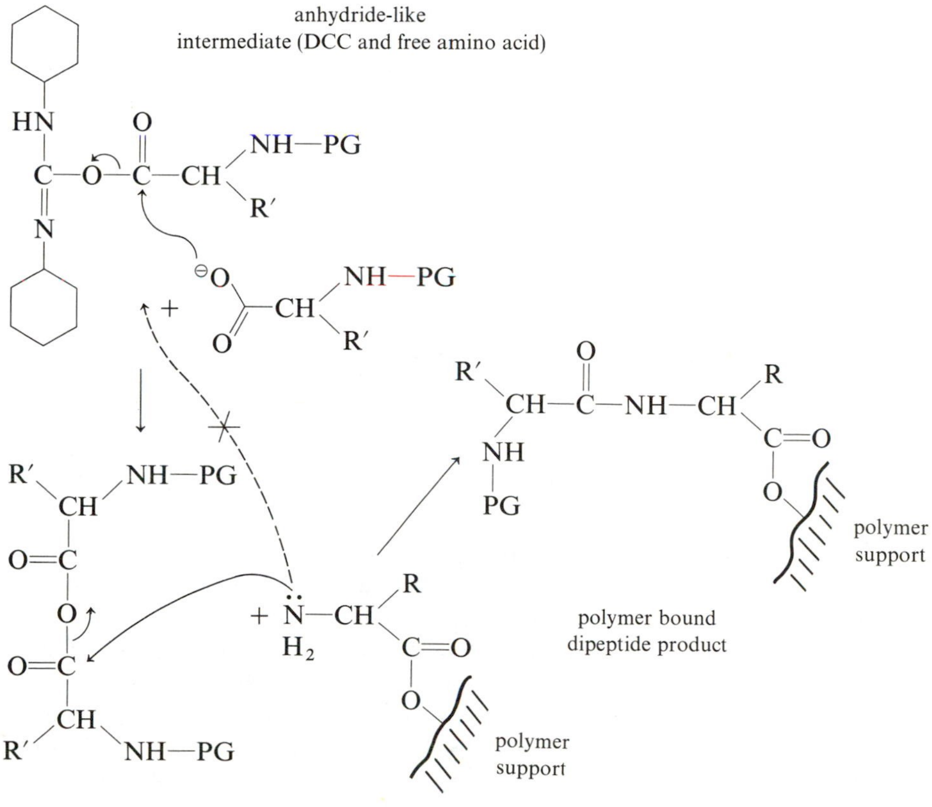

Any coupling method may be used, but that which has found the most use to date has been DCC. All of the amino acids, except asparagine and glu-tamine (whose amide side chains may be dehydrated to nitriles) have been successfully introduced into peptides with DCC. It is interesting to note that under the conditions of solid phase synthesis, the DCC coupling mechanism follows a different pathway than that observed in solution (9). This is understandable since the DCC and solution amino acid are in excess of the bound nucleophilic amino group. There is a greater probability of reaction between the anhydride-like intermediate and the carboxyl function of a second molecule of free amino acid than with the polymer bound amino function.

As the workup of each coupling reaction is so simple, it has been possible to automate the procedure, thus allowing for an even more rapid polypeptide synthesis. As such, the first chemical synthesis of an enzyme (bovine pancreatic ribonuclease, 124 amino acid residues) was accomplished by this method.

It should be pointed out that even with the simple methodology of the solid phase technique, problems are encountered. This is reflected by the fact that the ribonuclease enzyme synthesized by the Merrifield's method did not possess the same biological activity as the natural enzyme. Perhaps the most serious of these is the phenomena of "slippage," in which an amino acid may be misplaced or omitted from the polypeptide sequence, due to incorrect or a lack of coupling. A number of factors can contribute to this and most of these can be understood by remembering that the solid phase system is not a homogenous reaction medium. For example, during one coupling reaction, the bound amino nucleophile of a polypeptide chain may be sterically hindered by the resin thus preventing it from reacting with the free carboxylate, until later when a new (carboxylate) amino acid is introduced into the system. However, other bound polypeptide chains may not have been so prevented from reacting during the coupling reactions. The result is a mixture of protein products at the end of the synthesis which require purification by column chromatography. A partial solution to the problem has been the synthesis of small polypeptide chains which are then removed from the resin and latter coupled to give the desired protein.

As a result of such problems, other strategies using polymeric supports have been sought. While none has received the same attention as the Merrifield method, some have been successfully used for the synthesis of small polypeptides. An example is the use of the polymer formed from the Friedel–Crafts alkylation of polystyrene with 4-hydroxy-3-nitrobenzyl chloride (10). The first amino acid is attached to the support by reaction with DCC to form an active ester. A second amino acid then reacts with this active ester to form a peptide bond. The polymer is the leaving group. Thus, in contrast to the Merrifield synthesis, after each coupling reaction the product is obtained in solution and may be purified if necessary. Slippage is avoided, but each coupling requires re-addition of the polypeptide to the resin.

Attention should be given to the similarity between solid phase technology and the biological synthesis of proteins. In both cases, an amino acid is attached via the carboxylate function to a large macromolecular surface upon which the sequential addition of other amino acids and peptide bond formation occurs. In one case, the polymer, like the tRNA molecule at the peptidyl site, is the leaving group, while in the other case, the polymer, like the tRNA molecule at the aminoacyl site, remains bound to the chain after formation of the new peptide bond. Of course, the tRNA–ribosomal complex is of a more sophisticated nature than the polystyrene matrix.

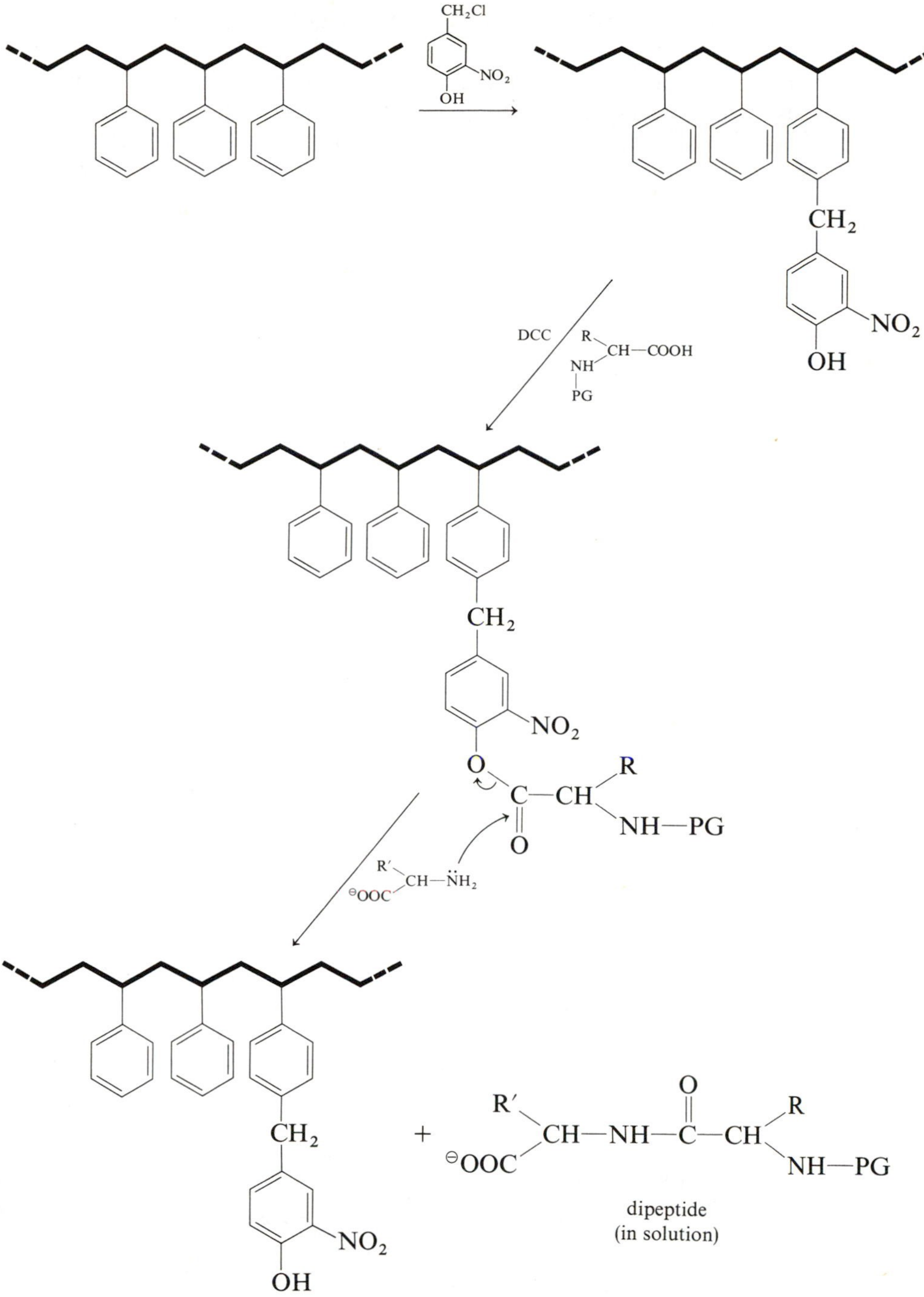

2.7 Asymmetric Synthesis of α-Amino Acids

In the course of chemical evolution, nature must have developed selective methods of amino acid synthesis and specific recognition. In this respect, what kind of chemical methods do we have presently to prepare amino acids in optically pure form and to selectively distinguish among enantiomers? We will therefore examine in this section two approaches to asymmetric synthesis of amino acids using the concept of asymmetric induction and specific metal ion complexation.

It is generally accepted that one of the remaining challenges in organic chemistry is to induce efficient asymmetric synthesis on a prochiral precursor, much as the enzyme does. One way to get around this problem is to use a chiral reagent that would produce diastereotopic interactions with a reactant molecule and lead to an asymmetric product.

The fact that α-amino acids are the constituents of proteins bestows on them a great importance. Eight amino acids are classified as "essential" because they cannot be synthesized by mammals and thus must come from food. They are isoleucine, leucine, lysine, methionine, valine, threonine, phenylalanine, and tryptophan. They are all of L-configuration and it is important to dispose of chemical methods giving access to these amino acids. Ten years ago, mainly biochemical methods were accessible, based on resolution of racemic mixtures.

2.7.1 Corey's Method

In 1968, a French team of chemists under the direction of H. B. Kagan reported the asymmetric synthesis of L-aspartic acid, starting with an optically active amino alcohol (11). The synthesis is outlined in Fig. 2.6.

The optically active unsaturated cyclic precursor is hydrogenated from the least hindered side of the double bond in very good yield creating a new asymmetric carbon. The stereochemistry of this reduction is insured by the presence of the bulky pseudo-axial phenyl ring. The chlorohydrate of monomethyl L-aspartate is obtained in 98% optically pure form.* One inconvenience to this approach resides in the loss of the starting asymmetric amino alcohol which is transformed to diphenyl ethane. In addition, the extension to the other α-amino acids appears uncertain.

---

* The percent optical purity of a compound is defined as

$$\% \text{ optical purity} = \frac{\text{specific rotation of the enantiomeric mixture}}{\text{specific rotation of one pure enantiomer}} \times 100$$

Expressed differently, it means that a racemic mixture which is 50% R-isomer and 50% S-isomer has 0% optical purity; and a 90% optical yield corresponds to a mixture of 95% of one isomer contaminated with 5% of its antipode.

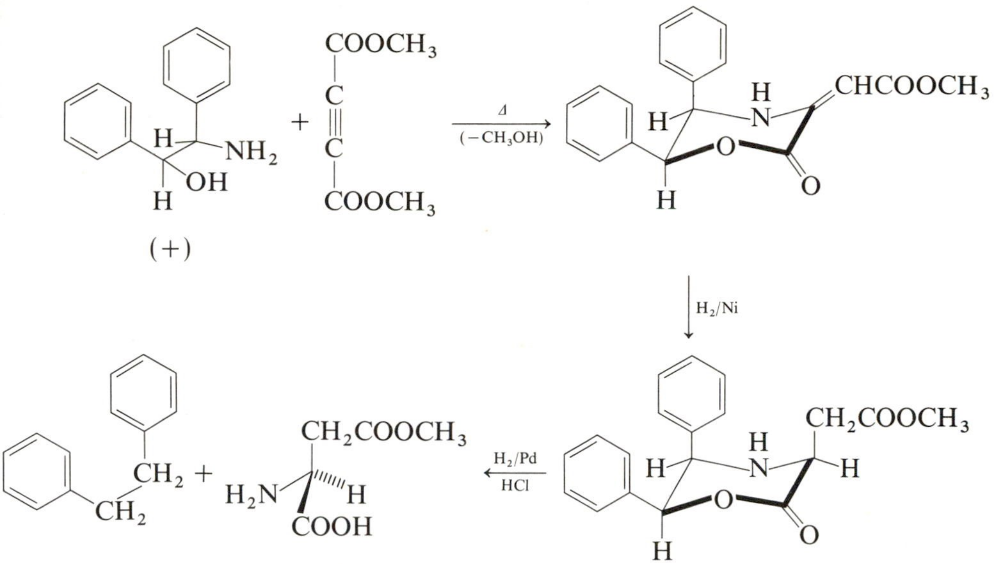

Fig. 2.6. Kagan's synthesis of L-aspartate monomethyl ester (11).

There was thus a need for a more general method where the asymmetric reagent will not be sacrificed and which insures the recovery of the starting material. To overcome these problems, E. J. Corey and collaborators (12) from Harvard University in 1970 proposed the methodology depicted in Fig. 2.7. The precursor of the α-amino acid is the corresponding α-keto acid. The α-keto acid is combined with a chiral reagent to form a ring of minimal size possessing a hydrazone function. Specific reduction of the double bond will produce the chiral carbon of the corresponding α-amino acid. Hydrogenolysis of this intermediate will generate the chiral α-amino acid and a chiral secondary amino alcohol, convertible to the original chiral reagent.

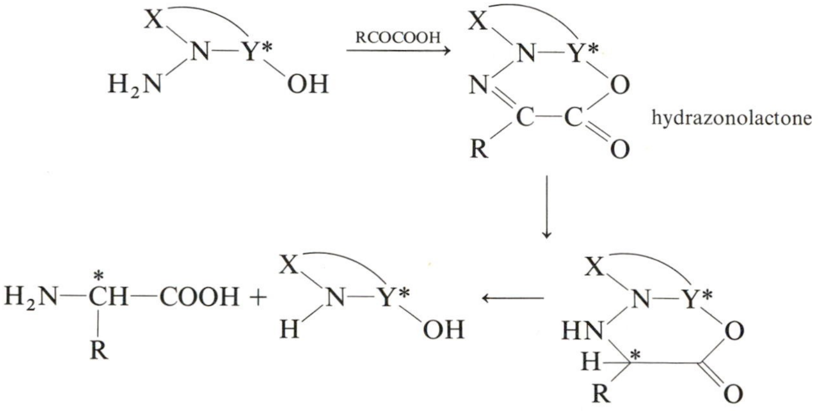

Fig. 2.7. Outline of Corey's methodology (12). *, chiral center.

To apply this methodology the chiral reagent chosen was a bicyclic in-
doline structure and its synthesis is given below:

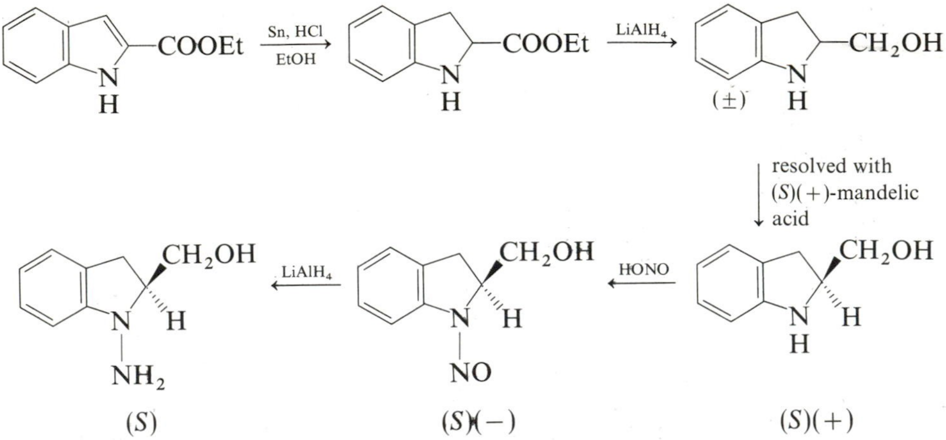

Its absolute stereochemistry was proved by correlation with L(−)-phenyl-
alanine. This simple rigid structure is necessary to insure a good steric control in the reduction of the C=N double bond by asymmetric induction.

Based on this model, the asymmetric synthesis of D-alanine was carried
out. All the steps proceed in good yield and the amino acid obtained has an
optical purity of 80%. Figure 2.8 outlines this synthesis.

The efficiency of the synthesis could in fact be improved by a modification
of the starting material, utilizing a more crowded chiral reagent with two
centers of asymmetry:

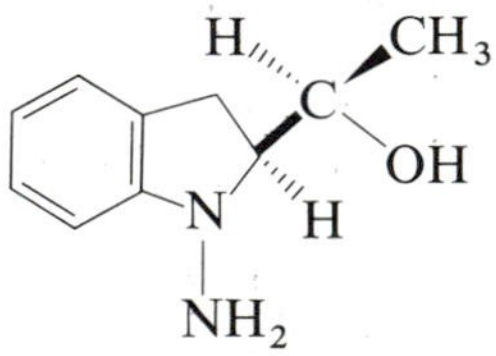

The optical yield of the conversion increased to 96% for D-alanine and to
97% for D-valine. Thus, the extra methyl in the front side of the corresponding
hydrazonolactone favors a more selective reduction of the double bond from
the back side of the tricyclic intermediate.

Another advantage of this approach is the possibility of preparing stereospecifically a series (by varying the R group of the keto acid precursor at will)
of α-deuterated amino acids by simply undertaking the chemical reduction
in heavy water.

2.7.2 Rhodium(I) Catalyst

Another elegant and useful recent method of production of optically active
amino acids is by homogeneous catalytic hydrogenation using rhodium(I)
complexes as the catalyst. Indeed, the discovery that the $[Rh(Ph_3P)_3Cl]$

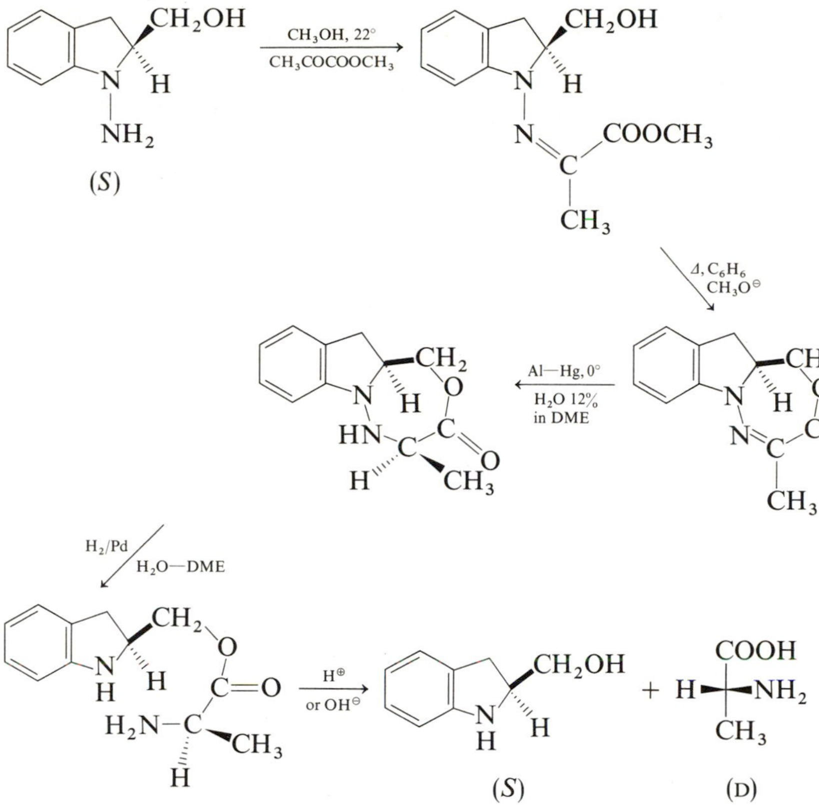

Fig. 2.8. Asymmetric synthesis of D-alanine (12).

complex (Wilkinson's catalyst) and related derivatives were efficient homo-geneous hydrogenation catalysts for many olefins provided a system which could potentially be modified into asymmetric catalysts.

The challenge was to find a catalyst which would stereospecifically reduce an alkene such as

Early attempts using unidentate phosphines gave only low optical yield. Bidentate systems are more promising but there is still a problem of flexibility and rapid interconversion:

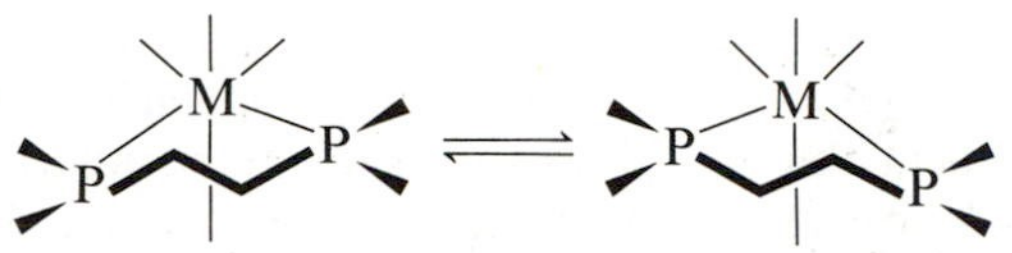

However, the incorporation of chiral bidentate phosphine and phosphite ligands should prevent this equilibrium. For example, if the aliphatic link

is substituted and thereby an asymmetric carbon center is produced, the chelate ring may be fixed into a single, static chiral conformation, with the requirement that the substituent be equatorially disposed. Also, a chiral substituent should transmit its chirality to the ring in such a way that the whole molecular framework is twisted into a single chiral conformation. A prochiral olefin would then coordinate this chiral metal complex in a preferential orientation, leading to an asymmetric reduction of the double bond. It was these principles that led B. Bosnich (13) from the University of Toronto, in 1971, to prepare the chiral ligand $(2S,3S)$-bis(diphenylphosphino) butane, abbreviated (S,S)-chiraphos. It was used to make the corresponding (S,S)-chiraphos chelate ring, M being a rhodium metal ion.

(S,S)-chiraphos chelated catalyst

The two methyl groups are equatorially oriented. By fixing the ring in one conformation only, a chirality is induced in the ring in the disposition of the phenyl residues to allow diastereotopic complex formation with olefins. The synthesis of the catalyst is summarized in Fig. 2.9.

Lithium diphenylphosphide displaces the tosylate groups by an S_N2 mechanism, Ni(II) ions are used to separate the desired product from side products via an insoluble Ni(II) complex and CN^- ions are finally used to liberate the metal. The (S,S)-chiraphos is a solid obtained in 30% yield, but which is oxidized slowly by air in solution. It is therefore immediately converted to the rhodium(I) complex by a displacement reaction involving a di-1,5-cyclooctadiene rhodium(I) precursor. The final product is obtained as an orange-red solid which is stable if kept under nitrogen at 0°–4°C. It is this species which is capable of hydrogenating a variety of olefins under catalytic conditions. The ratio of catalyst to substrate used is generally 1:100 and the reaction is carried out under nitrogen, at 25°C for 1 to 24 hr. The precursors for the α-amino acids are α-N-acylaminoacrylic acids and esters. Table 2.2 shows the optical yield of amino acids obtained. Only R-amino acids are obtained with the (S,S)-chiraphos catalyst. In principle, the corresponding antipode, the (R,R)-chiraphos catalyst should give natural S-amino acids, and indeed it does.

Table 2.2 shows that the yield of the reaction is sensitive to the nature of the N-acyl substituent, the β-vinyl substituent, and the polarity of the solvent. However, the reaction is always stereospecific in that only cis hydrogen addition to the olefins is observed. This was verified by undertaking the reaction with $D_2(^2H_2)$ instead of H_2. The resulting product is the [2R,3R-

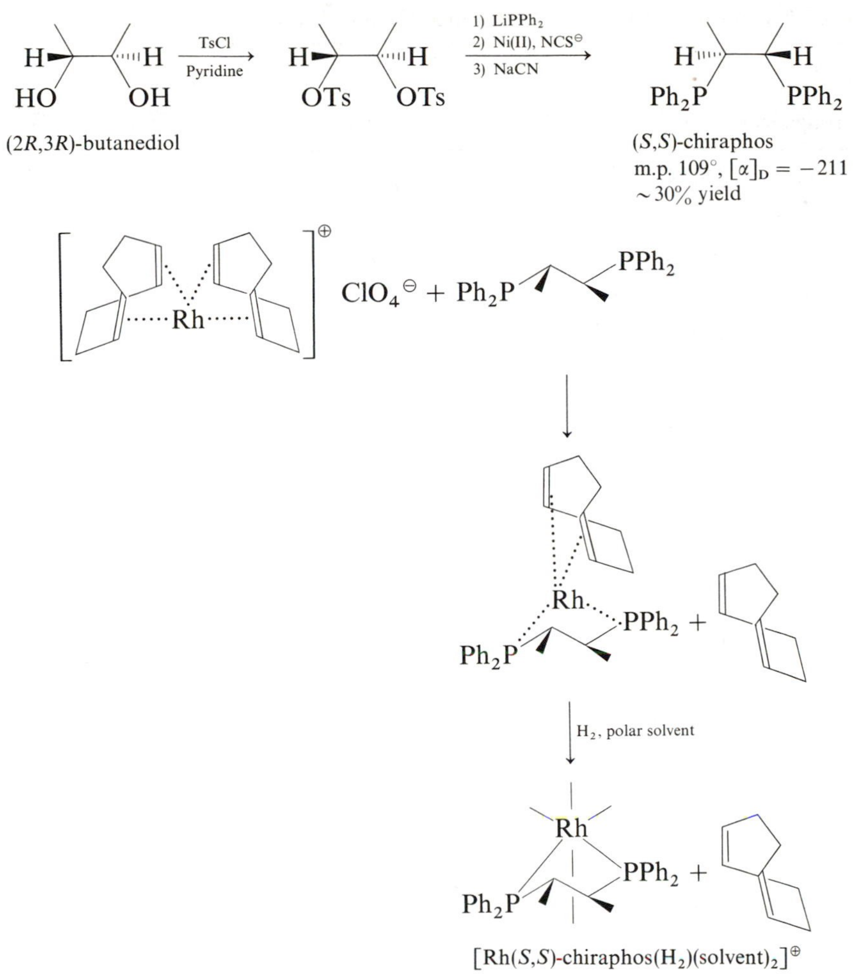
TsCl
Pyridine
1) LiPPh2
2) Ni(II), NCS⊖
3) NaCN
(2R,3R)-butanediol
(S,S)-chiraphos
m.p. 109°, [α]D = −211
~30% yield
ClO4⊖ + Ph2P
PPh2
H2, polar solvent
[Rh(S,S)-chiraphos(H2)(solvent)2]⊕

Fig. 2.9. Preparation of Rh(I) catalyst (13).

$^2H_2$]-*N*-benzoylphenylalanine from the corresponding acrylic acid precursor:

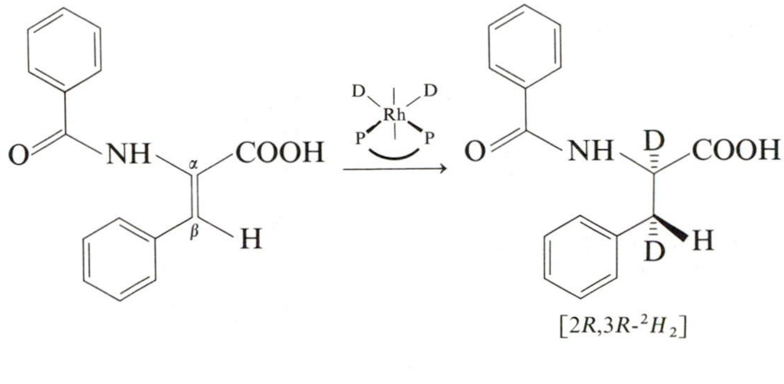
NH
COOH
α
β
[2R,3R-2H2]

Table 2.2 Optical Yield (%) of R-Amino Acid Obtained (13)

Starting olefin	Amino acid	Solvent	
		THF	EtOH
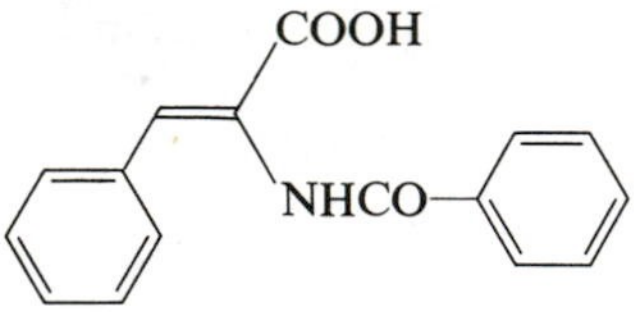 (structure: COOH, NHCO-phenyl, benzyl)	Phe	99	95
(structure: COOEt, NHCO-phenyl, benzyl)	Phe	83	—
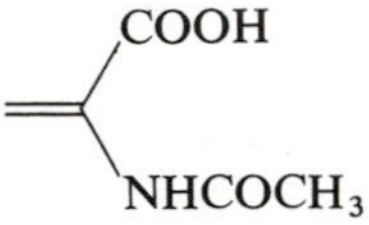 (structure: COOH, NHCOCH₃)	Ala	88	91
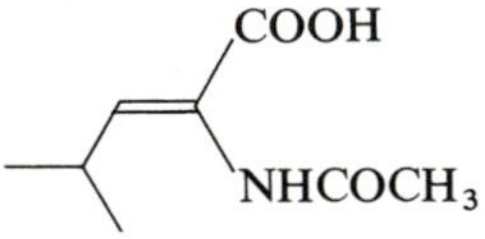 (structure: COOH, NHCOCH₃, isopropyl)	Leu	100	93
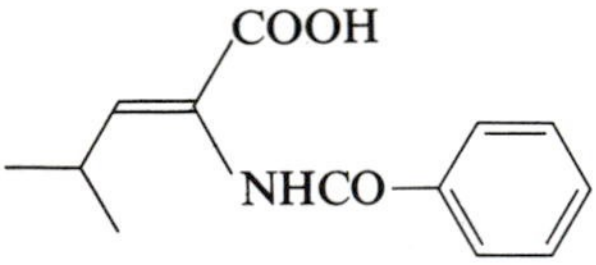 (structure: COOH, NHCO-phenyl, isopropyl)	Leu	87	72

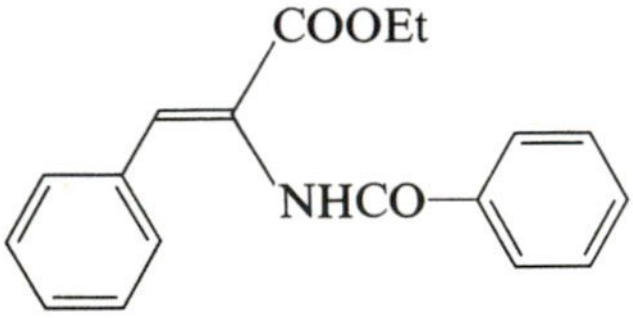

This asymmetric deuteration is believed to proceed in two steps involving first the addition of a hydride to the olefin to produce a rhodium alkyl bond. This is rapidly cleaved by the insertion of a proton into the metal alkyl bond:

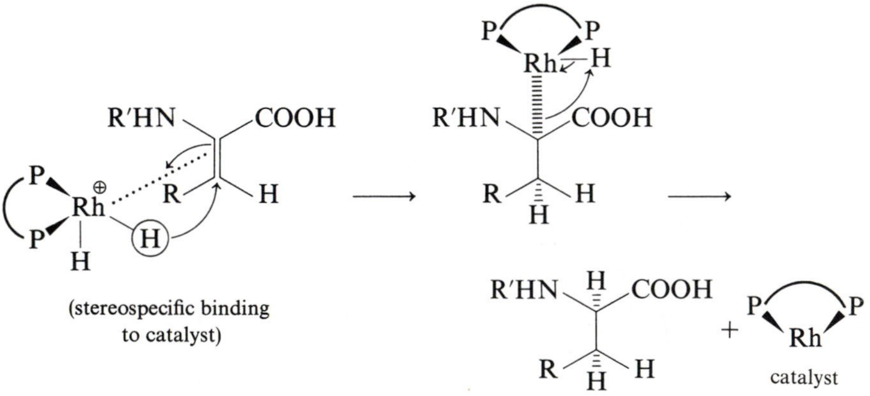

Despite the two-step mechanism, the process appears to be completely stereospecific. Furthermore, if a leucine precursor is deuterated in an analogous fashion, the deuterium appears only at the α and β centers. The γ-position carries no deuterium, implying, as expected, that no double bond migration occurs during the reduction. An important gain which is intrinsic to this mechanism is that the optical purity of the β-center is linked to that of the α-center.

The reasons for the effectiveness of the present catalyst in asymmetric hydrogenation of amino acid precursors can be summarized as follows: (1) the conformational rigidity of the chelated diphosphine ligand; (2) the dissymmetric orientation of the phenyl groups, held so by the 5-membered puckered chelate ring. These factors are the major source of the diastereotopic interaction with prochiral substrates to give: (a) a *cis*-endo stereospecific hydrogenation, producing both α- and β-carbons chiral with $^2H_2$*; (b) only *R*-amino acid, irrespective of the solvent.

The success of this approach is helpful for the preparation of specific β-deuterated amino acids. The following transformation is an outline of the method for racemizing the α-carbon center of *N*-benzoylphenylalanine via an oxazoline intermediate. The two diastereoisomers can be separated by conventional means.

$[2R,3R\text{-}^2H_2]$

$\xrightarrow{\text{Ac}_2\text{O}}$

1) Pyr/D$_2$O
2) OD$^\ominus$/CH$_3$OD/D$_2$O
3) HCl/H$_2$O

$[2R,S,3R\text{-}^2H_2]$

---

* One should realize that the catalyst approaches the prochiral double bond from the α-si, β-re face. (For an introduction to the nomenclature used to differentiate the two carbons of a prochiral double bond see Section 4.2.1.) Deuterium (^{2}H) and tritium (^{3}H) atoms are heavy isotopes of hydrogen (^{1}H) but for the clarity of the structural representations only, the symbols D and T are used throughout this text.

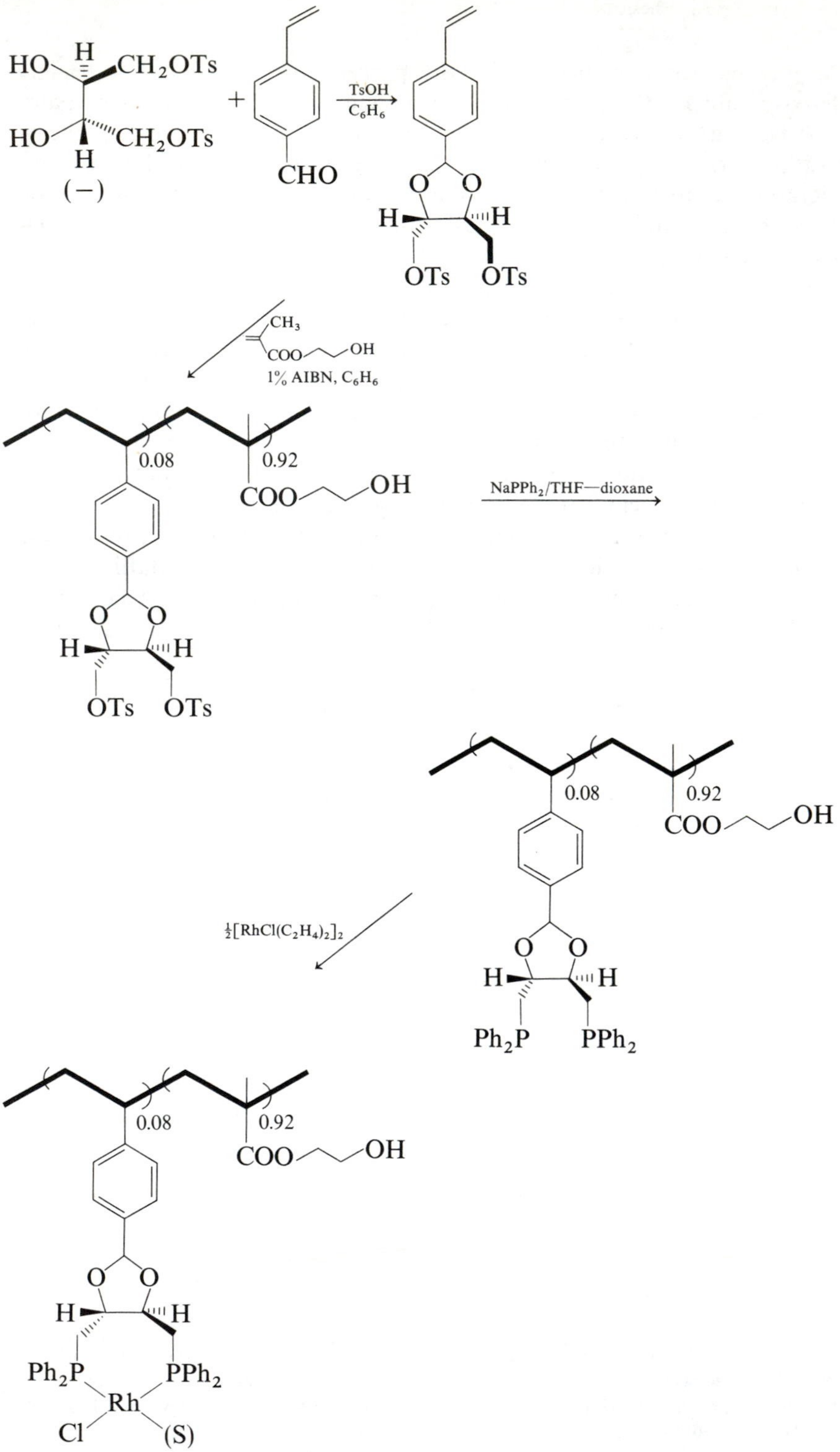

Fig. 2.10. Synthesis of a polymer-supported Rh(I) optically active catalysis (15). AIBN, azabiisobutyronitrile, a free radical initiator for polymerization. S, solvent.

Thus, as a bonus, this approach allows the rational design of selectively mono- or dideuterated amino acids and analogues. These chirally labeled molecules can be used to study the mechanistic details of biosynthetic pathways involving amino acid precursors. This approach has been recently extended (14) to allow the synthesis of chiral methyl lactic acid in which the three methyl hydrogens are replaced by a proton, deuteron, and triton.

In 1978, J. K. Stille and his group proposed an interesting extension of the concept of asymmetric synthesis via rhodium complexation by attaching the metallic site to an insoluble polymer (15). The main advantage of this modification is the possibility of recovering the optically active phosphine-rhodium complex catalyst.

One important and challenging problem in polymer-supported catalysis is the proper choice of the polymer matrix and the synthesis of the catalyst site in the matrix. The method used here consists of the introduction of a reactive site on a cross-linked polystyrene bead followed by the reaction of an optically active phosphine-containing ligand at the site. The approach is summarized in Fig. 2.10. The reaction of $(-)$-1,4-ditosylthreitol with 4-vinylbenzaldehyde affords 2-p-styryl-4, 5-bis(tosyloxymethyl)-1, 3-dioxolane, which is copolymerized radically with hydroxyethyl methacrylate to incorporate 8 mole % of the styryl moiety in the cross-linked copolymer. Further treatment with sodium diphenylphosphide to react with all the hydroxyl functions plus the tosylate groups gives after neutralization a hydrophilic polymer bearing the optically active 4,5-bis(diphenylphosphinomethyl)-1,3-dioxalane ligand. Exchange of rhodium(I) onto the polymer with $[(C_2H_4)_2$-$RhCl]_2$ gives the desired polymer-attached catalyst that swells in alcohol and other polar solvents to allow the penetration of the substrate.

Typical hydrogenations are carried out at 25°C with 1 to 2.5 atm of hydrogen and the catalyst and with an olefin to rhodium ratio of about 50. This way, α-N-acylaminoacrylic acids in ethanol are converted to the amino acid derivatives. The optical yields are comparable to those obtained with the previously mentioned homogeneous catalyst. The same absolute (R) configuration of the products is observed. The main advantage is that the insoluble catalyst can be reused many times. It may be recovered from the reaction mixture by filtration, under an inert atmosphere, with no loss of catalytic activity or optical purity in the hydrogenated product.

The ultimate variant of this approach was obtained recently by G. M. Whitesides, from M.I.T. (16, 17). He constructed an asymmetric hydrogenation catalyst based on embedding an achiral diphosphine-rhodium(I) moiety at a specific site in a protein. In this case the protein tertiary structure provides the chirality required for enantioselective hydrogenation.

The well-characterized protein avidin, composed of four identical subunits, each of which binds biotin and many of its derivatives, was used. For this, a hydroxysuccinimide substituted biotin was converted to a chelating diphosphine and complexed with rhodium(I) by the following sequence:

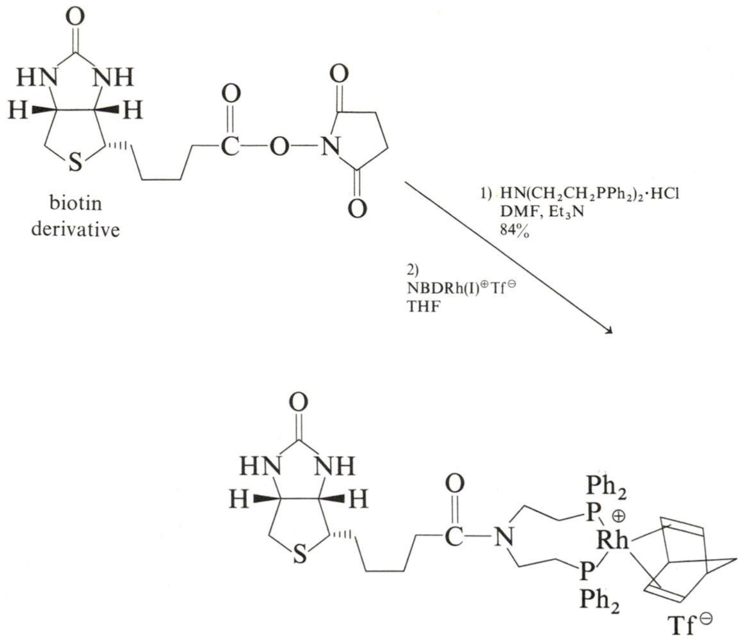
biotin
derivative
1) HN(CH2CH2PPh2)2·HCl
DMF, Et3N
84%
2)
NBDRh(I)⊕Tf⊖
THF
Tf⊖

$Tf^{\ominus}$ = triflate ion ($CF_3SO^{\ominus}$)
NBD = norbornadiene

In this procedure the diphosphine intermediate serves as the basis for the elaboration of a water-soluble rhodium-based homogeneous hydrogenation catalyst. The enantioselectivity of the catalyst was tested by the reduction of α-acetamidoacrylic acid to *N*-acetylalanine. The presence of avidin resulted in a definite increase in activity of the catalyst and in the production of the *S*-enantiomer (natural amino acid) in 40% excess.

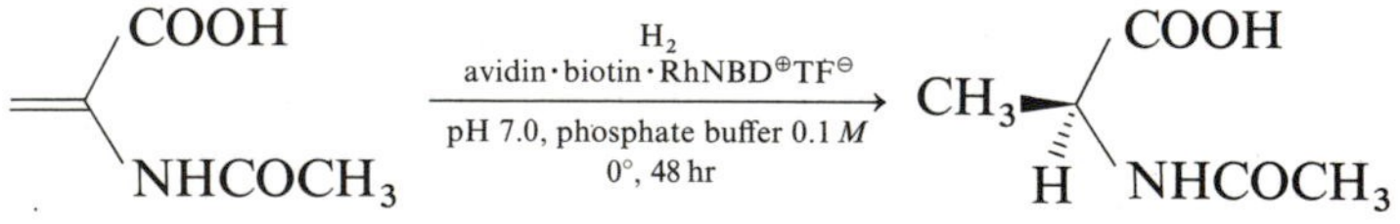
COOH
NHCOCH3
H2
avidin·biotin·RhNBD⊕TF⊖
pH 7.0, phosphate buffer 0.1 M
0°, 48 hr
CH3
COOH
H
NHCOCH3

S-isomer

The presence of other enzymes such as lysozyme or carbonic anhydrase had no significant influence on enantioselectivity.

Hence, Whitesides showed that it is possible to carry out in aqueous solution homogeneous hydrogenation using a diphosphinerhodium(I) catalyst associated with a protein. In addition, the chirality of the protein is capable of inducing significant enantioselectivity in the reduction. Finally, as the author pointed out, this technique developed to bind transition metals to specific sites in proteins may find use in biological and clinical chemistry unrelated to asymmetric synthesis (16).

Chapter 3
Bioorganic Chemistry of the Phosphates

"The whole is more than the sum of the parts."

Aristotle

Too often a detailed description of the synthesis and properties of peptides is given without consideration of the analogous, but equally important, synthesis and properties of phosphodiesters, and vice versa. Indeed many of the problems and strategies (i.e., use of DCC, common protecting groups, polymeric synthesis, etc.) are similar, if not the same, yet they are never presented "side-by-side." It is with this purpose in mind that this chapter is written. Again, for the sake of comparison, biological synthesis of the phosphate bond is also presented. As such, a chemical and biological comparison (bioorganic) of the two functionally important classes of macromolecules, the proteins and the nucleic acids, are presented. Of course, the picture is completed by examining two mononucleotides which are essential to the biological process: nucleoside triphosphates and cyclic nucleotides. This emphasizes that, as with the amino acids, not only polymers are important to the biological system. Again, a novel comparison of the chemical and biological synthesis is presented. This includes material which has been made available as recently as 1980.

Nonetheless, all this does require some overlap with elementary biochemistry. The chapter begins with an introduction to DNA and RNA. As with protein synthesis, this is described as briefly as possible, in chemical terms, omitting much of the descriptive material that is available in most biochemistry texts. Finally, the chapter ends with a brief presentation of the prebiotic origin of biopolymers.

3.1 Biological Role of Phosphate Macromolecules

In order to maintain the life process, it is necessary to *pass on specific information* from one generation of organisms to the next. Such information is vital, for it allows the continuation of those traits or properties of the organism that insure survival. This information must be stored, to be used at an appropriate time, much as data may be stored and played on a tape recorder. As such, the job of information storage and processing must have a molecular basis. Indeed, it is not difficult to imagine some of the properties that are requisite for a molecular biological "tape."

Considering that a great deal of information must be stored on this molecule (i.e., the type of organism, physical characteristics, chemical pathways, etc.), it is expected that it must be a biopolymer. Perhaps a protein could function as a storage molecule? Probably not, as proteins already play an important structural and functional (enzyme catalyst) role in the cell. Such an important job as data storage must require a unique macromolecular structure which cannot possibly be confused with routine cellular processes. This special biopolymer is expected to be rather homogenous in structure as it must fulfill one crucial role. It is not expected to be as diverse as enzyme structure for the latter are capable of participating in a wide variety of chemical reactions. On the other hand, it must have a heterogenous component in order to carry the different information necessary to the cell. This biopolymer is expected to be of a rigid or defined shape as it must be able to interact with the cellular apparatus when it becomes necessary to transmit the stored information. A "floppy" molecule containing acyclic polymeric chains which can assume any one of a number of available conformations would not be expected to interact well, perhaps even in a cooperative fashion, with the ordered configurations of the cellular components. Specific information must be passed on in a precise fashion. Remember, protein synthesis, for example, occurs on a "surface," in an ordered sequential fashion, and not in solution in a random manner (see Section 2.5).

An ordered structure suggests the presence of five- and six-membered rings, and not chains. Perhaps the simplest to imagine are hydrocarbons such as benzene, naphthalene, or indene. However, these molecules are quite hydrophobic. This is disadvantageous as biological processes are carried out in aqueous media. Furthermore, hydrocarbons cannot participate in a variety of noncovalent bonding interactions: hydrogen bonds and in particular, electrostatic bonds.

The molecules which function as the monomeric units for genetic information storage are nitrogen heterocycles: derivatives of *purine* and *pyrimidine*. The polymeric molecule which functions to both store and pass on
genetic information is *deoxyribonucleic acid* (DNA). The related polymer,
ribonucleic acid (RNA) helps in the passage of this genetic information. It
acts as the messenger which converts a specific genetic message to a specific
amino acid sequence. The common purine bases of DNA and RNA are
adenine and *guanine*. The common pyrimidine bases of DNA are *cytosine*
and *thymine* (5-methyluracil), while for RNA they are cytosine and *uracil*
(Fig. 3.1). The purines and pyrimidines occur in combination with the sugar
ribose (in RNA) or *deoxyribose* (in DNA) via the anomeric carbon of the
latter. The glycosidic linkage occurs through the ring nitrogen of the base:
either nitrogen-9 of the purines or nitrogen-1 of the pyrimidines. This
chemical combination of a sugar and base (with elimination of water) is
referred to as a *nucleoside* (Fig. 3.2). One important exception is the nucleoside
pseudouridine (ψ: found in tRNA) which is connected to the ribose sugar via
an atom of carbon. Thus, for example, by the rules of sugar chemistry, the
nucleoside adenosine is also referred to as 9-β-D-ribofuranosyladenine.

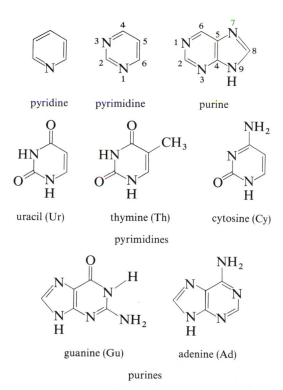

Fig. 3.1. Purines and pyrimidines of the DNA and RNA hereditary molecules (predominant tautomer at pH 7 is indicated).

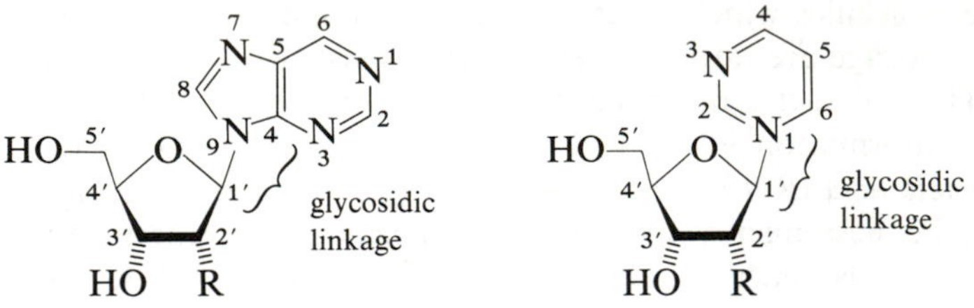

purine nucleoside   pyrimidine nucleoside

R	Purine or pyrimidine base	Nucleoside
OH	Cytosine	Cytidine (C)
H	Cytosine	Deoxycytidine
OH	Thymine	Thymidine (T)
H	Thymine	Deoxythymidine
OH	Uracil	Uridine (U)
H	Uracil	Deoxyuridine
OH	Adenine	Adenosine (A)
H	Adenine	Deoxyadenosine
OH	Guanine	Guanosine (G)
H	Guanine	Deoxyguanosine

Fig. 3.2. Structure and nomenclature of the common nucleosides.

Other bases are also present in the DNA and RNA polymers, but to a much lesser extent. Examples include 5-methyl cytosine (present in the DNA of some plants and bacteria), 5-hydroxymethyl cytosine (present in the DNA of T-even phages; bacterial viruses) and 4-thiouracil (present in some bacterial tRNA).

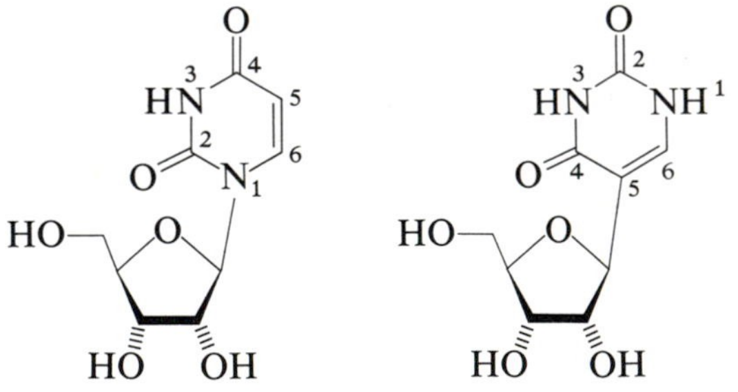

uridine (U)   pseudouridine (ψ)
(β-D-ribofuranosyluracil)   (5-β-D-ribofuranosyluracil)

It is possible to esterify a sugar hydroxyl with either phosphoric acid, pyrophosphoric acid or triphosphoric acid to give the nucleoside mono-, di-, or triphosphates, respectively. The example of adenosine triphosphate (ATP) has already been encountered. The combination of a base, sugar, and phosphate is referred to as a *nucleotide*. Just as the amino acid is the monomeric unit of the protein polymer, the nucleotide (nucleoside monophosphate) is the monomeric unit of the DNA and RNA polymers. Further, just as the monomeric amino acids are linked via an amide (peptide) linkage, the monomeric nucleotides are linked via a phosphoester linkage. A simple comparison is as follows:

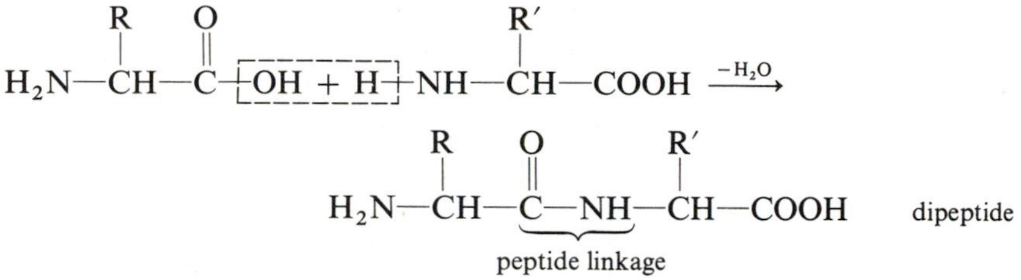

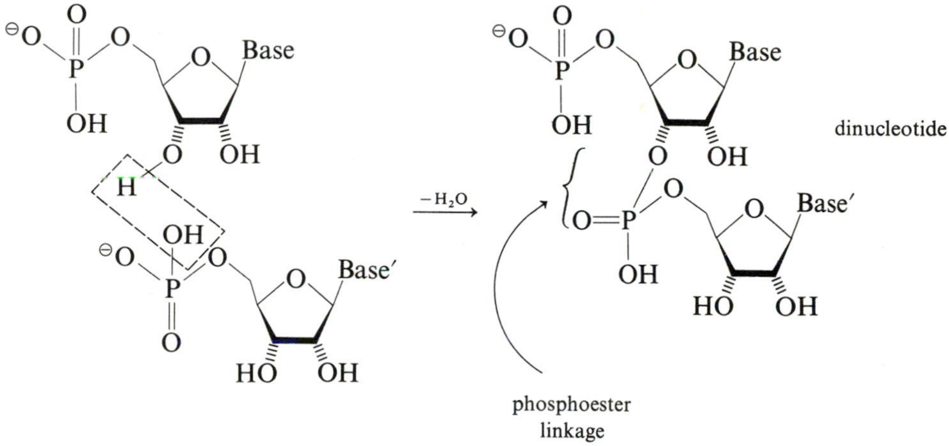

peptide linkage

dipeptide

dinucleotide

phosphoester
linkage

Base refers to a purine or a pyrimidine throughout this chapter

In both DNA and RNA, the nucleosides are then connected to inorganic phosphate via the 3'-hydroxyl (not 2'-) of one sugar and the 5'-hydroxyl of a second sugar. The phosphate exists as a phosphodiester, having one negative charge (pK_a of a phosphodiester ~ 2.0) at physiological pH. The cation is usually a metal (magnesium) or protonated amine (polyamines, histone protein).

It is important to realize that *it is the purine or pyrimidine base which carries the genetic information, just as the side chain of an amino acid determines its functional chemistry.* The DNA hereditary molecule is organized in the cell into units called *genes*. These are in turn located within structures called *chromosomes*, the latter being found in the nuclei of plant and animal cells. It is the gene which contains the information necessary for the determination of a specific trait: color of eyes, hair, height, sex, etc. However, the gene is a rather complex concept to describe at the molecular level, for the number of molecular events required for the exhibition of a particular trait may be numerous. Noting that any genetic trait will involve protein synthesis (either structural proteins or enzymes), it becomes possible to define a simpler unit: the *cistron*. The cistron is defined as that part of DNA which carries the genetic information (codes) for the synthesis of one polypeptide chain. A chromosome contains many hundreds of cistrons. The total DNA content of the cell is referred to as the *genome*.

The genetic information is passed from the parent cell to the daughter cell by *replication* (synthesis) of DNA. The genetic information is stored in DNA until it is needed, at which time it is converted to the appropriate message for the synthesis of a protein of specified sequence via the process of *transcription*. The genetic message is transcribed to the RNA (messenger RNA) biopolymer. This in turn interacts with the appropriate specific aminoacyl-tRNA's resulting in the sequential coupling of amino acids. The transformation of the genetic information from RNA to a specific amino acid sequence is referred to as *translation*. The terms replication, transcription, and translation may be understood by considering the analogy of printing a book. When a book goes to print, many replicas or copies are made; hence replication. When a passage is copied by hand from the book, it is transcribed; hence transcription. When the book is written in another language: it is translated; hence translation (from nucleotides to amino acids). The entire process is described below, and only a few more words will be said shortly to relate this to the molecular level.

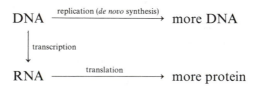

3.2 General Properties

3.2.1 Purines and Pyrimidines

These nitrogenous bases tend to be poorly soluble in water, but become more so upon their conversion to nucleosides and nucleotides. They possess high melting points: most above 300°C.

The oxygen-containing bases are capable of lactam–lactim tautomerism. In most instances, the resonance energy of the amide group(s) outweighs the resonance stabilization of the aromatic ring, so that the lactam form is the predominant tautomer. This is illustrated for uracil:

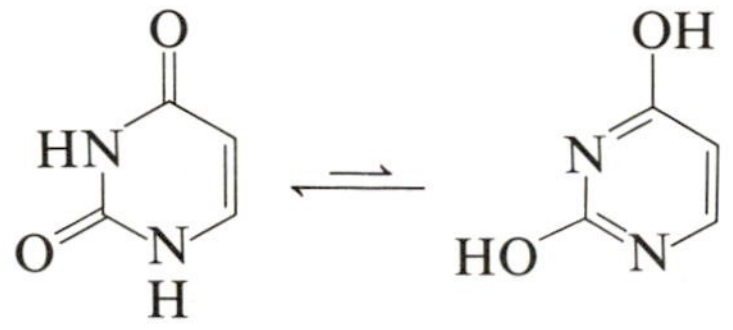

uracil (lactam)   uracil (lactim)

The purines and pyrimidines are weak bases, with pK_a values of ~ 9.5 for ring nitrogen atoms. For example, the pK_a values for the dissociation of the protons at N-1 or N-3 of uracil (in terms of their dissociation behavior, these protons are equivalent) or N-9 of adenine are 9.5 and 9.8, respectively:

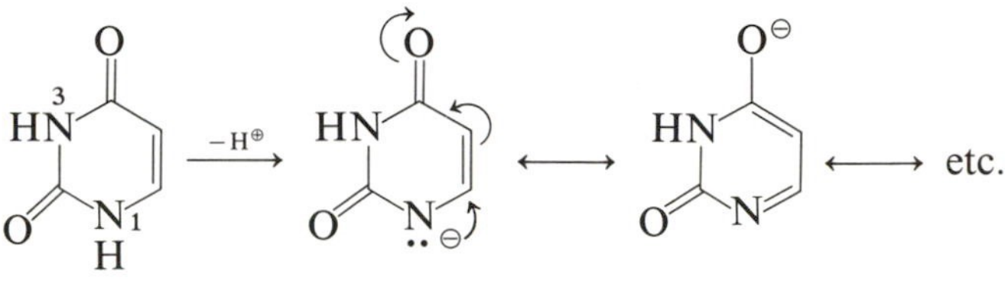

uracil

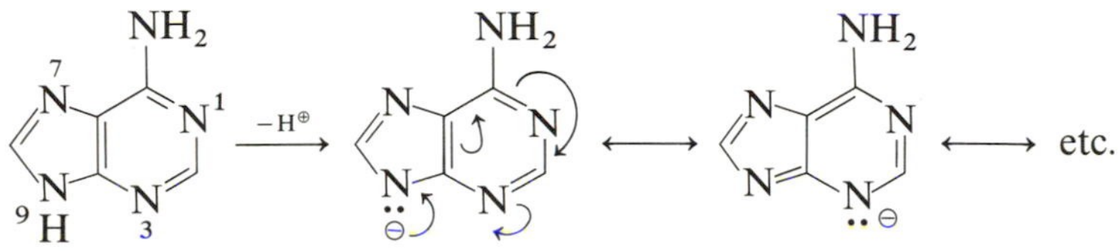

adenine

This basicity may be accounted for by resonance stabilization of the anion, and the sp^2 hybridization of the ring nitrogen atom. Remember, the greater the "s" character of the atom in question, the more acidic the atom.

The pK_a values of the exocyclic amino functions are also not typical of primary amines. Thus, they are not protonated at physiological pH and are not even the first nitrogen to be protonated as was thought for many years (see Table 3.1). Such behavior has already been discussed (compare aniline with cyclohexylamine, Table 2.1) and is accounted for by delocalization of the nitrogen nonbonding electrons into the ring system.

Of course, the presence of the nitrogen atoms as part of the ring structure, as well as exocyclic, will affect not only dissociation behavior, but overall chemical reactivity. The course of electrophilic and nucleophilic substitution reactions will be determined by the presence of these nitrogen atoms. For example, the two heterocyclic nitrogen atoms in pyrimidines result in an electron distribution about the ring which makes C-5 electron rich relative to the other carbon atoms and thus most susceptible to electrophilic addition.

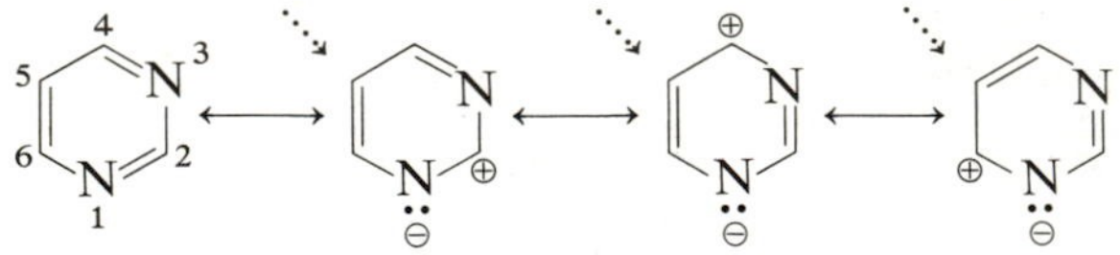

Similarly, for the purines, C-8 is the most electron rich position relative to C-2 and C-6, and thus the most susceptible to electrophilic substitution. Obviously, the carbon atoms at positions 4 and 5 cannot participate in an electrophilic substitution.

Table 3.1. The pK_a Values of the Most Common Purines and Pyrimidines

adenosine

guanosine

uridine

cytidine

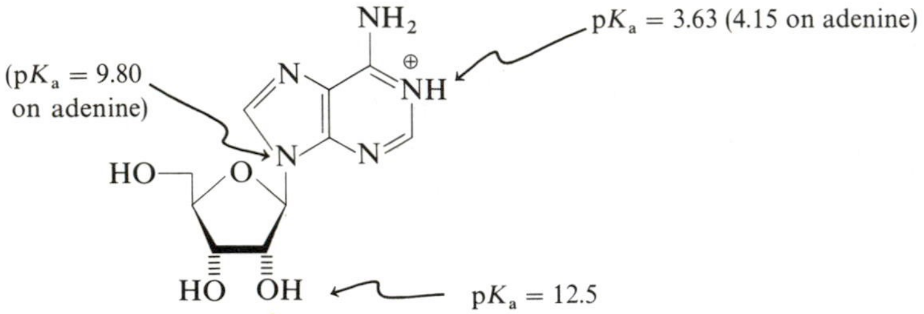

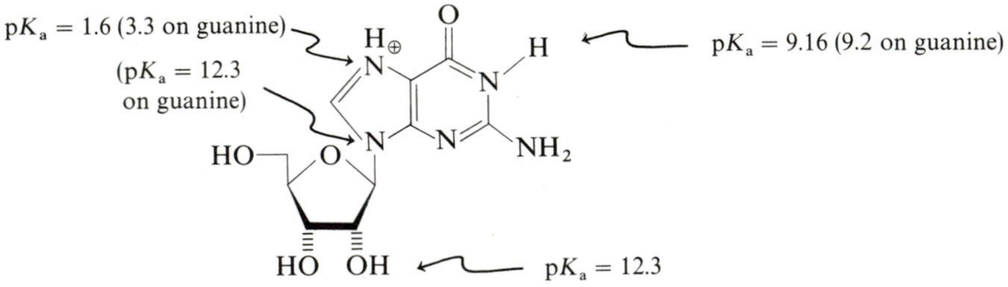

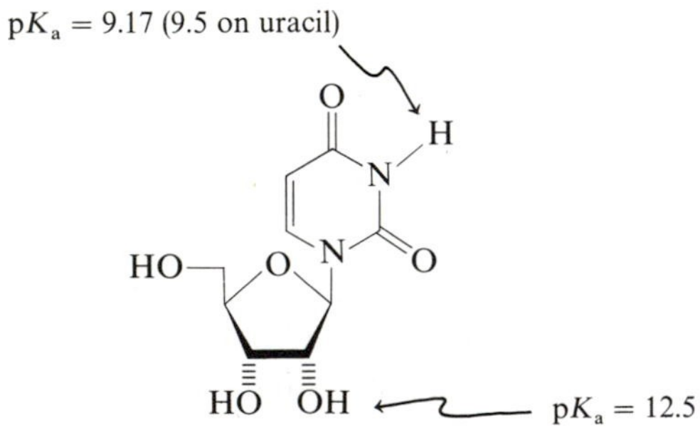

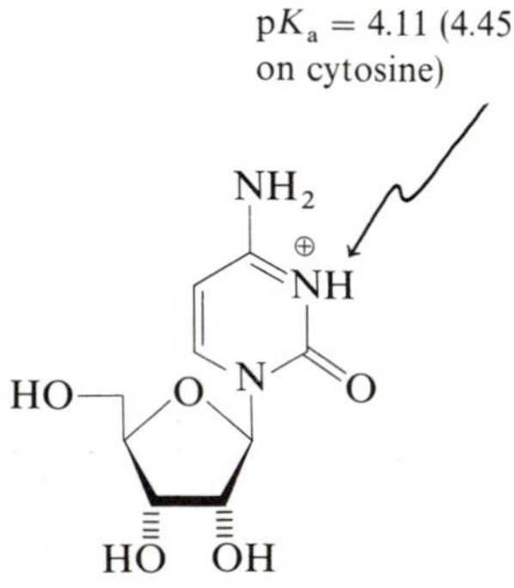

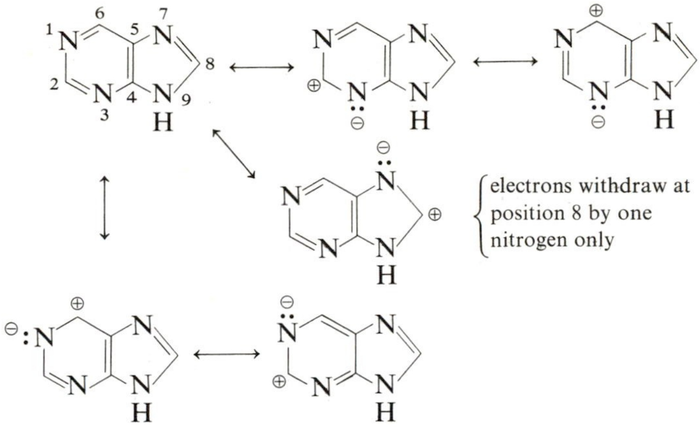

In addition, the sp^3 hybridized nitrogen at position 9 is capable of donating electron density to C-8:

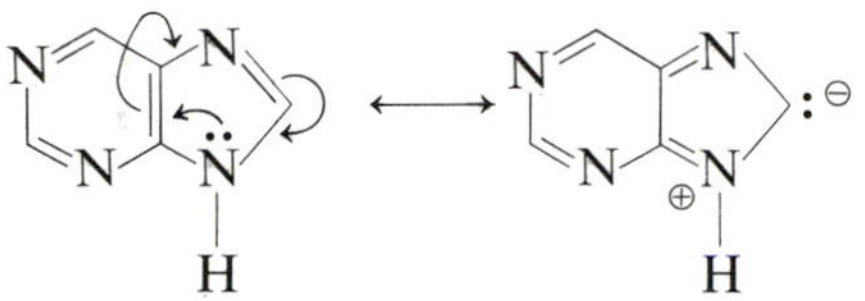

Thus, via electrophilic substitution, a number of biological analogues of the purines and pyrimidines may be prepared. In particular, halogenation allows the preparation of an intermediate that may undergo further derivatization. Some examples are presented:

(1) *Chlorination of Uridine*

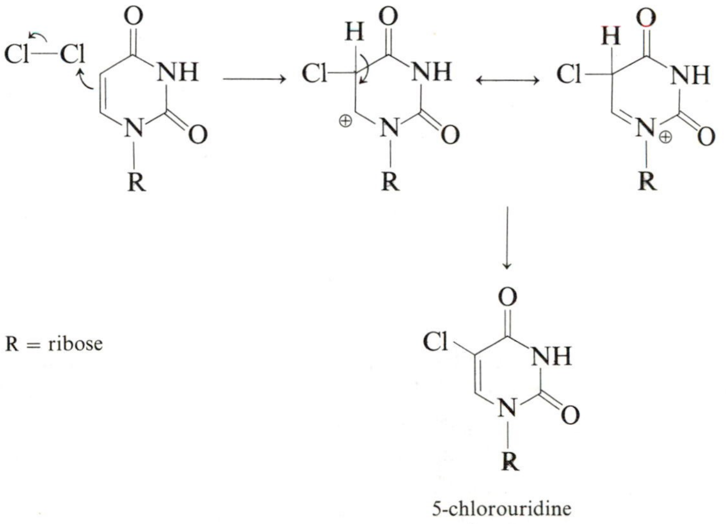

R = ribose

5-chlorouridine

(2) Bromination of Adenosine

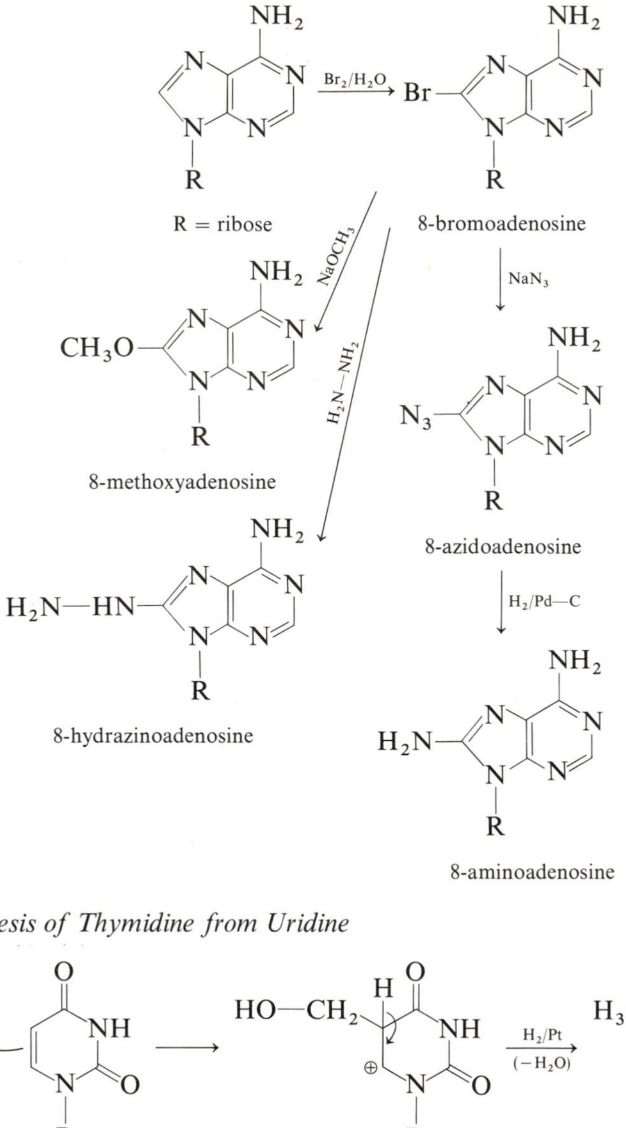

R = ribose

8-bromoadenosine

8-methoxyadenosine

8-azidoadenosine

8-hydrazinoadenosine

8-aminoadenosine

(3) Synthesis of Thymidine from Uridine

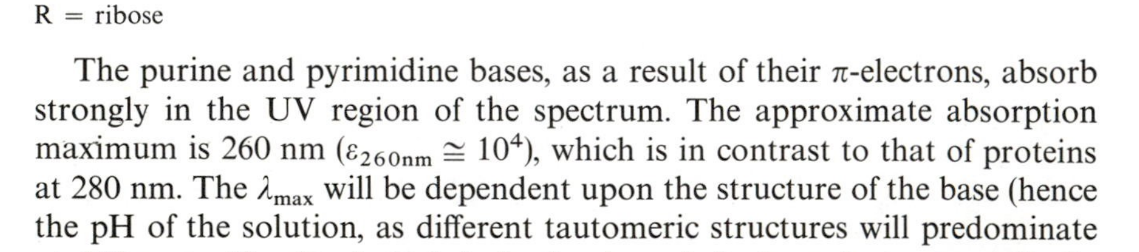

R = ribose

The purine and pyrimidine bases, as a result of their π-electrons, absorb strongly in the UV region of the spectrum. The approximate absorption maximum is 260 nm ($\varepsilon_{260nm} \cong 10^4$), which is in contrast to that of proteins at 280 nm. The λ_{max} will be dependent upon the structure of the base (hence the pH of the solution, as different tautomeric structures will predominate at different pH values), and derivatization of the base, but not markedly upon the structure of the sugar. This is of synthetic usefulness as purine and pyrimidine derivatives may be characterized by their UV absorption maxima, and chromatograms may be monitored by use of a UV lamp. For example, *N*-benzoyl guanosine (prepared by benzoylation of the base and sugar with benzoyl chloride in pyridine, followed by removal of the sugar benzoyl groups with sodium methoxide) has:

R = ribose
   guanosine

$\lambda_{max}^{EtOH} = 254$ nm

N-benzoyl guanosine

$\lambda_{max}^{EtOH} = 238, 257, 265, 295$ nm

The UV spectra of the common nucleosides can be found in most biochemistry texts.

3.2.2 DNA and RNA

Both DNA and RNA have been described as polymers whose monomeric units are nucleotides, connected together by a phosphoester linkage. This may present a picture of a long, single-stranded chain. However, both DNA and RNA may be double-stranded, or exhibit a *double-helix* structure. Such is not totally unusual, for the structural protein collagen exists as a triple helix: three polypeptide chains wrapped around one another and held together by interchain hydrogen bonds. Here also, hydrogen bonds serve to hold two strands together: either two strands of DNA, RNA, or a so-called DNA–RNA hybrid. The hydrogen bonds arise between the purine and pyrimidine bases of the two strands in a very specific fashion. That is, the bases of one strand align to the bases of the second strand such that a purine base is always hydrogen bonded to a pyrimidine base, and vice versa. More precisely, adenine and thymine (uracil), which are capable of forming two hydrogen bonds with each other, are paired with each other, just as guanosine and cytosine, which are capable of forming three hydrogen bonds with each other, are paired together. Such specific hydrogen bonding (*Watson–Crick pairs*) is important for replication purposes, as the two strands are complementary to each other.* (See Fig. 3.3.) This will be discussed further in the biological synthesis of the phosphoester linkage.

---

* It should also be noted that the extensive hydrogen bonding potential of the nucleotides provides the possibility for numerous types of hydrogen bonding in addition to the classical Watson–Crick base pairing scheme. For the most part, these interactions are found with synthetic polynucleotides. However, non-Watson–Crick base pairing, and hydrogen bonds involving the sugar–phosphate backbone are important elements in the structure of tRNA's.

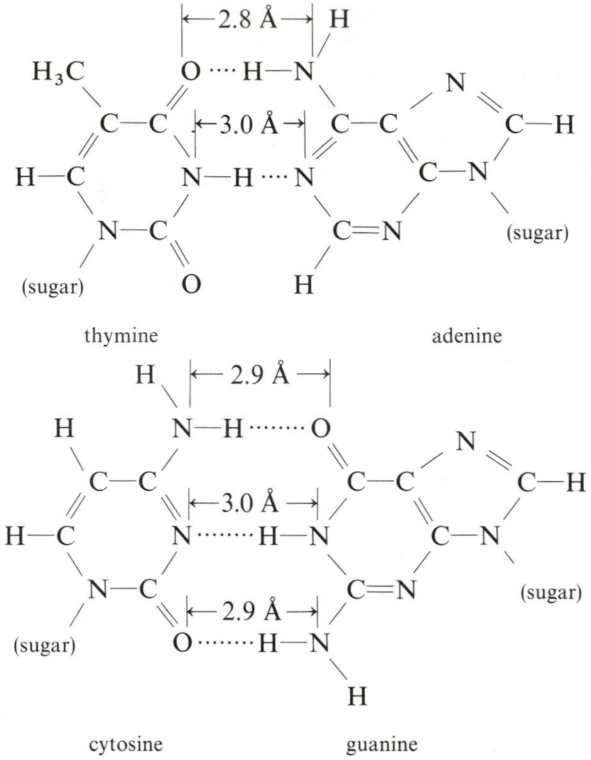

Fig. 3.3. Hydrogen bonding between the purine and pyrimidine bases of double-stranded DNA. Adenine is always paired with thymine while guanine is always paired with cytosine.

The geometry of the glycosidic linkage (C—N) is such that the planar (sp² hybridized) purine or pyrimidine is approximately perpendicular to the (approximate) plain of the sugar ring (Fig. 3.4). As a result, the bases of the double helix may be "stacked" one upon the other, so that hydrophobic bonds may exist between the bases and further stabilize the helical structure.

Again, the geometry of the sugar ring is approximately planar or precisely, it is in the "half-chair" form, with C-2′ and C-3′ out of the plain created by the furanose oxygen and C-1′ and C-4′. C-1′ and C-4′ are able to lie in the same plain as the furanose oxygen atom since the nonbonding electron lobes of the latter provide minimal eclipsing interactions.

The DNA double-helix consists of two strands of DNA wrapped one about the other, such that there are ten base pairs stacked one upon the other, per helical turn (18, 19). This is analogous to protein molecules: the DNA core will be of a hydrophobic nature, while the exterior phosphodiester backbone will face toward the solvent. Of course, during the synthesis of new DNA, it becomes necessary to unravel these strands so as to expose that portion of the molecule containing the genetic information. Further-

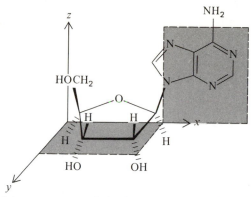

Fig. 3.4. The spatial arrangement of the furanose sugar and the base (adenine) of a nucleoside, about the glycosidic linkage.

more, the two chains of the double helix are "anti-parallel": they run in opposite directions. That is, at each end of the double helix there will be a 3'-hydroxyl function on one strand, and 5'-phosphate function on the other strand. These strands are very long. For example, in T_2 phage, the molecular weight of the DNA is 3.2×10^7, while in *E. coli* (a bacterium whose chromosome is a single molecule of DNA) the molecular weight is 2.0×10^9. The structure of the information storage from that of DNA is most often a double-stranded chain. One exception is the DNA of the phage ϕX174 (which attacks *E. coli*) which is single-stranded.

Polynucleotides, like their monomeric units, absorb UV light. They exhibit the phenomena of *hypochromism*: a solution of polynucleotide will have a lesser optical density of absorbance than will an equimolar solution of the monomeric units. Double-stranded polynucleotides exhibit an even lesser absorbance than single stranded polynucleotides. This simply reflects the fact that the more ordered the bases are in solution (i.e., the better they are able to undergo cooperative interactions), the lesser is the observed extinction coefficient. This allows the monitoring of the cooperative unwinding (hydrogen bond breaking) that occurs upon heating of a double-stranded polynucleotide by means of "melting curves." Experimentally, this is a plot of absorbance as a function of temperature (Fig. 3.5). The midpoint at which the transition occurs is referred to as the "melting temperature" (T_m). The T_m will tend to reflect the ratio of cytosine and guanine, relative to adenine and thymine, since the former base pairs are more stable.

While tRNA contains a significant amount of double-stranded helix, its structure is still different from that of helical DNA and DNA–RNA hybrids. It is of smaller molecular weight ($\sim 25,000$) than typical polynucleotides and assumes a "cloverleaf" structure (20) which contains a significant amount of single-stranded structure. The four "arms" of the molecule form two double-helices oriented at about 90° to each other and the three-dimensional structure looks like an L-shaped molecule (21).

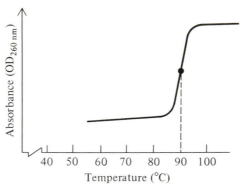

Fig. 3.5. Melting curve of DNA.

As has been briefly noted, all tRNA molecules possess certain common structural features. The first is the same trinucleotide sequence at the strand with the 3′-hydroxyl terminus (cytidine·cytidine·adenosine). The amino acid is esterified at the 3′(2′)-hydroxyl of this adenosine. At the opposite end of the tRNA molecule, within the "loop" is a nucleotide triplet that serves as a recognition site for binding to messenger RNA (mRNA) on the ribosome surface. This nucleotide sequence is different in all aminoacyl-tRNA's. Hence, if a triplet nucleotide on the mRNA polynucleotide (which is synthesized according to the genetic message dictated by the DNA molecule) has a sequence that is "complementary" to the tRNA nucleotide triplet, then binding will occur. *Complementary* refers to hydrogen bonding as described for two strands of DNA: adenine opposite uracil, and guanine opposite cytosine. The triplet sequence that codes for a specific aminoacyl-tRNA, and hence a specific amino acid of a protein, on the mRNA molecule is referred to as a *codon*. The complementary triplet on the tRNA molecule is referred to as the *anticodon*. Almost all tRNA molecules contain a common nucleotide sequence (guanosine·thymidine·pseudouridine·cytidine) which functions as a binding site to the ribosome.

3.2.3 Phosphates

The importance of phosphate in DNA and RNA macromolecules has been described. It serves as an integral part of the molecular "backbone" as well as the means by which the monomeric units are connected. It has important binding properties, being able to participate in strong (electrostatic) bonding interactions with metal and amine cations.

Free inorganic phosphate may be represented as a tetrahedral structure with three negative charges distributed over four oxygen atoms. Binding of appropriate ligands localizes these charges. The tetrahedral structure, the evenly distributed charge-density (per oxygen atom), and the pK_a values for the proton dissociation of phosphoric acid (H_3PO_4: 2.1, 7.2, and 12.3) may be considered as a consequence of the electrostatic repulsion between each

structure of inorganic phosphate ($PO_4{}^{3\ominus}$, or P_i)

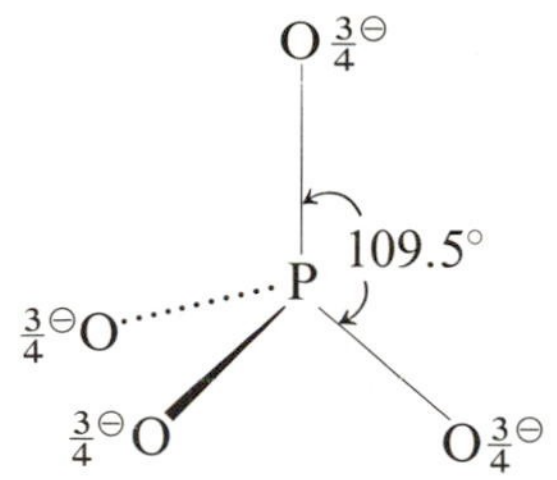

charge of the trianionic species. As such, the phosphoryl function (P=O) is not a localized entity in phosphate salts or all organophosphates, where it can migrate between two or more oxygen atoms. This is similar to the delocalization of the carbonyl function over two oxygen atoms in carboxylic acids. Here also, esterification is required to localize the carbonyl function. In many respects, carboxylic acid derivatives and phosphoric acid derivatives will be seen to have similar properties.

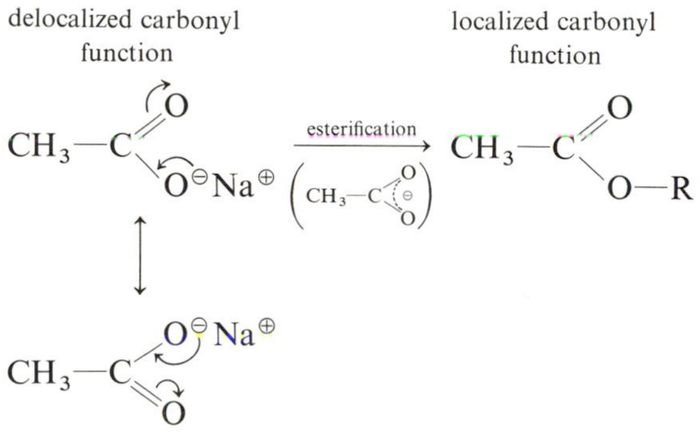

delocalized carbonyl function

localized carbonyl function

sodium acetate

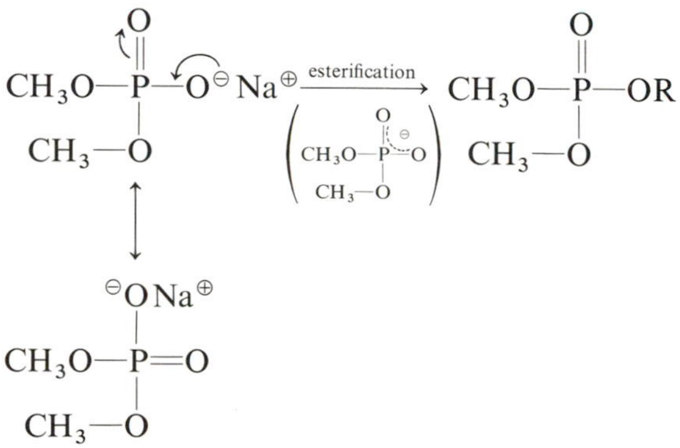

sodium dimethyl phosphate

However, the phosphoryl function does not consist of a π-bond between two p orbitals as does the carbonyl function. The phosphoryl π-bond involves an overlap between the p orbital of the oxygen atom and the d orbital of the phosphorus atom. This bond is a hydrid structure, even when localized, as is illustrated for triphenylphosphine oxide:

$$(Ph)_3P{=}O \leftrightarrow (Ph)_3\overset{\oplus}{P}{-}\overset{\ominus}{O} \quad \text{(ylid form)}$$

The phosphoryl function has significant double bond character, but the ylid is also important.

Phosphate readily forms a number of covalent compounds, ranging from simple esters (trimethyl or triethyl phosphate) to complex macromolecules (DNA or RNA). Many are of biological importance, and the role of phosphate is by no means confined to DNA and RNA structure. In fact just as a number of amino acids do not have to be part of a protein structure in order to have a biological function, many monomeric nucleotides are also of biological importance. Examples include ATP, and the biological regulatory molecules cyclic adenosine and guanosine monophosphate (cAMP and cGMP) (Section 3.4.2). Furthermore, a number of nonnucleoside phosphates are also of biological importance. Examples include sugar phosphates and creatine phosphate; an alternative energy store containing an acid labile phos-

$$\begin{array}{c} CH_3 \diagdown \quad \diagup CH_2{-}COO^{\ominus} \\ N \\ | \\ \overset{\oplus}{C}{=}NH_2 \\ HN \diagup \quad O^{\ominus} \\ \diagdown P \diagup \\ O \diagup \quad \diagdown O^{\ominus} \end{array}$$

creatine phosphate

phoramidate linkage. However, no further discussion will be given to the role of nonnucleoside phosphates, as these are overshadowed by their nucleotide counterparts and have been adequately described in numerous biochemistry texts. Of course, phosphate readily forms inorganic salts which also can be an important constituent of the biological organism (i.e., blood and urine phosphorus).

3.3 Hydrolytic Pathways

Most chemical reactions involving phosphates in the biological system are either the addition (phosphorylation) or removal (hydrolysis) of phosphate. In the case of biological phosphorylations, the phosphate source or donor is the energy store ATP (more shall be said about this in Section 3.4.1) and the reaction is catalyzed by an enzyme often referred to as a kinase. A simple example is the phosphorylation of the sugar glucose:

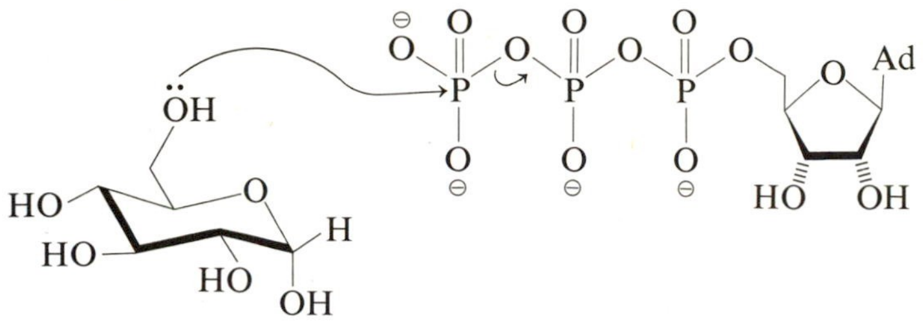

α-D-glucose

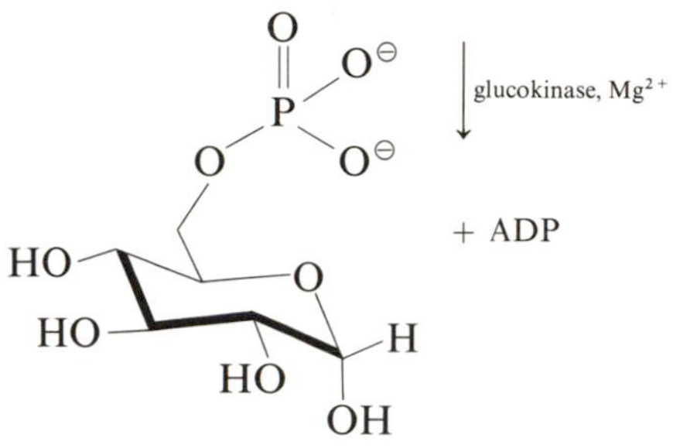

α-D-glucose 6-phosphate

Acid and base-catalyzed hydrolysis of carboxylate esters usually proceeds

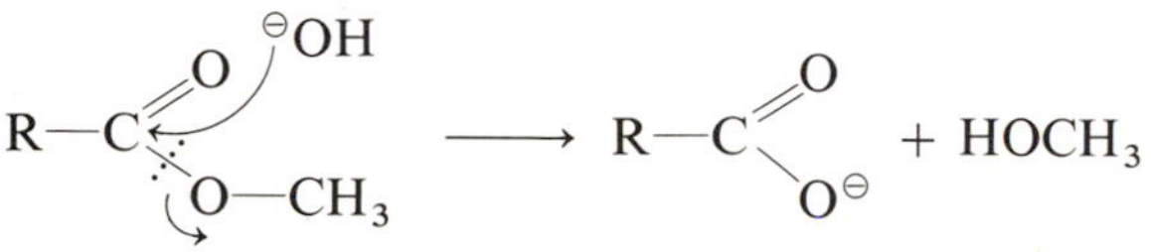

acyl-oxygen fission

by *acyl fission*. However, a few cases of *alkyl fission* are known. For example, dimethyl ether will form upon reaction of methoxide with methyl benzoate.

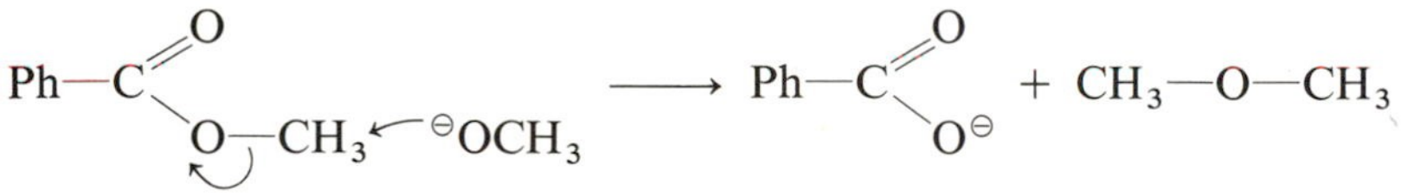

Acyl fission merely regenerates starting materials. The acid-catalyzed hydrolysis of *tert*-butyl acetate proceeds by an S_N1 process (stable carbonium intermediate) with alkyl fission.

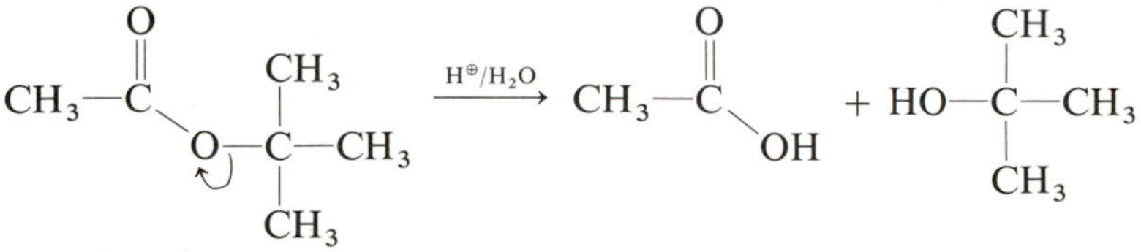

Neutral phosphate triesters will readily undergo acid or base-catalyzed hydrolysis with alkyl and/or *phosphoryl fission*. For example, trimethyl and triethyl phosphate will undergo hydrolysis in neutral water via an S_N2

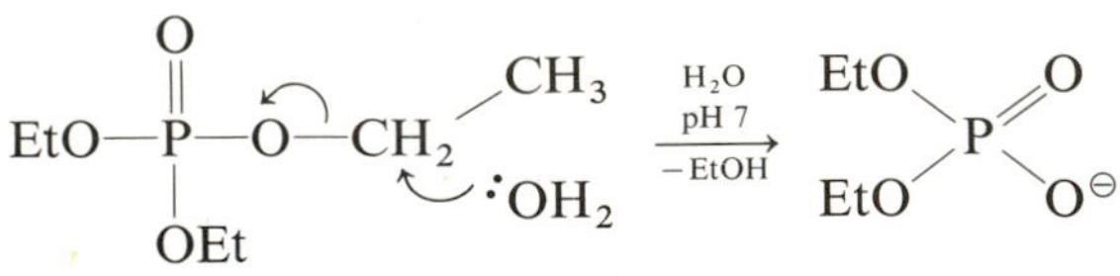

mechanism with alkyl fission. Acid-catalyzed hydrolysis also proceeds with alkyl fission.

The base-catalyzed hydrolysis of trimethyl phosphate proceeds by phosphoryl fission and allows a preparative synthesis of dimethyl phosphate.

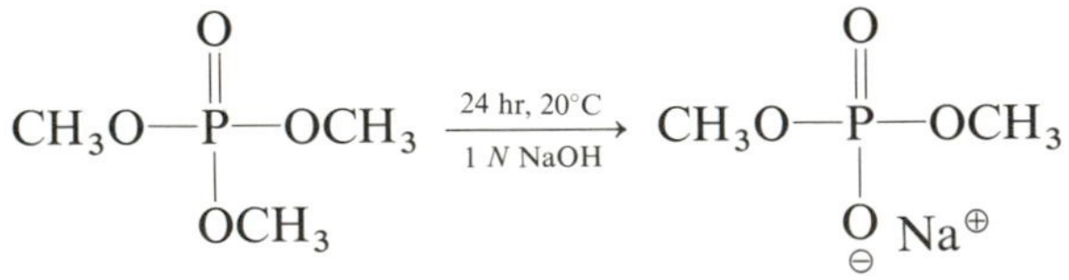

The anionic product causes electrostatic repulsion of any incoming hydroxide, so that monomethyl phosphate will not readily form. The half-life of this latter reaction is 16 days at 100°C, and the mechanism involves exclusively alkyl fission as the phosphorus atom is less accessible:

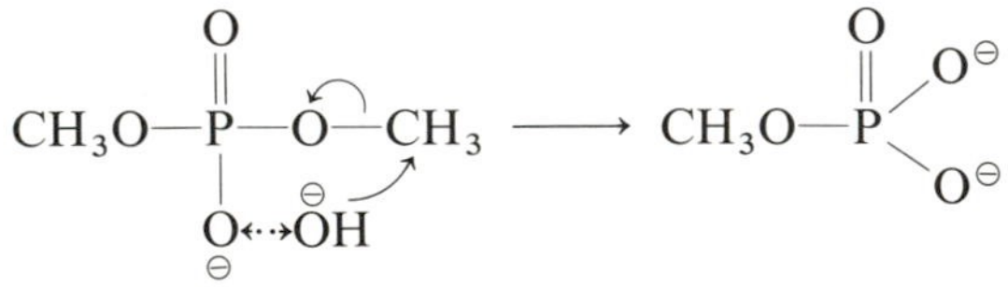

In the case of the base-catalyzed hydrolysis of diphenyl phosphate, only phosphoryl fission can occur. This proceeds very slowly (the half-life for the hydrolysis of diphenyl phosphate in neutral water, at 100°C, has been estimated at 180 years!) enhanced only by the better leaving ability of phenoxide, relative to alkoxide. Indeed, appropriately substituted diphenyl phosphates (i.e., bis-2,4-dinitrophenyl phosphate) with more acidic leaving groups will hydrolyze at a significantly greater rate. This data suggests that DNA, which functions as a "tape" for the storage of genetic information, is resistant to decomposition by hydrolysis. Indeed, after one hour at 100°C, in 1 N sodium hydroxide, no degradation of DNA is observed.

On the other hand, 2-hydroxyethyl methyl phosphate will easily undergo base hydrolysis: the half-life of the reaction is 25 minutes, at 25°C in 1 N sodium hydroxide. This may be accounted for by the intermediate formation of a strained five-membered cyclic phosphate by nucleophilic attack of the

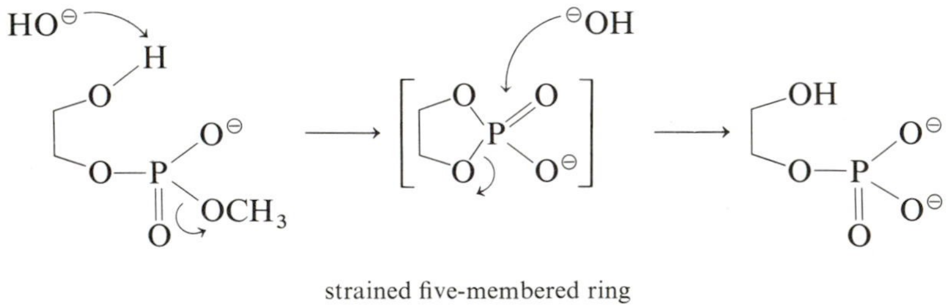

strained five-membered ring

vicinal hydroxyl (*anchimeric assistance*) function. An indication of the strain present in a five-membered cyclic phosphate lies in the observation that the rate of hydrolysis of ethylene phosphate is 10^6 times that of its acyclic analogue, dimethyl phosphate. Further, its hydrolysis is considerably more exothermic reflecting the inherent strain energy present in the cyclic structure:

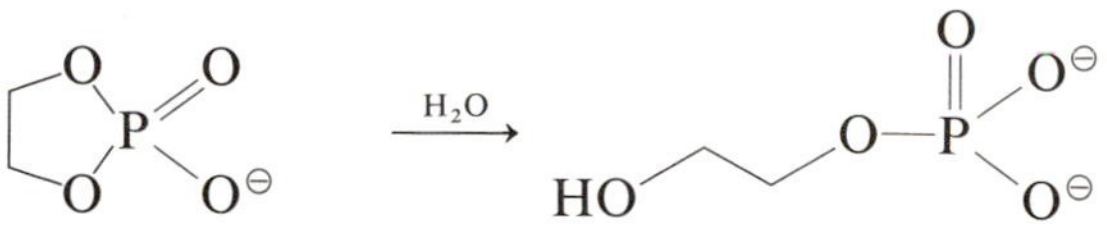

ethylene phosphate

$\Delta H = -26.8$ kJ/mol
$(-6.4$ kcal/mol$)$

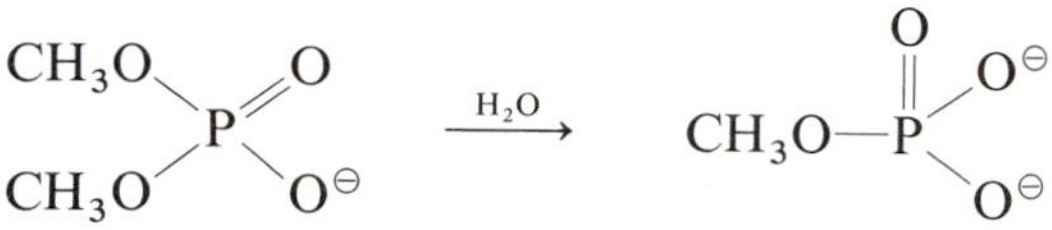

dimethyl phosphate

$\Delta H = -7.5$ kJ/mol
$(-1.8$ kcal/mol$)$

These data suggest that RNA, which functions as a carrier of genetic information and thus has a high turnover rate, is unstable or readily able to undergo degradation. Indeed, RNA will be degraded in $0.1 \, N$ sodium hydroxide, at room temperature, to give a mixture of 2′- and 3′-phosphates:

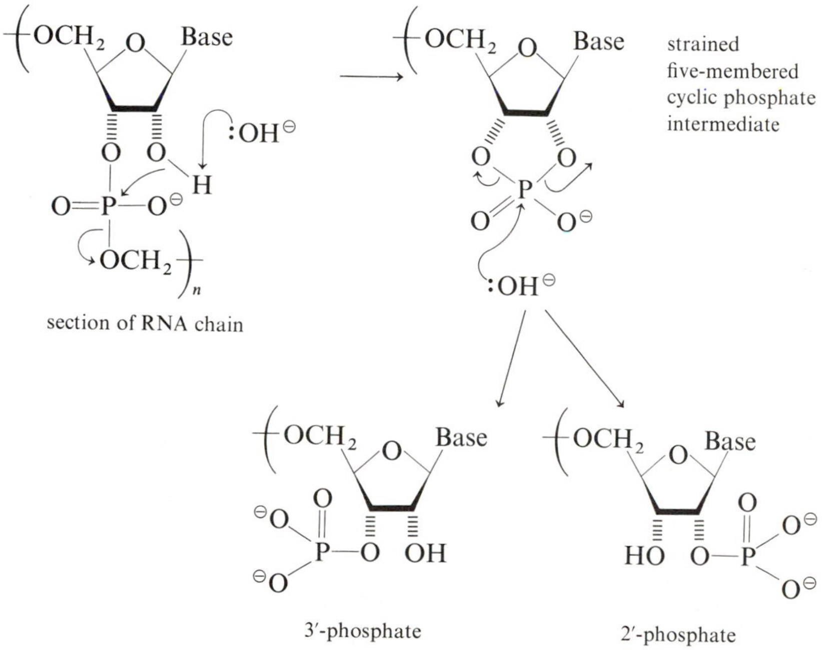

section of RNA chain

strained five-membered cyclic phosphate intermediate

3′-phosphate

2′-phosphate

Thus far, hydrolysis mechanisms have been written as if they proceed by a simple displacement process. However, just as the hydrolysis of carboxylate esters proceeds by a tetrahedral intermediate (as has been already noted), phosphate ester hydrolysis proceeds by a dsp³ hybridized "pentacoordinate" intermediate. The geometry of this intermediate is that which is representative of five electron pairs about a central (phosphorus) atom: trigonal bipyramidal. In fact, a number of stable pentacoordinate phosphorus compounds adopt such a geometry. For example, gaseous phosphorus pentachloride (PCl_5) exists as a trigonal bipyramid. It is important to note, as shall be seen shortly, that while it has five equivalent (chlorine) ligands, the dsp³ hybridization scheme provides five nonequivalent bonding lobes. As such, two of the chlorine–phosphorus bond lengths (designated as *apical* chlorines; 0.22 nm) are longer than the remaining chlorine–phosphorus bond lengths (designated as *equatorial* chlorines; 0.20 nm). The three equatorial chlorine atoms lie in a plane, while the two apical atoms are perpendicular to that plane.

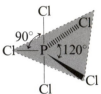

trigonal bipyramidal structure of gaseous phosphorus
pentachloride

During a hydrolysis reaction of an initially tetrahedral phosphate ester the entering group would approach from the distance, perpendicular to three ligands so that it will be perpendicular to the plane of the equatorial atoms in the pentacoordinate intermediate. Such is expected, for the new bond that is forming will, at least at first, be long in nature, which is compatible with the long bond of an apical ligand. This is analogous to the initial long bond formation that occurs in S_N2 displacements. Similarly, the leaving group will depart from an apical position only, going through a long bond geometry during its departure. This is illustrated for the hydrolysis of a phosphomonoester:

tetrahedral
phosphomonoester

$$\text{R—OH} + HPO_4^{2-}$$

"long-bond"
formation

departing
electron density

A consideration of the hydrolysis of methyl ethylene phosphate indicates that the hydrolysis of organophosphates does not necessarily proceed by a simple displacement mechanism. If this were the case, then exclusive opening of the strained ring would be expected to occur:

Notice that the five-membered ring in the pentacoordinate intermediate is represented as spanning an apical and equatorial position. This is the most favorable geometry since the strain in five-membered rings arises from angular distortions. A five-membered ring spanning two equatorial positions or two apical positions would be too unstable.

In fact, acid-catalyzed hydrolysis of methyl ethylene phosphate leads to significant displacement of the methoxy function:

70% 30%

This observation is accounted for by the phenomenon of *pseudo-rotation* (22, 23). Pseudo-rotation is the process of phosphorus–ligand bond deformation which occurs in pentacoordinate intermediates, such that an appropriate leaving group that could not leave prior to the process is made able to do so. Pseudo-rotation will occur as dictated by the five "preference rules":

1. Hydrolysis of phosphate esters proceeds by way of a pentacoordinate intermediate that has the geometry of a trigonal bipyramid.
2. During the hydrolysis process, groups will enter and leave from apical positions only.
3. Because of strain, a five-membered ring will not be formed between two equatorial or two apical positions, but instead between one apical and one equatorial position.
4. The more electropositive ligands tend to occupy equatorial positions, while the more electronegative ligands tend to occupy apical positions.
5. Ligands can exchange positions by the process of pseudo-rotation.

By pseudo-rotation two apical ligands become equatorial and simultaneously two equatorial lignads become apical. The driving force for the process is transforming a good leaving group from an equatorial to an apical geometry and thus allowing it to leave. Such bond deformations are not surprising when it is remembered that covalent bonds are not static entities, but instead are able to undergo various bending and stretching modes, as indicated by infrared spectroscopy. During pseudo-rotation, the two apical bond lengths may be thought of as being "squeezed together" to form two new equatorial bonds, while two of the three equatorial bonds (the third bond remains "neutral" or of constant bond length during the process: this is referred to as the "pivot") may be thought of as being "pushed out" (lengthened) in response to form two new apical bonds. This is illustrated (arrows indicate increasing or decreasing bond length):

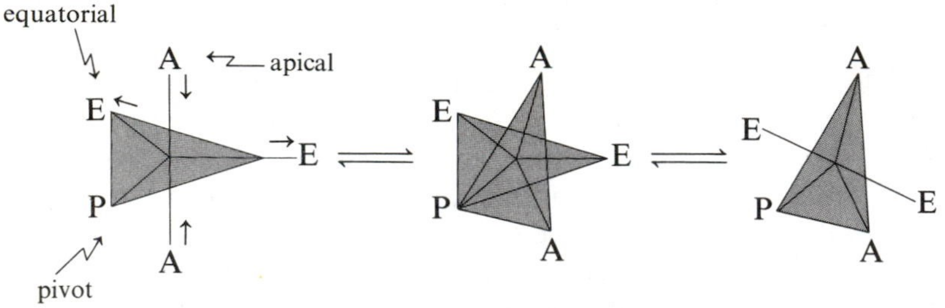

Hence, via pseudo-rotation, the methoxy function of methyl ethylene phosphate is able to leave.

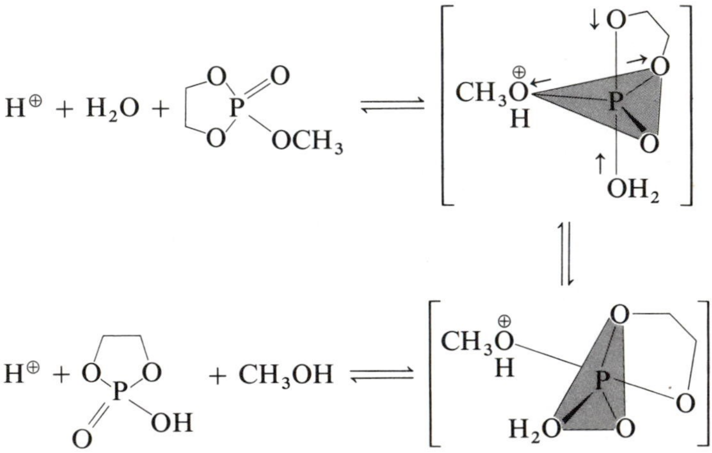

Hydrolysis of the phosphonate analogue proceeds exclusively with ring opening:

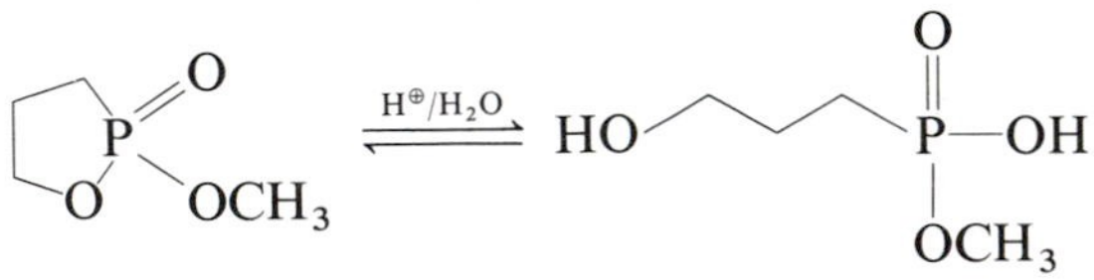

This is a consequence of the fourth preference rule which states that electropositive ligands prefer to be equatorial (close to phosphorus) and electronegative ligands prefer to be axial (away from phosphorus). Phosphorus, itself a nonmetal, is fairly electronegative and so prefers to be away from other electronegative atoms. As such, the carbon atom of the cyclic phosphonate will acquire an equatorial geometry and the oxygen atom an apical geometry:

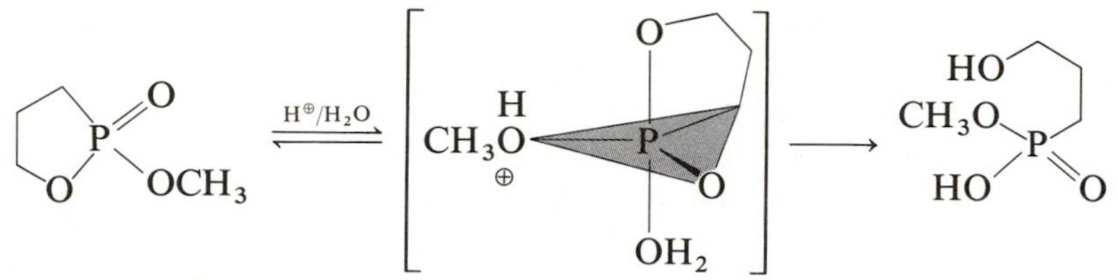

Once the pentacoordinate intermediate is formed it cannot undergo a pseudo-rotation as this would require the oxygen and carbon atoms of the ring to exchange positions. Inhibition of pseudo-rotation is the result and thus direct displacement by water is the only available pathway.

It has already been noted that within the organism hydrolysis of organo-phosphates is an important reaction pathway.

Having considered some of the mechanisms of hydrolysis, it is of interest to compare these with the active site mechanism of a few enzymes which catalyze such reactions. The enzyme ribonuclease A (RNase A)* (bovine pancreas, MW = 13,680, one polypeptide chain consisting of 124 amino acid residues) catalyzes the degradation of RNA by a two-step mechanism: transesterification followed by hydrolysis of the five-membered cyclic inter-

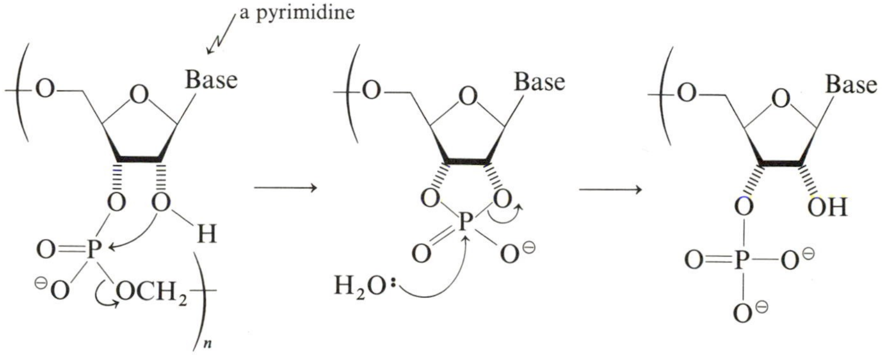

mediate. The enzyme will not cleave each phosphodiester linkage of the RNA polymer, but will instead attack only along certain points of the chain. The susceptibility of a particular phosphorus atom to attack by the enzyme is dependent on the purine or pyrimidine base present at the sugar esterified to the phosphate by a 3'-hydroxyl. Further, the base type is dependent on the source of the enzyme. For example, bovine pancreatic RNase will attack the phosphoester linkage if the nucleoside at the 3'-linkage contains a pyrimidine. On the other hand, RNase from the bacteria *B. subtilis* will attack the phosphorus linkage if the nucleoside at the 3'-linkage contains a purine.

---

* The fundamental basis of enzyme function will be presented only in Chapter 4. However, Sections 4.1 and 4.2 could be read at this point for a better understanding of RNase A mode of action.

Various studies indicate the participation of at least three amino acid residues in the active site chemistry of ribonuclease: two histidines and one lysine. RNA hydrolysis (Fig. 3.6) proceeds by two steps: transesterification followed by hydrolysis. Note that at physiological pH, one of the two imidazole rings is protonated while the other is not. The imidazole rings

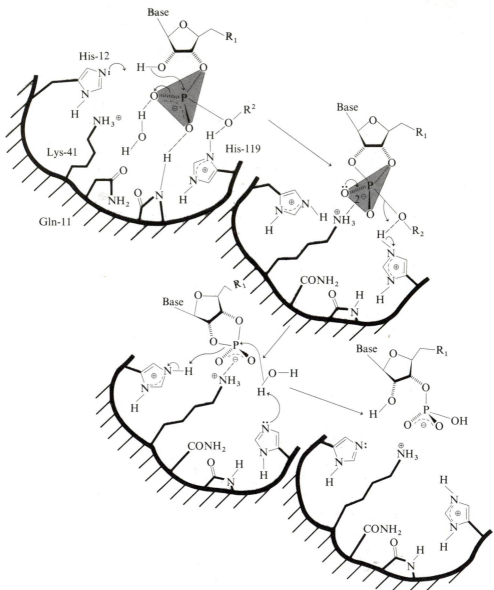

Fig. 3.6. Mechanism for the ribonuclease catalyzed hydrolysis of RNA (24). Amino acid residue number refers to its position relative to the N-terminal amino acid of the polypeptide (protein) chain.

function as a general-base–general-acid catalyst while the cationic lysine probably aids in the stabilization of the pentacoordinates intermediate.

This lysine residue apparently is not involved in the initial binding of the phosphate moiety to the enzyme (24). The phosphate substrate is instead bound by means of two hydrogen bonds between the phosphoryl oxygens and an amide backbone hydrogen and a hydrated glutamine residue. These hydrogen bonds will also increase the electrophilicity of the phosphorus atoms. Another important hydrogen bond occurs between the protonated imidazole ring and the oxygen atom of the substrate that will be cleaved from the phosphorus atom. In addition to a binding function, this hydrogen bond will polarize electron density in a direction that will facilitate complete proton transfer from the imidazole ring during hydrolysis of the substrate as will be seen below.

The mechanism of hydrolysis is straightforward and resembles that for the hydroxide catalyzed hydrolysis of RNA; with exception of specificity as only certain phosphodiester linkages are cleaved to give 3'- (and not 2'-) monophosphates (Fig. 3.7). However, with a knowledge of pseudo-rotation, two stereochemical mechanisms become possible for each step of the enzyme catalyzed reaction.

The two mechanisms are referred to as "in-line" or "adjacent": the type depending upon the relative geometry of the incoming and leaving groups during the displacement process. Referring to the plane of the three equatorial ligands, the stereochemical mechanism is in-line if the entering groups (nucleophile) enters on one side of the plane, while the leaving group departs on the other side. On the other hand, if the entering group (nucleophile) enters on one side of the plane, while the leaving group is situated in the plane the mechanism is adjacent. According to the preference rules, the adjacent mechanism will require a pseudo-rotation to transform the leaving group from an equatorial to apical geometry.

Transesterification begins with an unprotonated imidazole at the active site abstracting a proton from the 2'-hydroxyl (general-base catalysis) of the RNA substrate. As an alkoxide ion is generated it becomes a potent nucleophile which attacks the phosphorus atom. If the leaving group is in an apical geometry, it may do so, being protonated simultaneously by a second imidazole (imidazolium cation) situated below the plane of the three equatorial ligands. This in-line mechanism requires the participation of two histidine residues, as it would be physically impossible for the imidazole ring which removes the 2'-hydroxyl proton to extend below the plane and protonate the leaving group (see Fig. 3.6). If the leaving group is in an equatorial geometry, it must first pseudo-rotate, or change bond length. However, it would still be situated close enough to the incoming 2'-hydroxyl to be protonated, after the pseudo-rotation, by the same imidazole that abstracted the 2'-hydroxyl proton from the substrate. This adjacent mechanism requires an additional step, (a pseudo-rotation) but only the participation of one histidine residue.

Much the same arguments apply for the hydrolysis of the 2',3'-cyclic

in-line mechanism   adjacent mechanism

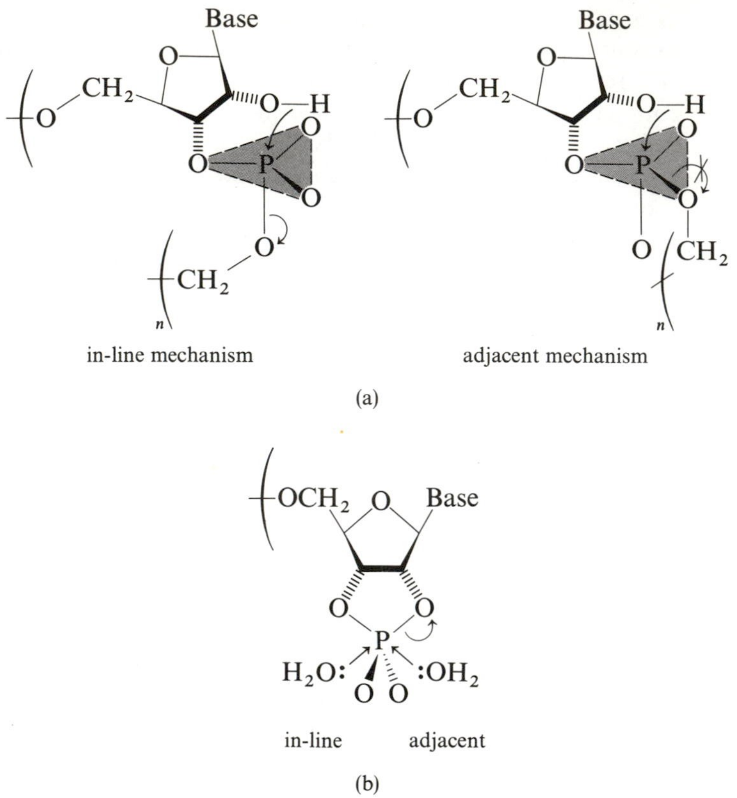

(a)

in-line   adjacent

(b)

Fig. 3.7. Differences between in-line and adjacent mechanism of phosphodiester bond hydrolysis. (a) First step (transesterification) in the RNase catalyzed hydrolysis of RNA illustrating two possible stereochemical pathways. The in-line mechanism allows a direct displacement while the adjacent mechanism requires a pseudo-rotation. (b) Second step (hydrolysis) in the RNase catalyzed hydrolysis of RNA illustrating two possible stereochemical pathways. Again, the in-line mechanism allows a direct displacement to form the 3'-phosphate, while the adjacent mechanism requires a pseudo-rotation (26).

phosphate. Here again, an imidazole ring functions as a general-base catalyst with abstraction of a proton from water, and depending on an in-line or adjacent mechanism, a second or the same imidazole can protonate the leaving group. Note that after hydrolysis is complete, the active site is fully regenerated and the initially unprotonated and protonated imidazole rings are again formed.

By a series of elegant experiments it has been determined that both steps of the RNase mechanism proceed by an in-line mechanism. The detailed mechanism was elucidated by D. Usher from Cornell University (25, 26).

Key to the solution is the isolation of two diastereoisomers of uridine 2',3'-cyclic phosphorothioate. Both isomers were used for hydrolysis in

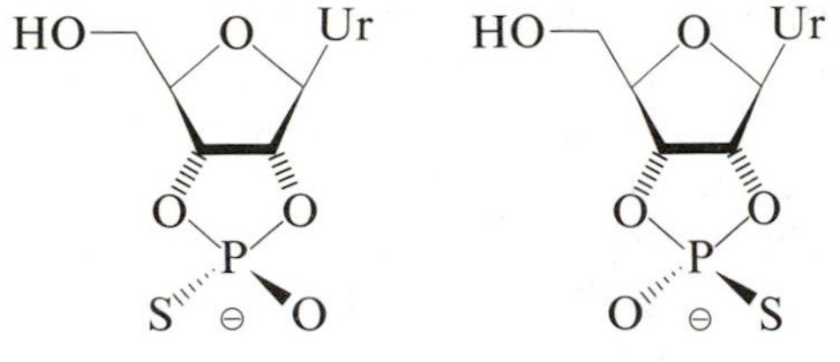

isomer A   isomer B

$H_2{}^{18}O$ by RNase A. The two mechanisms (adjacent or retention and in-line or inversion) are presented below for isomer A. Since the subsequent chemical

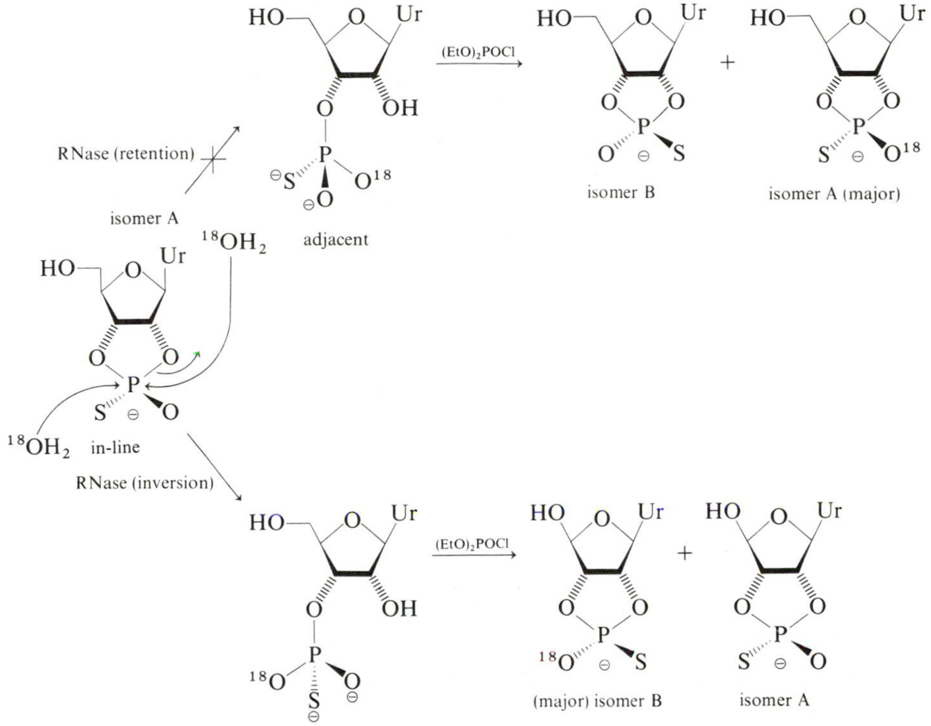

ring closure with diethyl phosphorochloridate in pyridine is known to proceed with inversion, the stereochemistry of the first step (the enzymatic one) can be determined by isolation of the epimeric cyclic diesters and measuring their ^{18}O content. As expected, ^{18}O was not present in the isomer A used in the enzymatic step, but in its epimer B. Therefore, the loss of ^{18}O of the nucleotide by a reaction that involves inversion requires that the introduction of ^{18}O be with inversion. These data preclude pseudo-rotation as a step in bond cleavage and help to identify the position that H_2O must occupy in models derived from crystallographic studies in RNase A.

A similar mechanistic study on DNA-dependent RNA polymerase from *E. coli* using phosphorothioate analogues has been carried out by the group of F. Eckstein (27).

A number of other enzymes which catalyze the hydrolysis of phosphoesters are of biological importance. These include cyclic purine phosphodiesterase (little is known about its active site chemistry at present, but more shall be said about its biological role shortly) and the phosphatases. Acid and alkaline phosphatase catalyze the hydrolysis of phosphomonoesters to the corresponding alcohol and inorganic phosphate. Their pH optimums are 5.0 and 8.0, respectively; hence their names. Both form covalent enzyme–substrate intermediates:

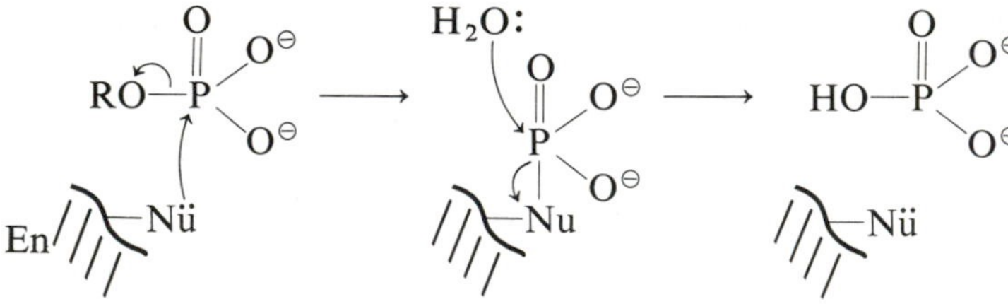

In the case of acid phosphatase, the nucleophile is an imidazole (the intermediate phosphoramidate is susceptible to acid pH) and in the case of alkaline phosphatase, the nucleophile is a serine. Alkaline phosphatase is known to contain a zinc atom which probably functions as a binding site and a Lewis acid for the phosphoryl function. This is understandable when it is remembered how difficult the hydrolysis of phosphomonoesters is at alkaline pH.

In conclusion, no enzymatic reaction has so far been reported which would call for the involvement of pseudo-rotation (28).

3.4 Other Nucleoside Phosphates

As noted above, nucleoside phosphates (nucleotides) are not important only because of their participation in biological polymers. A number of monomeric nucleotides are important to such diverse functions as energy storage (ATP), regulation (cyclic nucleotides), and cofactor chemistry (NAD$^+$ and NADP$^+$, see Chapter 7).

3.4.1 ATP (Adenosine Triphosphate)

As has been noted, ATP functions as a biological energy store, thus enabling

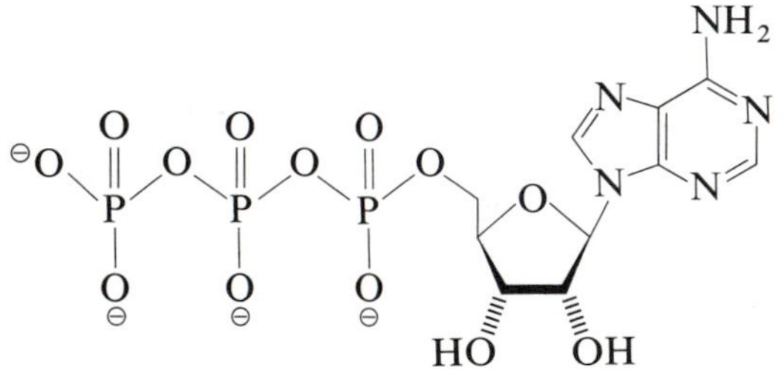

ATP

the synthesis of a number of biochemically important compounds. For example, the synthesis of the methyl donor S-adenosyl methionine from ATP and methionine has already been described (Chapter 2). The "high-energy" content arises from the presence of the anhydride structure in the triphosphate chain; the adenosine portion of the molecule functions as a recognition and binding site for the various ATP-utilizing enzymes. As such, simpler phosphate anhydrides (i.e., acetyl phosphate, pyrophosphoric acid) also possess a similar energy content. The reactivity and energy content of anhydrides may be attributed to "competing resonance," or two functions (phosphoryl, carbonyl, etc.) competing for the nonbonding electrons from one oxygen atom. Upon hydrolysis, this competition is relieved and a greater resonance

stabilization is observed. Hence, product formation is favorable, relative to the reactant anhydride. The function (carbonyl, phosphoryl, etc.) will be more electrophilic in the anhydride than in the ester analogue, and thus more reactive, since in the ester there will be one oxygen atom per function and not one oxygen atom shared by two functions. In the case of phosphoric anhydrides, product formation will be favorable since it will reduce electrostatic repulsion along the di- or triphosphate chain. Multiple anionic charges will no longer be covalently linked together.

At physiological pH (7.35), three of the four ionizable protons will be lost. The fourth has a pK_a value of 6.5 and so it will be mostly ionized. In the cell, this polyanionic species will tend to bind magnesium and so exist as a 1:1 complex with the latter. *In vitro*, ATP will also bind other divalent metal ions, such as calcium, manganese, and nickel. In addition to the two phosphate oxyanions, chelation of the metal ion can also involve the adenine base (i.e., N-7 of the imidazole ring). The metal ion can serve as an electrophilic catalyst (Lewis acid) for the hydrolysis of ATP. Of course, the presence of the metal ion bound to the phosphate chain would partially neutralize the total negative charge, thus making it easier for the approach of a negative nucleophile, such as hydroxide.

The ΔG^0_{hydro} at pH 7.0 for hydrolysis of the γ-(terminal) and β-phosphate is approximately -31.2 kJ/mol (-7.5 kcal/mol). This value is pH dependent and defined under standard conditions. However, within the cell the value has been estimated at approximately -50 kJ/mol (-12 kcal/mol). At any rate, ATP is, for the purpose of chemical reactions within the biological system, a "high energy" compound. While the definition is rather arbitrary, a high energy compound may be defined as one that has, at physiological pH, a ΔG_{hydro} in excess of -29.4 kJ/mol (-7 kcal/mol) or ΔH_{hydro} in excess of -25 kJ/mol (-6 kcal/mol).

ATP can transfer this potential energy to a variety of biochemically important compounds. For example, it has already been demonstrated that the energy for peptide bond formation (via intermediate acid-anhydride formation) may be acquired from the ATP molecule. Similarly, it will be shown shortly (see below) that cyclic nucleotides, which are also high energy, may be synthesized from an ATP (or GTP) precursor.

Guanosine triphosphate (GTP) is another high energy compounds which has the same structure as ATP, except the adenine base is replaced by guanine. While of less use than ATP in biological systems, it still finds some use in energy-requiring processes, such as peptide bond synthesis on the ribosome.

The biosynthesis of adenosine and guanosine triphosphates proceeds by phosphorylation of the low energy monophosphate precursors. For example, GTP may be synthesized from guanosine monophosphate (GMP), at the expense of two ATP molecules:

$$GMP + ATP \rightleftharpoons GDP + ADP$$
$$GDP + ATP \rightleftharpoons GTP + ADP$$

These reactions are catalyzed by the enzymes nucleoside monophosphokinase and nucleoside diphosphokinase, respectively. Note that these reactions are reversible, so that ATP may be synthesized at the expense of GTP or another nucleoside triphosphate. The precursor ADP (adenosine diphosphate) may also be synthesized from the reaction of AMP with ATP, catalyzed by the enzyme adenylate kinase:

$$AMP + ATP \rightleftharpoons 2\,ADP$$

All the above interconversions are thermodynamically acceptable as they involve the formation of an anhydride structure at the expense of one which already exists. Perhaps as this suggests, the total ATP, ADP, and AMP in the cell is in a steady-state concentration. It should be realized that ATP may be synthesized by phosphorylation of ADP with other phosphate donors besides the nucleoside triphosphates and that within the cell pathways exist by which the energy from the breakdown of sugars (glucose) is utilized for this ATP synthesis.

The chemical synthesis of adenosine and guanosine triphosphate may be undertaken in a fashion analogous to their biosynthesis: phosphorylation of the low energy monophosphate precursors. Phosphorylation is accomplished with phosphoric acid and a coupling (dehydrating) agent, DCC. Use of carbodiimides has already been discussed in connection with peptide bond formation. It will be recalled that reaction proceeds by way of an anhydride intermediate. Here, addition of a carboxylic acid to the carbodiimide is replaced by a phosphoric acid, as is illustrated for the conversion of AMP to

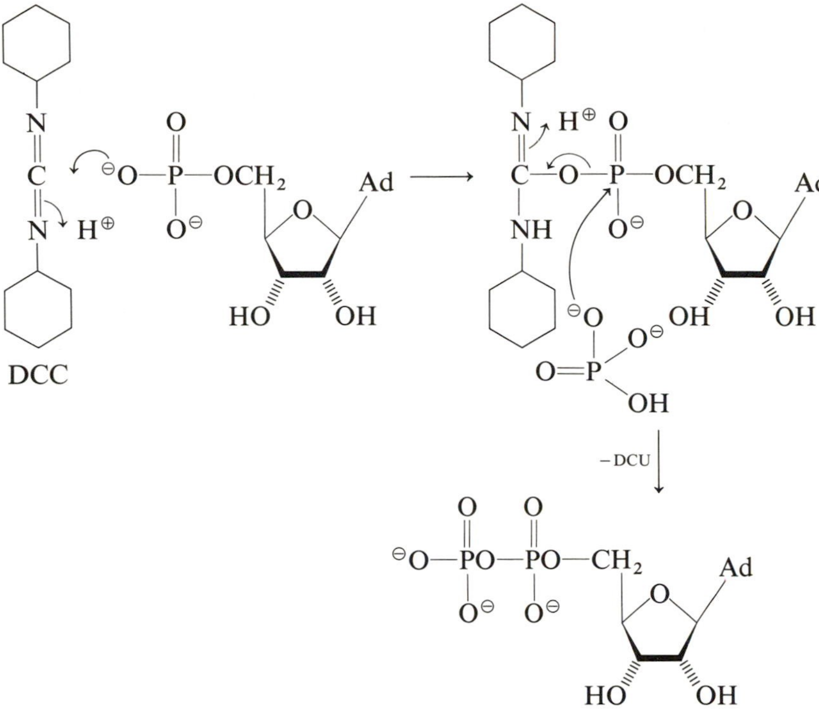

DCC

ADP

ADP. In the presence of excess phosphoric acid, the ADP product will be phosphorylated to form ATP. Of course, this reaction does not proceed with the same selectivity as the enzyme catalyzed phosphorylation, so that even under optimum reaction conditions, product mixtures of AMP, ADP, ATP, and some higher polyphosphates such as adenosine tetraphosphate are obtained. Nonetheless, the major product is ATP, which may be purified by ion-exchange chromatography:

$$\text{AMP} + \text{H}_3\text{PO}_4\,(85\%) \xrightarrow[\text{pyridine, N(Bu)}_3]{\text{DCC, 20°C}} \text{ATP} + \text{others}$$

One reason that ATP is the predominant product may be its existence in the reaction medium as a cyclic phosphate which decomposes upon workup:

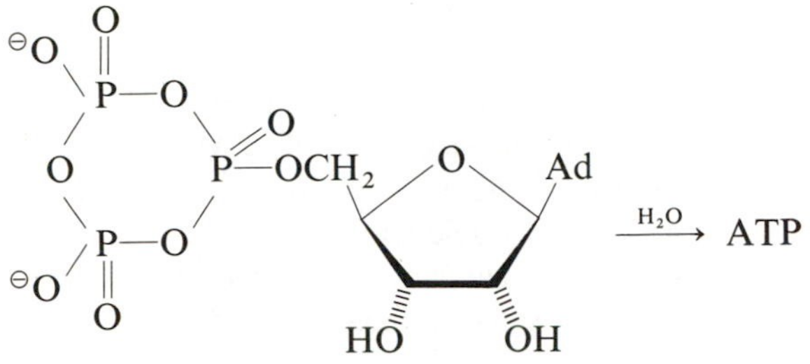

$\xrightarrow{\text{H}_2\text{O}}$ ATP

Such an intermediate would not be readily phosphorylated even in the presence of DDC. Indeed, ATP in the presence of DCC will form such a cyclic intermediate. This provides a synthetic route for ATP analogues modified at the γ-phosphorus.

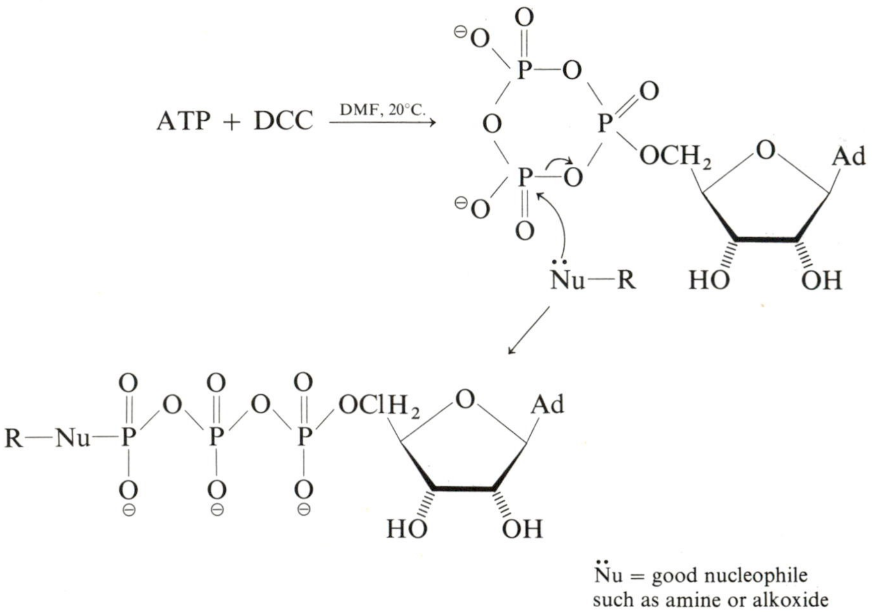

Nu = good nucleophile
such as amine or alkoxide

A more selective and thus more commonly used synthesis of nucleoside triphosphates involves coupling of a nucleoside monophosphate with inorganic phosphate (29).

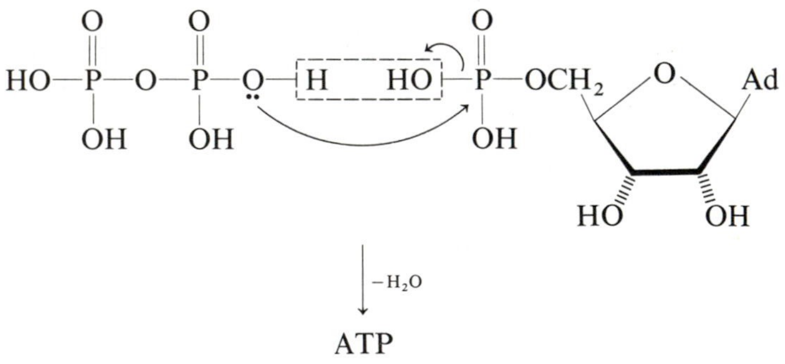

ATP

Again, the hydroxyl of the nucleoside monophosphate is a poor leaving group, or the process is not thermodynamically favorable. Conversion to a good leaving group is essential. Most popular is the formation of the imidazolidate by reaction of the nucleoside monophosphate with 1,1'-carbonyldiimidazole. This can be envisaged to proceed by way of a mixed anhydride analogous to that observed during peptide formation with EEDQ (see Section 2.6.3). The product anhydride will now readily undergo displacement with pyrophosphate. Unfortunately, significant 2',3'-cyclic carbonate formation can occur due to the entropically favorable *cis*-diol arrangement:

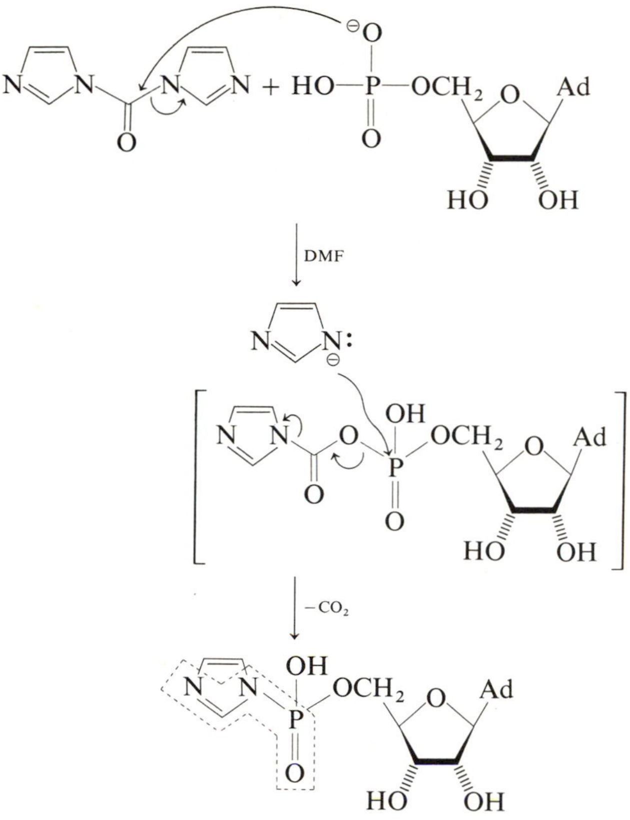

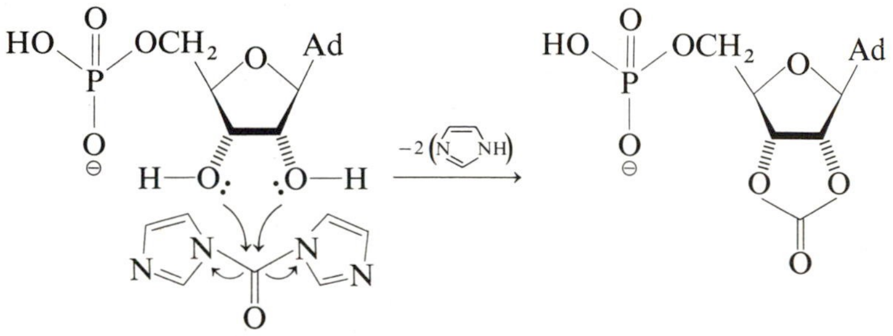

This may be overcome by forming the imidazolidate of the pyrophosphate and reacting this with the nucleoside monophosphate. Alternatively, the cyclic carbonate may be hydrolyzed by treatment with aqueous triethylamine. The imidazolidate method is presently the method of choice for the synthesis of ATP and GTP analogues modified at the base and/or sugar portion. One example is the synthesis of the triphosphate of the antibiotic cordycepin (3'-deoxyadenosine). This triphosphate is a potent inhibitor of RNA synthesis.

An alternative large-scale synthesis of ATP is one that was recently developed by G. M. Whitesides, M.I.T., which utilizes acetyl phosphate as the phosphate donor and *immobilized enzymes* (see Section 4.7) as a catalyst (30). The reaction occurs under mild conditions (approximately neutral pH, room temperature), the insoluble polymeric catalyst is easily removed (centrifugation) and need be present only in small amounts, relative to the substrate and the reaction is more specific than a nonenzymatic synthesis (i.e., adenosine tetraphosphate is not formed as a by-product). The net reaction is:

$$3\left(CH_3-\overset{O}{\overset{\|}{C}}-O-\overset{O}{\overset{\|}{P}}\overset{O^{\ominus}}{\underset{O^{\ominus}}{<}}\right) + HO-\underset{HO\quad\ OH}{\overset{O\quad Ad}{\diagup}}$$

$$\downarrow \text{3 enzymes, pH} = 6.7\text{–}6.9, 20°C$$

$$3\left(CH_3-\overset{O}{\overset{\|}{C}}-O^{\ominus}\right) + ATP$$

The reaction pathway in detail is as follows:

1) Adenosine + ATP $\xrightarrow{\text{adenosine kinase}}$ AMP + ADP

2) AMP + ATP $\xrightarrow{\text{adenylate kinase}}$ 2 ADP

_____

Total: Adenosine + 2 ATP $\xrightarrow{\hspace{2cm}}$ 3 ADP

3) 3 ADP + 3 acetyl phosphate $\xrightarrow{\text{acetate kinase}}$ 3 ATP + 3 acetate

While the equations do not necessarily indicate this, only a small amount of ATP (less than 1/1000 of the amount produced by weight) needs to be added in order to initiate the reaction, as the reaction will be perpetuated by the product ATP. The three enzymes are immobilized to a cross-linked polyacrylamide gel by reaction of their primary amino functions with "active esters" on the polymer surface.

Finally, mention must be made of the recent synthesis of ATP chiral at the γ-phosphorus by the group of J. R. Knowles at Harvard University (31). This was accomplished by reaction of adenosine diphosphate (ADP) with a phosphorylating agent isotopically labeled to form a $[\gamma\text{-}^{16}O,^{17}O,^{18}O]$-ATP product, as outlined below:

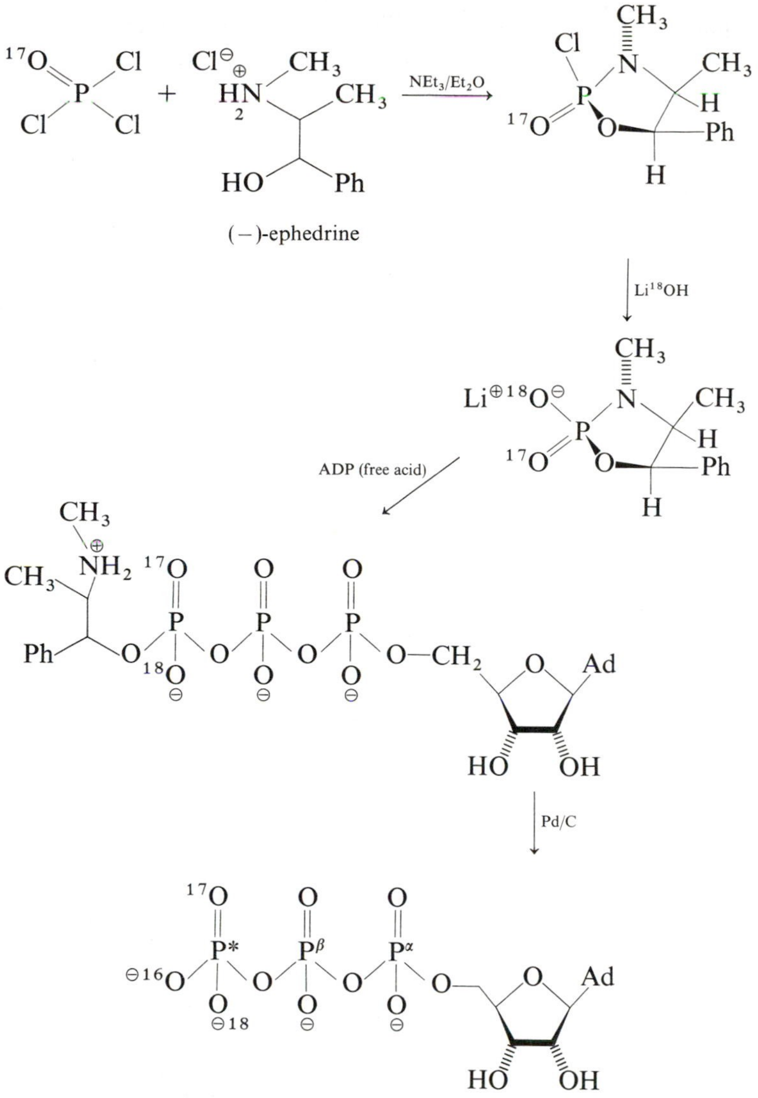

Reaction of ^{17}O-labeled phosphorus oxychloride with $(-)$-ephedrine produces a reactive five-membered ring whose chlorine atom may be displaced in excellent yield by reaction with ^{18}O-labeled lithium hydroxide. The product is an excellent phosphorylating agent by virtue of the strain of the five-membered ring, and the acid lability of the phosphoramidate linkage. The benzylic linkage of the acylic phosphomonoester is now susceptible to catalytic hydrogenolysis, to give rise to the chiral ATP product.

The synthetic ATP is an excellent probe for the stereochemistry and hence reaction mechanism of phosphorylating (kinase) enzymes (see page 109). At least two pathways are possible for the enzyme catalyzed donation of the γ-phosphate of ATP to a substrate. This may simply proceed via direct displacement on the enzyme surface, with an overall inversion of configuration in the case of a chiral γ-phosphate:

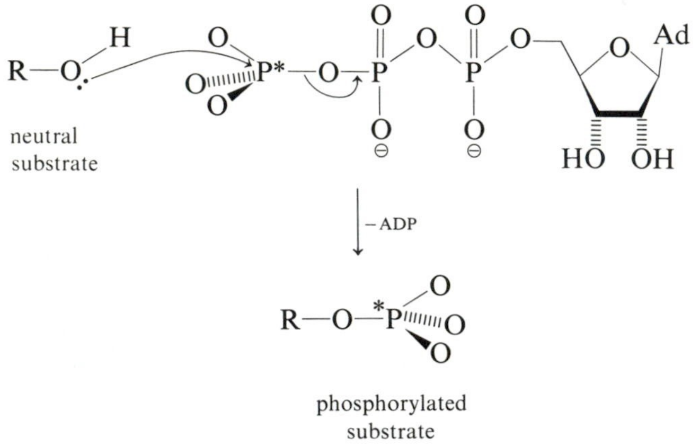

neutral
substrate

$-$ADP

phosphorylated
substrate

Alternatively, the enzyme may first be phosphorylated by the ATP to form a transient covalent enzyme–substrate intermediate, which then suffers displacement by the substrate. The net result is a retention of configuration or two inversion processes:

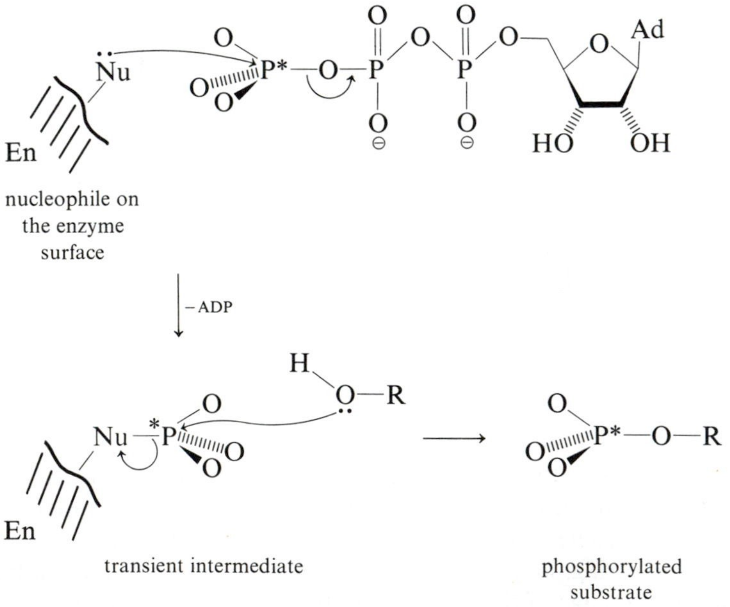

nucleophile on
the enzyme
surface

$-$ADP

transient intermediate   phosphorylated
                                                  substrate

In the case of the enzymes glycerol kinase, pyruvate kinase and hexokinase inversion of configuration was observed at the chiral γ-phosphorus (32).

3.4.2 cAMP and cGMP

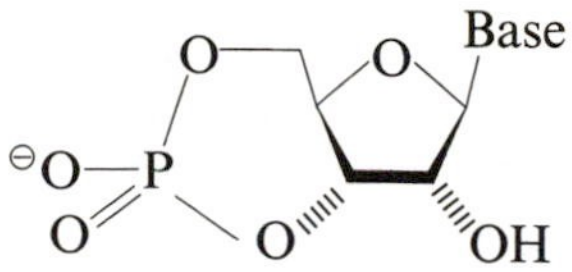

Base = adenine or guanine

Cyclic nucleotides, 3′,5′-cyclic adenosine monophosphate (cAMP) and 3′,5′-cyclic guanosine monophosphate (cGMP), function to regulate cell-to-cell communication processes (33). Cellular communication follows primarily three pathways. The first involves the transmission of electrical impulses via the nervous system. The second involves chemical messengers or hormonal secretions. The third involves *de novo* protein synthesis. All three processes are usually in response to some demand or stimulus and involve, at least to some extent, regulation by cyclic nucleotides:

For any given process, cAMP and cGMP usually operate in a "mirror" fashion. Hence, for example, if cAMP tends to regulate the process by "turning off" a particular synthesis, cGMP will most likely function to turn this synthesis "on." While such a "Yin–Yang" hypothesis appears operative in those cases studied it should be pointed out that less is known about cGMP than cAMP. It is known that both cAMP and cGMP do function as *second messengers* to bring about a biochemical response to an environmental (chemical) stimulus. The classical example is the release of the hormone adrenaline into the bloodstream upon being frightened. This results in increased blood sugar, a response that would not be possible without the second messenger cAMP.

The means by which cyclic nucleotides exert their influence involves a class of enzymes; the "protein kinases." As the name implies, these are enzymes that catalyze the phosphorylation of a protein (usually an enzyme) substrate. Just how protein kinases control various processes may be understood by returning to the problem of adrenaline secretion in response to fright.

The enzymes that synthesize cAMP and cGMP are adenyl cyclase and guanyl cyclase, respectively. These enzymes consist of two subunits: a regulatory subunit to which an effector (hormone, neurotransmitter) and guanosine 5′-triphosphate (GTP) bind and a catalytic subunit where the actual synthesis of the cyclic nucleotide takes places. These enzymes are bound to the membrane of the cell such that the regulatory subunit can bind a particular effector on the outside of the cell, but the catalytic subunit can synthesize cyclic nucleotides inside the cell (see Fig. 3.8). Hence, upon being frightened and releasing adrenaline into the bloodstream, this circulating hormone can bind to molecules of adenyl cyclase whose regulatory subunits specifically recognize that particular molecular structure (i.e., in the plasma membrane of liver cells). Once bound to this subunit along with GTP, conformational changes occur which are transmitted to the catalytic subunit, and cause the latter to synthesize cAMP. The endogenous cyclic nucleotide then binds to a

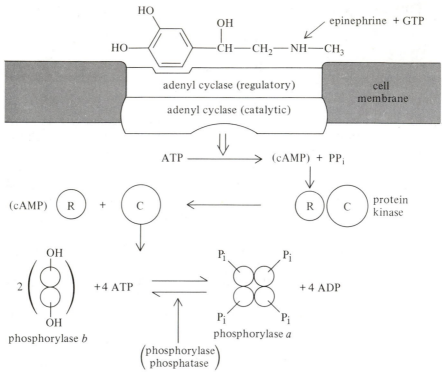

Fig. 3.8. The role of cAMP as an intermediate to environmental stimulus (hormone secretion) and the onset of physiological response. Adrenaline and GTP bind to a specific membrane bound regulatory subunit of adenyl cyclase. This elicits a conformational change such as to activate the catalytic subunit and stimulate cAMP synthesis. cAMP activates a protein kinase by causing its dissociation into an active catalytic subunit (R = regulatory subunit; C = catalytic subunit). This in turn catalyzes the phosphorylation of two serine residues of an inactive phosphorylase b enzyme. Two dimers combine to form an active tetramer: phosphorylase a. Phosphorylase catalyzes the breakdown of the sugar storage form: glycogen (phosphorolysis) to glucose 1-phosphate.

molecule of protein kinase. The following sequence of events is universal to all cyclic nucleotide mediated processes, but again emphasis will be given to the particular case of adrenaline release. Protein kinases also consist of two subunits: a regulatory subunit which binds the cAMP molecule and catalytic subunit which catalyzes the phosphorylation of a specific protein substrate.* However, binding of cAMP causes the regulatory subunit to dis-

---

* Recent work has shown this to be an oversimplification (34). cAMP-dependent protein kinase is a tetrameric protein consisting of two catalytic and two regulatory subunits, designated as R_2C_2. Further, cAMP-dependent protein kinases may be classified as Type I and Type II. The major difference between the two lies in the ability of the regulatory subunit of Type II kinase to be phosphorylated by the catalytic subunit. On the other hand, cGMP-dependent protein kinase is a dimeric protein consisting of two identical subunits. Each unit has a cGMP binding site, and a catalytic site within a single polypeptide chain.

sociate from the catalytic subunit. Presumably because the catalytic subunit is now exposed to the aqueous media or conformational changes occur which convert it to a geometry favorable for catalysis, the catalytic subunit is now an active entity. It will phosphorylate a specific protein. In this case, the protein is a dimeric enzyme: phosphorylase *b*, which itself has little enzyme activity. However, once phosphorylated by protein kinase (in this specific case referred to as phosphorylase kinase), the dimeric enzyme associates into an active tetramer, phosphorylase *a*. This tetramer catalyzes the breakdown of the sugar storage form, glycogen, into monomeric glucose. Hence, in response to epinephrine (adrenaline), blood sugar is elevated. However, this does not come about by the direct message received from the epinephrine hormone, but instead by the second messenger cyclic nucleotide.

There are numerous other examples of cyclic nucleotide mediated processes which fall into the three categories noted above: hormonal secretion, neuronal transmission, and protein synthesis. For example, cAMP, again via a protein kinase, activates the enzyme tyrosine hydroxylase.

Tyrosine hydroxylase, which catalyzes the conversion of the amino acid tyrosine to dopa (dioxyphenylalanine), is the rate-limiting step in the biosynthesis of the neurotransmitters dopamine and norepinephrine. A *neurotransmitter* is a substance that is released at the junction of two nerve cells or a nerve and muscle cell (a so-called *synapse*) when it is required that an electrical (nerve) impulse be transmitted from one cell to the next. This is a chemical alternative to an electrical discharge traveling between the two cells, as the neurotransmitter is only released upon arrival of the electrical impulse at the synapse and once the neurotransmitter binds to the second cell, a potential difference is created which results in the generation of an electrical impulse in the second cell. Interestingly, creation of this *action potential* in certain nerve cells appears to require cyclic nucleotide mediated protein kinase.

cAMP also activates the enzyme RNA polymerase I (mammalian), as always by protein kinase mediated phosphorylation. This enzyme catalyzes the synthesis of the mRNA polymer which is a process essential to the synthesis of proteins of correct amino acid sequence. Other enzymes that are known to be mediated by cyclic nucleotides include glycogen synthetase (synthesis of glycogen), phosphofructokinase (phosphorylation of fructose 6-phosphate), ornithine decarboxylase (decarboxylation of the amino acid ornithine), and ATPase (hydrolysis of ATP).

It is noteworthy that the cyclic phosphate ring of cAMP and cGMP is a high energy structure (35). It has a ΔG_{hydro} at pH 7.3 of approximately -12.5 kJ/mol $(-3.0$ kcal/mol) greater than the biological energy store ATP. It is the enthalpy term which accounts mostly for this free energy value; the ΔH_{hydro} of cAMP and cGMP are -59.2 (-14.1) and -44.1 kJ/mol $(-10.5$ kcal/mol), respectively. Of course, hydrolysis represents the conversion of cyclic nucleotide to a 5'-acyclic structure:

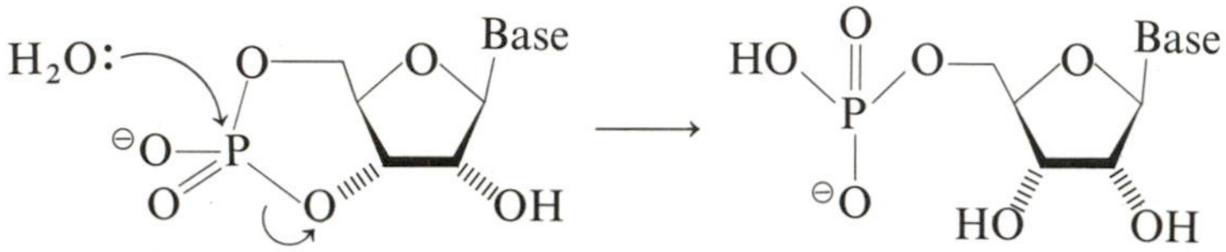

Base = adenine or guanine

The importance of these high energy cyclic structures to the biological function of cyclic nucleotides is not yet certain. However, it has been suggested that such is necessary for the binding of cyclic nucleotides to the regulatory subunit of protein kinase via a covalent intermediate. For example, a carboxylate residue of the regulatory subunit might open the cyclic phosphate to form a high energy intermediate. This in turn could break an ionic attraction between the regulatory and catalytic subunits, thus causing dissociation of the two structures:

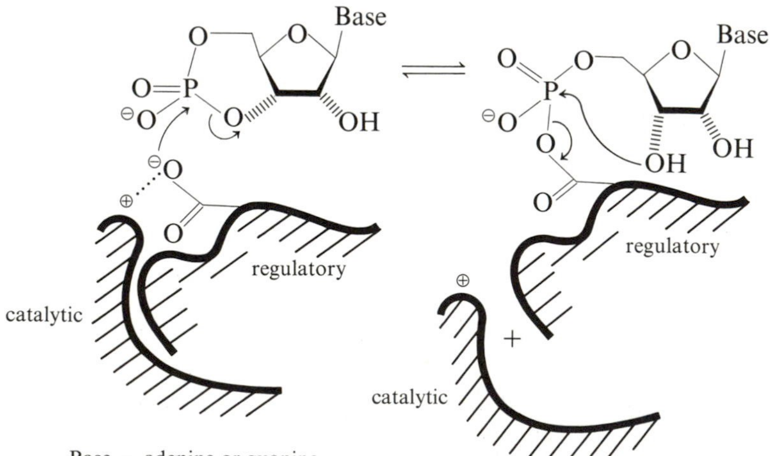

Base = adenine or guanine

It is known that six-membered cyclic phosphates are not typically high energy structures. That is, the six-membered cyclic phosphate, unlike the five-membered ring, is not strained (angular distortions). This is indicated by the enthalpy data for ethylene, trimethylene, and tetramethylene phosphoric acid:

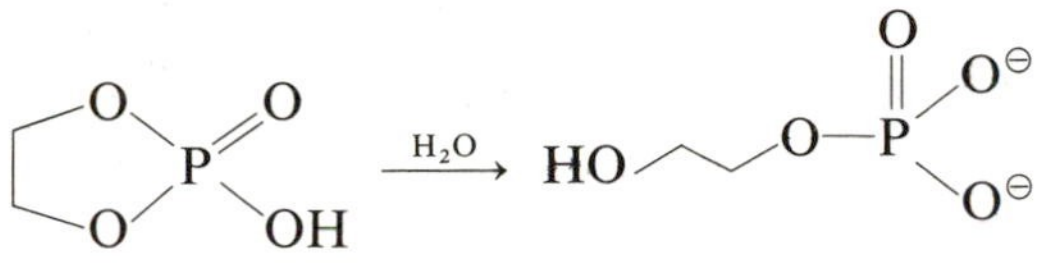

$\Delta H_{\text{hydro}} = -26.9 \text{ kJ/mol} (-6.4 \text{ kcal/mol})$

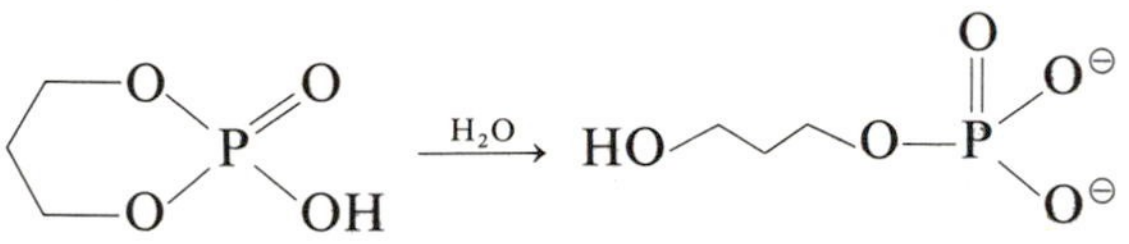

$\Delta H_{\text{hydro}} = -12.6 \text{ kJ/mol} (-3.0 \text{ kcal/mol})$

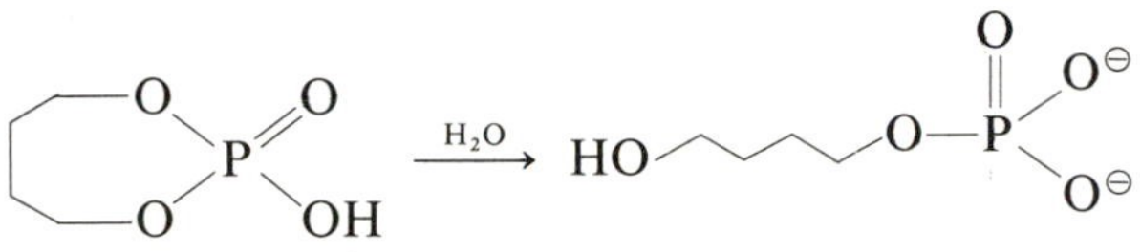

$\Delta H_{\text{hydro}} = 9.2 \text{ kJ/mol} (-2.2 \text{ kcal/mol})$

Further, the base is not responsible for this high energy content. It is the presence of the ether oxygen of the ribofuranoside ring *trans* fused to the six-membered cyclic phosphate which is believed to be responsible for the high energy content of the cyclic phosphate. Hence, replacement of the furanose ring with a pyranose (glucose) or a cyclopentane structure (i.e., the ether oxygen with a methylene group) results in a cyclic phosphate of significantly lower energy.

The biosynthesis of cAMP and cGMP then requires a high energy precursor: ATP or GTP. The enzymes which catalyze this reaction, adenyl and guanyl cyclase, have already been described above.

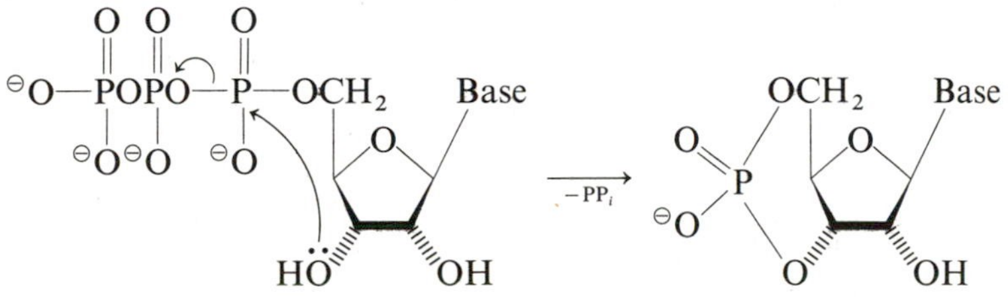

Base = adenine or guanine

The chemical synthesis of cyclic nucleotides may be undertaken in a fashion analogous to their biosynthesis: cyclization of an adenosine anhydride adduct. As the cyclic phosphate product is high energy, use of an anhydride intermediate, as may be obtained by reaction of a carbodiimide, will insure that cyclization is a thermodynamically allowable process. Hence, under conditions of high dilution and in the presence of DCC, adenosine

5′-phosphate will undergo cyclization to form cAMP. However, significant pyrophosphate formation can occur:

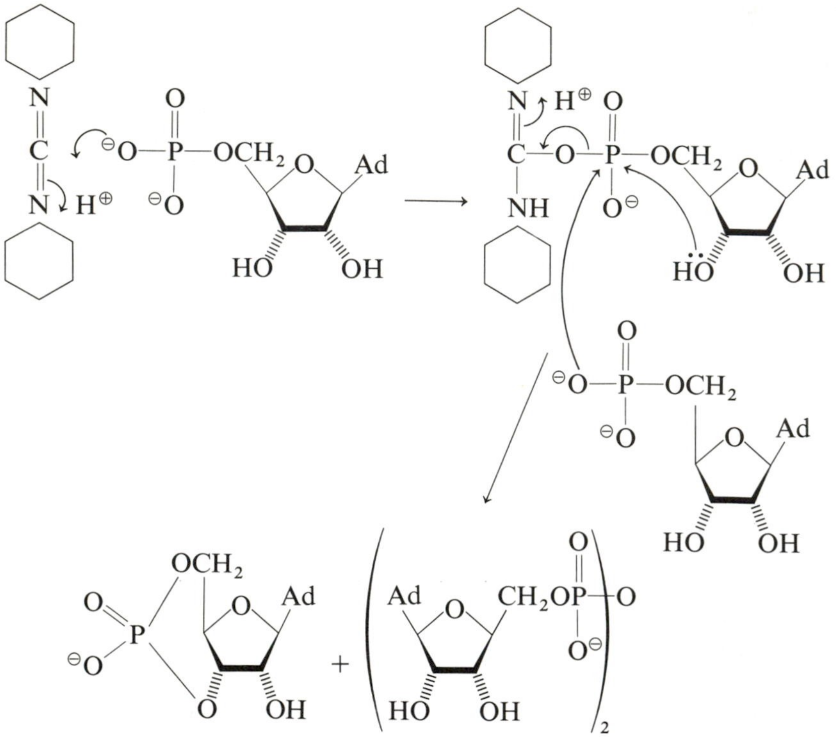

An alternative synthesis, which eliminates pyrophosphate formation, promotes cyclization by generation of a potent alkoxide nucleophile. The base used for this purpose is one of the most potent known: potassium *tert*-butoxide in dimethyl sulfoxide. Of course, the presence of a good leaving group is also necessary:

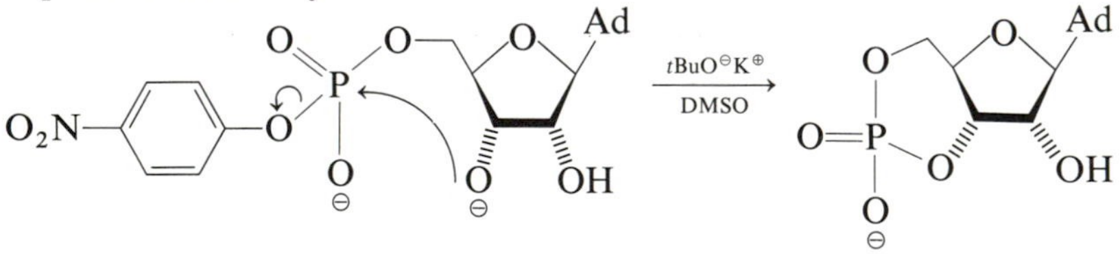

A similar strategy was used for the synthesis of a phenyl phosphotriester of cyclic uridine monophosphate (cUMP):

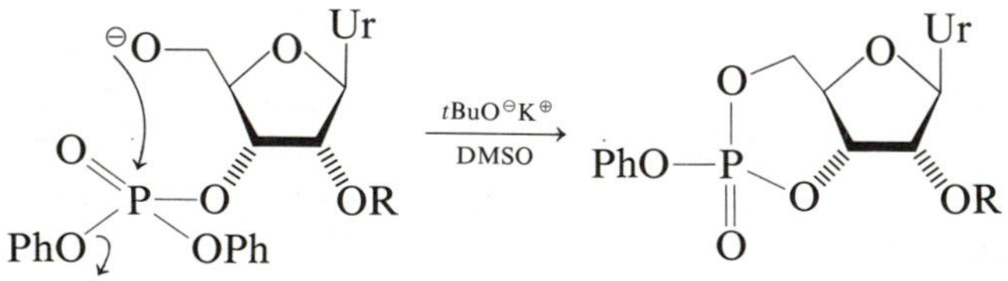

R = protecting group

The cyclic product is extremely sensitive to moisture (hydrolysis) and was more readily isolated as the cyclic diester. An interesting correlation has been observed: the phenyl triesters of high energy cyclic phosphates (cyclic nucleotides, ethylene phosphate) are very reactive species, susceptible to decomposition. On the other hand, the phenyl triesters of low energy cyclic phosphates are stable, crystalline compounds. Hence, for example, the phenyl triester of trimethylene phosphoric acid is a crystalline compound readily prepared in high yield by reaction of 1,3-propanediol with phenyl phosphorodichloridate.

A similar preparation of the phenyl triester of cAMP has not yet been reported.

3.5 Biological Synthesis of Polynucleotides

The synthesis (replication) of DNA will have to occur such that two new helices of double-stranded DNA of the same base sequence or genetic information as the parent molecule will form. By such a process, two daughter cells may arise from a given parent cell. This is made possible by prior separation of the parent double-stranded helix, with each of the separated strands serving as a *template* for the synthesis of a new helix. Had the two strands been covalently linked, the energy required for their separation would be considerable. Preservation of the base sequence is maintained because of the highly specific hydrogen bonding pattern between the purine and pyrimidine bases. That is, for example, adenine on one side of the double-helix will always see and hydrogen bond with thymine on the other side of the helix. If the two strands separate, the adenine from one strand will always interact with a thymine during the synthesis of a new complementary strand. Similarly, the thymine that was opposite to the adenine on the parent double-helix will, after separation, interact only with an adenine during the synthesis of a new complementary strand. Thus, with each of the separated strands from the parent double-helix acting as a template, two new strands of double-helical DNA may be synthesized of exactly the same base sequence as the parent molecule. This mechanism of DNA synthesis is referred to as a *semiconservative* mechanism of replication, for the original helix is "half-conserved" (Fig. 3.9). That is, each of the double-helix products contains one of the parent strands.

Of course, as the new bases are added to the template strand, they must undergo polymerization to form the complementary chain. This requires the formation of a 3',5'-phosphodiester linkage between two monomeric units. The enzyme that catalyzes this polymerization reaction is DNA polymerase. As with the peptide bond, formation of the phosphodiester linkage requires work. Hence, the monomeric units cannot be added sequentially as their monophosphates, but instead must first be activated as their triphosphates. To ensure a correct sequential polymerization, the enzyme requires the presence of the parent strand as a template. In addition, the enzyme requires a DNA *primer* from which to synthesize the new chain. Hence, a totally new strand is not synthesized, but instead a primer chain is elongated. This is not difficult to understand, because if the enzyme only required a DNA template, it might begin synthesis of the new strand at some point along the template instead of at the beginning (i.e., in proximity of the primer). This is especially true since the polymerase enzyme does not bind the entire DNA molecule to the active site, but instead just a small portion. The reaction catalyzed is as follows:

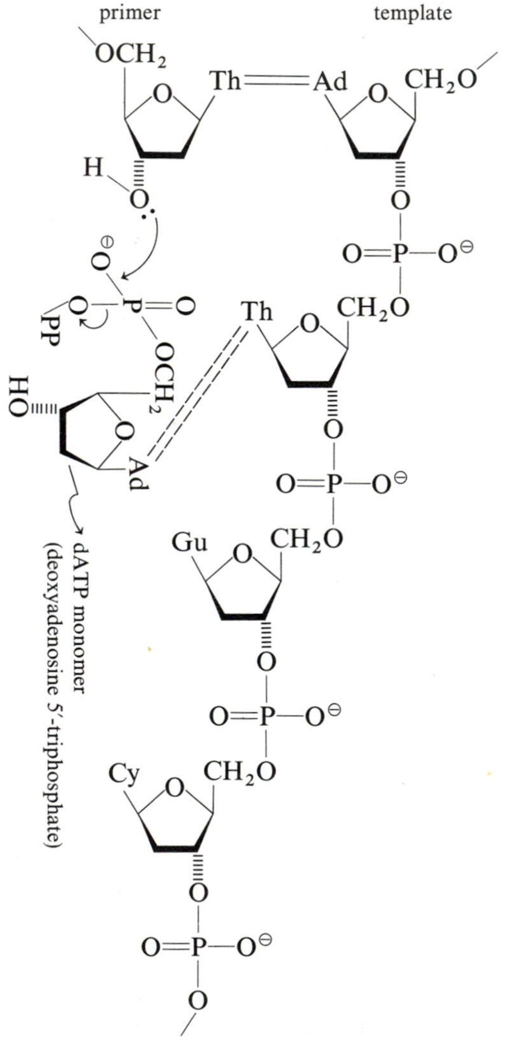

or

$$\text{primer DNA} + n(\text{dATP, dGTP, dCTP, dTTP})$$

$$\text{DNA polymerase} \Big|\; \text{Mg}^{2+},\; -n\,\text{PP}_i$$

$$\text{primer DNA-}[(\text{dAMP})_n\text{-}(\text{dGMP})_n\text{-}(\text{dCMP})_n\text{-}(\text{dTMP})]_n$$

The 3′-hydroxyl of the primer will attack the α-phosphorus of an appropriate (as dictated by the template) triphosphate, thus ejecting pyrophosphate and forming a phosphodiester linkage. The 3′-hydroxyl of the newly added monomer is now in a position to attack another molecule of triphosphate and so the polymerization can continue along the template. The chain will terminate such that the template will end with a 5′-phosphate but the new strand will end with a 3′-hydroxyl.

DNA polymerase exists in different forms in order to carry out different polymerization functions. While perhaps suggestive to the contrary, these different forms are not the result of subunit structure, at least in the case of the bacterial enzyme. Three forms of the enzyme from the bacterium *E. coli* have been characterized and are simply designated polymerase I, II, and III. DNA polymerase I functions principally as a repair enzyme, while DNA polymerase III is a replicating enzyme. The function of DNA polymerase II is not yet certain. The mammalian enzyme also exhibits multiple forms.

Actually, the process of DNA replication is more complicated than described above. Approximately twenty proteins are believed to be involved in the replication process, including those which separate the parent DNA strands, add and later remove small primer fragments, and snip out incorrect bases and repair the damaged region. Further, it appears that synthesis of the new strand along the template may not occur as one continuous step,

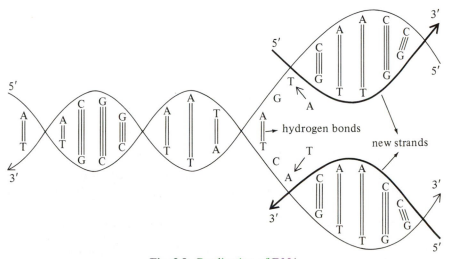

Fig. 3.9. Replication of DNA.

but instead by synthesis of smaller strands (*Okazaki fragments*) which are later joined together by a DNA ligase enzyme. The primer for these fragments may be short chains of RNA which are later replaced by DNA, synthesized by DNA polymerase I. Hence, while the above picture is basically correct, nucleophilic displacement by the 3'-hydroxyl of a growing DNA chain on a deoxynucleoside triphosphate as catalyzed by DNA polymerase III, it is indeed a simplified version of the story. A more complete picture is beyond the scope of this book.

Nonetheless, just as consideration has been given to drugs which inhibit protein synthesis, mention must be made of drugs that are able to inhibit DNA synthesis. These are particularly important to the problem of *cancer chemotherapy*, although they have the undesirable property of not being able to inhibit the synthesis of DNA in the cancer cell without affecting DNA synthesis in the normal cell. Their value does lie in the fact that in many cancers (e.g., leukemia) the rate of cancer cell proliferation greatly exceeds normal cell growth. Such drugs are one of two categories: nucleoside and nonnucleoside analogues.

Perhaps best known of the nucleoside analogues are the arabinose compounds: Ara(C) (cytosine arabinoside, 1-β-D-arabinofuranosylcytosine) and Ara(A) (adenine arabinoside, 9-β-D-arabinofuranosyladenine). Both compounds are triphosphorylated [converted to Ara(CTP) and Ara(ATP) *in vivo*] upon which they may exert a dual action on DNA polymerase. They may inhibit DNA polymerase (for example, Ara(CTP) inhibits DNA polymerase II and III) and they may be incorporated into newly synthesized DNA. In fact, Ara(C) but not Ara(A) is also incorporated into RNA. Such uptake into DNA is lethal since a subtle difference will develop in the DNA geometry as a result of the cytosine base adapting a new glycosidic bond angle due to steric hindrance from the 2'-hydroxyl function.

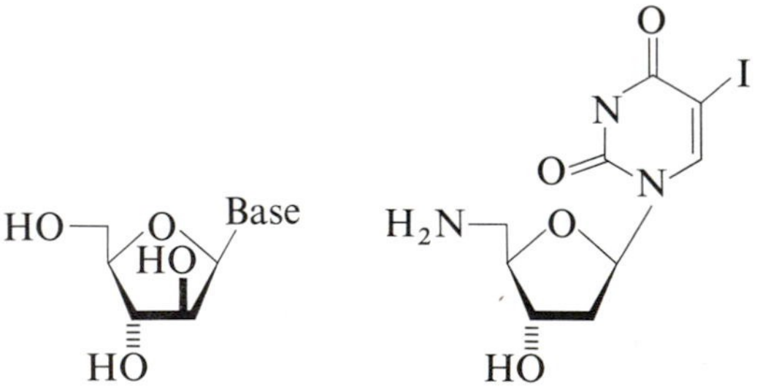

Base = cytosine: Ara(C)
Base = adenine: Ara(A)

5-iodo-5'-amino-2',5'-dideoxyuridine (IAdU)

One highly specific drug* has recently been developed: 5-iodo-5'-amino-2',5'-dideoxyuridine (IAdU). It inhibits herpes simplex virus (type-I) DNA

* Most recently (36) another drug [9-(2-hydroxyethoxymethyl)guanine] has been described which inhibits herpes simplex virus replication without affecting any virus infected cells. The basis of selectivity appears to depend upon the ability of the drug to be phosphorylated (by a virus-coded thymidine kinase) to compounds which are toxic to the viral genetic material.

replication without affecting the mammalian host DNA replication process. Unfortunately, this specificity has not been extended to oncogenic (cancer causing) viruses, but work is under way with this compound and related derivatives [i.e., the 5′-amino analogue of Ara(C)] which may lead to similar specificity. The mechanism of selectivity appears to originate at the level of the synthesis of the triphosphate substrate. Only viral infected cells appear able to prepare IAdUTP, which then becomes incorporated into the viral DNA. Such DNA will contain labile phosphoramidate linkages which presumably eventually interfers with the replication process (i.e., perhaps by decreasing the stability of the phosphodiester backbone).

Examples of nonnucleoside drugs which inhibit DNA synthesis by binding to the DNA double-helix include the acridines (e.g., proflavine) and various antibiotics (e.g. mitomycin C, adriamycin, daunomycin). The ability of the acridines to bind to DNA and RNA has led to their use as biological stains for these molecules. Binding is achieved by "intercalation"; the planar ring system of proflavine, for example, intercalates (squeezes) between the stacked base pairs of the double-helix. Such a "sandwich" of DNA with bound molecules would be expected to geometrically distort (i.e., lengthen) the double-helical structure; this has been observed.

proflavine   mitomycin C

R = OH, adriamycin

= H, daunomycin

A similar mechanism of binding is operative for the antibiotics. As such, all three possess anti-tumor properties.

Synthesis of the 3′,5′-phosphodiester linkage of RNA follows similar principles to DNA synthesis. Again, the 3′-hydroxyl of a growing (RNA) chain attacks an appropriate monomeric triphosphate, as dictated by the template, to form the phosphodiester linkage, with the subsequent ejection of pyrophosphate. The reaction is catalyzed by the enzyme RNA polymerase.

The required template is a strand of DNA, for, as it will be recalled, during the process of transcription, the genetic message from the DNA storage form is transferred to the RNA polymer (so-called mRNA). Only one of the two DNA strands serves as a template. However, a RNA primer is not required to initiate synthesis. Instead, the enzyme binds to a specific region of the DNA strand and from this point proceeds with RNA synthesis; the so-called *promoter region*. The reaction catalyzed may be represented as follows:

$$\text{DNA template} + n\,(\text{ATP, GTP, CTP, UTP})$$

$$\text{RNA polymerase} \quad \Big| \quad Mg^{2+}, -n\,PP_i$$

$$\text{RNA template} \cdot [(\text{AMP})\text{-}(\text{GMP})\text{-}(\text{CMP})\text{-}(\text{UMP})]_n$$

Unlike DNA polymerase, RNA polymerase has a complicated subunit structure.

3.6 Chemical Synthesis of Polynucleotides

The chemical synthesis of a polydeoxyribo- or polyribonucleotide is a complex problem (37–39). Problems that were faced in the consideration of the synthesis of a polypeptide must again be dealt with by the chemist in the laboratory. These include a sequential coupling method as opposed to a random synthesis, the need for protecting groups to prevent unwanted side reactions and a high yield for each protection, coupling, and deprotection step. Here, the problem of protecting groups is more acute since it may be necessary to distinguish between two functions that are almost chemically identical. This is particularly so in the case of a ribonucleotide synthesis, where it may be necessary to protect one secondary hydroxyl (2'-) so that an adjacent secondary hydroxyl (3'-) may be phosphorylated. In fact, once protected, this arrangement (1,2-*cis*-diol) may not be stable. Especially in the case of a small protecting group, isomerization from the wanted 2'-protected nucleoside to the unwanted 3'-protected nucleoside can occur:

2'-acetyl isomer   3'-acetyl isomer

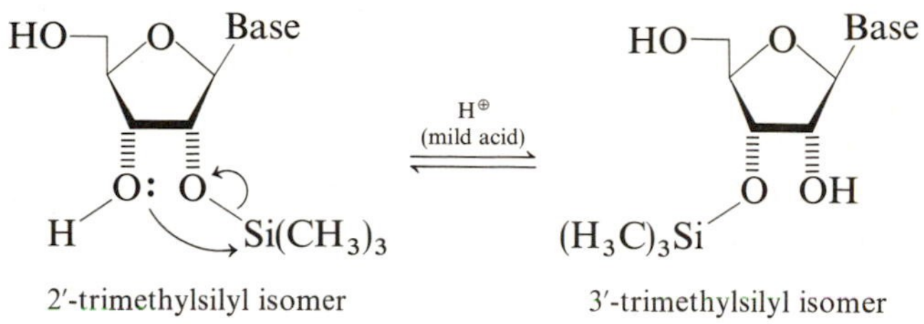

2′-trimethylsilyl isomer   3′-trimethylsilyl isomer

Remember, both DNA and RNA polymers possess a 3′,5′-phosphodiester backbone. Two potentially reactive species that require protection so that the desired bond formation will occur are hydroxyl functions on the sugar and exocyclic amino functions (if present) on the base. Once this is achieved, then supposedly the desired bond formation can occur:

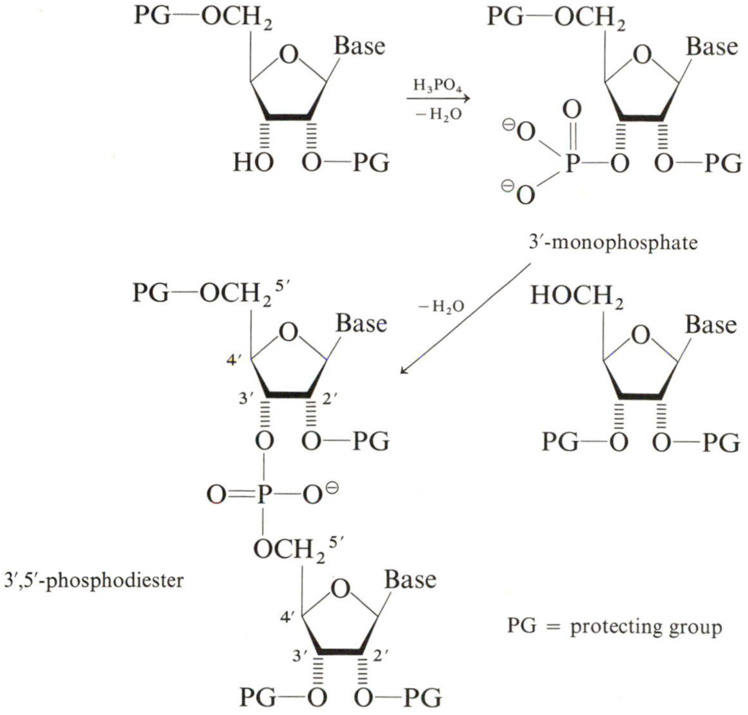

Of course, as with peptide bond formation, work must be done and so an activation process is necessary. A phosphodiester synthesis would not be possible by merely mixing phosphoric acid with the appropriately protected nucleosides. Finally, as will be discussed later, it may even be desirable to consider protection of the phosphate function. While this is not necessary (and was not done in the initial nucleotide synthesis), it is advantageous and is presently the most accepted protocol.

3.6.1 Base Protecting Groups

Protection of the amino function of adenine, guanine, or cytosine may be necessary to prevent phosphoramidate formation during a phosphorylation. In rarer cases (especially with guanine) it may also serve the purpose of introducing hydrophobicity in an otherwise too polar nucleoside.

(1) *Acyl Function*

Once acylated, the amino function is rendered chemically unreactive or nonnucleophilic. However, the amino acyl function often have to be introduced in a two-step procedure: acylation of the entire nucleoside followed by selective deacylation of the sugar. Further, the glycosidic linkage of the *N*-acyl derivative is usually more susceptible to hydrolysis. Typical

acyl functions include acetyl (CH_3—C—), benzoyl (Ph—C—), isobutyryl

(($CH_3)_2$CHC—), and anisoyl (CH_3O—⟨ ⟩—C—). For example, the synthesis of *N*-benzoyl adenosine:

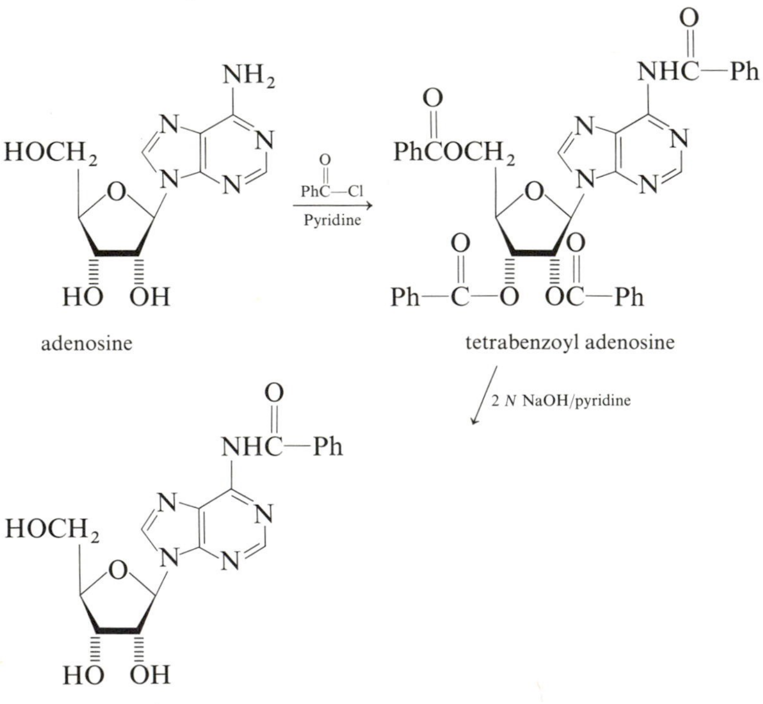

adenosine   tetrabenzoyl adenosine

N-benzoyl adenosine

The synthesis of *N*-benzoyl guanosine is similar except that the hydroxide is replaced by methoxide. However, *N*-benzoyl cytidine may be prepared by direct benzoylation of the base under conditions which do not affect the sugar. This reflects the better nucleophilicity of the cytidine amino function, relative to the purines.

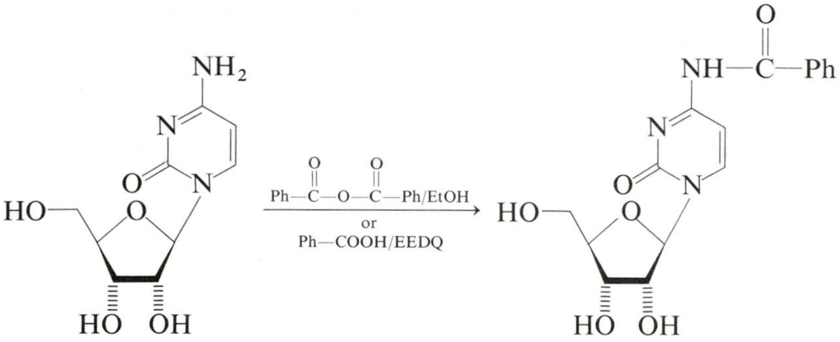

Removal of the acyl functions is typically accomplished, with concentrated ammonium hydroxide or aqueous sodium hydroxide. Further, the benzoyl group on the base may be selectively removed (leaving behind a benzoylated sugar) under neutral conditions by use of hydrazine hydrate $(NH_2NH_2 \cdot 0.5 H_2O)$. This preference for the base benzoyl function is accounted for by a concerted mechanism:

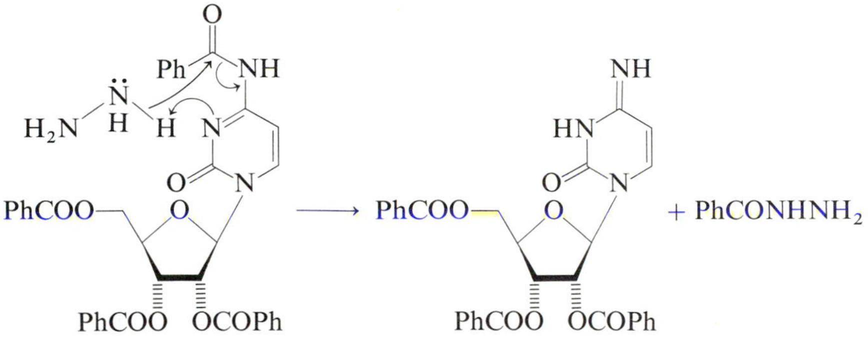

The *tert*-BOC and CBz functions have not yet found significant use in the protection of base amino functions. The only carbamate that has found extensive use is the isobutyloxycarbonyl function. This may be introduced by reaction of the nucleoside with isobutyryl chloroformate followed by deacylation of the sugar moiety:

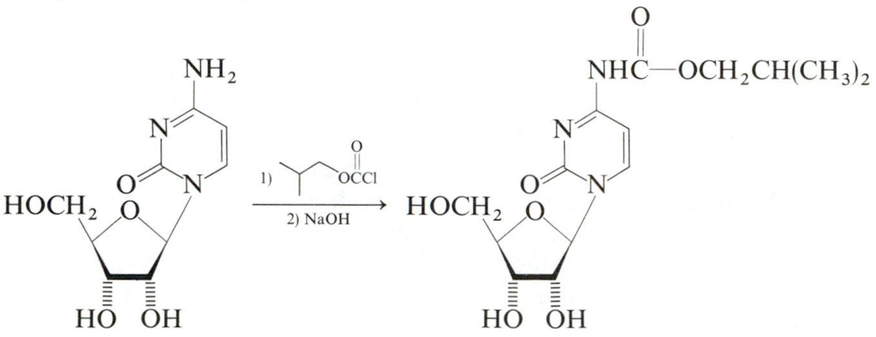

The advantage of this function is that the carbamate (i.e., less electrophilic carbonyl) is more resistant to hydrazine than the corresponding amide linkage. It may also be removed by treatment with concentrated ammonium hydroxide.

(2) N-Dimethylaminomethylene Function

Treatment of an exocyclic amino function with dimethylformamide dimethyl acetal in dimethylformamide solvent gives the N-dimethylaminomethylene derivative.

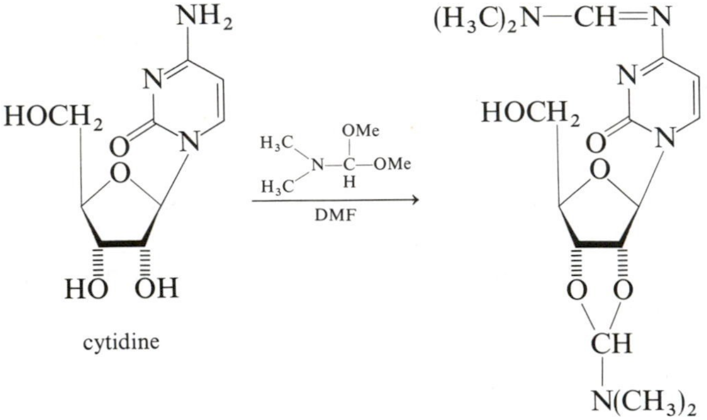

cytidine

The reaction involves nucleophilic displacement which may be thought of as proceeding via a reactive intermediate:

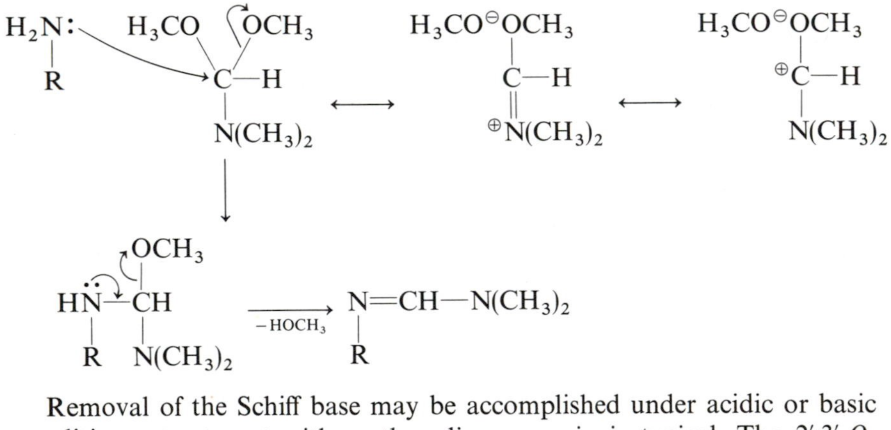

Removal of the Schiff base may be accomplished under acidic or basic conditions: treatment with methanolic ammonia is typical. The $2',3'$-O-dimethylaminomethylene function on the sugar is even more sensitive, being removed merely by the addition of water. Nonetheless, the sensitivity of the Schiff base to both acid and base can be disadvantageous if it is desired to carry out further operations on the protected nucleoside under one of these protic conditions. Further, thymine and uracil bases, if present (i.e., as part of an oligonucleotide) can be methylated. This is illustrated for 1-methyluracil:

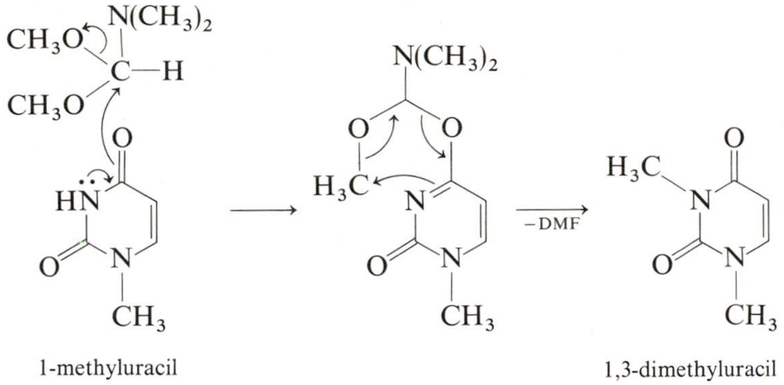

1-methyluracil   1,3-dimethyluracil

This may be avoided by the sterically bulky neopentyl analogue.

3.6.2 Sugar Protecting Groups

Of the three groups to be protected (amino, phosphate, and hydroxyl), protection of the hydroxyl is most important if unwanted by-product and isomer formation is to be minimized. As such, the number of protecting groups available for the hydroxyl function are numerous. This is desirable since it may be necessary to use two different hydroxyl protecting groups for a given molecule. This is usually done to distinguish the 5′-hydroxyl (primary) function from the secondary hydroxyl(s). More difficult is to distinguish, in the case of ribonucleosides, between the 2′- and 3′-secondary hydroxyl functions, which have similar chemical reactivity.

(1) *Monomethoxytrityl Function*
The *p*-methoxytrityl and to a lesser extent the trityl function are currently used primarily as protecting groups for the primary 5′-hydroxyl of nucleosides. Under mild reaction conditions, only minor tritylation of a secondary hydroxyl or base amino function will occur. If it seems surprising that tritylation of the exocyclic amino function is difficult, especially since the trityl group has been noted to be useful for the protection of the amino function of amino acids, it should be remembered that the base amino function (with its delocalized nonbonding electron pair) is significantly less nucleophilic than a primary alkyl amine. The secondary hydroxyl(s) will

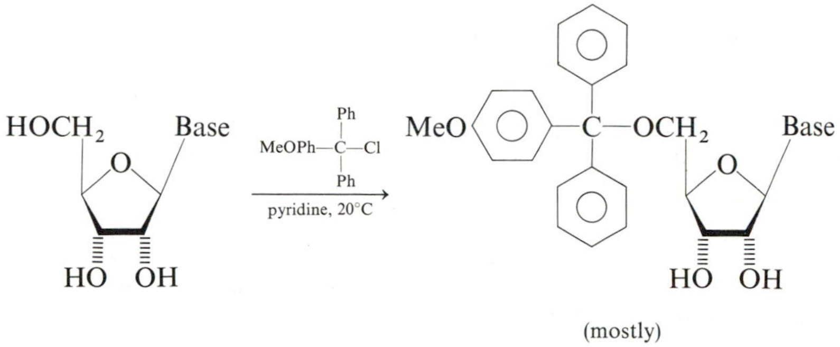

(mostly)

only react to a significant extent under harsh reaction conditions because of the steric bulk of the triphenylmethane function. Reaction occurs by an S_N1 displacement. The rate of reaction with the nucleoside increases with the introduction of the methoxy function (just as does the sensitivity to acid), so that the rate of reaction is trityl < p-methoxytrityl < di-p-methoxytrityl. This might suggest the di-p-methoxytrityl function to be the ideal protecting group. However, with this increased rate of reaction, some selectivity for the primary hydroxyl is sacrificed. Further, the di-p-methoxytrityl function is so acid labile that it may be partially removed during routine workup (i.e., purification on silica gel).

As with the amino acids, the trityl function may be removed from the nucleoside by treatment with acid (i.e., 80% acetic acid, pyridine–acetic acid buffer). Again, this proceeds by way of a stabilized carbonium ion intermediate. Further stabilization is achieved by the introduction of p-methoxy functions. Thus, the acid lability of the tritylated nucleoside is increased approximately by a factor of ten for each p-methoxy group present. In the case of nucleosides and nucleotides, removal of the p-methoxytrityl function

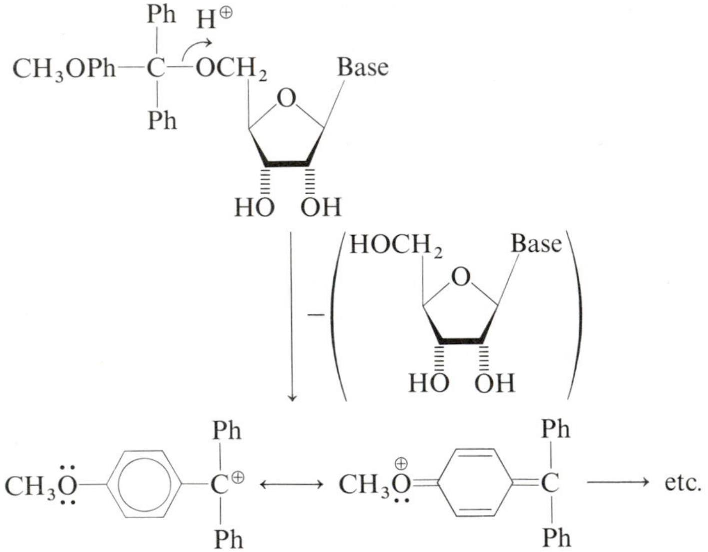

by hydrogenolysis is sluggish.* Further, unwanted side reactions, such as hydrogenation of pyrimidine bases, can occur.

(2) *Acyl Function*

Acylation of the primary and secondary hydroxyl(s) of nucleosides is most often accomplished by treatment of the nucleoside with the acyl anhydride

---

* During the final preparation of this text, it has been reported [M. H. Caruthers *et al.* (1980), *Tetrahedron Lett.*, 3243] that the trityl function may be removed under neutral conditions by treatment with $ZnBr_2$. The zinc ion supposedly labilizes the ether linkage by chelation with the 5'- and ring furanose oxygens.

or chloride in pyridine. Acylation of an exocyclic amino function, if present on the base, can also occur, depending upon its reactivity and the reaction conditions (see above). In addition to the typical acyl functions (formyl, acetyl, benzoyl, chloroacetyl, etc.) which are most often removed under alkaline conditions, a number of acyl groups have been developed which may be removed under conditions compatible with specific nucleotide synthesis.

Substituted acetyl groups have been prepared in order to devise a more labile protecting group than the parent acetyl function. Examples include the trifluoracetyl ($CF_3\overset{\overset{O}{\|}}{C}$—), phenoxyacetyl ($PhOCH_2\overset{\overset{O}{\|}}{C}$—), and methoxyacetyl ($CH_3OCH_2\overset{\overset{O}{\|}}{C}$—) functions. These substituents increase the electrophilicity of the carbonyl function via inductive pull ($I\ominus$).

Two groups of protecting groups have been developed which can be removed under neutral conditions by treatment with hydrazine hydrate. These functions are substituted γ-keto acids or acrylic acids. The general reaction with hydrazine is illustrated:

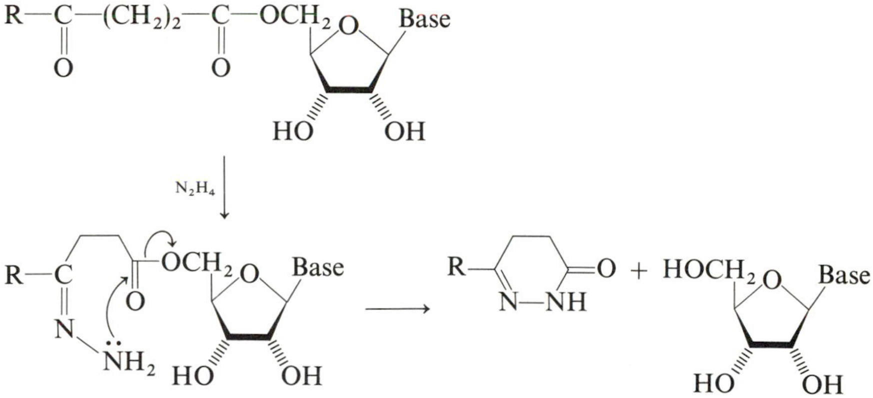

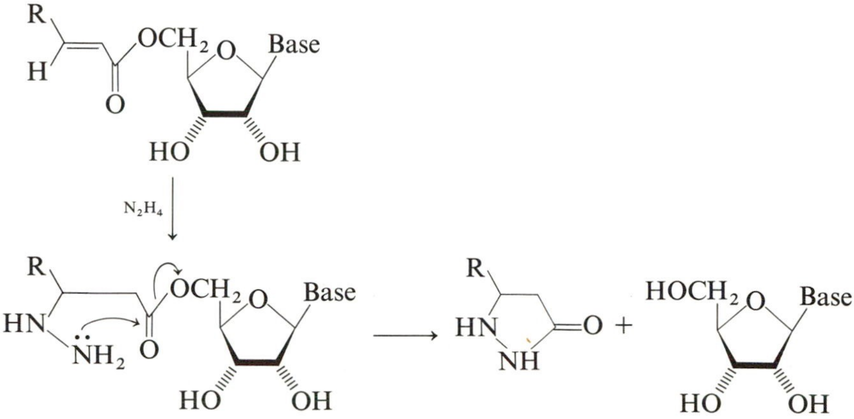

For example, 3'-O-benzoylpropionylthymidine has been prepared by the following reaction sequence:

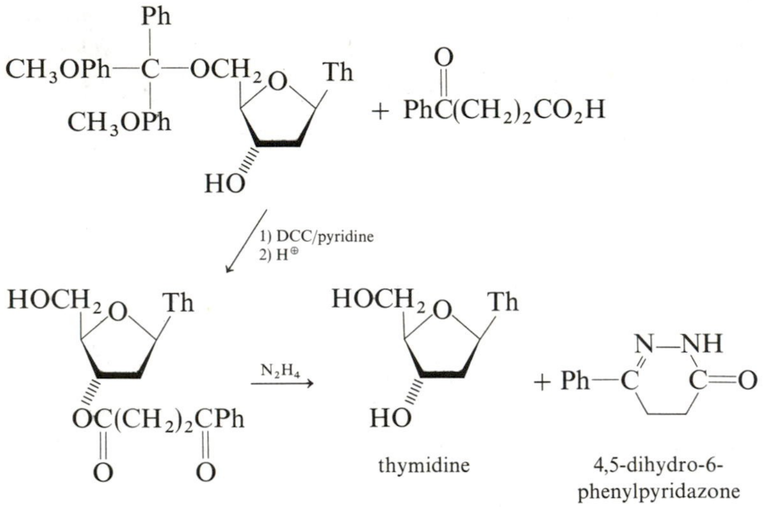

$$
\text{CH}_3\text{OPh}\!-\!\overset{\displaystyle\text{Ph}}{\underset{\displaystyle\text{CH}_3\text{OPh}}{\text{C}}}\!-\!\text{OCH}_2 \quad \text{Th} \qquad + \ \text{PhC(CH}_2)_2\text{CO}_2\text{H}
$$

(with HO on sugar ring)

1) DCC/pyridine
2) H⊕

$$
\text{HOCH}_2 \quad \text{Th} \qquad \xrightarrow{\text{N}_2\text{H}_4} \qquad \text{HOCH}_2 \quad \text{Th} \qquad + \ \text{Ph}\!-\!\text{C} \overset{\displaystyle\text{N}\!-\!\text{NH}}{\underset{}{\text{C}}}\!=\!\text{O}
$$

OC(CH₂)₂CPh (with two C=O)

HO

thymidine

4,5-dihydro-6-phenylpyridazone

The levulinyl function ($\text{CH}_3\overset{\text{O}}{\overset{\|}{\text{C}}}\text{(CH}_2)_2\text{CO}\!-\!$) is even more sensitive to hydrazine than the benzoyl analogue. This has been attributed to a loss of conjugation between the phenyl and carbonyl functions in the transition state for hydrazine addition. With the methyl function, no such conjugation was present prior to the hydrazinolysis.

Various acrylic acid derivatives include the

acrylyl $(\text{CH}_2\!=\!\text{CH}\!-\!\overset{\text{O}}{\overset{\|}{\text{C}}}\!-\!)$

crotonyl $(\text{CH}_3\text{CH}\!=\!\text{CH}\overset{\text{O}}{\overset{\|}{\text{C}}}\!-\!)$

methoxycrotonyl $(\text{CH}_3\text{OCH}_2\text{CH}\!=\!\text{CH}\overset{\text{O}}{\overset{\|}{\text{C}}}\!-\!)$

and

phenoxycrotonyl $(\text{PhOCH}_2\text{CH}\!=\!\text{CH}\overset{\text{O}}{\overset{\|}{\text{C}}}\!-\!)$

functions. These are typically introduced as their acid anhydrides.

Carbonate formation has also been used to protect hydroxyl functions. For example, isobutyl chloroformate will react preferentially with the primary hydroxyl of thymidine:

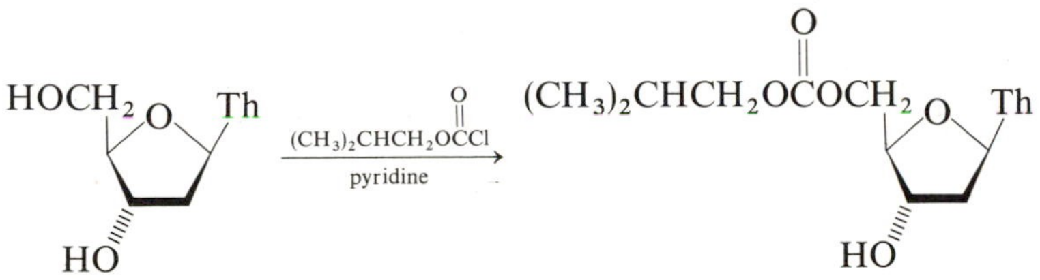

The carbonate so formed is stable to conditions of acidity that allow removal of a *p*-methoxytrityl function. Hence, a synthesis of 3'-*O*-*p*-monomethoxy tritylthymidine is as follows:

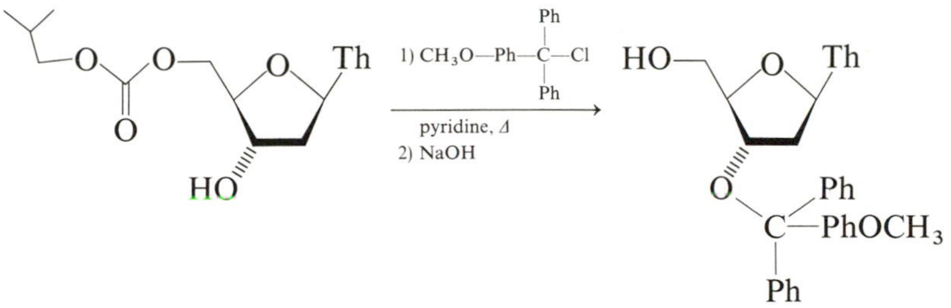

Of poorer selectivity, but of some use, is *p*-nitrophenyl chloroformate. In the case of a *cis*-diol arrangement, chloroformates or diphenyl carbonate will react to form a cyclic carbonate:

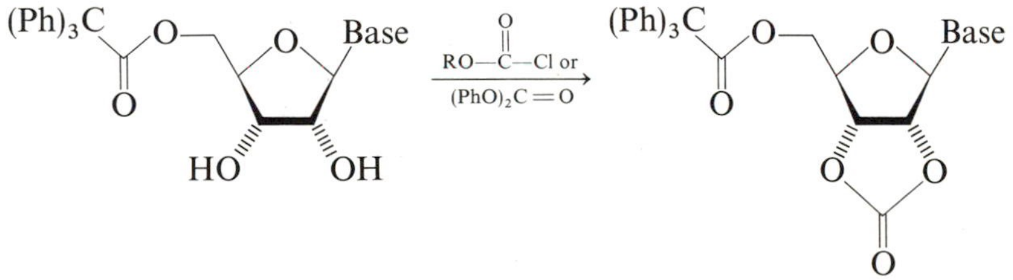

The cyclic carbonate may also be removed by treatment with mild base.

(3) *t-Butyldimethylsilyl Function*

Initially, silylation of carbohydrates and nucleosides was employed as a technique for preparation of samples for gas chromatographic and mass spectral analysis. However, silyl compounds, especially the *tert*-butyldi-methylsilyl function (*tert*-BDMS—), are presently finding use as protecting groups for the hydroxyl function. Bulky functions are preferred as they offer greater selectivity for primary hydroxyls, and are reasonably resistant to isomerization, even in the case of ribonucleosides where a 1,2-*cis*-diol arrangement is present. The reaction of thymidine with some silyl reagents is illustrated:

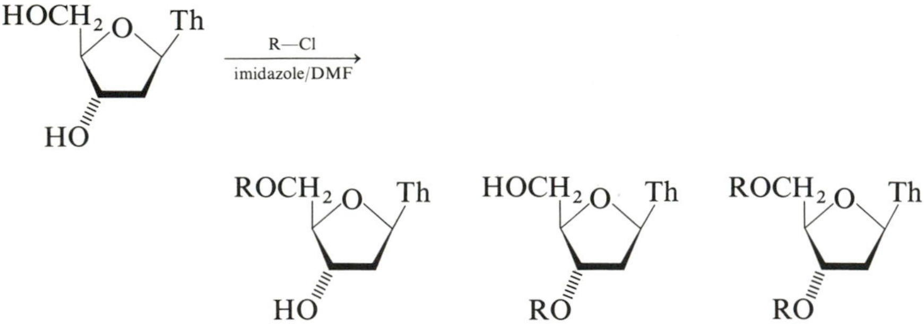

R	% yield		
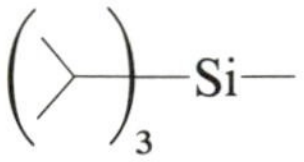 triisopropylsilyl	82	2	4
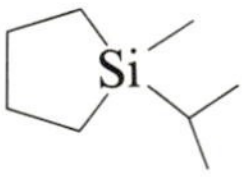 tetramethyleneisopropylsilyl	35	6	46
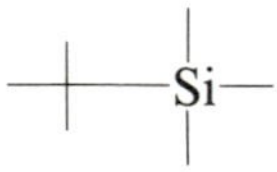 *tert*-butyldimethylsilyl	73	1	15

All the silylating agents possess properties which make them desirable for hydroxyl group protection. They show varying degrees of preference for the primary hydroxyl function, but will in addition, in the presence of excess silylating agent, silylate any secondary hydroxyls present. However, significant silylation of base amino functions will not occur. The silyl function may be removed by acid hydrolysis (80% acetic acid) or under neutral conditions by treatment with fluoride anion (most commonly, tetrabutylammonium fluoride, but other cations such as cesium or tetraethylammonium may be used). Silyl ethers are reasonably susceptible to acid cleavage. Further, especially in the case of the *tert*-BDMS function, primary silyl ethers will be removed preferentially. Thus, it is possible to prepare a 3'- or 5'-monosilylated deoxyribonucleoside with minor side product formation. Because of the hydrophobic nature of the silylated nucleosides, any side products can be removed by typical organic techniques (silica gel chromatography). Fluoride in anhydrous solvent systems is a good nucleophile. The silyl function is stable to base and hydrazine hydrate.

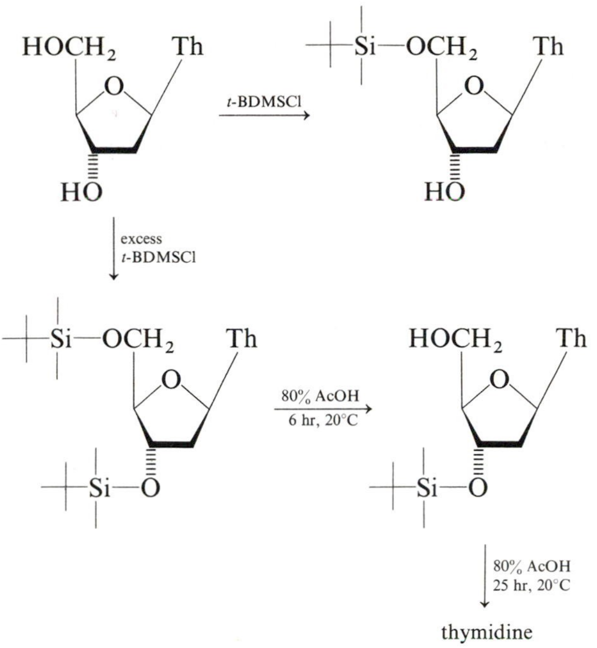

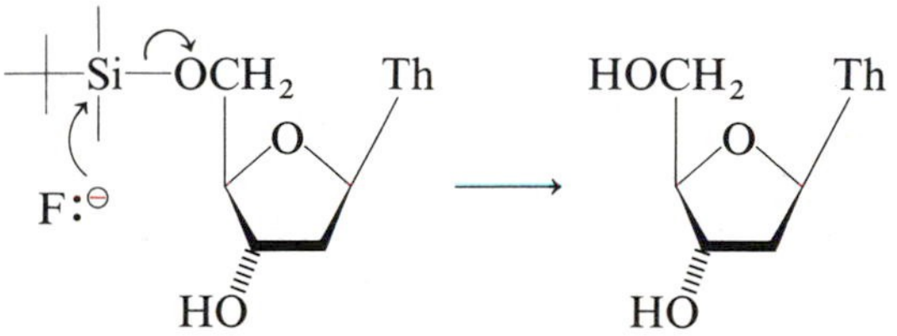

The *tert*-BDMS group is also useful for the synthesis of specific protected ribonucleosides. Again, it will preferentially silylate the primary hydroxyl. Addition of more silylating agent will lead to a mixture of 2',5'- and 3',5'-disilyl products. Depending upon reaction conditions and the base, the 2',5'-isomer will be formed in equal or greater amount than the 3',5'-disilyl compound. Such is useful, for the synthesis of a ribonucleotide requires protection of the 2'-hydroxyl function, but a free 3'-hydroxyl for phosphodiester bond formation. The two isomers may be separated by silica gel chromatography. Addition of still more silylating agent leads to total silylation of the nucleoside:

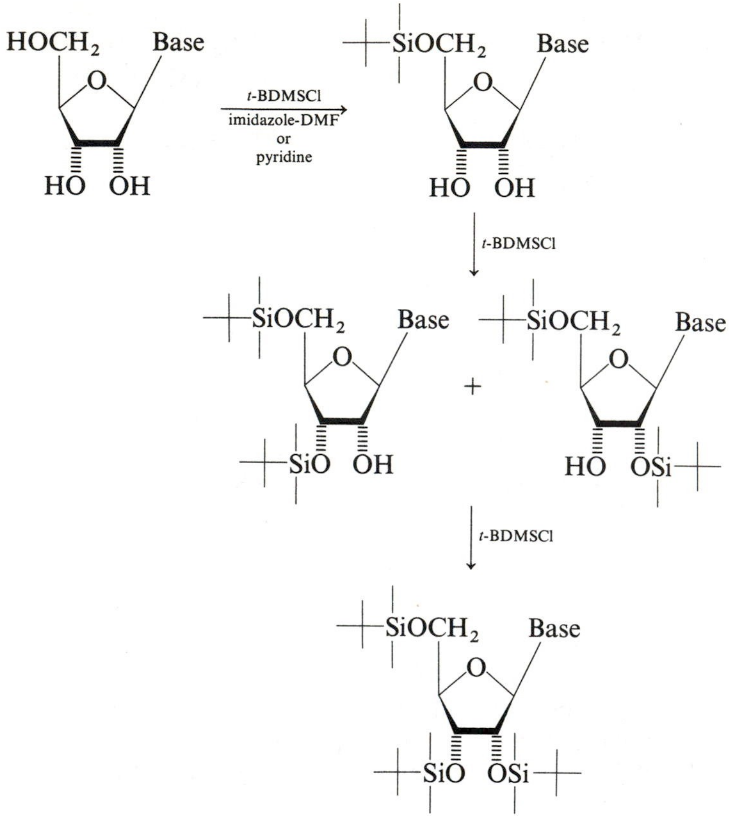

HOCH$_2$ Base   —SiOCH$_2$ Base

$$\xrightarrow[\substack{\text{imidazole-DMF} \\ \text{or} \\ \text{pyridine}}]{t\text{-BDMSCl}}$$

HO   OH   HO   OH

$$\downarrow t\text{-BDMSCl}$$

—SiOCH$_2$ Base   +   —SiOCH$_2$ Base

—SiO   OH   HO   OSi—

$$\downarrow t\text{-BDMSCl}$$

—SiOCH$_2$ Base

—SiO   OSi—

Remembering that the primary silyl function is the most susceptible to hydrolysis, it is possible to prepare the 5′-, 2′-, or 3′-*tert*-BDMS nucleoside, as well as the 2′,3′-, 2′,5′-, and 3′,5′-disilyl compounds. Actually, since the *p*-methoxytrityl function is more sensitive to acid hydrolysis than the primary silyl function (among other reasons), for the purpose of nucleotide synthesis, the 5′-*O*-*p*-methoxytrityl-2′-*tert*-BDMS-protected nucleosides have been prepared.

Silylation apparently requires not only a base as an acid scavenger, but a nucleophilic catalyst as well. Reaction rates are: triethylamine < pyridine < imidazole.

(4) *Tetrahydropyranyl Function*

The use of dihydropyran for the protection of alcohols is a well-known reaction. Both addition and removal of the group occurs in the presence of acid:

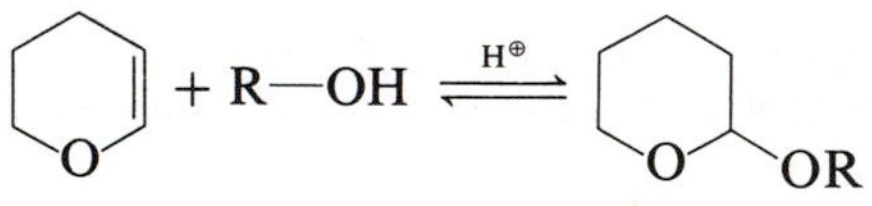

$$\text{(2,3-dihydropyran)} + R\text{—OH} \underset{}{\overset{H^\oplus}{\rightleftharpoons}} \text{(THP ether)}$$

2,3-dihydropyran   THP ether

The THP ether product is stable to base. The THP function (see below) has been used for the synthesis of 2′-protected ribonucleosides. As was noted (see above) the 2′-protected hydroxyl is vital for the synthesis of the 3′,5′-phosphodiester linkage. However, a complicating factor can be the presence of an asymmetric center in the THP protected product.

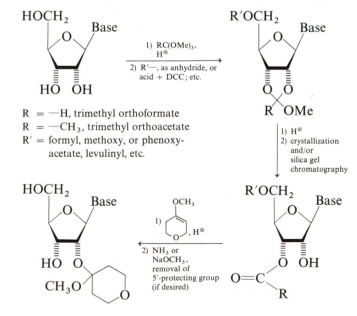

For the purposes of workup, it will appear as if two compounds are present with different R_f values, melting points, etc. This has been overcome by replacing 2,3-dihydropyran with 5,6-dihydro-4-methoxy-2H-pyran. The THP ether product is symmetrical.

Synthesis of the useful 2′-protected nucleoside has been undertaken by way of a 2′,3′-cyclic orthoester intermediate. The orthoester protects the 2′, and 3′-hydroxyls so that the primary hydroxyl can then selectively be protected, if desired. Hydrolysis of the fully protected nucleoside with acid

R = —H, trimethyl orthoformate
R = —CH₃, trimethyl orthoacetate
R′ = formyl, methoxy, or phenoxy-
acetate, levulinyl, etc.

(e.g., formic or acetic acid) yields a mixture of 2'- and 3'-acylated isomers. Separation is usually possible by crystallization or column chromatography. For example, 3'-O-acetyladenosine will selectively crystallize in the presence of 2'-O-acetyladenosine when both are dissolved in hot absolute ethanol. The separated 3'-isomer can then be tetrahydropyranylated (as will the primary hydroxyl, if it is not protected) and the acyl function removed by treatment with base to give a free 3'-hydroxyl-2'-protected nucleoside. Depending on the selection of protecting group for the primary hydroxyl, it will also still be protected (e.g., levulinyl, THP) or free (e.g., formyl, methoxy, or phenoxyacetate).

(5) *Other Protecting Groups*

Perhaps classical for the protection of 1,2- and 1,3-*cis*-diols is the isopropylidene (acetonide) function. This acid labile ketal may be formed by reaction of a ribonucleoside with acetone in the presence of an acid (HCl, *p*-toluenesulfonic, H_2SO_4) and water scavenger (2,2-dimethoxypropane, ethyl orthoformate). Other similar protecting groups include the benzylidene and cyclohexylidene functions.

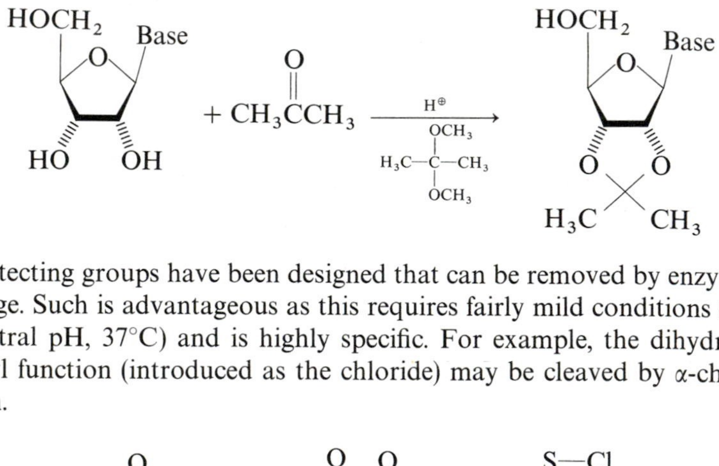

Protecting groups have been designed that can be removed by enzymatic cleavage. Such is advantageous as this requires fairly mild conditions (close to neutral pH, 37°C) and is highly specific. For example, the dihydrocinnamoyl function (introduced as the chloride) may be cleaved by α-chymotrypsin.

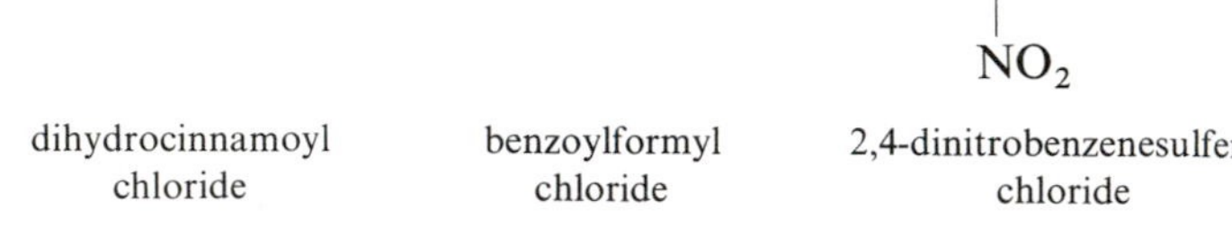

dihydrocinnamoyl chloride	benzoylformyl chloride	2,4-dinitrobenzenesulfenyl chloride

Other protecting groups include the benzoylformyl function (introduced as the chloride and removed in pyridine water) and the 2,4-dinitrobenzenesulfenyl function (introduced as the chloride and removed with thiophenol). The photolabile *o*-nitrobenzyl function, which has been noted as a protecting group for amino acids (see Chapter 2, p. 64), has found use as a 2'- hydroxyl protecting group.*

3.6.3 Phosphate Protecting Groups

With familiarity of the various protecting groups available for the hydroxyl and base amino functions, it is possible to prepare nucleoside intermediates with reactive groups available only for the formation of a 3',5'-phosphodiester linkage. Unwanted reactions (3',3'- or 5',5'-phosphodiester formation, phosphoramidate formation, etc.) cannot possibly occur with the selection of appropriate intermediates. A representative example would be the condensation between adenosine and uridine:

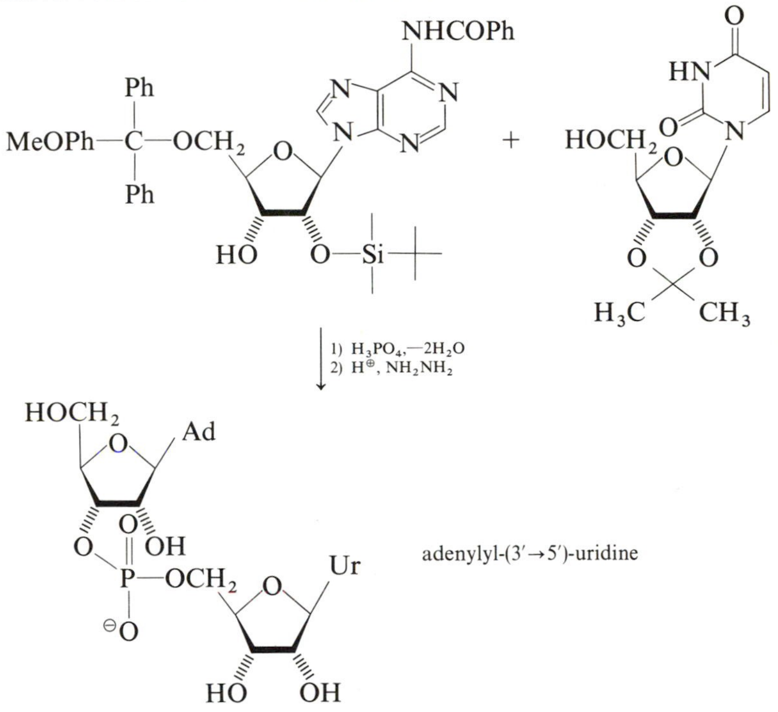

adenylyl-(3'→5')-uridine

Once product formation is complete, the protecting groups may be removed under mild conditions that will not affect the phosphodiester linkage. Such is the basis for the synthesis of polynucleotides via the *phosphodiester approach*. The product is a phosphodiester with a free, potentially troublesome, anionic charge. Further, with increasing size of the polynucleotide chain, the number of negative charges present in the product will also increase. Thus, depending on reaction conditions, these potentially nucleophilic charges can participate in unwanted side reactions. Further, such a

---

* For the most recent review concerning the use of photoremovable protecting groups in organic (amino acids, nucleotides) synthesis, see reference 362.

multicharged species will be too polar to allow purification by typical organic techniques such as silica gel chromatography. Instead, lower capacity ion-exchange (i.e., DEAE-cellulose) chromatography must be employed. The phosphodiester approach is a small scale synthetic technique. However, neutralization of this charge by esterification with an appropriate protecting group prior to phosphorylation of the nucleosides eliminates the above mentioned problems. In this case, the product of the condensation reaction is a phosphotriester. The *phosphotriester approach* is a large-scale synthetic technique. Some of the protecting groups used for the phosphate function are described below.

(1) *2,2,2-Trichloroethyl Function*

Introduction of this protecting group into an appropriate phosphorylating agent usually occurs by reaction of 2,2,2-trichloroethanol with a phosphorochloridate or phosphorochloridite:

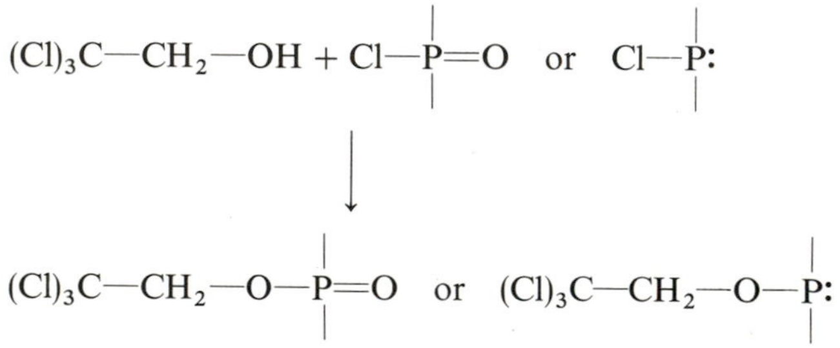

$$(Cl)_3C-CH_2-OH + Cl-\overset{|}{\underset{|}{P}}=O \quad \text{or} \quad Cl-\overset{|}{\underset{|}{P}}:$$

$$(Cl)_3C-CH_2-O-\overset{|}{\underset{|}{P}}=O \quad \text{or} \quad (Cl)_3C-CH_2-O-\overset{|}{\underset{|}{P}}:$$

It is the protected phosphorylating agent which is reacted with the nucleosides to form the 3′,5′-phosphodiester linkage. Once product formation is completed, the trichloroethyl protecting group may be removed by several means. Classic is the removal by zinc. Alternatively, tetrabutylammonium fluoride may be used:

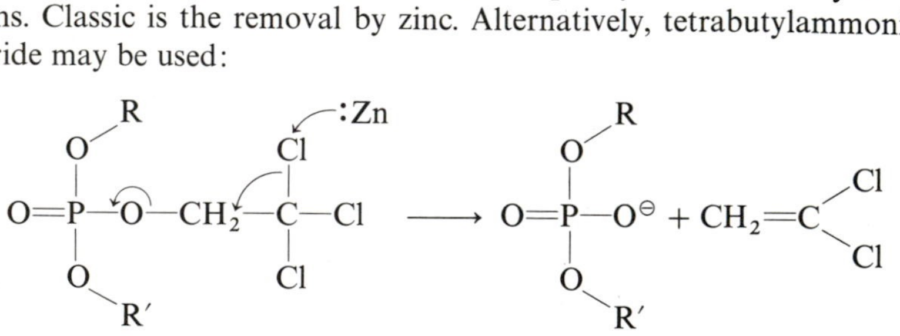

3′,5′-phosphotriester product

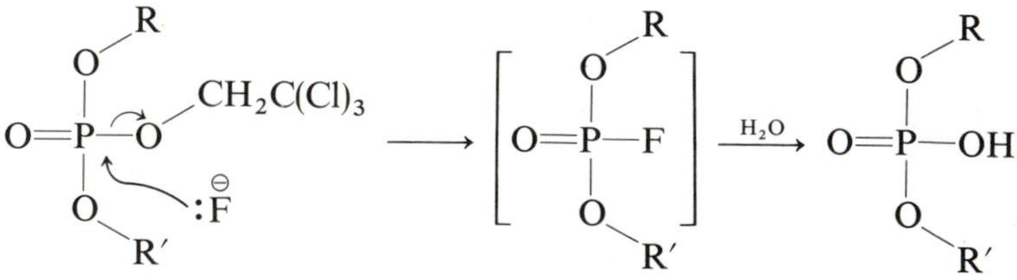

intermediate phosphorofluoridate

Remember, in nonaqueous solvents (e.g., tetrahydrofuran) fluoride is a good nucleophile. The inductive pull of the three chlorine atoms makes the trichloroethyl function the preferred leaving group. The highly reactive phosphorofluoridate intermediate decomposes on chromatographic workup. Cleavage has also been undertaken with sodium hydroxide.

(2) β-Cyanoethyl Function
Barium cyanoethyl phosphate, which is readily available, may be used for the phosphorylation of nucleosides. After product formation, this group may be removed by mild base or fluoride anion. Both bring about a β-elimination:

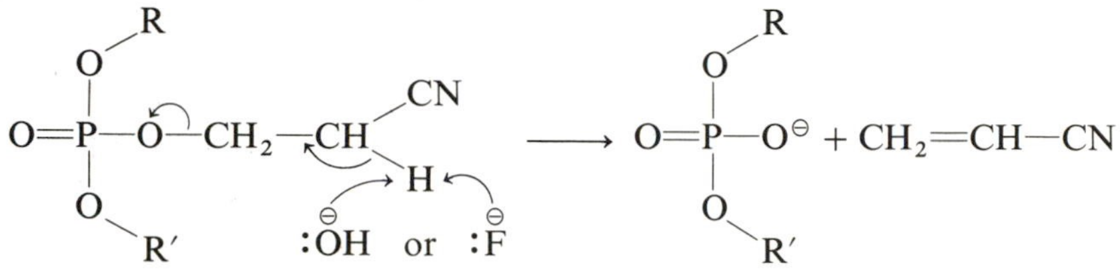

It should be noted that in anhydrous solvent systems, fluoride is both a good base and nucleophile.

(3) Aryl Function
The most popular aryl protecting groups have been introduced into an appropriate phosphorylating agent by the reaction of phenol or o-chlorophenol with a phosphorochloridate:

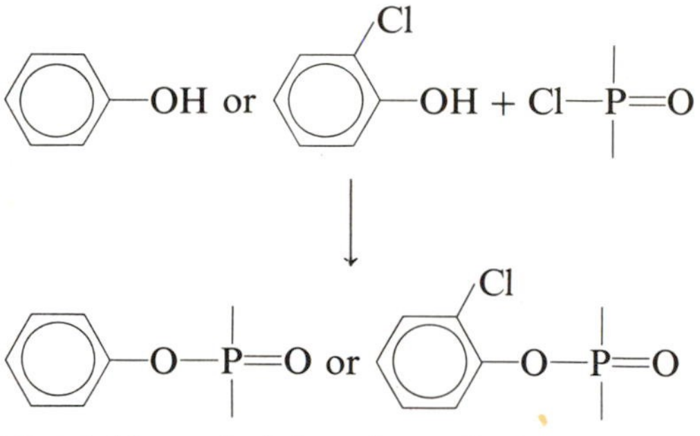

As with the 2,2,2,-trichloroethyl function, displacement of the protecting group may be undertaken with fluoride and sodium hydroxide. Obviously,

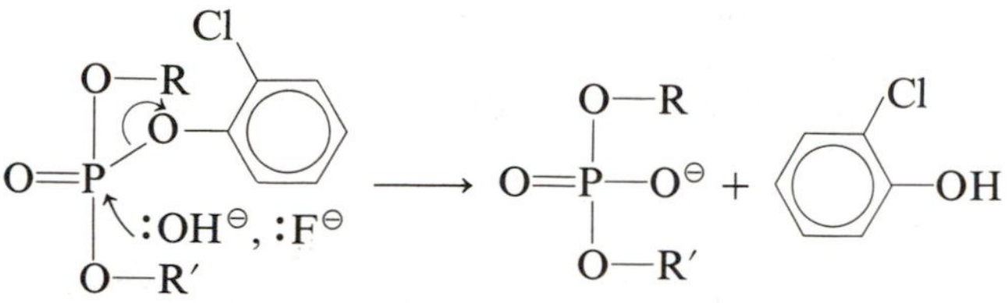

the chlorophenyl function is a better leaving group than phenyl itself. Also, attack by fluoride would again proceed by a phosphorofluoridate intermediate.

An unwanted side reaction of the triester approach has presented itself during polynucleotide synthesis (i.e., beyond the dinucleotide stage): internucleotide cleavage. This has been observed to a significant extent during the removal of aryl protecting groups with either hydroxide or fluoride:

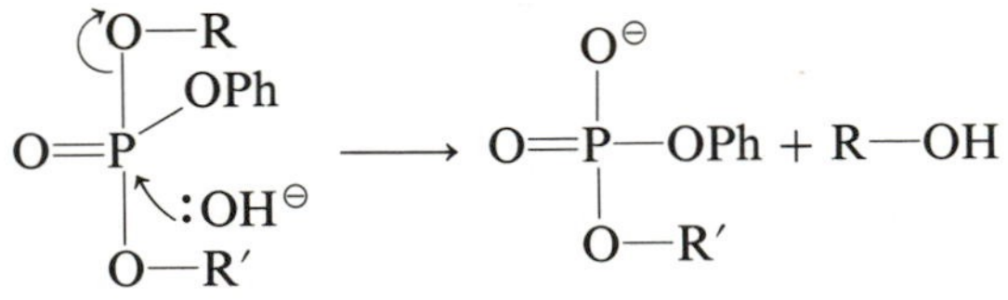

internucleotide cleavage

Use of oximes to remove the protecting groups has been observed to reduce internucleotide cleavage:

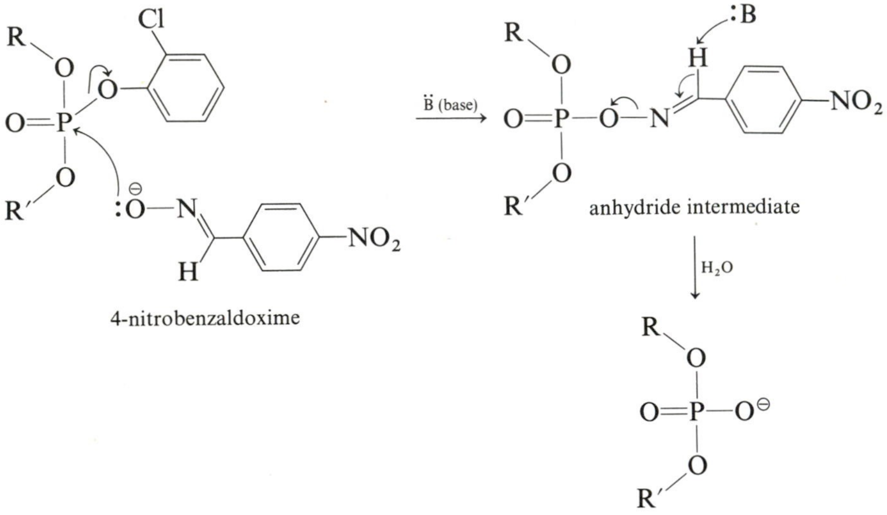

4-nitrobenzaldoxime

anhydride intermediate

The reaction is carried out in aqueous dioxane. Remember, the aldoximes are good nucleophiles, as dictated by the α-effect. Recently, the methyl group has found use as its removal by alkyl fission with good nucleophiles (thiophenoxide, butylamine) also reduces internucleotide cleavage (365).

(4) Anilidate Function

The acid sensitivity of phosphoramidates might suggest that they would be potentially useful as protecting groups. Indeed they have been so used, a well-known example being the anilidate function which may be introduced and removed as follows:

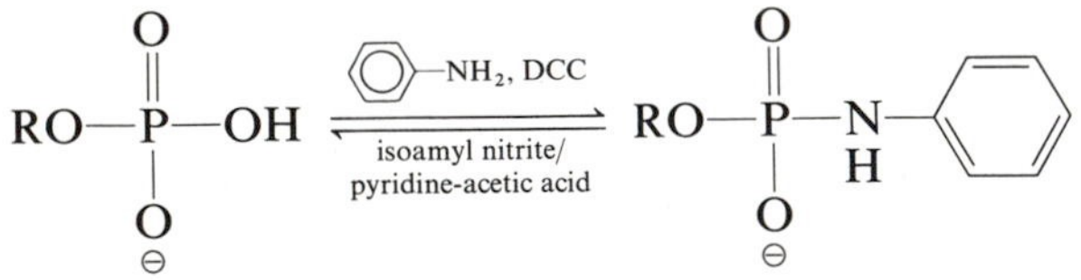

Alternatively, the anilidate function may be introduced as part of an unsymmetrical phosphorylating agent. The synthesis of such phosphoroamidochloridates is becoming of importance as a means to procure isotopically labeled and diastereomeric nucleotides.*

R = H, NO$_2$

phosphorylating agent which can
react with the free hydroxyl of
a nucleotide

3.6.4 Phosphodiester Bond Formation

As has been noted for the synthesis of the peptide bond, cyclic nucleotides and nucleoside triphosphates, synthesis of a phosphodiester requires that work must be done and therefore the expenditure of energy. This is reflected in the free energy of hydrolysis of phosphomono- and diesters, which approximates $-12.6\,(-3.0)$ and -25 kJ/mol $(-6.0$ kcal/mol), respectively. Thus, as with amino acids, two general approaches are possible. The first is to react an activated nucleotide with the appropriate hydroxyl function of a nucleoside to form the phosphodiester linkage. Alternatively, it is possible to react a nucleotide (nucleoside monophosphate) and nucleoside together in the presence of a "coupling agent" which activates the phosphate *in situ*. A similar approach may be used for the synthesis of the nucleotide (nucleoside monophosphate, phosphomonoester linkage) that is to enter into the phosphodiester linkage. Most often, phosphodiester formation is undertaken between a nucleoside 3'-phosphate and a nucleoside with a free 5'-hydroxyl function. This allows phosphorylation of a more reactive primary hydroxyl function.

(1) *Phosphorochloridates*

Analogous to peptide bond formation by reaction of an amine with an acyl chloride is phosphorylation by reaction of a nucleoside with a phosphorochloridate. Perhaps one of the simplest examples of this is reaction of a

---

* The synthesis of the two diastereomers of protected thymidine-3'-phosphorothioate has been reported recently (366). Key to this preparation was the reaction:

R = nucleotide
R' = phosphate protecting group

ribonucleoside with phosphoryl chloride. Under appropriate reaction conditions (trimethyl or triethyl phosphate solvent, 0°C: dioxane-pyridine, room temperature) phosphorylation will occur mostly at the more reactive (less sterically hindered) 5'-hydroxyl function:

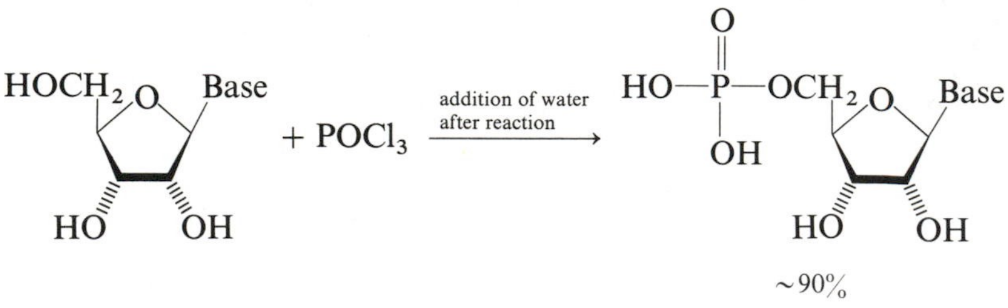

$\sim 90\%$

Of course, by reaction with an appropriately protected nucleoside, it is possible to phosphorylate the secondary hydroxyl function:

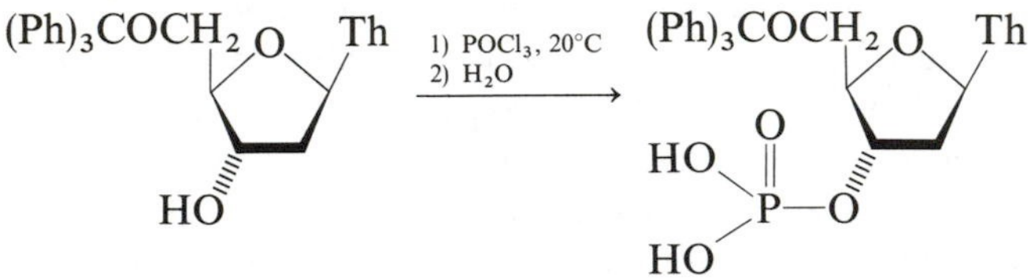

Replacement of the two chlorine atoms with bulkier substituents produces an even more selective phosphorylating agent. For example, under reaction conditions in which phosphorus oxychloride gives product mixtures (pyridine solvent), diphenyl phosphorochloridate and di(2-*tert*-butylphenyl) phosphorochloridate will give approximately 90% and 100% phosphorylation at the primary hydroxyl, respectively:

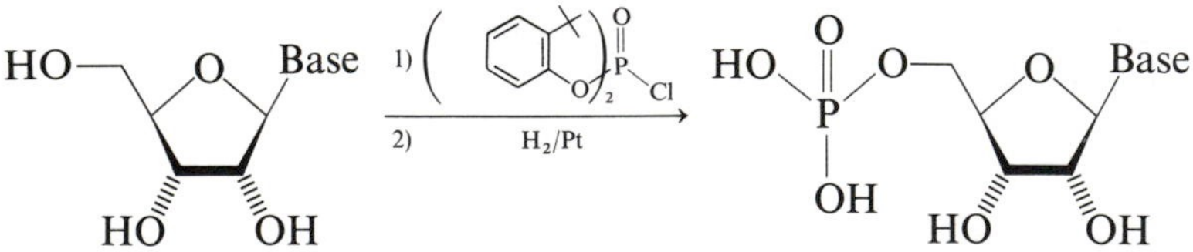

Such specificity for the primary hydroxyl indicates that preparation of desired 3'-phosphomonoesters may be difficult. For example, diphenyl phosphorochloridate will not react with 2',5'-bis-O-methoxytetrahydropyran-4-yl uridine in a pyridine solvent, at room temperature. The free 3'-hydroxyl is too hindered to react with this bulky phosphorylating agent. However, replacement of pyridine with a more potent nucleophilic catalyst, 5-chloro-1-methylimidazole will allow product formation to occur. Other phosphorodichloridates have been examined for the purpose of synthesis of nucleoside monophosphates. Perhaps one which has found the most use for the preparation of nucleoside 3'-phosphates is 2,2,2-trichloroethyl-2-chlorophenyl-phosphorochloridate. While phosphorylation will occur in pyridine, use of 1-methylimidazole provides a quicker reaction rate. This is important, for if the synthesis of an oligonucleotide is to be considered, then a rapid preparation of the monomeric units is important. The phosphorylation of a 5'-protected thymidine is illustrated:

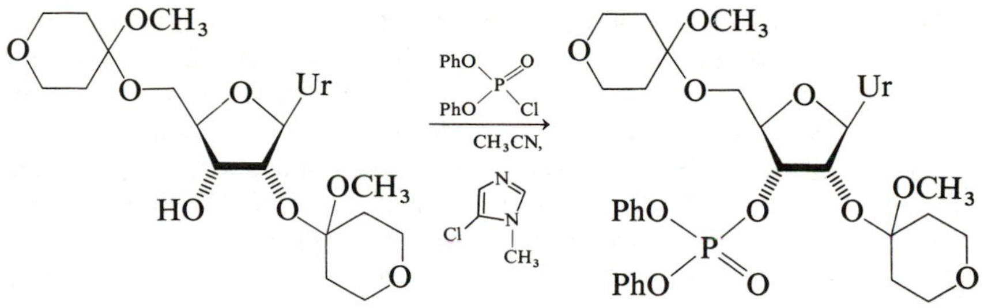

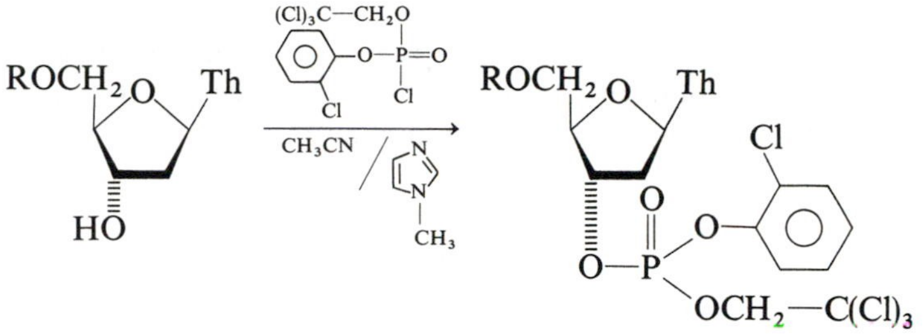

The trichloroethyl function may be selectively removed (zinc) leaving a protected nucleoside 3'-phosphate. This can be converted to a phosphodiester by reaction with an appropriate nucleoside, in the presence of a condensing (coupling) agent:

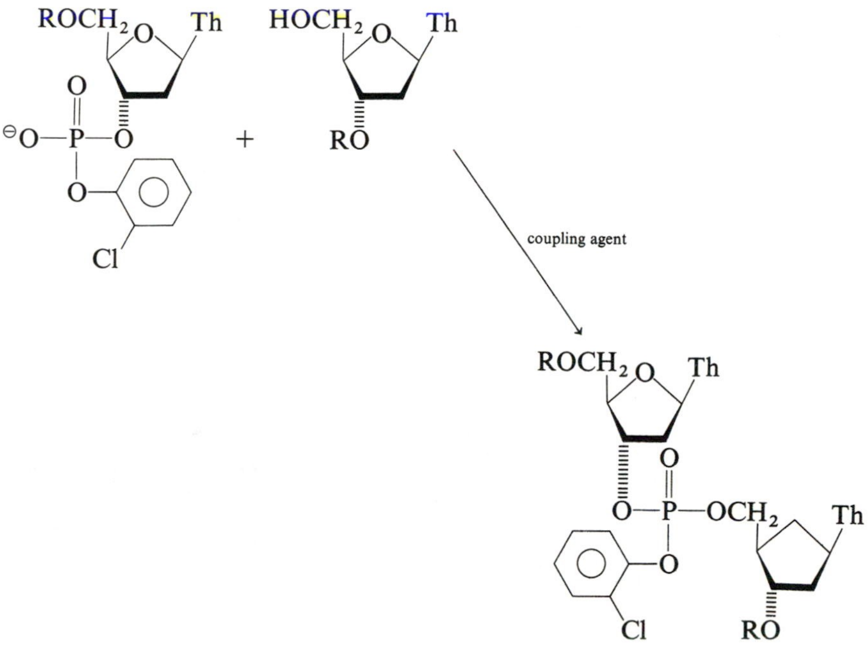

Notice that the actual phosphodiester bond formation did not involve the reaction of a phosphorochloridate with the 5'-hydroxyl of the nucleoside. This approach has been seldom used for diester synthesis as these reactions tend to be sluggish and can be complicated by comparison with other methods. More favorable is the reaction of a nucleoside phosphorochloridite with a second molecule of nucleoside, as shall be discussed shortly. Alternatively, phosphodiester formation may be accomplished via *in situ* activation of a nucleotide with a condensing agent. Some of these reagents are described below.

(2) *Carbodiimides*
Reaction of DCC with organic acids to form an anhydride intermediate has already been discussed in some detail. Much the same mechanism is operative in phosphodiester bond formation. This is illustrated for the synthesis of the dinucleotide thymidylyl-(3' → 5')-thymidine:

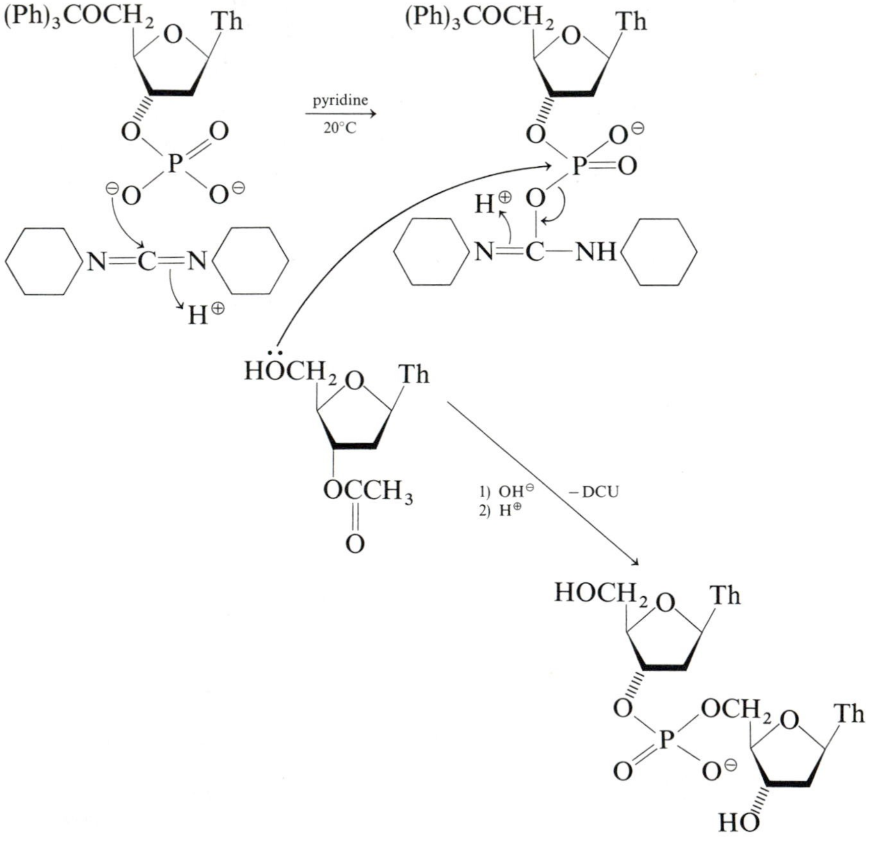

The prerequisite thymidine 3'-phosphate may be synthesized from reaction of a phosphorochloridate (see above), or by the reaction with DCC. Even

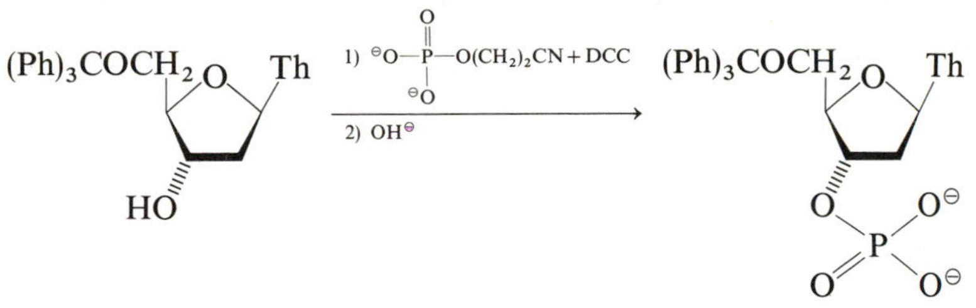

though β-cyanoethyl phosphate is readily available, this preparation is seldom used. In fact, because DCC reactions tend to be slow, suffer numerous side reactions, and can be only used for the phosphodiester approach (the phosphotriester approach cannot be used as a phosphodiester will not react with carbodiimides), they are seldom used presently for the synthesis of polynucleotides. Some possible carbodiimide side reactions, after anhydride formation are illustrated:

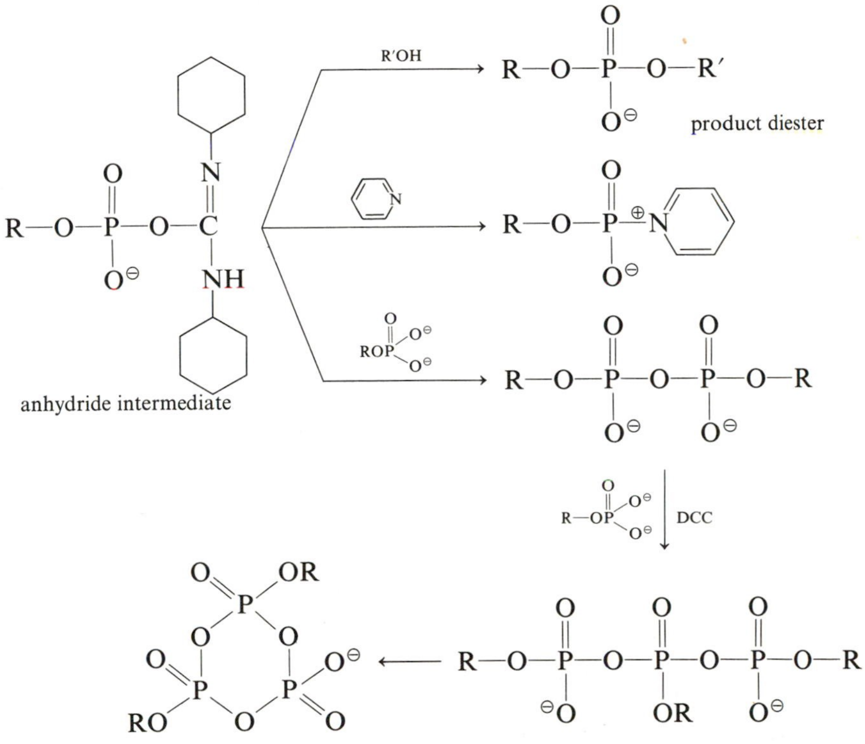

(3) *Aromatic Sulfonyl Chlorides*

These reagents are more effective than carbodiimides for the purpose of phosphodiester bond formation. Sulfonyl chlorides are more reactive, give better yields, and may be used for the phosphotriester approach of poly-nucleotide synthesis. Reaction may be envisaged to proceed by way of an anhydride intermediate:

$$R-O-\overset{\overset{O}{\|}}{\underset{\underset{PG}{O}}{P}}-O^{\ominus} \quad + \quad SO_2-Cl \quad \longrightarrow \quad R-O-\overset{\overset{O}{\|}}{\underset{\underset{PG}{O}}{P}}-O-\overset{\overset{O}{\|}}{\underset{\underset{O}{\|}}{S}}-Ph$$

| nucleoside monophosphate | *p*-toluenesulfonyl |
| PG = protecting group | chloride |

$R'-CH_2\ddot{O}H$

$$R-O-\overset{\overset{O}{\|}}{\underset{\underset{PG}{O}}{P}}-O-CH_2-R' + Ph-SO_3{}^{\ominus}$$

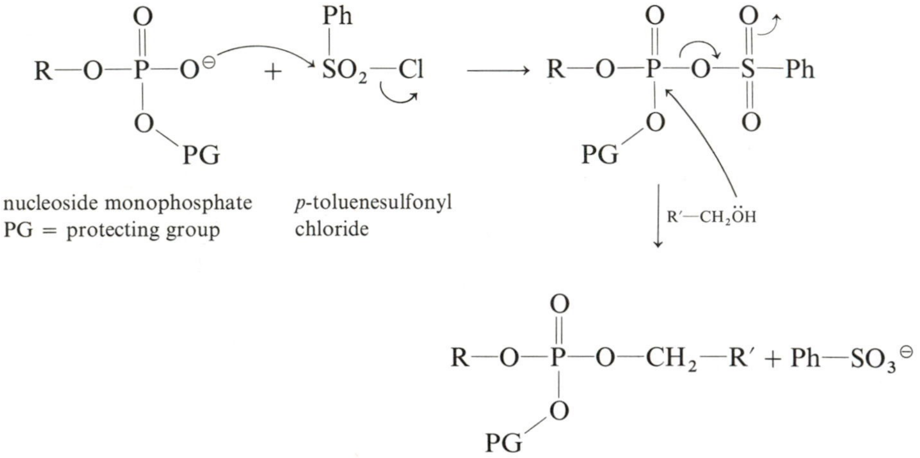

Initially, use was made of *p*-toluenesulfonyl chloride. However, the primary hydroxyl of the nucleoside may attack the phosphorus or sulfur atom of the anhydride intermediate, giving extensive sulfonation. This problem has been reduced by the introduction of steric bulk at the *ortho* positions of the benzene ring, thus favoring attack at the phosphorus atom. The com-monly used derivatives are mesitylenesulfonyl chloride (MS) and 2,4,6-triisopropylbenzenesulfonyl chloride (TPS).

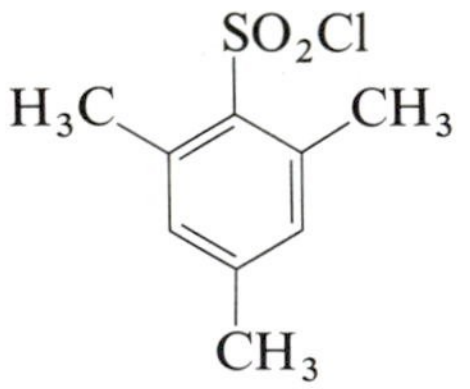

mesitylenesulfonyl chloride (MS) 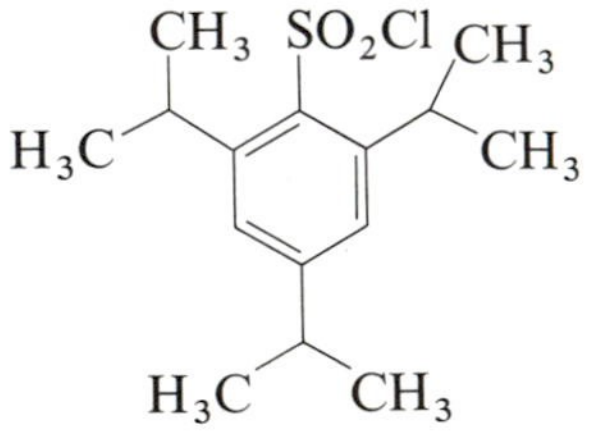   triisopropylbenzene sulfonyl chloride (TPS)

Thus, the first large (gram) scale synthesis of thymidylyl-(3' → 5')-thymi-dine was undertaken via the triester approach, using TPS. Note the steric selectivity of the anhydride intermediate for the primary hydroxyl, so that only 4% of the unwanted (3' → 3')-isomer is obtained.

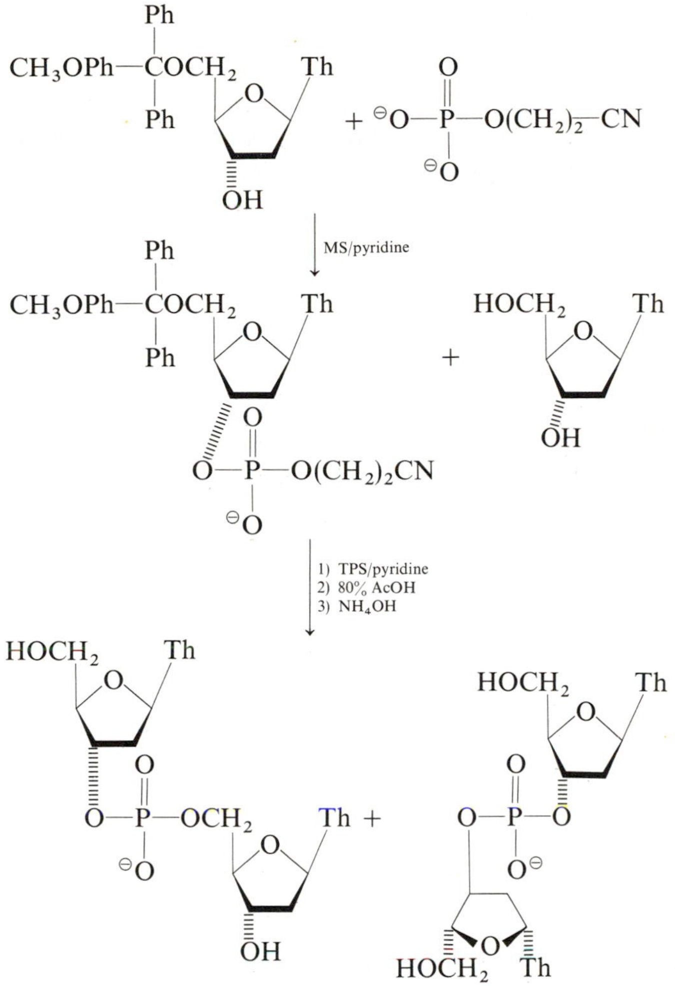

Even with TPS, a small amount of sulfonation is observed. Further, release of hydrogen chloride, even though scavenged by the pyridine solvent, can cause unwanted side reactions, especially if acid sensitive functions are present. Both problems have been solved by replacing the chlorine with imidazole, triazole and tetrazole leaving groups. Thus, the sulfonyl agents available for phosphodiester formation are as follows:

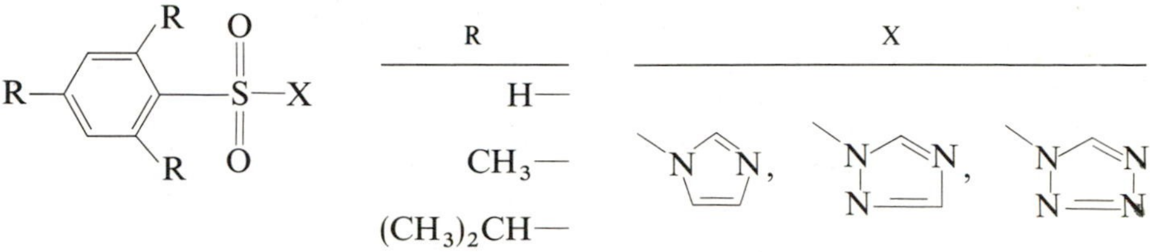

The rate of condensation follows the general order: imidazole < triazole < tetrazole.*

(4) 2,2,2-Trichloroethylphosphorodichloridite

Most recent to the technology of polynucleotide synthesis is the use of phosphorodichloridites (40, 41). This method makes use of an activated phosphorylating agent and thus does not require a condensing agent. However, the phosphorylating agent is not a phosphate, but instead a more reactive phosphite. This greater reactivity allows a one-step *in situ* conversion of a nucleoside with a free 3′-hydroxyl to a phosphomonoester, then to a phosphodiester by reaction with a second nucleoside with a free 5′-hydroxyl, and finally to a phosphate by a rapid oxidation with iodine. In fact, so great is the reactivity of the phosphorochloridite, that both phosphorylations are carried out at subzero temperatures. The entire operation is performed in less than one day. This is important if serious consideration is to be given to the synthesis of a polynucleotide. The one-step synthesis of protected thymidylyl-(3′ → 5′)-thymidine is illustrated:

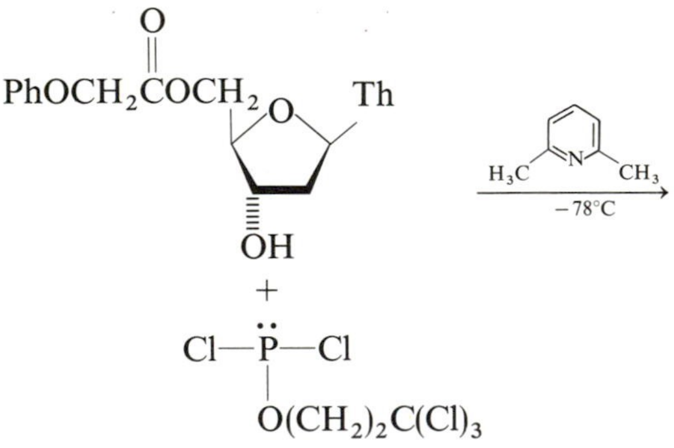

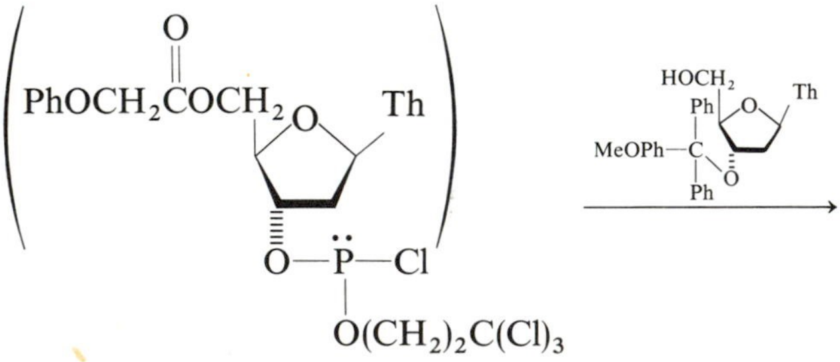

intermediate; not isolated

(continued on p. 167)

---

* Recently, it was reported that sulfonyl triazoles can give unwanted side reactions with protected guanosine and uridine (363).

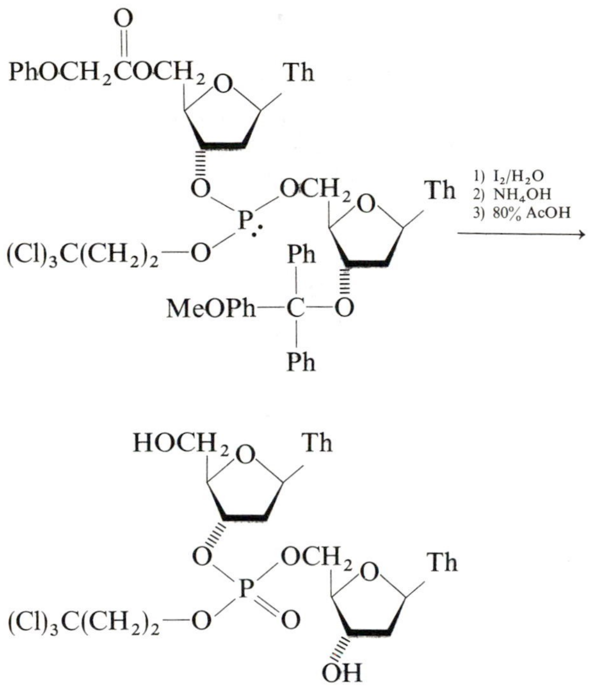

$PhOCH_2COCH_2$—O   Th

$(Cl)_3C(CH_2)_2$—O $\overset{..}{P}$ OCH_2—O   Th
1) I_2/H_2O
2) NH_4OH
3) 80% AcOH

Ph

MeOPh—C—O

Ph

$HOCH_2$—O   Th

$(Cl)_3C(CH_2)_2$—O   P OCH_2—O   Th

OH

The reaction is accomplished by adding a tetrahydrofuran solution of the first nucleoside to a lutidine solution of the phosphorylating agent, allowing the two to react for a short time and then adding a tetrahydrofuran solution of the second nucleoside. Only after completion of the condensation is the oxidation undertaken by the addition of iodine crystals dissolved in water-tetrahydrofuran. This oxidation is also very quick. Of course this order of reagent addition will lead to formation of some unwanted $(3' \rightarrow 3')$- and $(5' \rightarrow 5')$- isomer formation. For example, if two molecules of the first nucleoside react with the phosphorodichloridite before addition of the second nucleoside to the reaction mixture, than $(3' \rightarrow 3')$-isomer formation will occur:

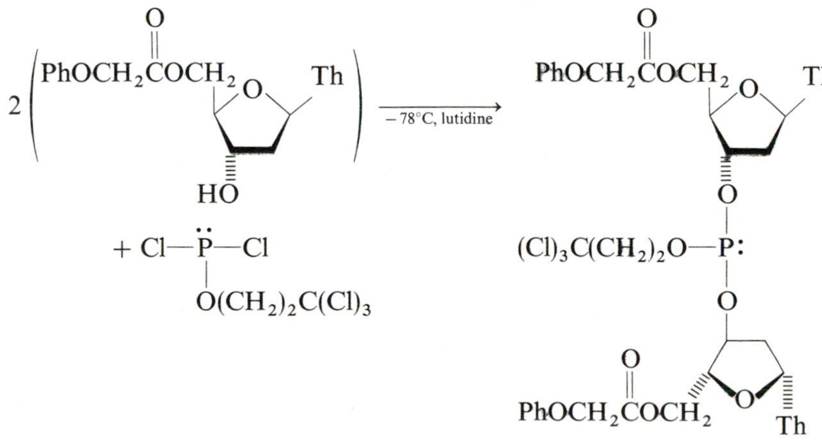

$2\left(PhOCH_2COCH_2\text{—O} \quad Th \atop HO \right)$ $\xrightarrow[-78°C, \text{ lutidine}]{}$ $PhOCH_2COCH_2$—O   Th

$+ Cl\text{—}\overset{..}{P}\text{—}Cl$
$O(CH_2)_2C(Cl)_3$

$(Cl)_3C(CH_2)_2O$—$\overset{..}{P}$:

O

$PhOCH_2COCH_2$ O   Th

These isomers are readily separated from the desired triester product by silica gel chromatography. A similar rapid synthesis of ribonucleotides is also possible, as is illustrated for the synthesis of uridylyl-(3′ → 5′)-uridine. Here, both the trichloroethyl and silyl protecting functions may be removed simultaneously by dissolving the triester product in a solution of tetrabutylammonium fluoride in tetrahydrofuran.

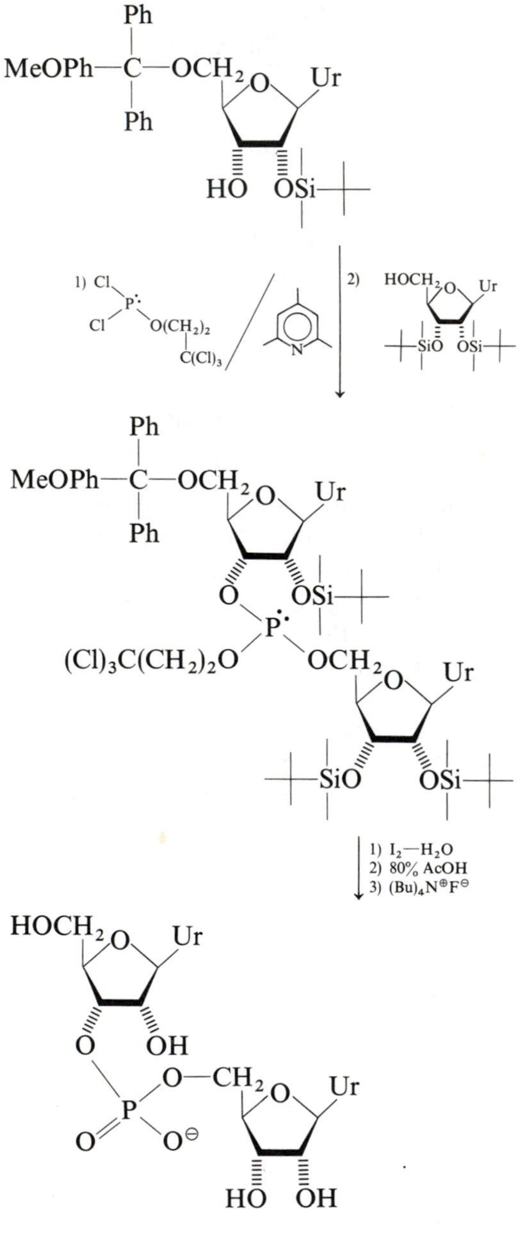

$$
\text{Ph}
$$

MeOPh—C—OCH$_2$—O—Ur

Ph

HO   OSi⟨

1) Cl–P(Cl)–O(CH$_2$)$_2$C(Cl)$_3$ / (lutidine)   2) HOCH$_2$—O—Ur, ⟩SiO OSi⟨

Ph

MeOPh—C—OCH$_2$—O—Ur

Ph

O   OSi⟨

(Cl)$_3$C(CH$_2$)$_2$O—P—OCH$_2$—O—Ur

⟩SiO   OSi⟨

1) I$_2$—H$_2$O
2) 80% AcOH
3) (Bu)$_4$N$^{\oplus}$F$^{\ominus}$

HOCH$_2$—O—Ur

O   OH

O—CH$_2$—O—Ur

O=P—O$^{\ominus}$

HO   OH

(5) *Solid Phase Synthesis*
The solid phase synthesis of oligonucleotides has not reached the level of sophistication as that for peptides. The design of an appropriate solid support has been difficult with such problems as undesirable polymer characteristics and irreversible adsorption of reagents onto the support. Only recently* have appropriate polymer matrices (e.g., HPLC silica) been described that appear to have potential for the synthesis of nucleotide polymers. In fact, a major impetus for the development of solid phase methodology for the synthesis of polydeoxyribonucleotide (DNA) sequences has been the industrial application of the so-called "biotechnology" or recombinant DNA technology.

3.7 Chemical Evolution of Biopolymers

Proteins (amino acid polymers) and nucleic acids (nucleotide polymers) are the essence of life. Enzymes are proteins that catalyze the chemical reactions necessary for life, while nucleic acids serve as a data bank for the storage of genetic information in the cell's nucleus. To end this chapter we will take a brief excursion into the origin of these biopolymers. For this we will ask the following fundamental questions. How did the chemical processes necessary for the maintenance of life begin? In other words, how did peptide bonds form during prebiotic life? What is the origin of macromolecules of biological interest? What is the origin of asymmetry and chirality in organic molecules? Some of these questions have been answered in part by chemists who have tried to duplicate the conditions of the primitive atmosphere prevailing on Earth at that time.

3.7.1 Prebiotic Origin of Organic Molecules

Chemical evolution started some 4.6 ± 0.1 billion years ago, and this process alone, as opposed to *biological evolution*, took about 1.5 billion years (42). We are interested here by the period of chemical evolution where organized organic molecules were formed and eventually "transformed" into living matter. Most scientists now believe that there were four stages in the evolution of life. The formation of small molecules (amino acids, nucleotides, sugars) occurred initially. These building blocks then formed macromolecules such as proteins and nucleic acids. The third stage involved the development

---

* The first of a series of papers from M. H. Caruthers' group (367) at the University of Colorado appeared describing the potential of silica gel as a support for solid phase synthesis.

of a cell-like structure that could reproduce itself. In the final step, this primitive cell evolved into a modern cell containing the genetic mechanism for protein synthesis.

It is believed that the prebiotic or primitive atmosphere on Earth at the time of the origin of life was strictly reductive; no oxygen was present. Oxygen appeared much later, largely as the product of photosynthesis by green plants (42). This reducing atmosphere contained such gases as CH_4, NH_3, N_2, CO, CO_2, H_2, and water. Much evidence now suggests that reactions between these molecules and inorganic components were activated by the energy of ultraviolet light, electric discharges, heat, radiation, and other forms of energy such as shock waves.

This sequence of events was first postulated in 1924 by a Russian chemist, A. I. Oparin, and a British chemist, J. B. S. Haldane. As chemists, we would like to see if it can be proved experimentally. In the 1950s, S. Miller, then at the University of Chicago, constructed an apparatus with the intention of proving this suggestion. His apparatus was designed to simulate the primitive atmosphere prevailing on Earth 4 billion years ago using an electric arc as energy source. He obtained a limited number of components of which 50% were of biochemical interest and about 1% of a racemic mixture of the following three amino acids: glycine, alanine, and aspartic acid. Today, seventeen different amino acids can be made by extension of Miller's original conditions. A few simple reactions are as follows:

$$CH_4 + H_2O \xrightarrow{h\nu} \ \ \underset{H}{\overset{H}{>}}C{=}O + 2\,H_2$$

$$CH_4 + NH_3 \xrightarrow{h\nu} H{-}C{\equiv}N + 3\,H_2$$

Hydrogen cyanide and formaldehyde are in turn precursors of more reactive compounds such as aminoacetonitrile and aminoacetamide. A combination of these molecules can lead to the formation of amino acids, purines, pyrimidines, and sugars. For example:

$$\underset{H}{\overset{H}{>}}C{=}O + NH_3 + HCN \xrightarrow{\Delta} H_2N{-}CH_2{-}CN + H_2O$$

aminoacetonitrile

$$\Big\downarrow H_2O$$

$$\text{poly-Gly} \xleftarrow{\Delta} H_3\overset{\oplus}{N}{-}CH_2{-}COO^{\ominus} \xleftarrow{H_2O} H_2N{-}CH_2{-}C\overset{\nearrow O}{\underset{\searrow NH_2}{}}$$

aminoacetamide

This transformation is reminiscent of the well-known Strecker amino acid synthesis:

$$R-CHO + NH_3 + HCN \longrightarrow R-\underset{\underset{NH_2}{|}}{C}H-CN + H_2O$$

The hydrolysis of the nitrile yields the corresponding amino acid.

J. Oro in 1960 first demonstrated the synthesis of a nucleic acid component (43, 44). By refluxing a concentrated solution of ammonium cyanide he obtained adenine in low yield. Later it was found that UV irradiation of a dilute HCN solution produced adenine and guanine. 4-Amino-5-cyano-imidazole and 4-aminoimidazole-5-carboxamide have been shown to be important intermediates in these reactions. Furthermore, NH_3 and HCN could be converted to formamidine ($H_2NCH=NH$).

An oversimplified pathway for the synthesis of a purine, adenine, from HCN and NH_3 is:

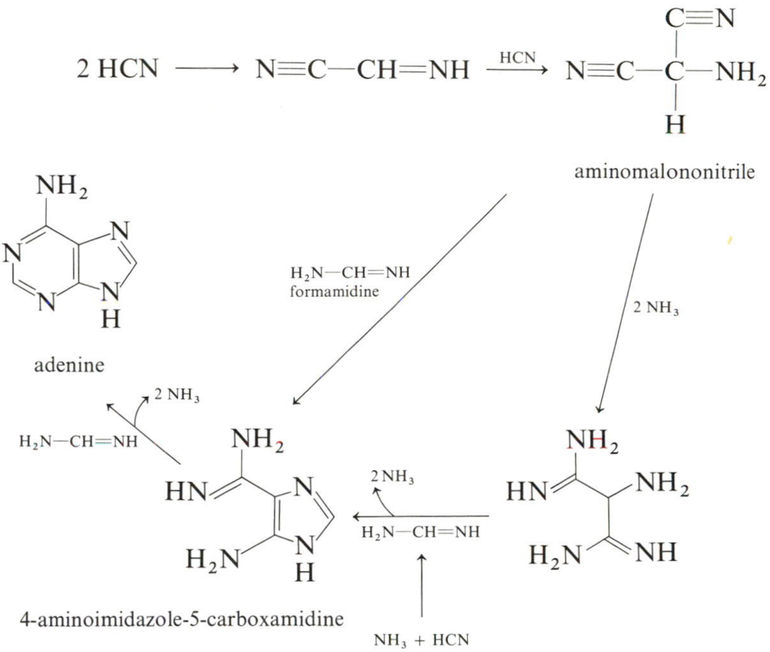

$$2\ HCN \longrightarrow N\equiv C-CH=NH \xrightarrow{HCN} N\equiv C-\underset{\underset{H}{|}}{\overset{\overset{C\equiv N}{|}}{C}}-NH_2$$

aminomalononitrile

adenine

$H_2N-CH=NH$ formamidine

2 NH_3

4-aminoimidazole-5-carboxamidine

2 NH_3

$H_2N-CH=NH$

$NH_3 + HCN$

In 1965, L. E. Orgel and co-workers from the Salk Institute found that in a sealed tube this transformation proceeds in 40% yield (44). Adenine could have been the first purine formed on Earth, and it is interesting to realize that this component is present in ATP and coenzyme A. Both are essential mediators of many biochemical reactions.

Consequently, the role of HCN in the origin of life is more important than previously thought and a second process called *oligomerization* (45) of HCN has also been postulated to occur in aqueous solution:

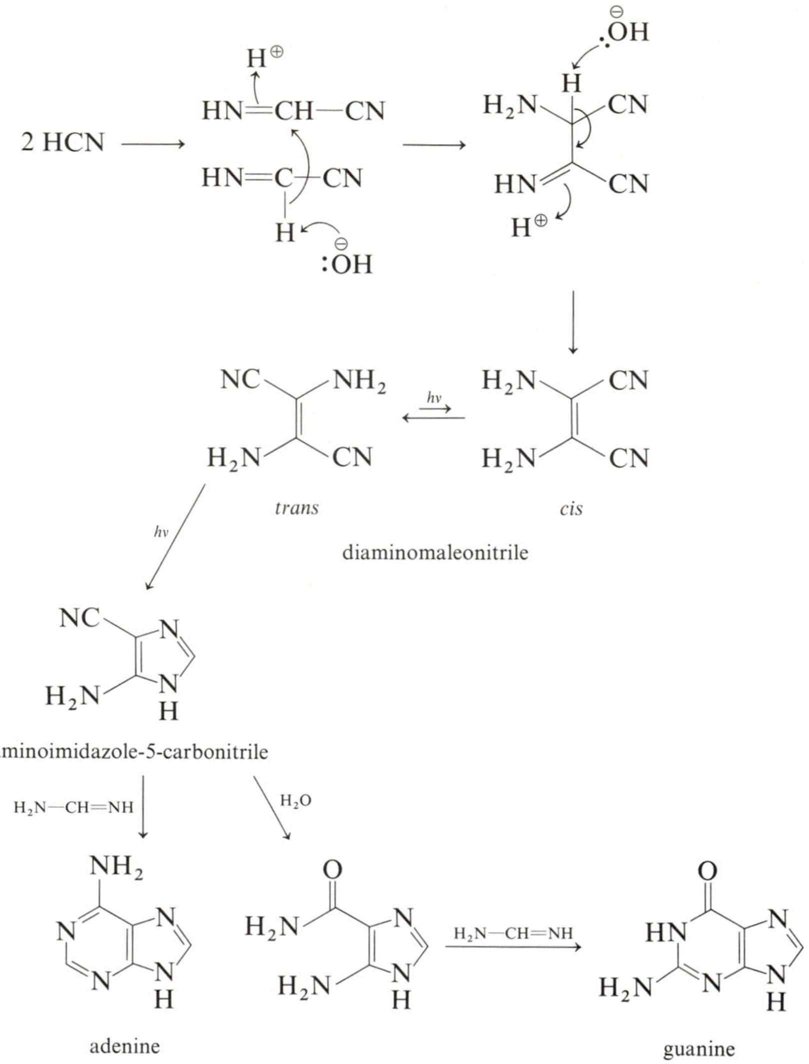

adenine   guanine

The above transformations show that HCN is converted to a tetramer which is further transformed photochemically to an imidazole ring system. In absence of oxygen, the reaction proceeds in essentially quantitative yield in dilute aqueous solution.

Hydrogen cyanide can also be converted to cyanoacetylene and hydrogen cyanate, both precursors of pyrimidines. These reactions were reproduced in the laboratory. In fact, in 1828, F. Wöhler made urea from hydrogen cyanate and ammonia, the first synthesis of an "animal substance" from inorganic materials. Very likely, all these processes occurred primarily in an aqueous environment where H^+ and OH^- ions acted as specific-acid or specific-base catalysts. It is particularly impressive that the three major classes of nitrogen containing biomolecules, purines, pyrimidines, and amino acids are formed by the hydrolysis of the oligomers formed directly from

$$HC\equiv C-C\equiv N \xrightarrow[\Delta]{CNO^{\ominus}} OCN-CH=CH-CN$$

cyanoacetylene

uracil   cytosine

dilute aqueous solution of HCN. A primitive Earth scenario for the synthesis of all these biomolecules would be the constant formation of HCN by the action of electric discharges and ultraviolet light on the atmosphere. The HCN would be dissolved in rain droplets and carried to the Earth's surface where it would oligomerize and then slowly hydrolyze to yield the bio-molecules which were incorporated into the primitive forms of life (369).

Knowledge of the reaction conditions on primitive Earth is essential to carry out laboratory investigations, and neither proteins nor nucleic acids form spontaneously in aqueous solution (47). Sugars must have been formed by the self condensation of formaldehyde, another possible precursor of life, via the photolysis of water in the presence of CH_4.

In summary, the prebiotic condensation of small molecules such as NH_3, H_2O, HCN, HCHO, and $HC\equiv C-CN$ provided the building blocks for the synthesis of polyamino acids or proteins and polynucleotides or nucleic acids. L. E. Orgel contends that living organisms are in their present state now because of this continuity in building block synthesis which prevailed at the origin of primitive Earth. He was also able to show that polyphos-phates, necessary for the synthesis of polynucleotides, can be formed by simply warming orthophosphate with urea and ammonium ions (44). Modern radio-frequency spectroscopes have revealed that most of these small mole-cules also occur in interstellar clouds and hence makes these suggestions all the more plausible.

Well before life arrived on Earth, however, a self-replicating process was necessary. How? It seems reasonable to suppose that specific nucleic acid–nucleic acid and nucleic acid–protein interactions were of fundamental im-portance for the replication of nucleic acids and the evolution of the genetic code (48). Such recognition processes are dependent on base sequence and amino acid sequence. According to R. D. MacElroy from NASA, these would probably play a key role in the formation of protein nucleic acid complexes, and may have been of fundamental importance in the early stages of macro-molecular evolution (49).

Only recently, G. L. Nelsestuen, at the University of Minnesota, proposed a bioorganic model which can account for a "code" which could describe the specific interaction of polynucleotides and proteins (50). He postulated the existence of a primitive hybrid polymer, or copolymer, composed of a ribonucleic acid chain (RNA) with amino acids attached covalently to the 2'-position of the ribose sugar ring. Such an organized "template" would be responsible for specific polypeptide–polynucleotide recognitions that might be at the origin of the present genetic code.

3.7.2 The Concept of a Dissymmetric World (46)

The amino acids obtained from Miller's experiments were all racemic. Furthermore, only racemic amino acids were found on the Murchison meteorite which fell in Australia in 1969. This presents an interesting question: How was optical activity introduced in complex molecules? Among the various hypothesis, S. Akabori (46) proposed the following transforma-tions for the synthesis of more complex polypeptides, which have been veri-

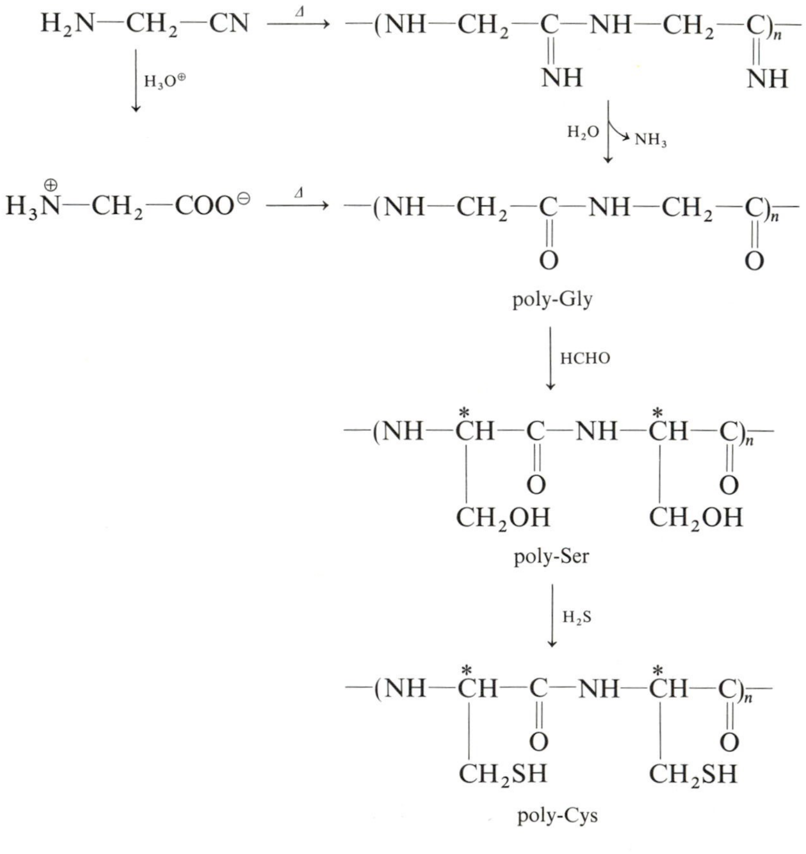

$$H_2N-CH_2-CN \xrightarrow{\Delta} -(NH-CH_2-\underset{\overset{\|}{NH}}{C}-NH-CH_2-\underset{\overset{\|}{NH}}{C})_n-$$

$$\downarrow H_3O^{\oplus}$$

$$\xrightarrow{H_2O} NH_3$$

$$\overset{\oplus}{H_3N}-CH_2-COO^{\ominus} \xrightarrow{\Delta} -(NH-CH_2-\underset{\overset{\|}{O}}{C}-NH-CH_2-\underset{\overset{\|}{O}}{C})_n-$$

poly-Gly

$$\downarrow HCHO$$

$$-(NH-\overset{*}{C}H-\underset{\overset{\|}{O}}{\underset{\overset{|}{CH_2OH}}{C}}-NH-\overset{*}{C}H-\underset{\overset{\|}{O}}{\underset{\overset{|}{CH_2OH}}{C}})_n-$$

poly-Ser

$$\downarrow H_2S$$

$$-(NH-\overset{*}{C}H-\underset{\overset{\|}{O}}{\underset{\overset{|}{CH_2SH}}{C}}-NH-\overset{*}{C}H-\underset{\overset{\|}{O}}{\underset{\overset{|}{CH_2SH}}{C}})_n-$$

poly-Cys

fied experimentally. It is important to note that cyanamide (NH_2—$C\equiv N$), a tautomeric form of carbodiimide (HN=C=NH), could have acted as a primitive condensing agent in the early synthesis of polypeptides. Most likely, the formation of very reactive molecules were necessary to bring about condensation reactions on the primitive Earth.

These processes resulted in racemic mixtures. However, the resolution of this mixture is believed to have occurred by spontaneous crystallization. This process most likely occurred by chance. Minerals such as natural dissymmetric quartz crystals and metal ions may have played a crucial role of optical selection by selective chelation of only one stereoisomer. After all, stereoselective polymerization of olefins by metal surfaces (Ziegler–Natta catalysts) is a well-documented industrial process for the synthesis of isotactic* polymers. We also know the importance of metal ion binding in many biochemical transformations. It is essential for the maintenance of the native structure of nucleic acids and numerous proteins and enzymes. Other physical forces through radioactive elements, γ-radiation, or from cosmic rays, may have also been involved in optical selection. For instance, recent experiments with strontium-90 indicate that D-tyrosine is destroyed more rapidly than the naturally occurring L-isomer. It is tempting to incorporate such factors into the origin of dissymmetry in life process (46).

Optical activity is considered essential for protein chain folding and life would probably not be possible without dissymmetric molecules. Molecules lacking reflective symmetry (no plane of symmetry) are designated as *dissymmetric*. It should be noted that asymmetric molecules (possessing only a C_1 axis element of symmetry) are a special class of dissymmetric molecules, but not all dissymmetric molecules will be asymmetric. *Stereoisomers* possess the same molecular bonding skeleton but differ in the absolute arrangement of the atoms in space. They are optically active and may be characterized as *chiral*. Stereoisomers that are related as object and nonsuperimposable mirror image are termed *enantiomers*. Those stereoisomers not so related are called *diastereoisomers*.

3.7.2.1 Template-Directed Synthesis

Is it possible to prepare, by a chemical reaction, a polynucleotide and mimic the synthesis of DNA? Indeed, such a model can be developed from "primary" templates. The preformed nucleic acid (template) can then be used to direct the synthesis of complementary molecules by using the well-established concept of Watson–Crick base pairing (see Chapter 4). Orgel and his colleagues (44) have demonstrated that polyuridine [poly(U)] can act as a template in aqueous solution to help the condensation of a 5′-activated nucleotide (AMP), in the presence of a water-soluble carbodiimide via

---

* An isotactic polymer structure occurs when the site of steric isomerism in each repeating unit in the polymer chain has the same configuration. This is illustrated in Chapter 4, p. 191.

specific complementary Watson–Crick pairs. This template directed synthesis has no effect on GMP. Most surprising is the finding that the complementary polynucleotide formed under these conditions has only 2′,5′-phosphodiester linkages whereas the naturally occurring polynucleotides all have 3′,5′-phosphodiester linkages. If a deoxy-2′ nucleotide analogue is used to avoid the nonnatural bond, 5′,5′-linkages are obtained instead, and only under special conditions can natural 3′,5′-bonds be formed.

If the template is made of a pyrimidine [poly(C)] doped with a small amount of a purine (like A), so that some regions have an alternation of a purine and a pyrimidine, the template will be a more efficient condensing agent of the complementary bases (adding pG and pGU dimers). Therefore, polynucleotides can lead to very specific complexing. This observation leads to interesting speculation for the origin of the genetic code. The new com-

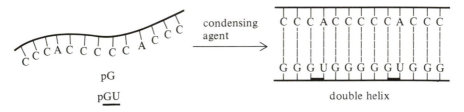

plementary chain is predominantly 2′,5′-linked, while the enzymatic conversion with RNA polymerase yields exclusively the 3′,5′-isomers.

This problem of 2′,5′- vs 3′,5′-linkages was elegantly solved by D. A. Usher (51). A dodecanucleotide, $(pA)_{12}$, with only 3′,5′-bonds was prepared by enzymatic degradation of a larger poly(A) molecule. A second oligonucleotide, $(pA)_6$-$(pA)_6$, was prepared chemically where only the central phosphodiester bond was 2′,5′-linked. In solution, both polymers have similar stability toward alkaline hydrolysis. However, if poly(U) is added to either of them, Usher found that the polymer having a 2′,5′-bond is rapidly hydrolyzed to give two $(pA)_6$ units. From this observation Usher estimated that when an organized double helix is formed, the 2′,5′-bond is 700 times less stable than the naturally occurring 3′,5′-bond.

It also offers the interesting speculation that a natural selective molecular environment could have favored 3′,5′-linkages during the process of "maturation" of DNA molecules. Later on, the arrival of an enzyme could have accelerated this process and eventually removed the ambiguity by converting ribose to deoxy-2′-ribose. Then the double-helix formation of DNA, as we know it now, would have been favored since only the most stable 3′,5′-linkage remained—an example of Darwinism at the molecular level. These observations also help to clarify the question of which came first, DNA or RNA? The exact chemical reasons why 3′,5′-bonds are favored over 2′,5′ can be understood from the discussion developed in Section 3.3.

Finally, we can also ask why ribose is the only sugar found in polynucleotides? No other sugars will allow such an efficient condensation of corresponding nucleotides and a deoxy-3′ nucleotide analogue does not condense. Obviously both 2′-OH and 3′-OH groups are needed for the polymerization to proceed. The reason being that a hydrogen bond between both groups, which are *cis* only in the ribose series, will increase the acidity of the OH group at the 2′-position. Another important observation is that

$$RO—\quad O\quad Base$$

the polynucleotide condensation is very specific for nucleotides having a ribose residue of the D-series. Orgel has provided an experimental demonstration of this. If a mixture of L- and D-ribonucleotides are added to poly(U) of the D-series, only the corresponding D-ribonucleotides can condense and form a complementary double helix. The reason lies in the fact than for helix formation, only a second strand of the *same* specificity can be compatible. This observation is important in connection with the evolution of optical activity in biological systems.

3.7.2.2 Later Steps in Chemical Evolution

The chance association of a number of prebiotically formed macromolecular structures could have yielded a favorable situation of enhanced survival value. As soon as a single asymmetric molecule was involved in a reaction, steric factors may have become important. Asymmetric synthesis was then introduced and molecular selection led to molecular evolution of more complex structures. Short polypeptides could have acted as the first primitive catalysts. With the arrival of long polypeptide chains, three-dimensional globular conformations, helped by both hydrophobic and electrostatic interactions of molecular components, were favored and later evolved into enzymes. Such macromolecular networks acquired *molecular information* necessary for self-replication. This is the minimum requirement for life and the development of primitive metabolisms. We can also imagine than favorable nonpolar interaction of lipids and fatty acids produced *micelle-like* aggregates (Section 5.2 for definition) which were eventually transformed into membranes to produce primitive cells. Oparin, in his theory of the origin of life on Earth, postulated that the association of fundamental chemical structures generated polymeric microspheres called *coacervates.* These charged droplets played an important role in the origin of life according to him.

More recently, the American biochemist S. W. Fox, University of Miami, has described experimental conditions in which the thermal condensation of amino acid mixtures gives polymeric substances. These mixtures of poly-peptides in salt water form *proteinoid* microspheres and exhibit many aspects of cell-like behavior in the presence of ATP. Indeed, the droplets of Oparin and Fox behave like thermodynamically open systems. This is one of the fundamental properties of living matter.

Many more questions have to be answered in the laboratory and for this, team efforts from polymer chemists, organic chemists, biochemists, and biologists will be necessary to remove some of the "mystique" still underlying the origin of life on Earth.

Chapter 4
Enzyme Chemistry

*"Imagination and shrewd guesswork are powerful
instruments for acquiring scientific knowledge quickly
and inexpensively."*

van't Hoff

An *enzyme* is characterized by having both a high degree of specificity and a
high efficiency of reaction. The factors involved in enzyme-catalyzed reactions
are the main subject of this chapter.

For this, hydrolytic enzymes will be used to present the concept of the *active
site*. However, an introduction to the general concepts of catalysis, which are
based on *transition state theory* is needed first. Proximity and orientation of
chemical groups will also be illustrated as factors responsible for the magnitude of enzyme catalysis. This will eventually allow the bridging of nonenzymatic heterogeneous catalysis and enzymatic catalysis.

4.1 Introduction to Catalysis

It was noted earlier that enzymes are proteins that function as catalysts for
biological reactions. When it is remembered that body temperature is 37°C
and that many organic reactions occur at temperatures well above this, the
need for these catalysts becomes apparent. It becomes of interest to understand how these proteins perform their catalytic function. The exact mechanism of enzyme action is a fundamental problem for the bioorganic chemist.
Most of the "action" occurs on the surface of the protein catalyst at an area
designated as the *active site* where chemical transformations follow the basic

principles of organic and physical chemistry. Several parameters operate simultaneously which must be sorted and examined individually by such techniques as model building. However, an appreciation of the catalytic conversion of reactant (substrate) to product first requires a basic understanding of the phenomena of *catalysis*. The word *substrate* is commonly referred to the chemical reactant whose reaction is catalyzed by an enzyme.

By definition, a *catalyst* is any substance that alters the speed of a chemical reaction without itself undergoing change. This is true of the enzymes, for they are of the same form (i.e., conformation and chemical integrity) before and after the catalytic reaction. This is mandatory for catalytic efficiency. If the enzyme were altered after the chemical reaction with a first molecule of substrate, then it would not be able to interact with a second substrate molecule. Of course, as catalysts, enzymes need only be present in small amounts.

A catalyst may either increase or decrease the velocity of a chemical reaction. However, in current usage, a catalyst is a substance that increases the reaction velocity; a substance that decreases the rate of a reaction is called an *inhibitor*. This definition also implies that a catalyst is not consumed during the course of the reaction, but serves repeatedly to assist molecules to react. In biochemistry many enzyme-catalyzed reactions require other substances which may also properly be called a catalyst, but which are consumed (or modified) in the course of the reaction they catalyze. They are the *coenzymes* and are often restored to their original form by a subsequent reaction, so that in the larger context the coenzymes are unchanged. The detailed chemistry of these substances will appear in Chapter 7.

In M. L. Bender's view, the function of a catalyst is to provide a new reaction pathway in which the rate-determining (slowest) step has a lower free energy of activation than the rate-determining step of the uncatalyzed reaction. Furthermore, all transition state energies in the catalyzed pathway are lower than the highest one of the uncatalyzed pathway. Figure 4.1 illustrates this with a free-energy diagram of an exothermic reaction.

According to transition state theory, the processes by which the reagents collide are ignored. The only physical entities considered are the reagents, or

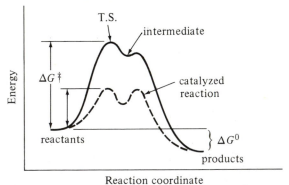

Fig. 4.1. Hypothetical free-energy diagram of a reaction.

ground states, and the most unstable species on the reaction pathway, the *transition state* (T.S.).

The importance of transition state theory is that it relates the rate of a reaction to the difference in Gibbs free energy ($\Delta G^{\ddagger}$) between the transition state and the ground state. This theory may be used quantitatively in enzymatic reactions to analyze structure reactivity and specificity relationships involving discrete changes in the structure of the substrate.

Catalysts may be classified as *heterogenous*, or not in solution. A typical example is the hydrogenation of ethylene in the presence of a metal catalyst such as palladium, platinum, or nickel. Here the catalyst functions to "bring together" the hydrogen and ethylene molecules so that they may react with each other. Whereas the reactants do have an affinity for the metallic surface (i.e., through the π electrons of the ethylene molecule), the product (ethane) does not, and so is "desorbed" to make room for more reactant molecules.

Catalysts may also be *homogenous*, or limited to one phase, in which a surface or phase boundary is absent. Two examples are the acid-catalyzed hydrolysis of esters, in which the proton is a catalyst:

$$H_2O + CH_3-\overset{\overset{\displaystyle O}{\|}}{C}-O-R \underset{}{\overset{H^{\oplus}}{\rightleftharpoons}} CH_3-\overset{\overset{\displaystyle O}{\|}}{C}-OH + ROH$$

and the benzoin condensation, in which cyanide is the catalyst:

$$2\ Ph-C\overset{\displaystyle O}{\underset{\displaystyle H}{\diagdown}} \xrightarrow{:\overset{\ominus}{C}N} Ph-\overset{\overset{\displaystyle O}{\|}}{C}-\overset{\overset{\displaystyle OH}{|}}{C}H-Ph$$

In the first case, the mechanism is as follows:

$$CH_3-\overset{\overset{\displaystyle O}{\|}}{C}-OR + H^{\oplus} \rightleftharpoons CH_3-\overset{\overset{\displaystyle \oplus OH}{\|}}{C}-OR \longleftrightarrow CH_3-\overset{\overset{\displaystyle OH}{|}}{\underset{\displaystyle \oplus}{C}}-OR$$

$$\curvearrowleft :OH_2$$

$$ROH + \qquad CH_3-\overset{\overset{\displaystyle \oplus OH}{\|}}{\underset{\displaystyle OH}{C}} \rightleftharpoons CH_3-\overset{\overset{\displaystyle OH}{|}}{\underset{\displaystyle OH}{C}}-OR \rightleftharpoons CH_3-\overset{\overset{\displaystyle OH}{|}}{\underset{\displaystyle \oplus OH_2}{C}}-OR$$

$$CH_3-C\overset{\displaystyle O}{\underset{\displaystyle OH}{\diagup}} + H^{\oplus}$$

The protonation of the ester makes the carbonyl carbon more electrophilic (note the carbonium ion resonance form) and hence more susceptible to nucleophilic attack by water.

The benzoin condensation proceeds as follows:

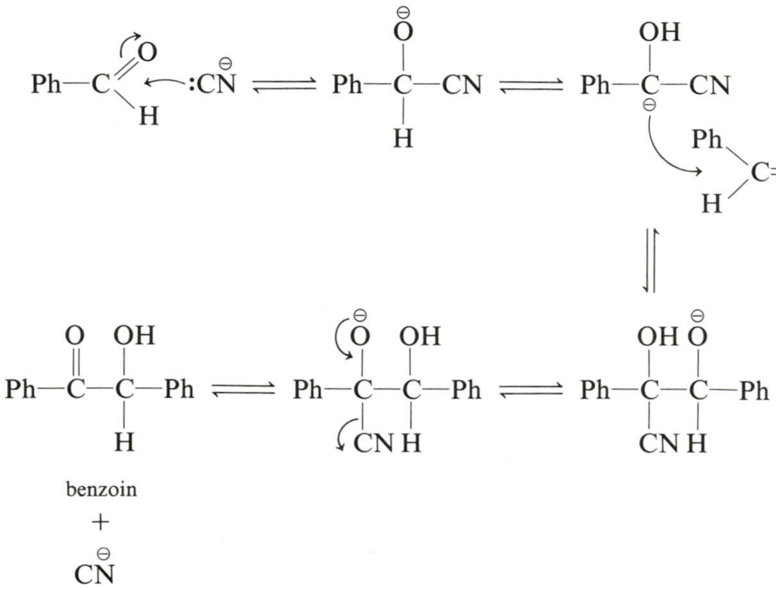

benzoin

+

$\overset{\ominus}{CN}$

The hydrogen of benzaldehyde is not acidic, so it cannot attack a second molecule of the same. However, by addition of cyanide (cyanohydrin formation), the hydrogen atom becomes acidic (the carbanion will be resonance stabilized by the cyano function) and the reaction proceeds.

Enzymes incorporate the features of both classes of catalysts. They may bring reactants together on a protein surface, or extract them from an aqueous phase into a hydrophobic environment. However, they may interact with the reactants in a manner such that the rate of chemical reaction is greatly improved. For example, as will be seen shortly, the enzyme catalyzed hydrolysis of an amide bond does not proceed merely by hydrolysis on the protein surface, but instead involves chemical interaction of the enzyme to form a more susceptible ester intermediate which then undergoes hydrolysis.

Homogenous catalysts may be further subdivided into *specific-acid, specific-base, general-acid, general-base, nucleophilic,* and *electrophilic catalysts*. A specific acid is merely a proton, a specific base is a hydroxyl ion. General-acids and -bases are any other acidic or basic species, respectively. The bromination of acetone illustrates these different catalysts. It may proceed by specific-acid catalyzed enolization (52):

$$CH_3-\overset{\displaystyle O}{\overset{\|}{C}}-CH_3 + H^{\oplus} \rightleftharpoons CH_3-\overset{\displaystyle \overset{\oplus}{O}H}{\overset{\|}{C}}-CH_2 \underset{\qquad}{\overset{-H^{\oplus}}{\rightleftharpoons}} CH_3-\overset{\displaystyle OH}{\overset{|}{C}}=CH_2$$

$$H \qquad\qquad Br-Br$$

$$\overset{-HBr}{\nearrow\!\!\!/\!\!\!/}$$

$$CH_3-\overset{\displaystyle O}{\overset{\|}{C}}-CH_2-Br$$

It may proceed by specific-base enolization:

$$CH_3-\overset{\displaystyle O}{\overset{\|}{C}}-CH_3 + O\overset{\ominus}{H} \rightleftharpoons CH_3-\overset{\displaystyle O}{\overset{\|}{C}}-\overset{\ominus}{\underset{\underset{Br-Br}{\downarrow}}{C}}\!\!H_2 \rightleftharpoons CH_3-\overset{\displaystyle O}{\overset{\|}{C}}-\underset{Br}{CH_2} + Br^{\ominus}$$

It may also proceed by general-acid and/or general-base catalysis. For example, in a sodium acetate buffer, both general-acid and general-base catalysis may occur.

$$AcO-H \qquad\qquad AcO^{\ominus}$$

$$CH_3-\overset{\displaystyle O}{\overset{\|}{C}}-CH_2 \longrightarrow CH_3-\overset{\displaystyle HO}{\overset{|}{C}}=CH_2 + AcOH$$

$$H \qquad\qquad\qquad \downarrow {\scriptstyle Br_2}$$

$$\overset{\ominus}{OAc} \qquad\qquad\qquad products$$

Nucleophilic catalysis may be defined as the catalysis of a chemical reaction by a (rate-determining) nucleophilic substitution. Thus, if a particular nucleophilic reaction is slow, then the reactant may undergo attack by a nucleophilic catalyst to give rise to an intermediate that is more susceptible to the desired nucleophilic displacement than the reactant. A classical example of nucleophilic catalysis is the acetylation of alcohols with acetic anhydride in pyridine:

$$R-OH + CH_3-\overset{\displaystyle O}{\overset{\|}{C}}-O-\overset{\displaystyle O}{\overset{\|}{C}}-CH_3 \xrightarrow{\text{(pyridine)}} CH_3-\overset{\displaystyle O}{\overset{\|}{C}}-O-R + CH_3COOH$$

At first glance, it might be thought that pyridine merely functions to scavenge any acetic acid. However, this is not correct:

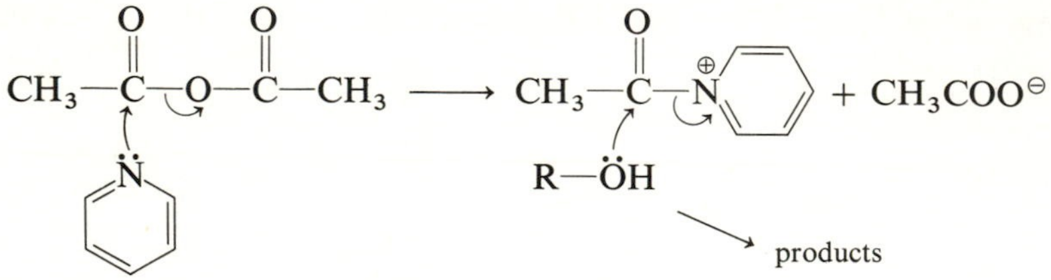

$$CH_3-\overset{O}{\underset{\;}{C}}-O-\overset{O}{\underset{\;}{C}}-CH_3 \longrightarrow CH_3-\overset{O}{\underset{R-\overset{..}{O}H}{C}}-\overset{\oplus}{N} \!\!\!\bigcirc + CH_3COO^{\ominus}$$

products

Without the pyridine, the acetylation of the alcohol would be much slower.

Another well-known example of nucleophilic catalysis is hydrolysis of *p*-nitrophenyl acetate, as catalyzed by imidazole:

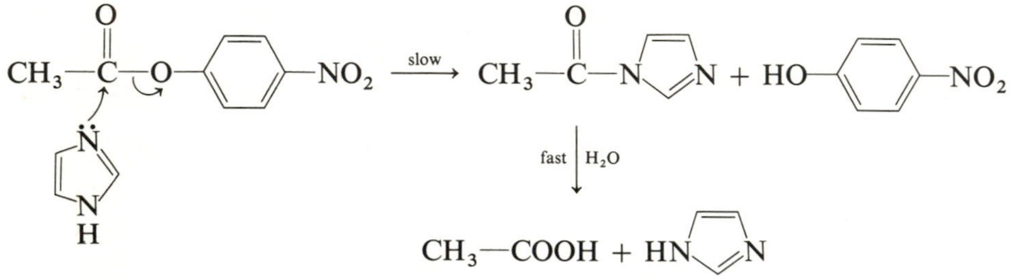

$$CH_3-\overset{O}{\underset{\;}{C}}-O-\bigcirc-NO_2 \xrightarrow{\text{slow}} CH_3-\overset{O}{\underset{\;}{C}}-N\!\!\bigcirc\!\!N + HO-\bigcirc-NO_2$$

fast $\Big\downarrow$ H$_2$O

$$CH_3-COOH + HN\!\!\bigcirc\!\!N$$

The imidazole is a nucleophilic catalyst for the reaction, but under basic conditions, an OH$^-$ ion can act as a proton acceptor to assist the nucleophilic catalyst (the imidazole) in attacking the substrate. Thus, the OH$^-$ ion is acting as a general-base. A transition state for the tetrahedral intermediate can be postulated:

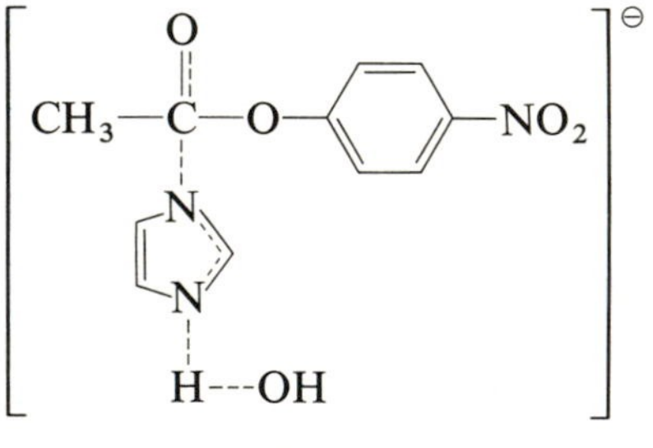

$$\left[\begin{array}{c} CH_3-\overset{O}{\underset{\;}{C}}-O-\bigcirc-NO_2 \\ | \\ N \\ \bigcirc \\ N \\ | \\ H---OH \end{array} \right]^{\ominus}$$

tetrahedral T.S. intermediate

This reaction is a typical example of nucleophilic catalysis. It is also called *covalent catalysis* when the substrate is transiently modified by forming a covalent bond with the catalyst to give a reaction intermediate.

Just as nucleophilic catalysis involves the addition of electrons from the catalyst to the substrate, *electrophilic catalysis* involves the abstraction of electrons or electron density from the substrate to the catalyst. Metal ions are excellent electrophilic catalysts. Especially in the chemistry of phosphates where anionic charges tend to repel nucleophiles from the phosphorus atom, electrophilic catalysts can be important. For example, the synthesis of 3',5'-cyclic guanosine monophosphate (cGMP, see Section 3.4.2) from guanosine triphosphate is markedly accelerated by divalent metal cations [e.g., Mg(II), Mn(II), Ba(II), Zn(II), Ca(II)].

The hydrolysis of ATP is also subject to metal ion catalysis. In both of the above cases, the metal chelates to the oxygen atoms of the triphosphate moiety, and withdraws electron density from the phosphorus atom, making the latter more electrophilic.

Enzymes may use any of the above mentioned modes of catalysis in order to catalyze a particular chemical reaction. For example, the imidazole ring of a histidine residue of the enzyme α-chymotrypsin (Section 4.4) can function as a general-base catalyst, while in the enzyme alkaline phosphatase, the same residue can function as a nucleophilic catalyst. Indeed, enzymes are complex catalysts which employ more than one catalytic parameter during the course of their action. It is by this successful integration of a combination of individual catalytic processes that a rate enhancement as high as 10^{14} may be achieved. Furthermore, it is this combination of factors which results in a specific catalyst.

To further illustrate the importance of imidazole in catalysis, let us examine its role as a basic catalyst which is extremely common in chemical reactions. Hydrolysis of carboxylic acid derivatives for instance is both assisted by general-acid and -base catalysis.

The example on the following page shows that hydroxyl ions and imidazole are proton acceptors and can participate in the rate-determining step of ester hydrolysis by general-base catalysis.

Why is the reaction accelerated by participation of a base? Many reasons can be given. Principally, the base (imidazole) accepts in the transition state (T.S.) a proton from the attacking water molecule so the oxygen of the latter acquires a greater share of electron density. Thus, the oxygen of water becomes more negatively charged and the ability to denote an electron pair to the carbonyl group is enhanced. The net result is that the presence of a base lowers the free energy of activation of the reaction. In the uncatalyzed reaction, another molecule of water, which is less basic and thus a less efficient catalyst, would accept the proton.

An important experimental question from the standpoint of catalysis is: How can general-base catalysis be distinguished from nucleophilic catalysis?

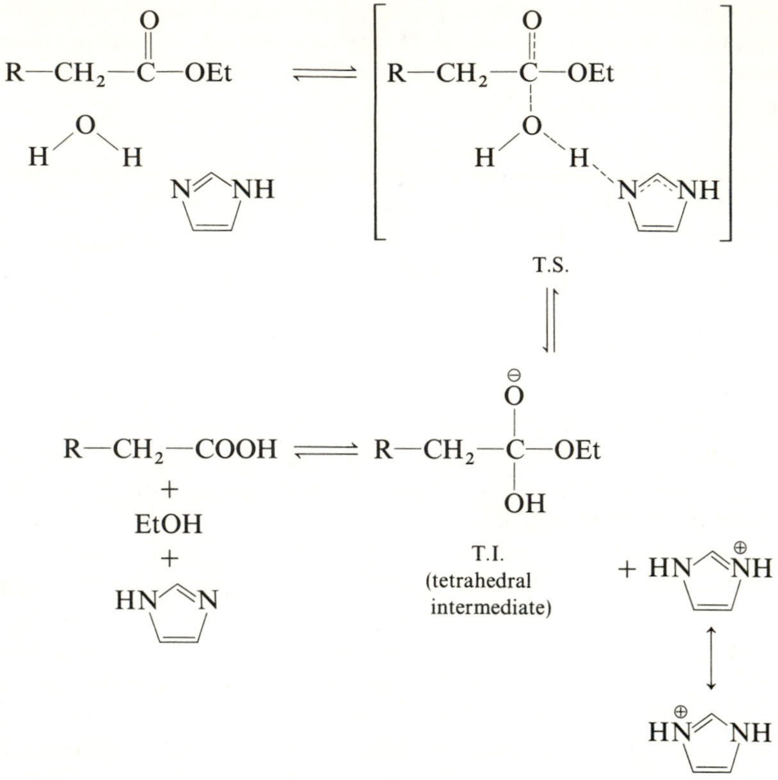

$$R—CH_2—\overset{\overset{\displaystyle O}{\|}}{C}—OEt \rightleftharpoons \left[R—CH_2—\overset{\overset{\displaystyle O}{\|}}{\underset{\underset{H\,\diagdown\!\!\diagup\,H}{O}}{C}}—OEt \right]$$

$$\underset{N\diagup\!\!\diagdown NH}{\overset{H\diagup\!\!\diagdown H}{O}} \qquad\qquad \underset{N\diagup\!\!\diagdown NH}{}$$

T.S.

$$R—CH_2—COOH \rightleftharpoons R—CH_2—\overset{\overset{\displaystyle \overset{\ominus}{O}}{|}}{\underset{\underset{\displaystyle OH}{|}}{C}}—OEt \qquad + HN\diagup\!\!\diagdown\overset{\oplus}{N}H$$

+

EtOH

+

$$HN\diagup\!\!\diagdown N$$

T.I.
(tetrahedral
intermediate)

$$HN\diagup\!\!\diagdown\overset{\oplus}{N}H$$

As seen in the above reaction sequence, water is involved in the transition state. Thus, a possible procedure is to perform the reaction in ^2H_2O. A $^2H—O$ bond is stronger than a H—O bond so that the energy of transfer of a deuterium ion is higher. Experimentally, base catalysis in ^2H_2O shows a difference of the order of 2 to 3 times smaller for the rate of reaction as compared to H_2O.

On the other hand, catalysis by a nucleophile will give the same rate in both solvents. An example is the hydrolysis of anhydrides by formate ions. Formate is less basic than acetate but is a better nucleophile. Furthermore, in a nucleophilic catalysis an unstable acyl-intermediate is formed and can be accumulated by having an excess of nucleophile.

In a less polar solvent (water plus alcohol), the individual groups are less solvated and they interact together more often. This situation involves *multiple catalysis* where the presence of both acid and base might have a cooperative effect greater than the sum of the individual effects. However, four molecules; substrate, nucleophile, general-acid, and general-base have to line up in the right way before such cooperativity of catalysis can take place. This situation is thus more unlikely to occur than in general-acid or -base catalysis where collision among three molecules has a higher probability of occurrence.

This situation is, however, more probable in an enzyme active site that combines several types of catalytic groups which operate simultaneously. It explains in part why enzymes are such efficient catalysts.

In cases where the nucleophile is on the same molecule, *intramolecular catalysis* takes place. We can illustrate this by the amino group participation in the hydrolysis of an ester. Four types of mechanism are possible (53).

(a) intramolecular nucleophilic catalysis:

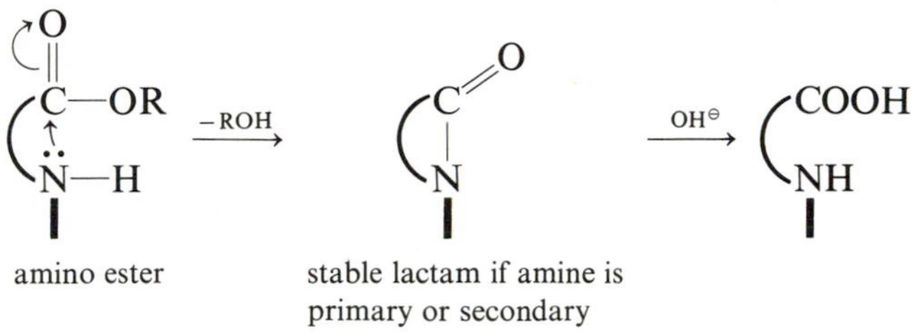

O
C—OR
N—H
−ROH
C
O
N
OH⊖
COOH
NH
amino ester
stable lactam if amine is primary or secondary

(b) intramolecular general-base catalysis:

O
C—OR
HO
H
NH

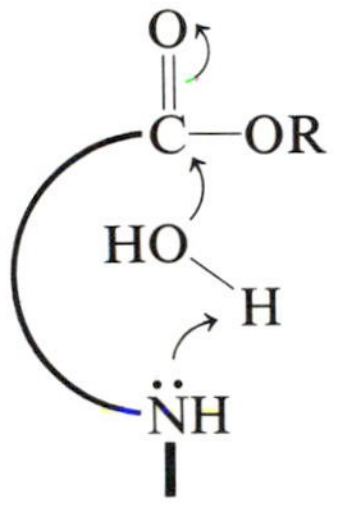

(c) intramolecular general-acid specific-base catalysis:

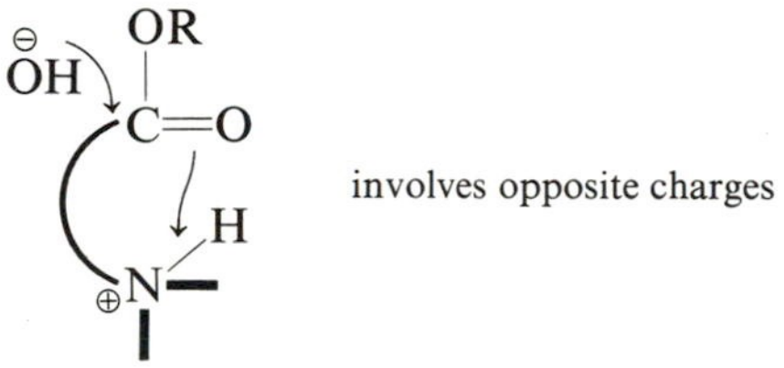

OR
OH
C=O
H
N
involves opposite charges

(d) electrostatic facilitation:

OR
HO
C=O
N
R‴
R′
R″
the only mechanism with a quaternary amine

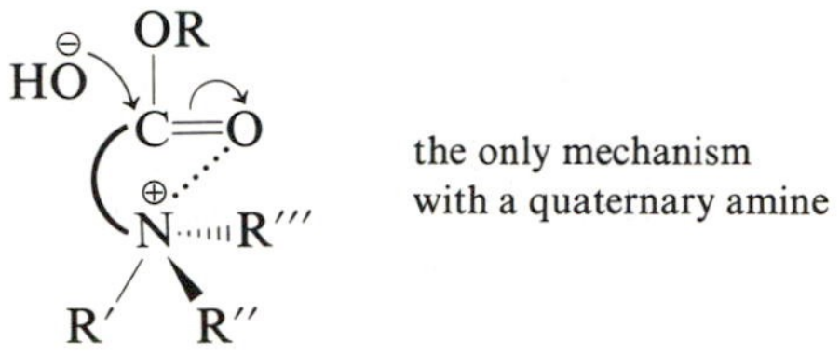

Mechanisms a, b, and c are kinetically indistinguishable but in principle b and c can be differentiated from a by deuterium solvent isotope effects. Mechanism c should be sensitive to ionic strength effects, but not b because it is electrically neutral. Mechanisms c and d are alike but the most favored one is c. However, with a good leaving group, process a is the most likely to occur (53).

Thus far we have considered only *homogeneous catalysis* in which all reactants and the catalyst are in the same phase (in solution). In *heterogeneous catalysis*, however, at least two phases are present in the reaction mixture. As mentioned earlier the most common types are systems with a solid catalyst in contact with substrates in the gaseous or liquid phase (356, 368).

Many analogies can be made between surface heterogeneous catalytic polymerization of olefins and the way that nature catalyzes chemical reactions. It is in this context that nonenzymatic and enzymatic catalysis can be bridged. Indeed, heterogeneous catalysis is similar to enzyme catalysis in many respect. A substrate molecule collides with an *active site* on the surface of the solid catalyst to form an adsorptive complex. The adsorbed substrate reacts in one or more steps under the influence of catalytic groups at the active site. Finally, the product molecules desorb (or escape) from the active site. Thus, the concepts of an active site and of complex formation between substrate and active site are common both to enzyme catalysis and heterogeneous catalysis. A comprehension of these concepts serves to bridge nonenzymatic and enzymatic catalysis. They are nonetheless fundamentally different in the sense that most enzymes have only one active site per molecule whereas heterogeneous catalysts have many active sites per particule. Furthermore, in heterogeneous catalysts, the sites are not necessarily identical.

The concept of constrained geometry is also important and this is why π-complexes of metal and olefins can lead to structured molecules. For example a model of cyclododecatriene-nickel shows that the nickel atom is situated exactly in the center of the ring. This picture represents a "lock and key" fit very precisely; another analogy with an enzyme–substrate complex.

Heterogeneous catalysts are exceedingly useful and important in the petroleum industry and in the synthesis of synthetic polymers and plastics. In this respect, a well-known catalyst for the polymerization of olefins is the Ziegler–Natta catalyst, developed in the 1950s. It is formed by the couple $TiCl_4 \cdot AlEt_3$. Ti(IV) is the coordinating transition metal and the organo-

metallic portion acts as an activator (a reductor) by alkylating the active
site titanium ion and removing one chlorine ligand:

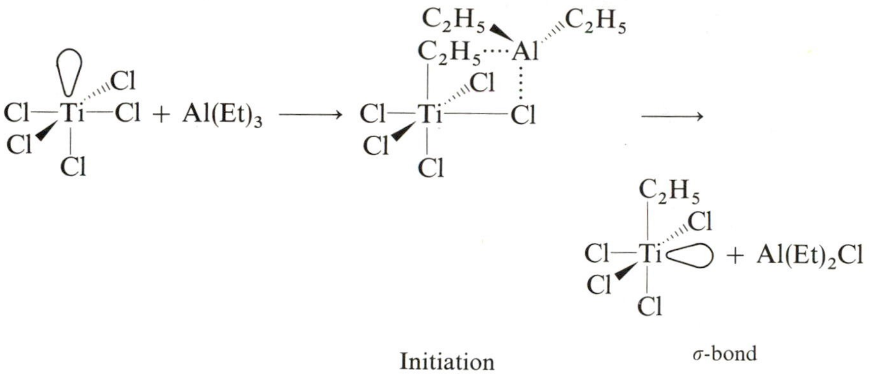

Initiation   σ-bond

active center   π-complex

regeneration   active center   transition state
of former active center   (σ-bond)

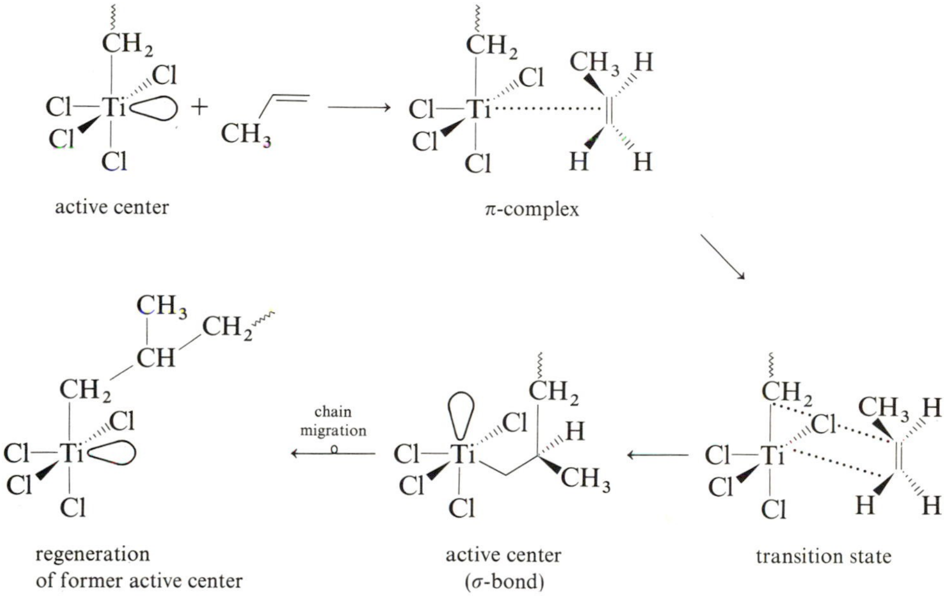

Propagation

  Titanium is the real catalyst; aluminium serves only at the beginning to
initiate the reaction by alkylation. The alkylated titanium surface-complex
produced in this way has a ligand vacancy (empty orbital) oriented in a
cavity for an olefin to make a π-complex (Fig. 4.2).
  Therefore, the titanium atom has just the right electronic structure to
form a strong covalent bond with an alkyl chain which becomes labile only
in the presence of an olefin to form a π-bond which allows an easy migration
of the ligand. This complex evolves to a new Ti—Cσ bond with the olefin

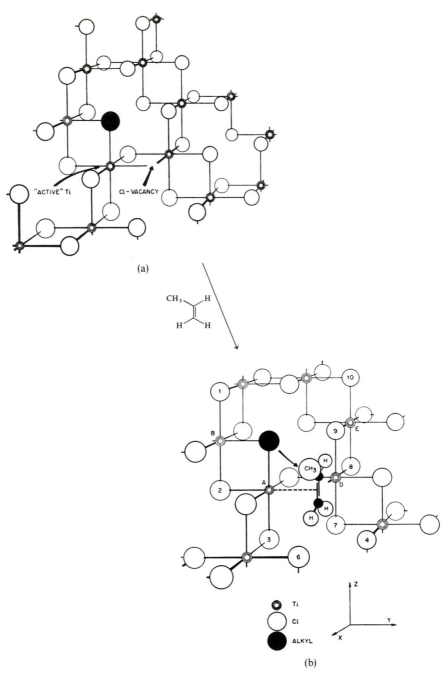

Fig. 4.2. Surface lattice structure of titanium trichloride showing (a) active site titanium ion with a ligand vacancy, and (b) a propylene molecule complexed to the titanium ion of the active site. The black sphere represent the growing polymeric chain (54). Copyright© 1962 by The Chemical Society.

and the extension of the chain occurs by ligand migration to regenerate the free ligand in the original orientation.

The reaction is stereospecific because each new olefin monomer always enters and reacts from the same ligand position on the titanium active site. A *stereoregular* polyolefin results with all substituents pointing in the same direction. It is an *isotactic* polymer. If the groups alternate on the chain it is a *syndiotactic* polymer. If the orientation is random, the molecule is said to be *atactic*.

isotactic polypropylene

syndiotactic polypropylene

atactic polypropylene

Thus, the polymerization of olefins under the influence of a metal catalyst can proceed with a high degree of specificity to give crystalline polymers in which the monomer units are all arranged in the same configuration. This demands that there be regio- and stereospecific interactions between the growing polymer, the catalyst, and the reacting monomer as the polymer is synthesized.

This unique example of solid surface catalysis has many analogies with an enzymatic reaction. In particular there is an active site located in a specific region of the metal surface to accommodate the "substrate," and the transformation is always stereospecific.

4.2 Introduction to Enzymes

In its broadest sense, an *enzyme* is a protein capable of catalytic activity. More specifically it can be defined as a polypeptide chain or an ensemble of polypeptide chains possessing a catalytic activity in its native form. It is a sophisticated copolymer made of amino acids monomers of the same configuration. The catalysis takes place in a specific region of the enzyme referred to as the *active center* or catalytic cavity. The active center consists of all the amino acid residues implicated in the mode of binding and the specificity of the substrate. On the other hand, the *active site* is composed only of the side chains of the amino acids which are directly implicated in the catalytic process. An integral part of the concept of the active site is that there are a small number of groups on the enzymes that produce its high catalytic activity and very often in a complementary way. The simple template "lock and key" theory, which assumes immobility, is only a first approximation for the binding of substrate to enzyme (55, 56).

The outstanding characteristic of enzyme catalysis is that the enzyme specifically binds it substrate and the reactions take place in the confines of the enzyme–substrate complex. Thus, to understand how an enzyme works, we do not only need to know the structure of the native enzyme but also the structure of the complexes of the enzyme with its substrates, intermediates, and products.

The amino acid sequence or primary structure of an enzyme dictates the secondary and tertiary (three dimensional) structure which is the folding of the peptide chain to form a globular macromolecule with a well-defined cavity to accommodate the substrate, and the coenzyme if needed. Enzymes have complex and compact structures where the polar amino acids are at the surface of the molecule directed toward the solvent while the nonpolar ones are generally oriented toward the interior of the molecule away from the solvent. The three-dimensional structure is maintained by a large number of intramolecular noncovalent apolar or hydrophobic interactions as well as by ionic interactions, disulfide bridges, hydrogen bond interactions, and sometimes salt bridges (57). *Hydrophobic forces* are probably the most important single factor to explain the large binding free energies which are observed in enzyme–substrate interactions.

Why are enzymes macromolecules? The reason is that the active center must be of a defined or highly ordered geometry if it is to contain all the binding and catalytic amino acid residues in a correct alignment for optimal catalysis. This imposes a heavy entropy demand upon the system which can be compensated for at the expense of another already ordered region of the biopolymer (58).

The most obvious means by which an enzyme might increase the rate of a bimolecular reaction is simply to bring the reacting molecules together at the active site of the enzyme. Therefore, the two most important questions with respect to this topic are, first, the magnitude of the rate enhancement which might be expected from such an approximation of the reactants, and second, the mechanism by which this rate enhancement is brought about. This chapter will try to clarify these points. Change in solvation may also provide a significant rate-acceleration effect in both intramolecular and enzymatic reactions. The nonpolar interior of an enzyme is similar to the low dielectric constant of an organic solvent. Electrostatic interactions are no longer shielded by solvent molecule and thus become stronger with the substrate. Before evaluating the factors responsible for enzyme specificity we should emphasize again the importance of chirality.

4.2.1 Molecular Asymmetry and Prochirality

Enzymes catalyze biochemical reactions stereospecifically. For this reason asymmetric syntheses are very common in nature and are often essentially unidirectional. Consequently, most natural products are optically active as a result of having been constructed by the catalytic action of three-dimensional enzymes. In its simplest terms, the substrate "fits" into the active site of the enzyme in a precise geometric alignment. For example, the enzyme triosephosphate isomerase catalyzes the conversion of achiral dihydroxycetone monophosphate to D-glyceraldehyde 3-phosphate (59).

The substrate here has a *prochiral* center* and one hydrogen is transferred specifically only from one face (the *re*-face) of the double bond to the carbonyl function. It is primarily the chirality of the enzyme which determines the correct course of the reaction. Another example of this is when these same two substrates are in the presence of the enzyme aldolase, to give fructose diphosphate, the H_S rather than the H_R-hydrogen is exchanged with water.

fructose 1,6-diphosphate   dihydroxyacetone monophosphate   D-glyceraldehyde 3-phosphate

---

* The concept of prochirality and a proposed system of pro-*S*, pro-*R* configurational nomenclature was definitively developed by K. R. Hanson in 1966. Hanson's general definition of prochirality is as follows: "If a chiral assembly is obtained when a point ligand in a finite achiral assembly of point ligands is replaced by a new ligand, the original assembly is prochiral." Point ligands are the chemical groups attached to the center of prochirality. From a practical point of view, a group designated as the pro-*R* (H_R) substituent is the one which upon replacement or modification leads to a chiral molecule with the *R* configuration (60). Hanson also proposed a systematic way of labeling the faces of trigonal atoms such as the carbon of the C=O group. A trigonal prochiral center is planar and is viewed from one side as in the drawing below. The three groups surrounding the carbon atom are given priorities a, b, and c, according to *R,S* nomenclature of Cahn–Ingold–Prelog.

re-face   *S*-configuration
attack

If the sequence a, b, c of priorities of the groups is clockwise, the face toward the reader is *re* (rectus), if counterclockwise, *si* (sinister). If necessary, "phantom atoms" attached to C and O must be added to assign priorities on the basis of the atomic numbers of atoms surrounding the carbon of the carbonyl function.

The concept underlying this specific behavior is called the *Ogston effect* (61). A. G. Ogston suggested in 1948 that in order for an enzyme to be capable of asymmetric attack upon a symmetric substrate, it must bind the substrate to at least three points. Only then stereoselective recognition or synthesis is possible.

Fundamentally, for a compound of the type

$$
\begin{array}{c}
y \\
| \\
x - C - x \\
| \\
z
\end{array}
$$

the enzyme can discriminate between the like groups x because the two transition states for attack of any chiral reagent on either x groups are diastereoisomers and consequently the rates of reaction proceeding through them will differ. This hypothesis laid the foundation for the concept of prochirality and provided the impetus for investigation of the stereochemistry of enzymatic reactions occuring at prochiral centers.

For example, yeast alcohol dehydrogenase (YADH) was the first alcohol dehydrogenase to be crystallized and the first one to be shown that direct hydrogen transfer between substrate and coenzyme can occur. Deuterium-labeled ethanol was used by F. Westheimer (62) in 1953 to prove that the process is sterospecific and only one of the enantiotopic protons attached to C-1 of the primary alcohol is transferred to C-4 of the 3-acetamido-pyridinium ring of NAD^+ coenzyme (RPPRA = the rest of the coenzyme). If *R*-deutero-1-ethanol is used, all the deuterium is found on the top face of the aromatic portion of the coenzyme (more details on this coenzyme will be given in Chapter 7). The new asymmetric carbon on NAD^2H has the *R* stereochemistry. If *S*-deutero-1-ethanol is used instead, all the label is found on the acetaldehyde and none on the reduced coenzyme. This is a particularly good example of the Ogston effect where coenzyme and substrate are held at the active site of the enzyme in a specific manner. The enzyme has the capacity to distinguish without ambiguity between the two identical hydrogens at the prochiral carbon of ethanol. Of course the reverse reaction is also stereospecific.

A more dramatic example of asymmetric synthesis is the photosynthetic process of plants which can convert solar energy into chemical energy with the assistance of chlorophyll molecules. In this complex process achiral carbon dioxide ends up as D-glucose.

In summary, a large number of natural products (hormones, vitamins, biopolymers, etc.) become chiral substances due to dissymmetric influences exerted by enzymes and other organic materials during biochemical pathways.

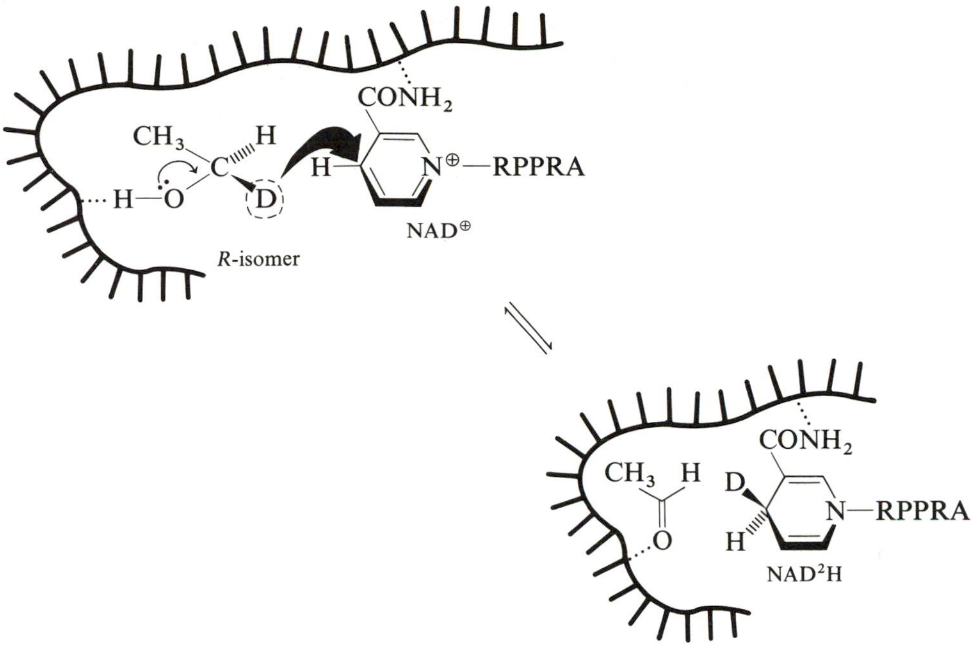

4.2.2 Sensory Responses at the Molecular Level

As humans, we have developed five perceptive senses: smell, taste, touch, sight, and hearing. Sensory response is a good example of adaptation at the molecular level and for this we shall examine the potential of the first two. Odor, like taste, is related to a direct contact of molecules with the olfactive (or gustative) epithelium. These membranes or surfaces contain chemo-receptors which upon excitation by a stimulant give characteristic organo-leptic sensations. The specific interactions between small molecules and receptors involve an equilibrium between absorption and desorption and can be studied by chemists using models.

For instance, R- and S-carvone are enantiomers; the R-isomer has a caraway odor whereas the S-isomer has a spearmint odor.

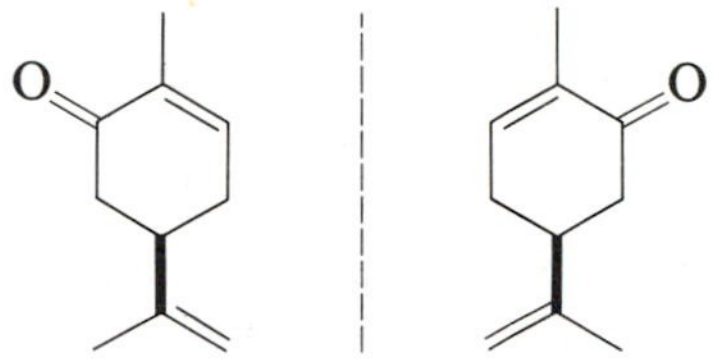

R-carvone   S-carvone

R-Carvone can be converted chemically to another naturally occurring terpene; (+)-limonene. This compound now has an "orange-like" odor.

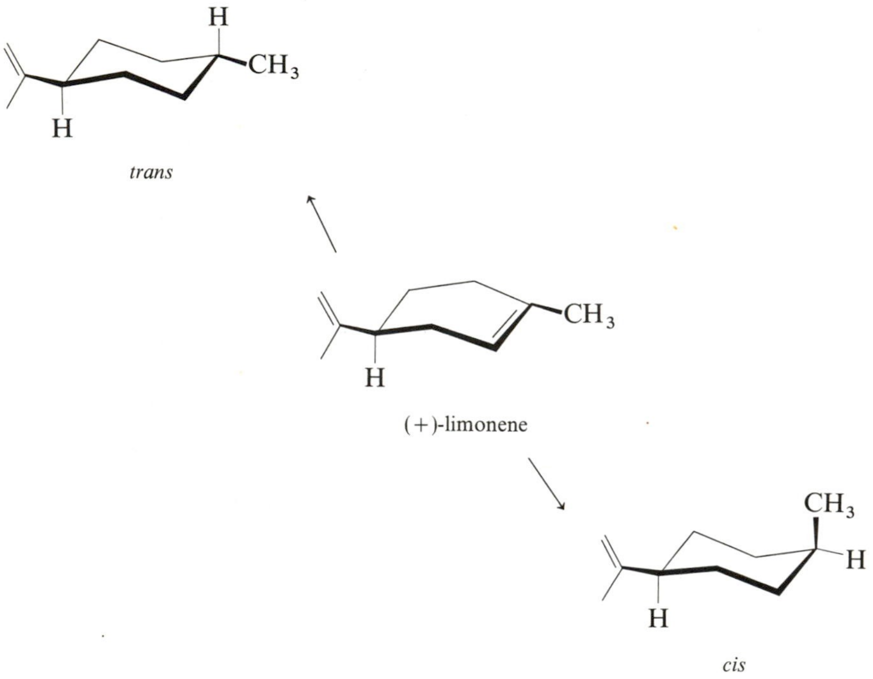

trans

(+)-limonene

cis

(+)-Limonene can be reduced to either *cis*- or *trans-p*-menth-8-ene. The *trans* compound conserves its orange-like odor but the *cis* has an unpleasant hydrocarbon-like odor. These examples illustrate that olfaction depends on the geometry of the molecule. Similar odors are probably related to similar molecular dimensions which are receptor specific. Our system has thus evolved to such a degree of molecular complexity that it can "smell the shape of a molecule."

Taste also registers differences between enantiomers. L-Glutamic acid imparts a meaty flavor and has been sold as taste intensifier for meats. The D-isomer, however, is almost tasteless. The proteins thaumalin (MW about 21,000) and monellin (MW about 10,700) have been observed to exhibit intense sweetness. An even more potent taste modifier is miraculin (MW about 44,000), found, like the other two, in African berries, which causes acids to taste sweet. Monellin contains 92 amino acid residues. Its intact tertiary (three-dimensional) structure is necessary to produce sweetness. The protein is in fact composed of two noncovalently bound chains of 50 and 42 residues. When separated, neither shows sweetness. This represents a case of molecular recognition at a conformational level.

A last but fascinating and practical example to illustrate again the importance of molecular geometry in life. Female insects (some male also) can secrete simple organic molecules such as brombycol, from *Bombyx mori*, or the juvenile hormone below, from the hemolymph of *Manduca sexta* larvae. Usually they have an isoprenoid origin. Juvenile hormones serve to keep the insects in the larvae stage or prevent insect eggs from hatching. They are produced by the corpus allatum and are involved in the control of morphogenesis and vitalogenesis. These substances are highly active chemical

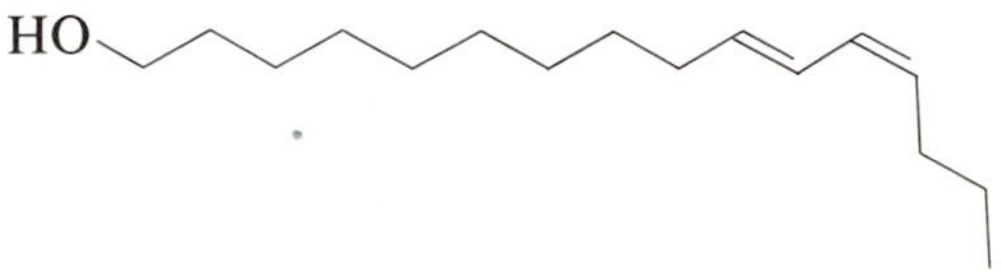

brombycol
(10-*trans*-12-*cis*-hexadecadiene-1-ol)

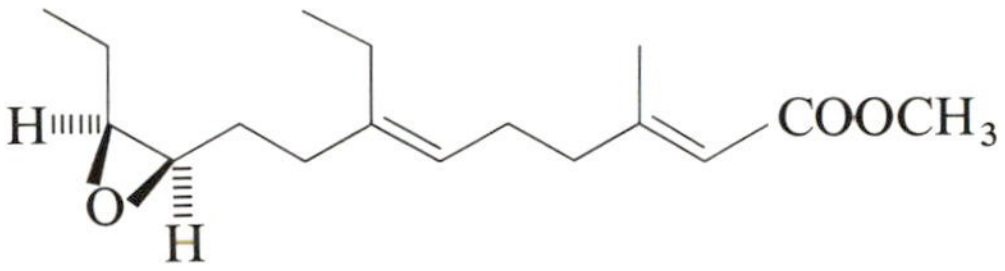

juvenile hormone

messengers and are the major forms of communication among insects. They can be percepted from very large distances by the antenna of the mate of the same species to attract them for reproduction purposes. Neurohormones or sex-attractant molecules such as brombycol are called *pheromones*. The other geometrical isomers are inactive for the same insect species. These findings can assist agricultural scientists in fighting insect destruction of crops (pest control) by biological means (bioselective insecticide) rather than chemical means which is more hazardous to the health of other animals and human beings. Synthetic analogues can also be used for mimicking their action.

4.2.3 Factors Responsible for Enzyme Specificity (55, 56)

It has been recognized that an enzyme has three levels of specificity: *structural specificity*, *regiospecificity*, and *stereospecificity*. An enzyme must first recognize some common structural features on a substrate (and a coenzyme) to produce a specific catalysis. Second, catalysis must occur at a specific region on the substrate (or the coenzyme) and the stereochemical outcome must be controlled by the enzyme.

Reduction of ketones by the reduced coenzyme nicotinamide adenine dinucleotide (NADH) and alcohol dehydrogenase provides again a good example (see Section 7.1.2 for pertinent applications in organic synthesis).

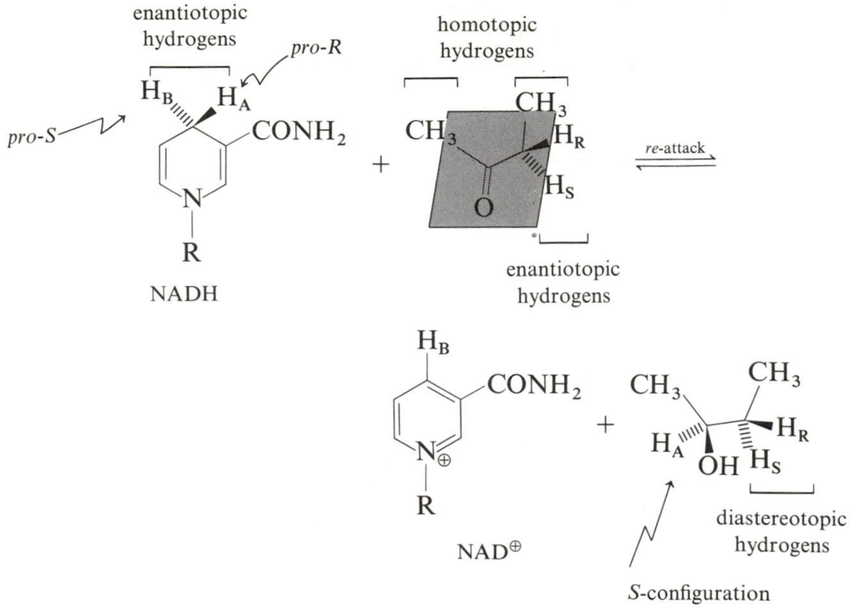

The carbonyl function of the ketone substrate has two faces and is thus said to be *enantiotopic*. In the presence of the enzyme, the coenzyme NADH delivers selectively the hydrogen H_A and only from the rear side of the carbonyl function, producing a chiral alcohol where the methylene hydrogens then become *diastereotopic*. The methyl hydrogens, however, cannot be distinguished by the enzyme and are called *homotopic*. In this example, the H_A (*pro-R*) hydrogen of the coenzyme migrated on the substrate, but other alcohol dehydrogenases have the opposite regiospecificity where the H_B (*pro-S*) hydrogen is used.

Enzyme-catalyzed reactions obey the principle of *microscopic reversibility*. This states that the mechanism of any reaction in the backward direction is just the reverse of the mechanism in the forward reaction. This thermodynamic concept is very useful for elucidating the nature of a transition state from knowledge of that for the reverse reaction. Certainly, enzymes are nature's most specific and powerful catalysts. No man-made catalysts have been capable of duplicating the great catalytic efficiency exhibited by enzymes under mild physiological conditions. Rate enhancements by factors of the order of 10^{10} to 10^{14} compared with similar nonenzymatic reactions have been observed.

One property which most profoundly distinguishes enzymes from other catalysts is their ability to bind their substrates in close proximity to each other and to the catalytic groups of the enzyme. Thus, enzymes accelerate reactions by juxtaposition of reacting atoms, the *proximity effect*, and by proper orientation of the relevant chemical groups. Certainly proximity and orientation effects are very important in catalysis. Random collision between molecules is too hazardous a process to bring about specific and efficient catalysis. Direction must be given. This direction can be brought about by a stereospecific complexing of the substrate. Correct stereochemistry between catalyst and substrate would then give a favorable entropy of activation, leading to efficiency of catalysis.

Generally speaking, the factors involved in the catalytic activity of an enzyme can be of four types. First, a *chemical apparatus* is needed at the active center which can deform or polarize bonds of the substrate to make the latter more reactive. Second, a *binding site* that can immobilize the substrate in the correct geometry relative to other reactive groups that are participating in the chemical transformation. Third, a correct and precise *orientation* of the substrate which permits each step of the reaction to proceed with minimal translational or rotational movement about the bonds of the substrate. Finally, the mode of fixation of the substrate must *lower the energy of activation* of the enzyme–substrate complex in the transition state. The proper distribution of charges at the active site and the geometry of the active center are among the factors responsible for a decrease in the overall entropy in the transition state. All these factors act, to a different degree, on the structure of the active site of an enzyme and cannot be separated from one another. Together they accelerate the rate of the enzymatic reactions and allow the enzyme to function as a powerful catalyst (77).

These factors can be evaluated experimentally with the aid of model systems. Rate enhancements have been obtained in several biomodel chemical reactions that are of similar magnitude to those observed in analogous enzyme-catalyzed reactions.

4.2.4 Intramolecular Catalysis

Intramolecular acid–base catalysis is an effective way of catalyzing reactions in organic systems. However, it would be useful to know the contribution of this to enzyme catalysis. There is a fundamental difference between enzyme and solution chemistry. The rate constants for solution catalysis are second order; the rate increasing with increasing concentration of catalyst. The reactions in an enzyme–substrate complex are first order, the acids and bases being an integral part of the molecule. The importance of acid and base catalysts in the latter system may be evaluated by synthesis of model compounds with the catalytic group as part of the substrate. This may be compared with the reaction rates for the corresponding intermolecular reactions.

Let us examine the following two examples:

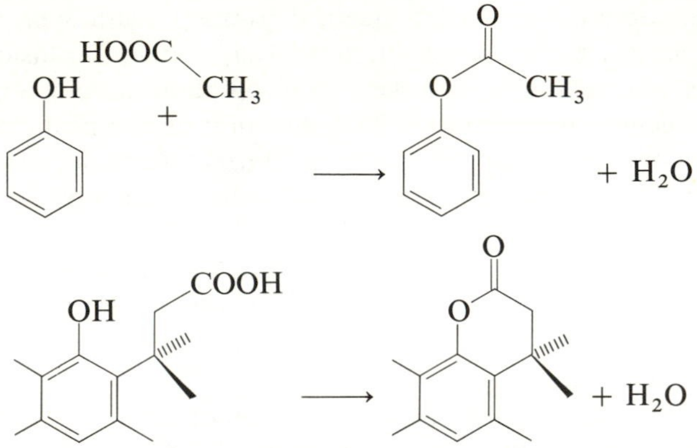

The intramolecular reaction shows an enhancement in rate relative to the intermolecular reaction by a factor of 10^{15} and the molecule lactonizes even in alkali! The aromatic methyl group fits between the geminal methyl groups of the side chain because of steric compression and "locks" the system. Cyclization also relieves the large ground state strain built into the molecule. Furthermore, the "effective concentration" of the carboxyl group in the proximity of the attacking hydroxyl group is very high.

As mentioned already, the high *effective concentration** of intramolecular groups is one of the most important reasons for the efficiency of enzyme catalysis. One function of an enzyme is thus to bring the substrate, by binding at the active site, into proximity with functional groups of the enzyme. This involves a change in *entropy* of the system. Hence, the catalytic consequence of an intramolecular reaction over its intermolecular counterpart is due to an entropic effect. The intermolecular reaction involves two or more molecules associating to form one, leading to an increase in "order" and a consequent loss of entropy.

Other factors can also be invoked, in particular solvation factors. An intramolecular nucleophile is less heavily solvated in the ground state than an intermolecular nucleophile in dilute solution. Furthermore, an intramolecular nucleophile is more rigidly held with respect to the reaction center. The relief of strain in an intramolecular reaction can also account in a large part for the rate enhancement.

Since the entropy of a molecule is composed of the sum of its translational, rotational, and internal entropies, T. C. Bruice in 1960 argued that the rate increases found in intramolecular reactions could be due to the

---

* The effective concentration of the group concerned is obtained by dividing the observed first-order rate constant for the intramolecular reaction by the second-order rate constant for the corresponding intermolecular process (349). Values in the range of 10^7 to $10^9 M$ are anticipated in enzymatic reactions.

restriction of unfavorable rotomer distribution (63). To prove this point, he studied the intramolecular displacement of *p*-bromophenol (to form an anhydride intermediate) in the following series of compounds as a function of increasing rigidity in the molecule:

	relative rate of diacid formation
	1
	200
	10,300
	53,000

Note that removal of one degree of freedom increases the rate by about 200. With the rigid molecules, the reactive groups are properly aligned for reaction, and the reaction rates are much larger. The resulting rate enhancement in these reactions is a direct consequence of the *proximity effect* or closeness of the reactive groups. This results in favorable changes in translational and rotational entropy of activation. Bruice believes that "freezing" internal rotation of the substrate accounts for the principal factor in catalytic efficiency of enzymes, plus the entropic effect.

In contrast, W. P. Jencks of Brandeis University argues that loss of *translational entropy* (less motion) is among the most important factors to explain the large rate increase in intramolecular catalysis. In enzymatic reactions this entropy loss is offset by a favorable binding energy of the substrate to the enzyme which provides the driving force for catalysis. In other words, besides a lowering of the degree of rotation and of the translational entropy, Jencks suggests that the concept of *intrinsic binding energy*, which results from favorable noncovalent interactions with the substrate at the site of catalysis, is to a large degree responsible for the remarkable specificity and the high rates of enzymatic reactions (64).

By studying intramolecular lactonization reaction, where rate enhancements of 10^3 to 10^6 over the bimolecular counterpart were found, D. E. Koshland (65) in 1970 proposed a new concept: *orbital steering*. His experimental evidence suggests that "steering" reacting atoms can lead to (or explain) large rate factors. Not only the reacting groups have to be close, but also they must be properly oriented. In this concept, both favorable proximity and orientation are important. The more rigid molecules show the largest rates. The loss of translational entropy is not as important as Jencks originally suggested, according to Koshland's views.

$$CH_3CH_2OH$$
$$+$$
$$CH_3COOH$$

relative rate
of
lactonization: 1   413   1660

18,700

In essence this hypothesis suggests that a major factor in the intramolecular enhancements in lactonization rates results from the fact that the approach of the attacking atoms is confined to certain pathways as compared to the random orientation occurring with bimolecular collisions. Koshland says that "orbital steering" may explain why enzymes are so efficient. It is likely that enzymes line up the bonding orbitals of reacting molecules and catalytic groups with a precision that is impossible to obtain in ordinary bimolecular collisions in solutions. An enzyme does not merely bind substrates into close contact (Bruice's proximity effect) with one another; there is an orientation factor that is related to the shape of the electron orbitals of the potentially reactive atoms. This would be the source of the unique catalytic power of enzymes, according to Koshland. The remarkable catalytic activity of enzymes, therefore, arises not only from their ability to juxtapose the reacting atoms, but also to "steer" the orbitals of these atoms along a path that takes advantage of this strong angular dependence. The way the bond angles are affected by "orbital steering" is illustrated with a bicyclic compound.

$\pm 10°$ of angular dependence for γ-lactone formation; a misalignment can cause a serious decrease in rate

However, Bruice (66) has argued that some of the rate enhancements observed in many model compounds are much larger than expected from orientation effects. For instance, "orbital steering" could not account for rate enhancements of up to 10^8.

In response to Bruice, Koshland presented other results using again bicyclic rigid molecules but with a sulfur atom instead of an oxygen as the nucleophile (67, 68).

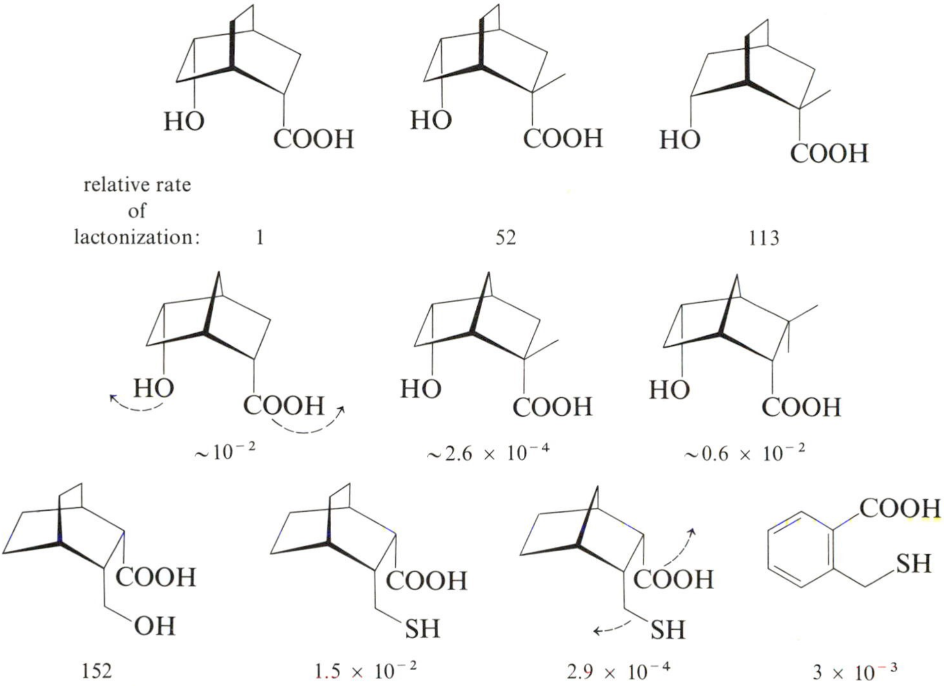

relative rate
of
lactonization: 1   52   113

$\sim 10^{-2}$ $\sim 2.6 \times 10^{-4}$ $\sim 0.6 \times 10^{-2}$

152   1.5×10^{-2} 2.9×10^{-4} 3×10^{-3}

The above results show that the values of the rate constants obtained are not always consistent with structural rigidity although the proximity factors are the same. Large differences in rate of 10^2 to 10^4 are seen between bicyclo-[2.2.2] and bicyclo[2.2.1] systems where the reactive groups in the latter system are only slightly further apart. So the orbital orientation must be different and responsible for the differences in rate.

In particular, thiolactonization is much lower in rate. The orientation is thus less favorable and this is expected because of the distortion in the overall shape of electron orbitals. Similarly, if the hydroxyl group of the active serine residue of the enzyme subtilisin is replaced by a —SH function, the activity of the protease is drastically reduced. This substitution (O → S) has a considerable effect (lowering) on the acyl-enzyme deacylation rate constant (69). Koshland argues that such inhibition would be expected if orbital orientation were of great importance.

Consequently, "orbital steering" explains some high rates of intramolecular reactions in terms of a favorable orientation of the orbitals that will undergo rehybridization in the transition state. However, it is not easy to see how this factor can be separated from other structural factors. There are presently more arguments against its overwhelming importance. For instance, the treatment is based principally on the effective concentration of a neighboring group to be 55 M, the concentration of water in water. This value is believed to be a serious underestimation of the contribution of translational entropy (Jencks' proposal) to effective concentration (70). Although some indispensable overlap of orbitals is needed in the transition state, it does correspond to an adjustment in orientation of no more than 10°. Such distortion would cause a strain of only 11 kJ/mol (2.7 kcal/mol) for a carbon–carbon bond (58).

Whether orientation effects are due to "orbital steering" or elimination of a conformation unfavorable for the reaction makes a conceptual difference only in that "orbital steering" would include a transition state effect whereas the latter would arise from restriction of possible ground states. What is unmistakably clear from available data is that constraining an intramolecular nucleophile close to the reaction center can produce large rate increases. Bruice's viewpoint is that the advantage of intramolecular reactions is entropic in nature because of the freezing out of degrees of freedom in the ground state. The Bruice–Koshland controversy forms part of the large question of why intramolecular reactions are so favorable. The truth is probably a blend of both effects, and other factors. In Section 4.6 we will see that proper orientation of lone-pair orbitals is an important stereoelectronic requirement to explain the rate enhancement in hydrolytic reactions.

Besides *proximity* effect and *orientation* effect, *steric compression* effect could be a third factor to consider. Nonetheless there can still be an enhancement of 10^2 to 10^4 in rate constants not accounted for by these factors. Among the likely candidates, electrostatic stabilization of the transition state and release of ground state strain should be mentioned. The notion of "freezing of substrate specificity" by Bender is also a factor for consideration. An example is the "aromatic hole" in α-chymotrypsin (see below) which allows a favorable steric situation for the amino acid side chain of the substrate.

In summary, entropy is one of the most important factors in enzyme catalysis. For chemical reactions catalyzed in solution, the catalyst and the substrate molecules have to come together in close contact and this involves a large loss of entropy. Enzymatic reactions, however, can be looked at as being restricted to an understanding of the chemistry going on at the enzyme–substrate complex where the catalytic groups are acting cooperatively on the same molecule. Thus, there is no loss of translational and/or rotational entropy in the transition state. As a consequence, the catalytic groups on an enzyme have very high effective concentrations compared with bimolecular reactions in solution. This results in a gain in entropy which is compensated by the enzyme–substrate binding energy. In other words, the rotational and translational entropies of the substrate are lost on formation of the enzyme–substrate complex and not during the subsequent catalytic steps (58).

4.3 Multifunctional Catalysis and Simple Models

Another fundamental idea that has been invoked to explain enzymatic catalysis is that such reactions utilize bifunctional or multifunctional catalysis. That is, several functional groups in the active center are properly aligned with the substrate so that *concerted catalysis* may occur.

One of the earliest examples (71) in solution chemistry is the study of the rate of mutarotation of *O*-tetramethyl-D-glucose in the presence of α-pyridone. A 10^{-3} *M* solution of α-pyridone was 7000 times more effective than an equivalent concentration of phenol and pyridine.

A concerted mechanism was proposed where the catalyst acts as a base and an acid simultaneously:

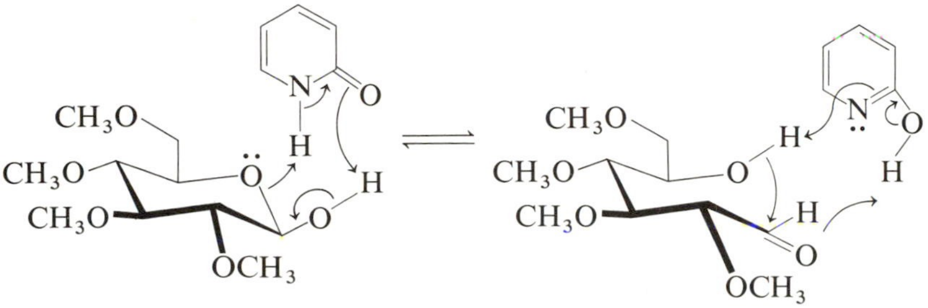

A key feature of enzyme catalysis is observed; regeneration of the catalyst (via the reversible pathway). 2-Aminophenol, which is both a stronger acid and base than α-pyridone, is not as potent a catalyst. In the presence of 2-aminophenol a concerted mechanism of catalysis is not possible.

Another classical example (72) of multifunctional catalysis is the hydrolysis of the monosuccinate ester of hexachlorophene:

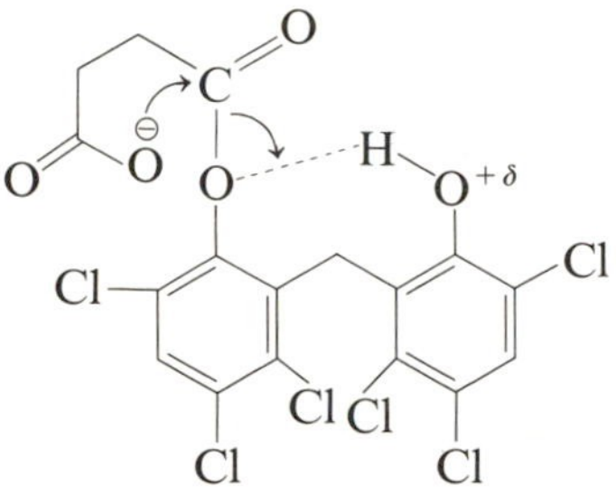

The proximity and participation of a hydroxyl group greatly accelerates the reaction rate. Similar, comparison of the corresponding monoacetate and diacetate shows that the monoacetate is hydrolysed 500 times faster that the diacetate below pH 5. A plot of the rate of reaction vs pH has a bell-shaped* profile and indicates intramolecular nucleophilic and general-acid hydrolysis. It is also indicative of ionic species involved with opposite pK_a values for maximal activity.

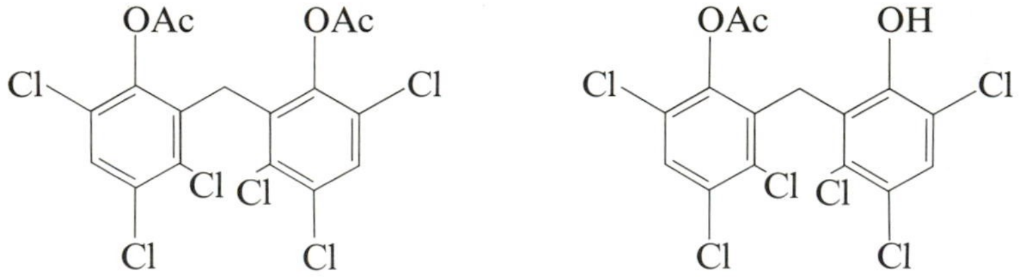

diacetate   monoacetate

Perhaps of more relevance to the understanding of enzyme function is a model for bifunctional catalysis in an aqueous milieu. One of the first re-actions to show a significant rate enhancement in water was the hydrolysis of an iminolactone. This work was undertaken by Cunningham and Schmir (73) who observed that phosphate buffer was at least 200 times as effective as imidazole buffer (even though both buffers have approximately the same pK_a) in catalyzing the expulsion of aniline.

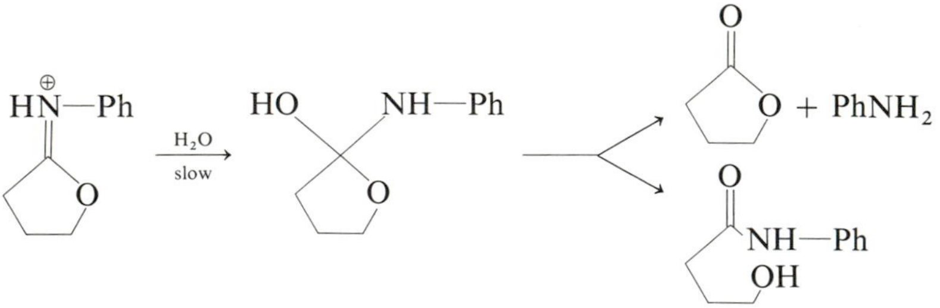

At basic pH, amide formation would be expected to be the preferred pathway, as the amide nitrogen would be a poor leaving group. This is observed in imidazole buffer. However, with phosphate, bifunctional catal-ysis allows ejection of aniline to become significantly more favorable:

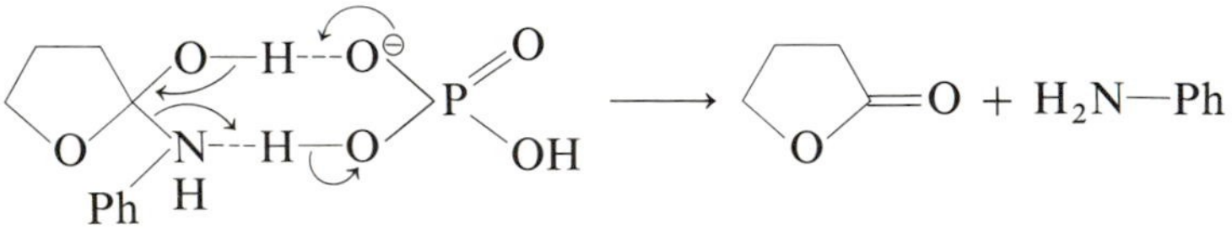

Similar catalysis was observed with bicarbonate and acetate buffers.

More recently, Lee and Schmir (74, 75) have studied similar bifunctional catalytic mechanisms in more detail with acyclic imidate esters. Again,

---

* A bell-shaped rate profile as a function of pH is characteristic of acid–base concerted catalysis.

hydrolysis proceeds via the rate-determining formation of a tetrahedral intermediate followed by breakdown to an amide or ester product:

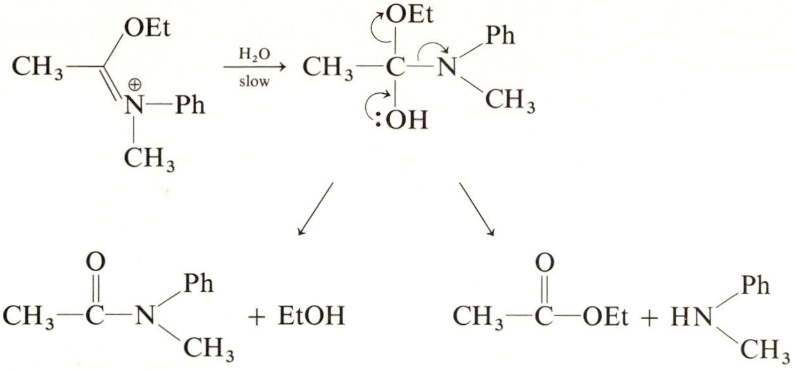

In the absence of buffer, the yield of amide will increase with increasing pH. At pH 7.5, 50% amide and 50% ester is formed. However, in the presence of various bifunctional catalysts amine formation will become the favored pathway. This is illustrated in Table 4.1.

With the exception of acetone oxime, all of the bifunctional catalysts contain an acidic and basic group in a 1,3-relationship. The cyclic transition

Table 4.1. Hydrolysis of Acyclic Imidate Esters in Various Buffers (74, 75)

Buffer	pH range studied	Rate ratio[a]
$HCO_3^{\ominus}$	7.49–9.73	290
$CH_3C(=N-OH)-CH_3$	7.73–8.85	8700
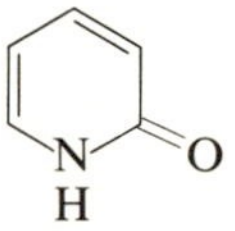 (pyridinone)	7.92–8.55	1460
$HPO_4^{2\ominus}$	6.96–9.11	52
$HAsO_4^{2\ominus}$	7.18–8.82	61
$H_2AsO_3^{\ominus}$	7.42–8.99	150
$CF_3C(O^{\ominus})(OH)CF_3$	7.13–8.90	47

[a] Refers to the ratio of amine formation for the bifunctional catalyst, and its monofunctional equivalent of the same pK_a.

state required for a concerted proton transfer would consist of an eight-membered ring. This is illustrated for phosphate:

$$\left[\begin{array}{c} CH_3 \\ \diagdown \\ EtO \diagup \end{array} C \begin{array}{c} O\text{---}H\text{---}O \\ \diagdown \quad\quad \diagdown \\ \diagup \quad\quad \diagup \\ N\text{---}H\text{---}O \end{array} P \begin{array}{c} \diagup\!\!\!\diagup O \\ \diagdown \\ O^\ominus \end{array} \right]$$
$$\begin{array}{cc} \diagup & \diagdown \\ Ph & CH_3 \end{array}$$

Once the cyclic transition state has been achieved, breakdown will occur in a fashion analogous to that described for the iminolactone, to give ejection of aniline and ester formation. Notice that while imidazole does possess an acidic and basic function in the 1,3-geometry, these groups are sterically unsuited for a cyclic proton transfer.

In 1972, Jencks (76) concluded that concerted bifunctional acid–base catalysis is rare or nonexistent because of the improbability of the reactant and catalyst meeting simultaneously. Furthermore, bifunctional catalysis does not necessarily represent a favorable process in aqueous solution even when a second functional group is held sterically in proper position to participate in the reaction. Caution should then be used in assuming that most enzymes are utilizing bifunctional or multifunctional catalysis. Nevertheless, this idea has played a leading role in concepts of enzyme catalysis.

4.4 α-Chymotrypsin

The catalytic efficiency of enzymes is a subject of great fascination, particularly when crystallographic structures are available and a great deal is known about the physical organic chemistry of the enzymatic mechanism. In this regard, the most studied enzyme of a group of enzymes called serine proteases is α-chymotrypsin. The term *serine protease* derives from the fact that this class of enzymes contain at their active site a serine hydroxyl group which exhibits unusual reactivity toward the irreversible inhibitor diisopropylphosphorofluoridate (DFP).

$$\begin{array}{ccccc} & & F & O & \\ & & \diagdown & \diagup\!\!\!\diagup & \\ CH_3 & & P & & CH_3 \\ \diagdown & \diagup & \diagdown & & \diagup \\ CH\text{---}O & & O\text{---}CH & \\ \diagup & & & \diagdown \\ CH_3 & & & CH_3 \end{array}$$

DFP

The highly reactive (P—F) linkage readily suffers displacement by nucleophiles such as the hydroxyl of the active serine of proteolytic enzymes. Other enzymes in this group include trypsin, thrombin, and subtilisin.

The mechanism of action of α-chymotrypsin is probably understood in more detail than any other enzyme at the present time. Its physiological function is to catalyze the hydrolysis of peptide bonds of protein foods in the mammalian gut. It is secreted in the pancreas as an inactive *zymogen* precursor, chymotrypsinogen, having a single polypeptide chain of 245 amino acids. Such an inactive form is necessary to prevent "self-digestion." The precursor is activated by trypsin hydrolysis (scission of the peptide backbone) at two specific regions to form active α-chymotrypsin having three peptide chains held by disulfide bridges. The three-dimensional structure was deduced by the X-ray diffraction study of D. M. Blow (77) and the mechanism of action was elucidated in collaboration with B. S. Hartley and J. J. Birktoft (78, 79), all at the MRC Laboratory of Molecular Biology, Cambridge, England.

In addition to the active serine-195 residue, the other active site residues are histidine-57 and aspartic acid-102, deduced from the X-ray work. The other histidine residue, His-40, is not implicated in the catalysis. The enzyme has a specificity for aromatic amino acids. Esters of aromatic amino acids are also good substrates for the enzyme and most of the kinetic data were obtained with ester substrates. The enzyme cuts up proteins on the carboxyl side of aromatic amino acids. After the formation of the Michaelis complex the uniquely reactive Ser-195 is first acylated to form an acyl-enzyme intermediate with the substrate.

The transformation of the Michaelis complex to the acyl-enzyme involves first the formation of a tetrahedral intermediate (T.I.) (see following section). Finally, attack by a water molecule hydrolyzes the acyl-enzyme which normally does not accumulate.

$$\text{En}-\text{OH} + \text{R}-\overset{\overset{\text{O}}{\|}}{\text{C}}-\text{X} \longrightarrow [\text{En}-\text{OH}\cdot\text{R}-\text{COX}] \longrightarrow \text{En}-\text{O}-\overset{\overset{\text{O}^{\ominus}}{|}}{\underset{\underset{\text{R}}{|}}{\text{C}}}-\text{X} \overset{\text{HX}}{\longrightarrow}$$

X = OR′, NHR′   Michaelis complex   T.I.
R = Phe, Trp, Tyr

$$\text{En}-\text{O}-\overset{\overset{\text{O}}{\|}}{\text{C}}-\text{R} \overset{\text{H}_2\text{O}}{\longrightarrow} \text{En}-\text{O}-\overset{\overset{\text{O}^{\ominus}}{|}}{\underset{\underset{\text{OH}}{|}}{\text{C}}}-\text{R} \longrightarrow \text{En}-\text{OH} + \overset{\text{O}}{\underset{\text{HO}}{\diagup}}\!\!\!\diagdown\!\text{C}-\text{R}$$

   acyl-enzyme   T.I.

It has long been thought that the basicity of the imidazole residue His-57 increases the nucleophilicity of the hydroxyl of Ser-195 by acting as a general-base catalyst.

$$\underset{\text{His-57}}{\text{H}-\text{N}\diagup\diagdown\text{N:}\frown\text{H}-\text{O}\diagup^{\text{Ser-195}}}$$

However, on the basis of the crystallographic structure for α-chymotrypsin, a new form of general catalysis was proposed. It is called the *charge-relay* system and originates with the alignment of Asp-102, His-57, and Ser-195, linked by hydrogen bonds (Fig. 4.3).

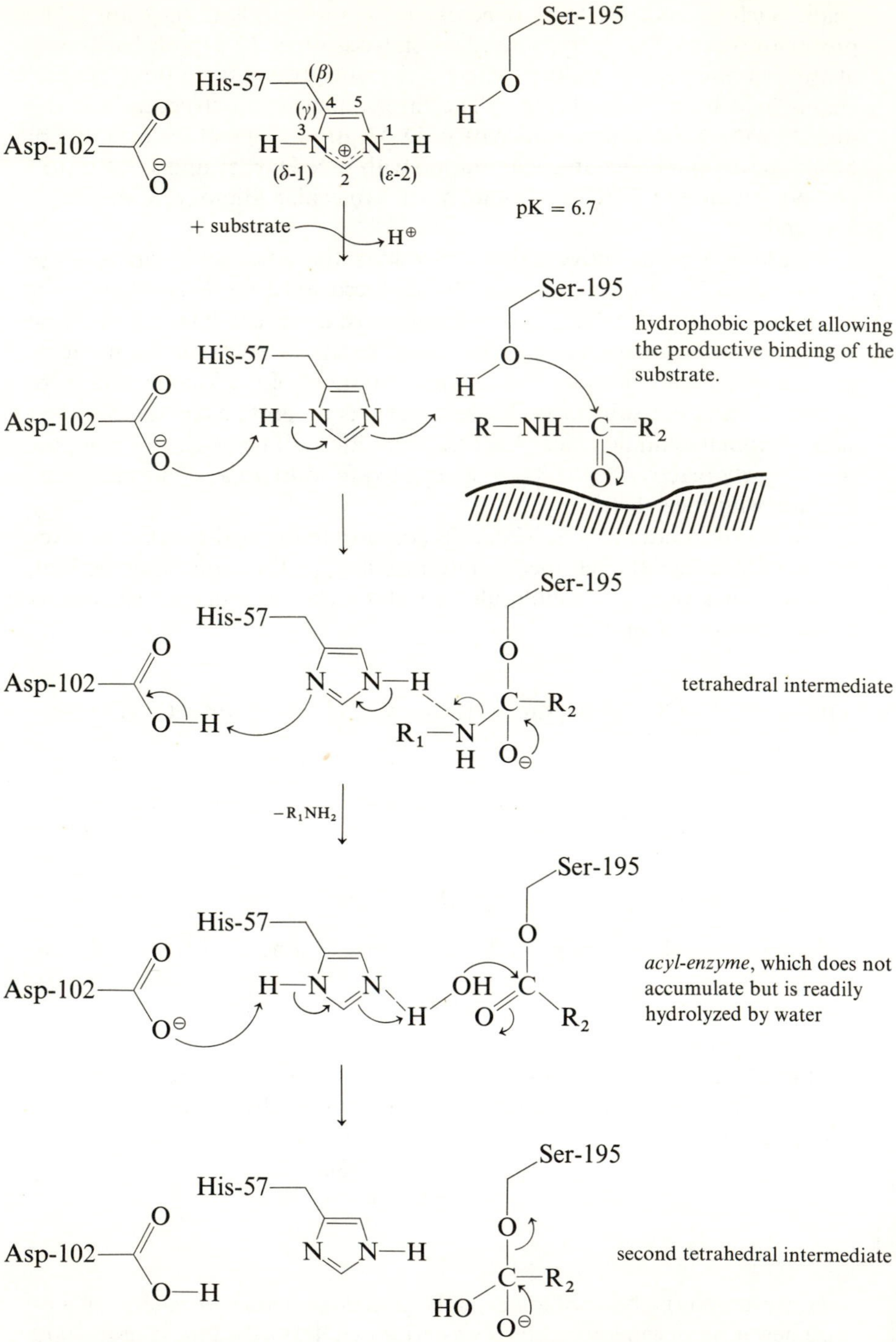

Fig. 4.3. The "charge-relay" system of α-chymotrypsin (position 3 of the imidazole group is often referred to as Nδl or π and position 1 as Nε2 or τ).

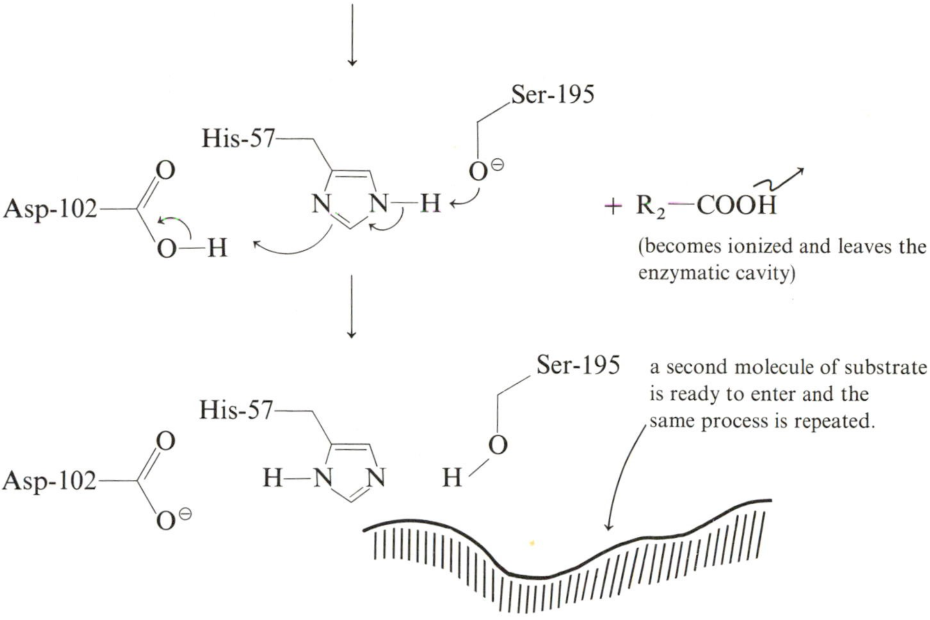

Fig. 4.3. (continued)

The acid group (Asp-102) is believed to be buried and its unique juxta-position with the unusually polarizable system of the imidazole ring (His-57), which in its uncharged form can carry a proton on either of the two ring nitrogens, seems to be the key to the activity of this enzyme, and other serine proteases (80). The function of the buried Asp-102 group is to polarize the imidazole ring, since the buried negative charge induces a positive charge adjacent to it. This gives a possibility of proton transfer along hydrogen bonds, to allow the hydroxyl proton of Ser-195 to be transferred to His-57. The active serine residue becomes then a reactive nucleophile capable of attacking the scissile peptide bond.

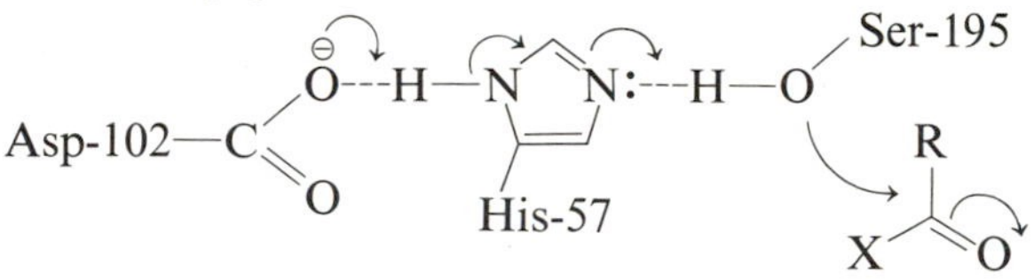

$$pK_a \sim 6.5 \text{ to } 7.0 \qquad X = NHR', OR'$$

The catalytic efficiency of α-chymotrypsin cannot be solely attributed to the presence of the charge-relay system. X-Ray work (81) has indicated the many parameters operative in the catalytic process. Nine specific enzyme substrate interactions have been identified in making the process more efficient. For example, stabilization of the tetrahedral intermediate, and thus lowering of the transition state energy barrier, is accomplished by hydrogen bond formation of the substrate carbonyl function with the amide hydrogen of Ser-195 and Gly-193. In chymotrypsinogen, the latter hydrogen bond is absent. Indeed, refinement of the X-ray structures of chymotrypsinogen and α-chymotrypsin indicates a difference between the structure of the catalytic triad in the zymogen and the enzyme. This conformational change in the overall three-dimensional structure of the enzyme may produce significant changes in the chemical properties of the catalytic residues which may play an important role in the amplification of enzyme activity upon zymogen activation.

Evidence has been accumulated for the existence of a tetrahedral intermediate and a concerted proton transfer mechanism from Ser-195 to His-57 and from His-57 to Asp-102 during the hydrolysis of p-nitroanilide peptide substrates catalyzed by a serine protease from *Myxobacter* 495 named α-lytic protease. It possesses only one histidine residue (82, 83). The rate-limiting step of the hydrolysis is the decomposition of the tetrahedral intermediate to acyl-enzyme and p-nitroaniline. The experimental results support the view that the two proton transfers which occur during this step take place in a concerted, not stepwise, manner.

Surprisingly, ^{13}C-NMR work (84) suggested that the group ionizing with a pK_a of about 7 is the buried aspartate and not the expected His-57, suggesting that the imidazole ring remains unprotonated during catalysis in the pH range 7–8. However, careful high resolution ^{1}H-NMR work by J. L. Markley's group (85) at Purdue University on chymotrypsinogen, α-chymotrypsin, and trypsin supports the previous conclusion of Robillard and Shulman (86) that the pK_a of His-57 is higher than that of Asp-102 in both the zymogen and enzyme. Based on this argument and others, Markley favors the following mechanism, a variant of the one in Fig. 4.3, for formation of the tetrahedral intermediate:

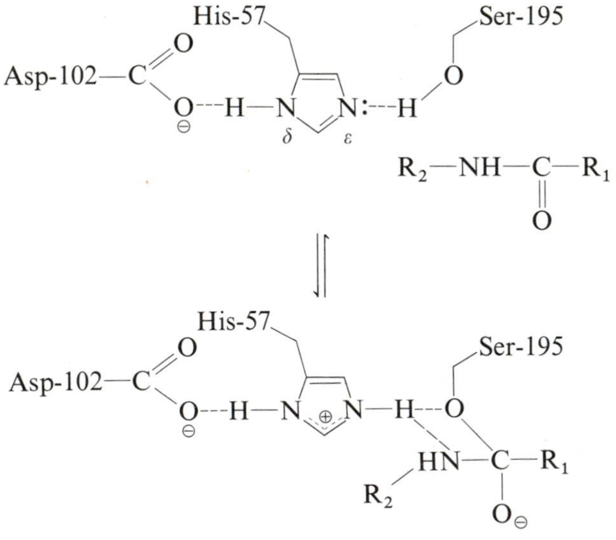

Therefore, both ^{1}H-NMR evidence clearly support the existence of a His-57–Asp-102 ion pair (86) and a concerted transfer of two protons becomes unlikely.

A recent X-ray work (87) on α-lytic protease showed that Asp-102 is in a strongly polar environment with a pK_a of 4.5. Furthermore, histidine, enriched in ^{15}N in the imidazole ring, has been incorporated into α-lytic protease, using a histidine auxotroph of *Myxobacter* 495 (88). The behavior of the ^{15}N-NMR resonances of this labeled α-lytic protease in the "catalytic triad" as a function of pH indicated clearly the presence of a hydrogen-bonded interaction between NH at the position-3 of histidine (Nδ1) and the adjacent buried carboxylate group of aspartic acid.

Furthermore, Komiyama and Bender (92) proposed that the proton abstracted from the hydroxyl group of the serine by the imidazolyl group of the histidine is donated to the nitrogen atom of the leaving group of the amide before the bond between the carbonyl carbon atom of the amide and the attacking serine oxygen atom is completed.

It would be most interesting to know the pK_a of the active site His-57 in the transition state of an α-chymotrypsin catalyzed reaction. Efforts in this direction comes again from the group of Markley (90). They have looked at the pH dependence of the ^{31}P-NMR chemical shifts of a series of diisopropyl-phosphyryl-serine (DFP-serine) proteases. The pK_a values obtained from these titration studies agree with earlier pK_a values derived from ^{1}H-NMR data for peaks assigned to the hydrogen at position Cϵ1-H (C-2) of His-57 of each of these DFP derivatives. Interestingly, the pK_a of His-57 of three enzymes studies (α-chymotrypsin, trypsin, and α-lytic protease) increases (by as much as 2.0 pK_a units for α-lytic protease) when Ser-195 is derivatized. Since inhibitors such as DFP can be considered as transition state analogues (91) of serine proteases, Markley argued that the increased pK_a values of His-57 observed in his derivatives may represent an approximation of the transition-state pK_a values. The fact that His-57 has a lower pK_a in free enzymes ensures that the imidazole residue is in the unprotonated form at physiological pH (90). After formation of the substrate–enzyme complex and as the transition state is approached, the pK_a of His-57 is raised, making the imidazole a more efficient base for accepting the hydroxyl proton of Ser-195. Obviously, biophysical techniques such as nuclear (^{1}H, ^{15}N, ^{13}C, and ^{31}P) magnetic resonance spectroscopy have played a key role in the evaluation of the correct mechanism of action of serine proteases.

Trypsin, a mammelian protease, and subtilisin, a bacterial protease, have both been shown to have a mechanism of action similar to α-chymotrypsin. While α-chymotrypsin and subtilisin have totally different foldings of their polypeptide backbones, the residues involved in catalysis (serine, histidine, aspartic acid) have the same spatial relationships. This similarity of active centers is a prime example of *convergent evolution* of active center geometries in enzymes (77).

4.4.1 Tetrahedral Intermediates

The existence of a tetrahedral intermediate in the enzyme-catalyzed reaction of an ester or an amide has been a long standing question of major concern in mechanistic enzymology. For this reason much attention has been given to the catalytic mechanism operative in serine proteases, based on the availability of much data from chemical and X-ray studies. In 1979 (89), tetrahedral intermediates were observed with the mammelian enzyme elastase. With specific di- and tripeptide *p*-nitroanilide substrates, tetrahedral intermediates have been detected, accumulated and stabilized at high pH (pH of 10.1) by using subzero temperatures ($-39°C$) and fluid aqueous/organic cryosolvents. When corrected for the effect of temperature and cosolvent, the rate of intermediate formation was in good agreement with results obtained at $25°C$ in aqueous solution by stopped-flow techniques. These recent findings are necessary prerequisites for eventual crystallographic cryoenzymological studies. The implication of an imidazole function in the mechanism of action of the serine proteases has led to extensive investigations of imidazole catalysis of ester hydrolysis. For instance, many model compounds indicated that histidine can act as a nucleophile in intramolecular reaction. An early example is the hydrolysis of *p*-nitrophenyl γ-(4-imidazolyl) butyrate (93):

acyl-enzyme analogue

At pH 7, $25°C$, the rate is comparable ($200 \ \text{min}^{-1}$) to the release of *p*-nitrophenolate by α-chymotrypsin ($180 \ \text{min}^{-1}$). The rate-limiting step here is the acylation step k_2. In the first approximation it is a satisfactory model to mimic the formation of an acyl-enzyme intermediate although His-57 in α-chymotrypsin acts as a general-acid–general-base catalyst and not as a nucleophilic catalyst.

As shown in the previous section, the geometrical arrangement of the triad Asp-His-Ser in the active site of serine proteases has been referred to as the charge-relay system. The ensemble is situated in a hydrophobic environment inside the enzyme and attack by the hydroxyl of the serine residue is assisted by general-base removal of the proton by the imidazole ring of the histidine residue. This results in the (partial) generation of a potent (unsolvated) alkoxy nucleophile.

G. A. Rogers and T. C. Bruice have made an extensive study with model compounds to evaluate the validity of the charge-relay system (94–97). The following model compounds show an enhancement in acetate hydrolysis of up to 10^4-fold relative to a mixture of phenylacetate and imidazole.

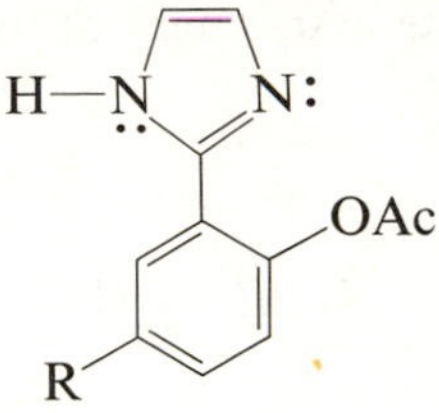

$$R = H, SO_3^{\ominus}$$

Three schemes of catalysis have been proposed, depending on experimental conditions.

(a) at low pH

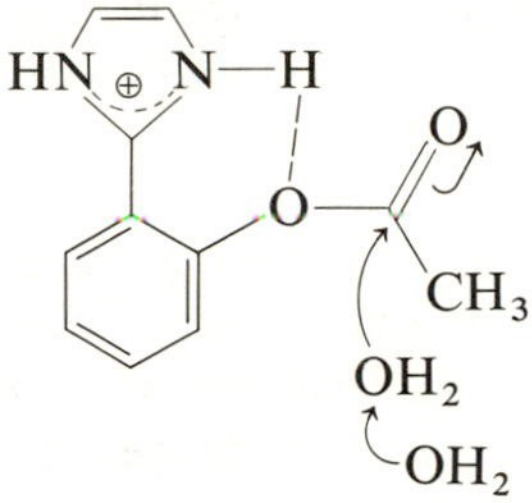

general-acid assisted
H_2O attack

(b) at neutral pH

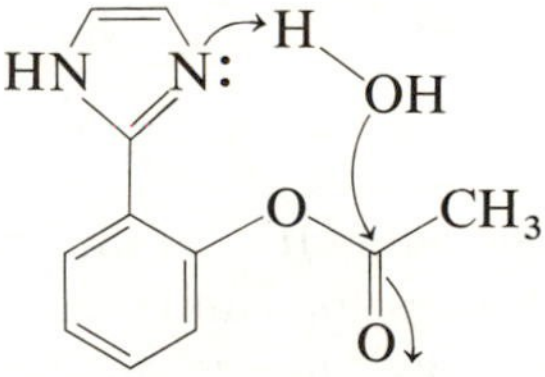

general-base assisted
H_2O attack

(c) at high pH

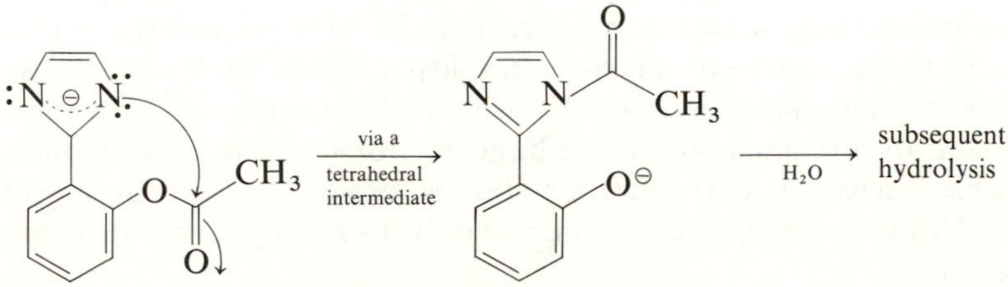

nucleophilic mechanism

At high pH the possibility of O → N acyl-transfer to imidazole anion occurs via a nucleophilic mechanism. Subsequent hydrolysis of the acyl-imidazole intermediate takes place because the rate of hydrolysis exceeds the thermodynamically unfavorable N → O transfer. This model does have some analogy with serine proteases mode of action although only the neutral pH mechanism is relevant. The introduction of a carboxylate group, however, provides a more accurate assessment of the charge-relay system:

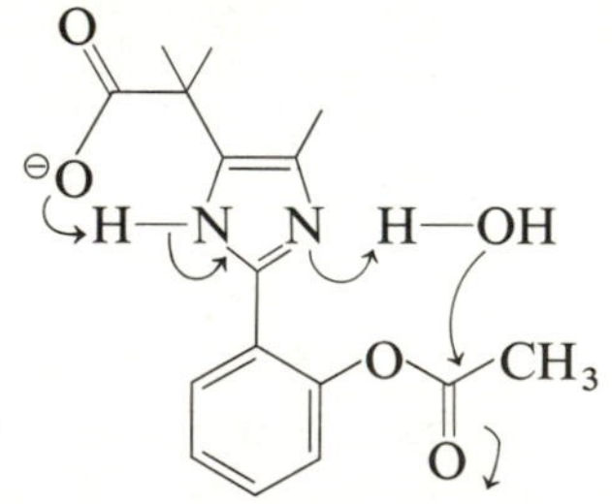

The introduction of the hydrogen-bonded carboxylate does enhance catalysis by the neighboring imidazole and assesses the role of tandem general-base hydrolysis. However, the enhancement in rate of ester hydrolysis is still poor (only by a factor of three). So it is negligible from the stand point of enzymatic catalysis.

What is the effect of solvent? Is water an integral part of the mechanism? Experiments with a less polar solvent did not alter the role of the carboxylate group. A hydrogen bond should be expected between the carboxylate group and an imidazolium ion. In fact, hydrogen bonding has been definitely established for the zwitterion:

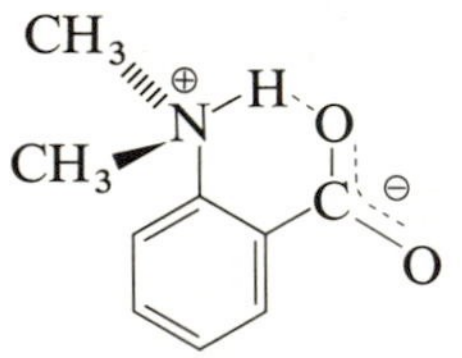

It was found that the reaction with OH⁻ at the bridged-proton is 10^3 slower than diffusion controlled values for simple anilinium ions (96). So if the concentration of water decreases in the medium (dielectric constant decreases), both dipole–dipole interactions and hydrogen bonding are expected to increase, a situation not encountered in Bruice's model compound.

Nevertheless, Rogers and Bruice's model does support the charge-relay hypothesis because larger rate enhancements have been reported when nearly anhydrous acetonitrile or toluene is employed. Under these conditions, the hydrogen bond system is "locked" inside and does not exchange with the medium. In this situation scheme b, general-base assisted water attack, is possible where a dipolar transition state is formed from a neutral ground state. This would happen in acetonitrile only if a nearby carboxylate (anion) is present.

The model also had the merit of allowing the first isolation of a tetrahedral intermediate in an acetyl-transfer reaction (94). The scheme is as follows:

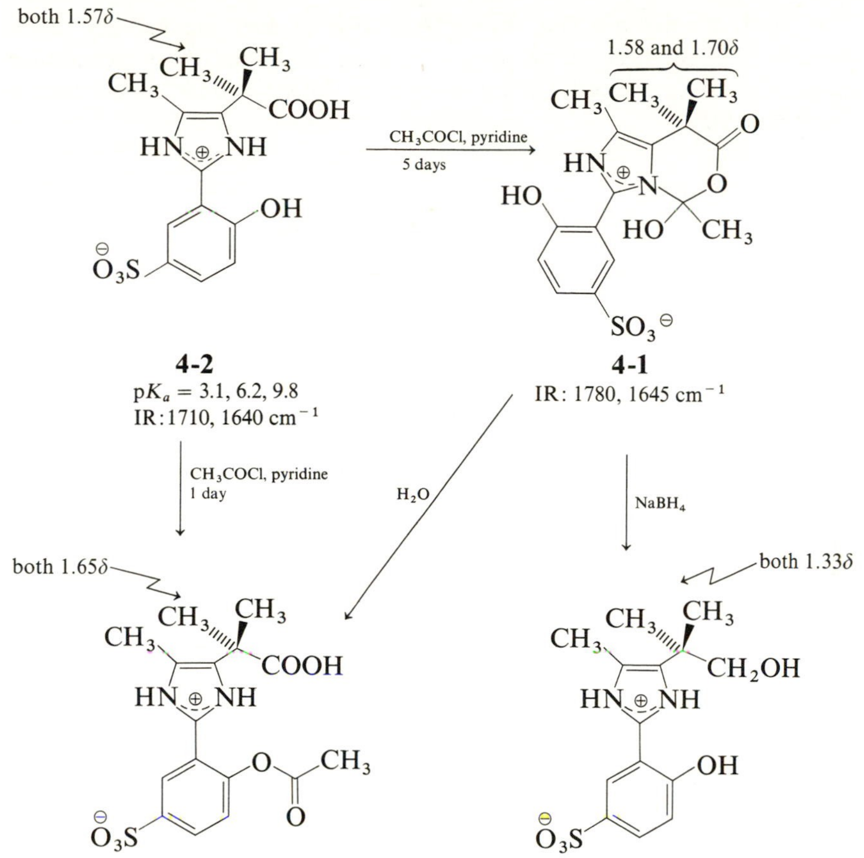

both 1.57δ —

CH₃ CH₃ ⁗ CH₃
CH₃ — ... — COOH
HN ⊕ NH
— OH
⊖O₃S

4-2
pK_a = 3.1, 6.2, 9.8
IR: 1710, 1640 cm⁻¹

$\xrightarrow[\text{5 days}]{\text{CH}_3\text{COCl, pyridine}}$

1.58 and 1.70δ

CH₃ CH₃ ⁗ CH₃
O
HN ⊕ N O
HO HO CH₃
SO₃⊖

4-1
IR: 1780, 1645 cm⁻¹

CH₃COCl, pyridine
1 day

H₂O

NaBH₄

both 1.65δ —

CH₃ ⁗ CH₃
CH₃ — COOH
HN ⊕ NH
O — CH₃
⊖O₃S O

4-3
IR: 1780, 1730, 1645 cm⁻¹
δ = chemical shift (in ppm)

both 1.33δ

CH₃ ⁗ CH₃
CH₃ — CH₂OH
HN ⊕ NH
— OH
⊖O₃S

4-4
pK_a = 5.8, 8.9

The conversion **4-1** → **4-3** involves a N-acyl imidazole intermediate. The evidence is the following. Competition experiments in the presence of an amine acting as a base indicates that the carboxyl group of **4-1** is attacked to give the corresponding amide of **4-3**. NMR spectroscopy shows the non-equivalence of the two geminal methyl groups in **4-1** but not in **4-2** or **4-3**. This is caused by the rigidity and the asymmetric nature of the tetrahedral intermediate. Furthermore, borohydride reduction of **4-1** to **4-4** shows the presence of a lactone function. No reduction occurred with compounds **4-2** or **4-3**.

The formation of **4-1** from **4-2** probably proceeds through a carboxyl attack on the N-acetyl pyridinium salt to give an anhydride intermediate before imidazole displacement. This represents the first isolation and un-equivocal characterization of a labile acyl tetrahedral intermediate **4-1** capable of a migration to give an acyl-transfer product **4-3**.*

---

* Recently, tetrahedral adducts between a strongly electrophilic ketone and tertiary amines have been isolated and characterized (358).

In 1977, Komiyama and Bender (98) also studied the validity of the charge-relay system in serine proteases. They examined the general base-catalyzed hydrolysis of ethyl chloroacetate by 2-benzimidazoleacetic acid as a model system.

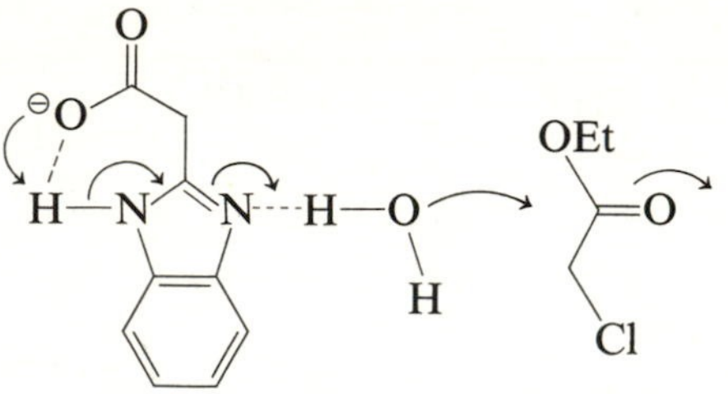

Compared to benzimidazole itself, an 8-fold increase in rate of hydrolysis is observed, implying general-base catalysis assisted by the carboxyl group.

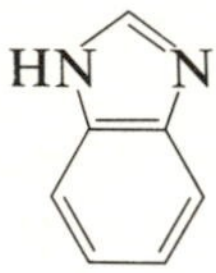

benzimidazole 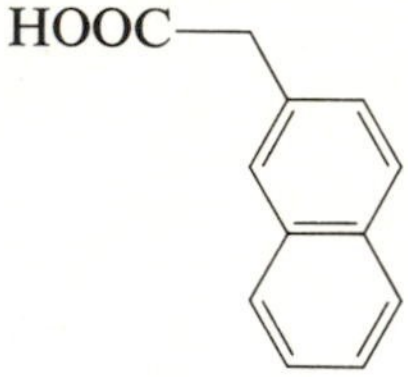   β-naphthyl   N-methylimidazole
                       acetic acid

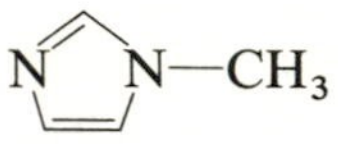

But the carboxyl group alone, as in β-naphthyl acetic acid, cannot catalyze the reaction. Therefore, the mechanism involves the cooperative function of the carboxylate, the imidazole and water.

N-Methylimidazole was also tested as a model comparable to methylated α-chymotrypsin. It was completely inactive toward the hydrolysis of ethyl chloroacetate and indicates that in the charge-relay system both nitrogens of the imidazole ring must be free to participate in the relay.

The properties of α-chymotrypsin methylated at His-57 were also examined to explain the mechanism of this enzyme which is about 10^5 times less active than α-chymotrypsin (99). The results suggest that general-base catalysis remains an integral feature of the hydrolytic mechanism of the modified enzyme. While only subtle alterations occur in the active site upon methylation of His-57, the transition state and the tetrahedral intermediate are destabilized relative to the native enzyme.

Hydroxyl group participation in hydrolysis of esters and amides also proceeds through the formation of a tetrahedral intermediate (T.I.).

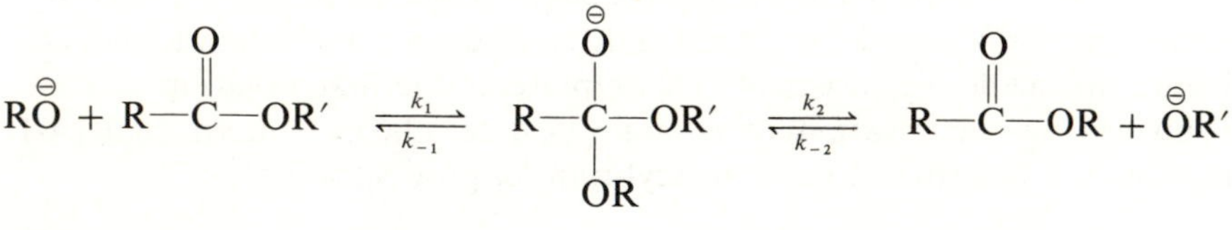

formation   T.I.   breakdown

The pK_a value of the leaving group (R'O$^-$) leads to two transition states:

$$\left[\quad RO\overset{-\delta}{}\quad \underset{\underset{R}{|}}{\overset{\overset{O}{\|}}{C}}\cdots\overset{-\delta}{OR'}\quad\right]\qquad \left[\quad RO\overset{-\delta}{}\cdots\underset{\underset{R}{|}}{\overset{\overset{O}{\|}}{C}}\cdots\overset{-\delta}{OR'}\quad\right]$$

    resembles product   resembles reactant
         T.S.   T.S.

If the transition state of the tetrahedral intermediate resembles the product, the rate limiting step is the breakdown of the tetrahedral intermediate. In other words, $k_2 < k_{-1}$, that is k_2 is the uphill process (slow) and k_1 is fast. This situation arises when RO$^-$ is a good nucleophile, or its pK_a is less than the pK_a of R'O$^-$.

In the other situation where the transition state resembles the reactant, the rate-limiting step is the formation of the tetrahedral intermediate where $k_2 > k_{-1}$ and k_1 is slow. In this case R'O$^-$ is a good leaving group and its pK_a is smaller than the pK_a of RO$^-$, the nucleophile. These principles are readily applicable to the hydrolysis of p-nitrophenyl esters by a methoxy group.

$$R\!-\!\underset{\underset{\underset{CH_3O}{\ominus}}{\uparrow}}{\overset{\overset{O}{\|}}{C}}\!-\!O\!-\!\!\left\langle\right\rangle\!\!-\!NO_2$$

It is a downhill process when the nucleophile is the conjugate base of a weak acid (methanol) and the leaving group is the conjugate base of a strong acid (pK_a is small). Consequently, the reverse reaction is much more difficult.

In summary, with a good leaving group (weak conjugate base) the rate limiting step (slow) in ester hydrolysis is the formation of the tetrahedral intermediate. With a poor leaving group the rate-limiting step is the breakdown of the tetrahedral intermediate.

Of course, hydroxyl group participation in catalysis has some analogy with the participation of the serine residue in serine proteases. For this reason model compounds have been prepared and studied.

An example is the following intramolecular transformation:

The rate of this reaction is 10^5 larger than for ethyl benzoate. The reaction is base-catalyzed and as expected in the presence of ^2H_2O instead of H_2O solvent, the rate is reduced by a factor of at least two $(k_H/k_D = 3.5)$. Thus, proton-transfer occurs in the transition state and the analogy with α-chymotrypsin acylation is apparent.

Another interesting simple model is the amide analogue (100):

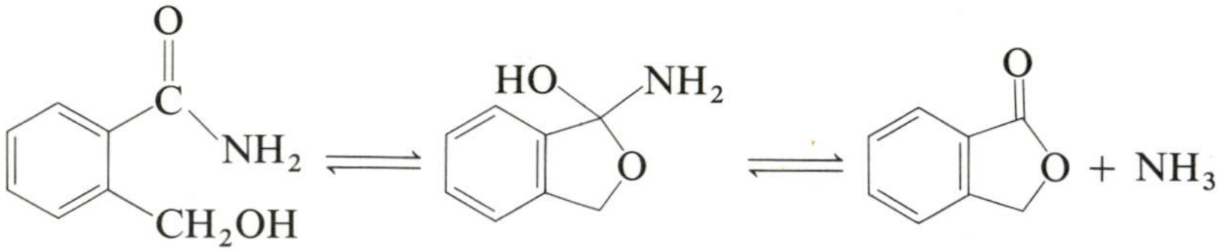

At low pH, protonation of the nitrogen occurs and $-\overset{+}{N}H_3$ becomes a good leaving group. At high pH, however, the hydroxyl group of the tetrahedral intermediate becomes a better leaving group. The reaction was studied in the stable imido form:

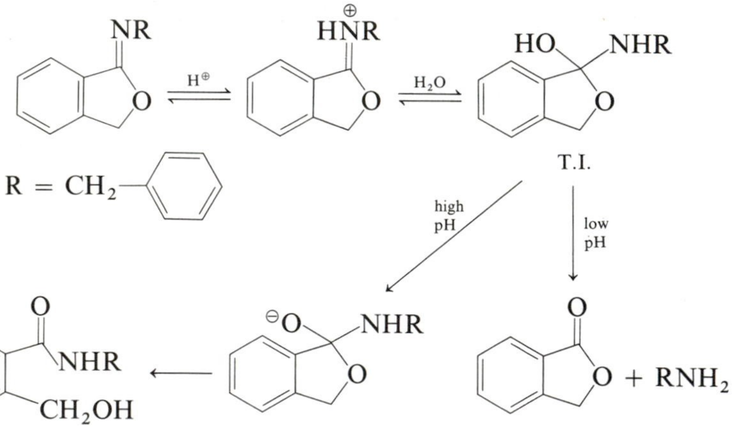

Other model compounds were synthesized with an amino group in proximity. At position 6 it was hoped that the amino group would help to decompose the tetrahedral intermediate and hence accelerate the rate of the reaction. However, the presence of the amino group resulted in a reduction in rate by a factor of ten.

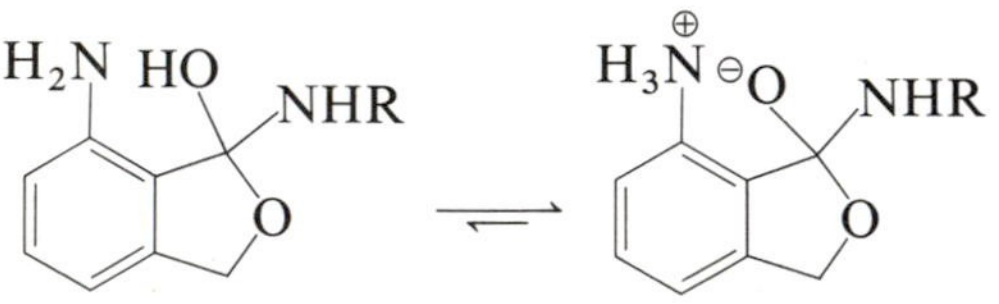

stable ion pair

Possibly the formation of an ion-pair attractive interaction stabilizes the tetrahedral intermediate from being broken down.

What would happen if the amino group were at position 3 instead?

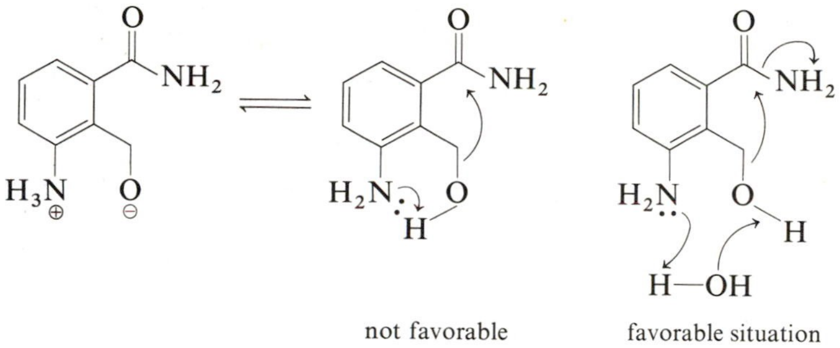

not favorable   favorable situation

The rate of the reaction becomes 10^3 faster than with the 6-amino analogue. In ^2H_2O a k_H/k_D value of 2.82 is obtained, suggesting an intramolecular base catalysis through solvent participation.

4.4.2 Absolute Conformation of Bound Substrate

What is the exact orientation of a substrate in the active center? What is the position of the scissile bond relative to the catalytic groups? Those are some of the questions asked by C. Niemann of Caltech and B. Belleau of McGill University.

One way to approach this problem would be through the use of simple model substrates for the enzyme α-chymotrypsin. In this regard, Hein and Nieman (101) first attempted the elucidation of the conformation of well-selected chymotrypsin-bound substrates by using a conformationally constrained molecule as a model for the active conformation of a typical open-chain substrate, N-acetyl L-phenylalanine methyl ester (L-APME). For

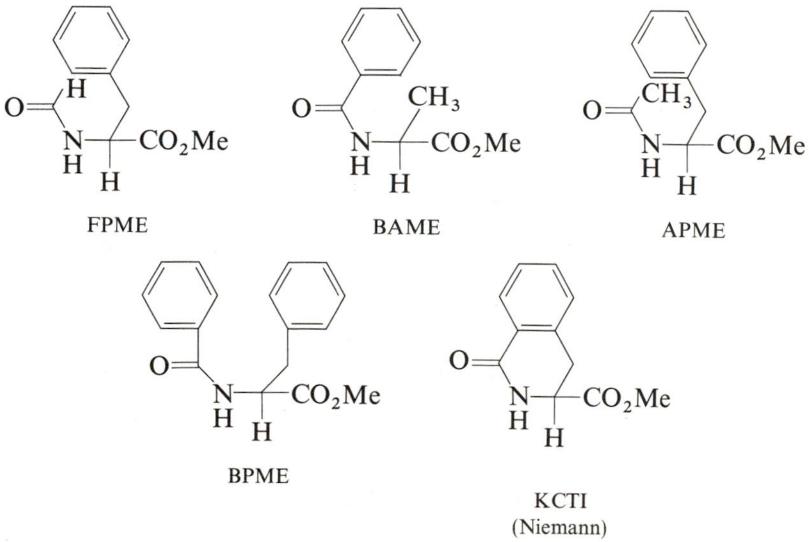

FPME   BAME   APME

BPME

KCTI
(Niemann)

this Niemann studied the kinetic behavior of D- and L-1-keto-3-carbomethoxy-1,2,3,4-tetrahydroisoquinoline (KCTI). It is a rigid analogue of L-APME as well as N-formyl phenylalanine methyl ester (L-FPME) and N-benzoyl L-alanine methyl ester (L-BAME), all of which are good substrates of α-chymotrypsin.

Although, α-chymotrypsin is sterospecific toward the L-isomer of most amino acid substrates, Niemann showed that the stereospecificity is reversed in the case of KCTI. The D-isomer of this conformationally restricted ester is hydrolyzed at a rate comparable to that of N-acetylated L-phenylalanine methyl esters while the L-isomer is hydrolyzed very slowly. Hein and Niemann (101, 102) pointed out that this anomaly is consistent with a requirement for the carboxylate group of D-KCTI to be in an axial conformation, a conformation which matches a probable conformation of open-chain L-amino acid ester substrates.

D-KCTI

Subtilisin, a serine protease of bacterial origin, also shows the same inversion of stereospecificity toward D-KCTI as α-chymotrypsin (103). This suggests that both enzymes have similar specificity with respect to the configuration of their substrates. This common specificity neatly confirms and reflects the close structural analogy between the primary binding sites of the two enzymes. This close structural analogy shows how α-chymotrypsin and subtilisin, which have completely different phylogenetic origins, actually "freeze" their substrates in the same active conformation.

The work of Silver and Sone (104) favors, however, the equatorial orientation of the ester group. Thus, the results of these studies gave rise to very intense controversies regarding the orientation (axial or equatorial) of the carbomethoxy function of D-KCTI, and hence of enzyme-bound L-APME (104, 105). Whether the orientation of this ester function is axial or equatorial has not been unambiguously resolved with Niemann's compound and its derivatives. The reason for this probably lies in the fact that in most of the compounds used the structural features of the normal substrates of α-chymotrypsin are so grossly altered as to cast strong doubts on the validity of the models.

However, one uniquely constrained substrate, 3-methoxycarbonyl-2-dibenzazocine-1-one, was synthesized in 1968 by Belleau and Chevalier from diphenic anhydride (106). This 2,2'-bridged biphenyl compound is a constrained analogue of N-benzoyl phenylalanine methyl ester (BPME).

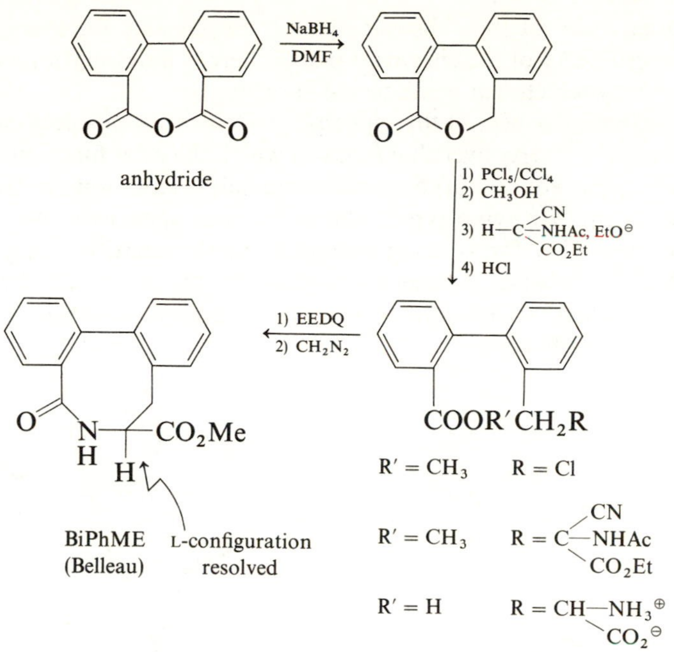

$$\xrightarrow[\text{DMF}]{\text{NaBH}_4}$$

anhydride

1) PCl$_5$/CCl$_4$
2) CH$_3$OH
3) H—C$\overset{\text{CN}}{\underset{\text{CO}_2\text{Et}}{\text{—NHAc, EtO}^\ominus}}$
4) HCl

$$\xleftarrow[\text{2) CH}_2\text{N}_2]{\text{1) EEDQ}}$$

N—CO$_2$Me
H H↑

BiPhME   L-configuration
(Belleau)   resolved

COOR′CH$_2$R

R′ = CH$_3$ R = Cl

R′ = CH$_3$ R = C$\overset{\text{CN}}{\underset{\text{CO}_2\text{Et}}{\text{—NHAc}}}$

R′ = H   R = CH—NH$_3^\oplus$
                        CO$_2^\ominus$

This biphenyl model compound was shown to possess the so-called *primary optical specificity* of α-chymotrypsin. That is, only the enantiomer related to L-phenylalanine was hydrolyzed by the enzyme. Generally speaking, there are three more levels of specific substrate recognition in enzyme catalysis. Let us consider a peptide bond in a polypeptide chain. The lateral side chain R$_2$ is responsible for the normal specificity of the enzyme. For α-chymotrypsin, R$_2$ is an aromatic side chain and the hydrophobic cavity a "aromatic hole" in the active center is there to accommodate the amino acid to be recognized by the enzyme. This is referred to as the *primary structural specificity.*

primary optical
specificity (D or L)

bond to
be cleaved

secondary
structural
specificity

—NH—CH—C—NH—C—C—NH—R$_3$

R$_1$ R$_2$

secondary
structural
specificity

primary
structural
specificity

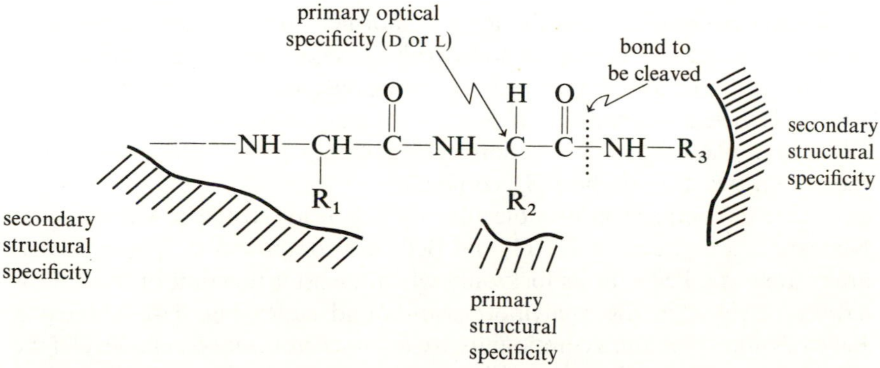

The influence of the adjacent amino acid residues R$_1$ and R$_3$, in the active center subsites, is also important to maintain favorable interaction and proper orientation of the substrate. This is responsible for the *secondary structural specificity* of the enzyme. Finally, there is a *tertiary* level of structural specificity which will be discussed shortly.

One fascinating aspect of the biphenyl model is that in solution it exists as two sluggishly interconvertible forms in which the ester function occupies either an outside (equatorial) or an inside (axial) orientation relative to the biphenyl system. α-Chymotrypsin shows marked specificity for the S,S_{eq} conformer, the other forms being essentially inert toward the enzyme. The rate of hydrolysis and Michaelis constant for the active conformer are virtually identical to those of the corresponding normal substrate N-benzoyl phenylalanine methyl ester.

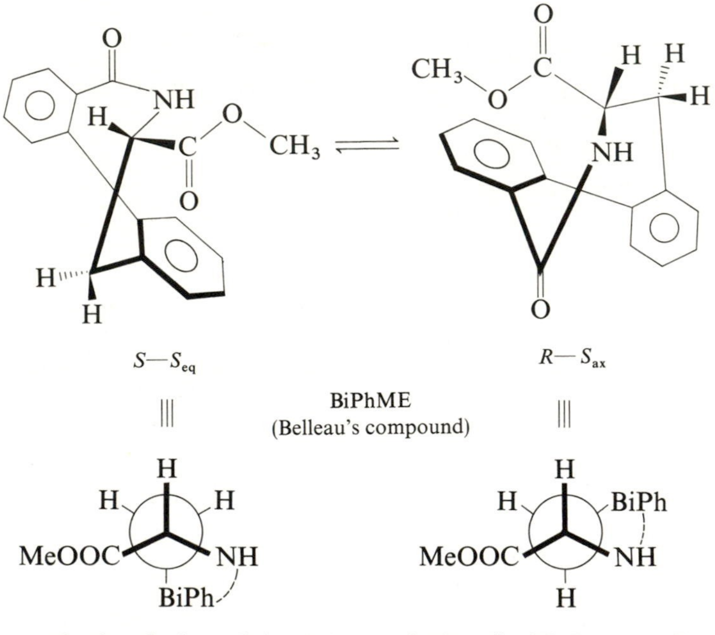

$S—S_{eq}$ $R—S_{ax}$

BiPhME
(Belleau's compound)

When a fresh solution of the R,S_{ax} conformer (in 95:5 water–dioxane) is incubated with α-chymotrypsin no hydrolysis occurs. However, with an aged stock solution in dioxane α-chymotrypsin-catalyzed hydrolysis takes place readily. Hence, isomerization to an enzymatically active conformer is gradually taking place upon destruction by dissolution in an organic solvent of the crystal structure of the R,S_{ax} conformer.

A careful comparison of molecular models leads to the conclusion that Niemann's compound (D-KCTI) and Belleau's compound (S,S_{eq} conformer) are related as a key is to its lock only when the ester function of D-KCTI is *axially* oriented in the α-chymotrypsin-bound state (Fig. 4.4). Therefore, Belleau's and Niemann's compounds are identical at the molecular level if the ester function is axially oriented in the Niemann's product. The versatility of Belleau's compound makes it a better model for analyzing the conformation of α-chymotrypsin-bound substrates.

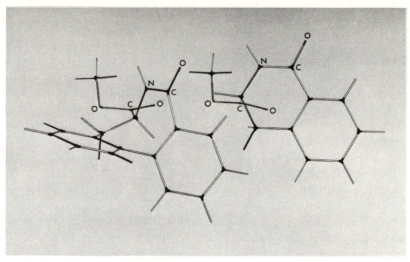

Belleau's compound   Niemann's compound
($S - S_{eq}$ form)   (D-isomer)

Fig. 4.4. Dreiding stereomodels showing that the D-configuration of Niemann's compound can be superimposed on Belleau's biphenyl model compound.

Also important is the finding that not only the R,S_{ax} conformer is inert to hydrolysis by α-chymotrypsin, but it also failed to inhibit enzymatic hydrolysis of the active S,S_{eq} conformer. In marked contrast, L-KCTI has been shown to strongly inhibit chymotryptic hydrolysis of D-KCTI. This pattern of competitive inhibition has also been demonstrated for other enantiomeric pairs of chymotrypsin substrates. To understand this behavior it should be realized that the two conformers of Belleau's compound differ in two important aspects: orientation of the carbomethoxyl group and the chirality of the biphenyl system. Consequently, it must be concluded that in its reaction with this constrained substrate, α-chymotrypsin displays specific recognition of molecular asymmetry. This is referred to as *tertiary structural specificity*. The specificity of the biphenyl compound thus serves to extend the concept that appropriately constrained substrates can serve as very useful tools.

Is this "molecular chiral specificity" the same for enzymes having such different phylogenic origins as papain (thiol protease of plants) and subtilisin (serine protease from bacteria)? The constrained biphenyl compound is naturally a good choice for answering this question because it is a molecular asymmetric analogue of BPME, which is a good substrate for all three proteases. It turns out that substilisin behaves exactly like α-chymotrypsin toward the biphenyl isomers. However, with papain, both R,S_{ax} and S,S_{eq} conformers are inactive. The inactivity of the compound toward papain might be due to the *cis* configuration of the acylamido portion of the substrate. In other words, papain would require a *transoid* configuration about the amide bond in order to hydrolyze its substrates. However, the situation is not as simple because *cisoid* and *transoid* 9- and 10-membered lactam substrate analogues were synthesized by Elie and Belleau (107). They found that the *trans* isomers have no detectable substrate activity toward any of the three proteases tested.

Although these negative results are somewhat disappointing (it is often the case with model compounds) they nevertheless suggest that the reactivity of a particular substrate seems to be determined principally by the conformation of the scissile bond relative to the overall shape of the substrate molecule. The crucial parameter is whether or not the scissile bonds are held in the correct orientation for catalysis.

As already mentioned, there exist a considerable cross-specificity among α-chymotrypsin, papain, and subtilisin. The results of such studies on chiral specificity will eventually help bring to light new aspects of the evolutionary divergences between mammalian, bacterial, and plant proteases. In addition, zymogen activation is often involved in the biosynthesis of proteases themselves as well as in a great variety of biological processes such as blood coagulation, complement reaction, hormone production, fibrinolysis, etc. Such precise and limited proteolysis by enzymes, exhibiting a generally broad primary specificity, illustrates also the overwhelming importance of the tertiary structural specificity of proteases in their interaction with their natural substrates (107).

In parallel fashion, one of the key problems in *molecular pharmacology* is the cross-specificity exhibited by different classes of receptors. As a result, drugs having the structural features indispensable for activity toward a given receptor often cause undesirable side effects through interactions with other related receptor sites. Knowledge of the *chiral specificity* of a particular receptor (where applicable) could allow the design of drugs with a much improved specificity for this receptor. Hopefully, a systematic examination of the chiral specificities of proteases could provide, at a simpler level, a more rational basis for the development of receptor-specific effectors because enzyme–substrate interactions are fundamentally of the same nature. Proteases offer the advantage of being less complex and more accessible than the often elusive receptors.

4.5 Other Hydrolytic Enzymes

Among the best known hydrolytic enzymes of which enough structural and mechanistic information are available are the exopeptidase carboxypeptidase A (see Chapter 6), ribonuclease A (see Chapter 3), and lysozyme. In this chapter we shall examine the chemistry of this last enzyme.

Lysozyme is an important enzyme that catalyzes the hydrolysis of a polysaccharide that is the major constituent of the cell wall of certain bacteria. The polymer is formed from $\beta(1 \rightarrow 4)$ linked alternating units of *N*-acetylglucosamine (NAG) and *N*-acetylmuramic acid (NAM) (Fig. 4.5).

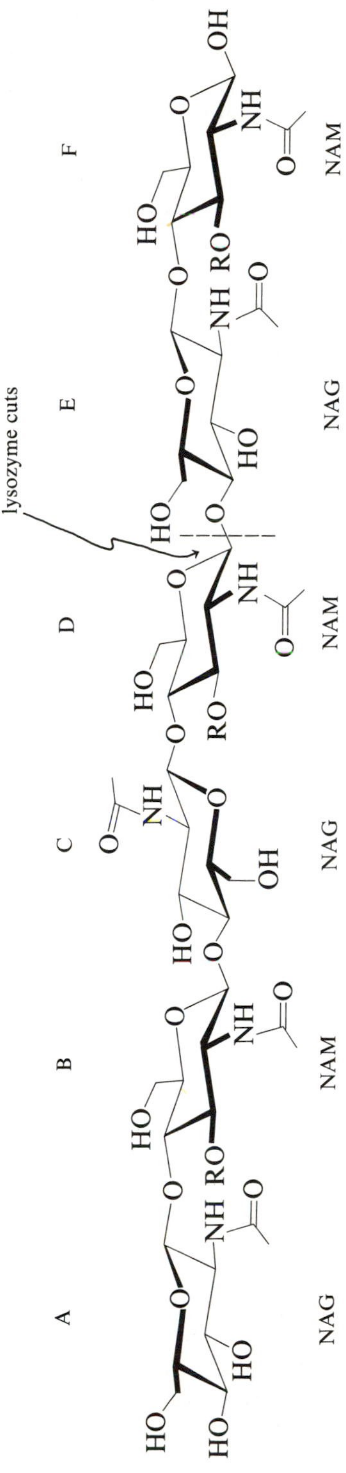

Fig. 4.5. Structure of an hexasaccharide that can bind at the active center of lysozyme. Upon binding to the enzyme, the sugar ring D of the substrate becomes distorted and catalysis proceeds through the promotion of an oxocarbonium ion (see p. 228). This results in a polar transition state. However, an important feature of enzyme is the capacity to stabilize (neutralize) the enzyme–substrate complex by *electrostatic inter-actions* with amino acid residues at the active site.

The enzyme is small, having a polypeptide chain of 129 amino acids. It was the first one, in 1967, to have its tertiary structure elucidated by X-ray crystallography (108). Unlike α-chymotrypsin, lysozyme has a well-defined deep cleft running down one side of the ellipsoidal molecule for binding the substrate.

The cleft is divided into six subsites, ABCDEF. NAM residues can bind only in sites B, D, and F, while NAG residues of synthetic substrates may bind in all sites. The bond that is cleaved lies between sites D and E.

The carboxyl group of Glu-35 in the unionized form and the carboxyl group of Asp-52 in its ionized form are the two functions implicated at the active site.

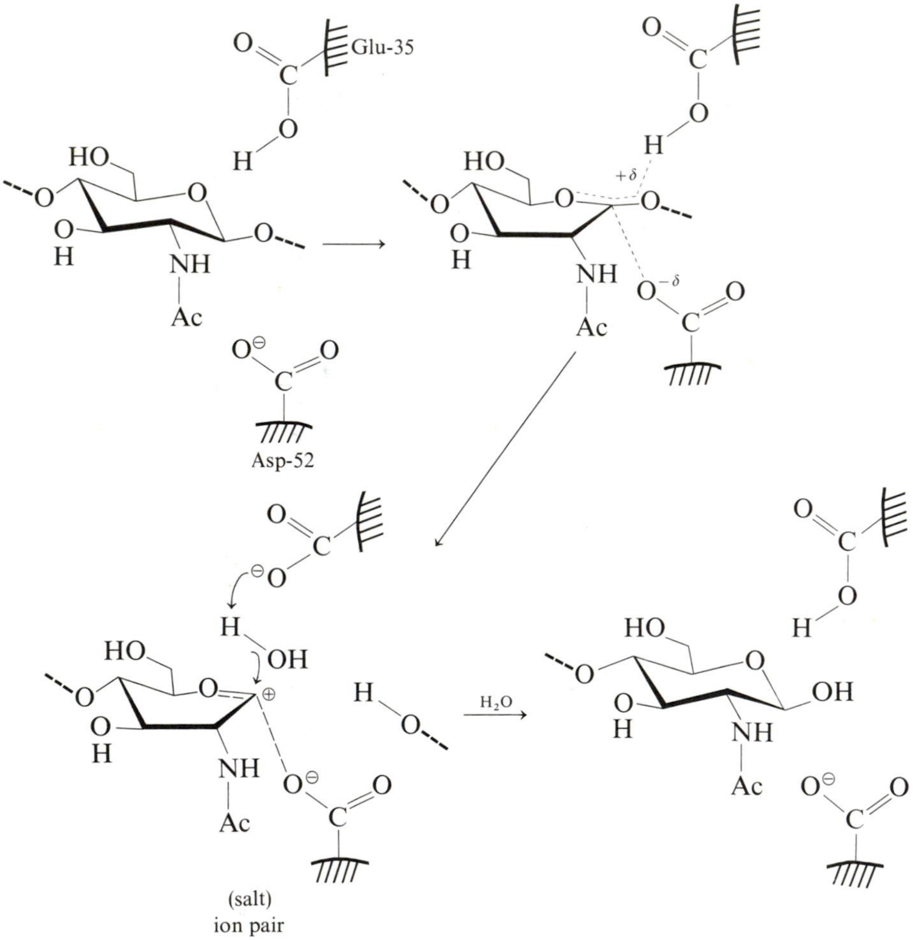

Glu-35 serves as a general-acid catalyst and Asp-52 stabilizes, by electrostatic interaction, the oxocarbonium ion intermediate. It helps the development of a positive charge at the anomeric carbon. Interestingly, the intermediate develops a half-chair conformation to relieve the strain created by the introduction of a transition state with sp^2 character (double bond) in the ring. Such structural change of the substrate allows departure of the leaving group (ring E) via stereoelectronic control (Section 4.6). The mechanism of action of lysozyme is also a good illustration of a concept stated in 1948 by L. Pauling: *The active site of an enzyme is complementary to the transition state of the reaction it catalyzes.* In other words, the enzyme has evolved to bind more efficiently a transition state intermediate between substrate and product than the substrate itself in its ground state (349).

A great deal of theoretical work has been carried out to evaluate the binding energy involved between substrates in enzymes. In this respect, the efforts of H. A. Scheraga's group from Cornell University have given promising results (110). They have investigated, by conformational energy calculations, the most favored binding modes of oligomers of β-D-N-acetylglucosamine (NAG) to the active site of lysozyme. In order to to this, both the substrate and the side chains of the enzyme were allowed to undergo conformational changes and relative motions during energy minimization. It was found that $(NAG)_6$ had a clear preference for binding to the active site cleft of lysozyme with its last two residues on the "left" side of the cleft region thus in agreement with earlier X-ray work (109). The calculations also showed that the presence of the N-acetyl groups on NAG oligomers is important to provide sufficient interactions with the enzyme and accounts for the fact that glucose oligomers bind with much lower affinity.

Study of substrates' distortion could of course, in model compounds, have important mechanistic consequences as well as an effect on reaction rate.

For example, the hydrolysis of the following acetals has been studied (111):

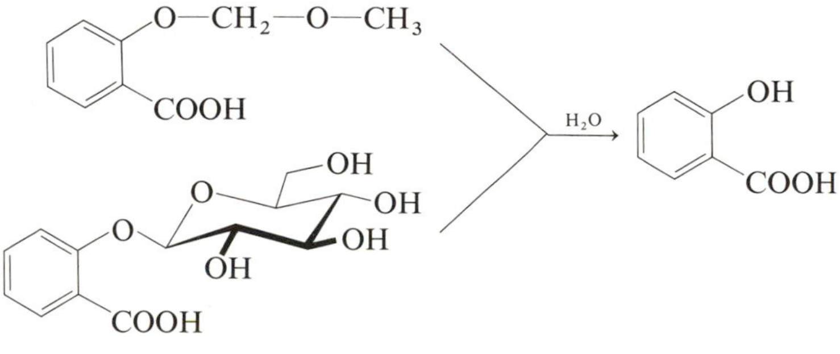

The presence of an o-carboxyphenyl facilitates the rate of hydrolysis by a factor of 10^4 in the glucose series and by 600 in the other simpler molecule relative to the *para* compound. But the participation of the carboxyl group in the mechanism of the latter example remains obscure. Again, reaction in ^2H_2O vs H_2O indicates that proton-transfer occurs in the critical transition state.

Three possible intramolecular mechanism are possible: two involve a nucleophilic attack by the carboxylate and one involves an intramolecular general acid catalysis.

(a) [structure: benzene ring with $O-CH_2-\overset{\oplus}{O}-CH_3$ substituent bearing H, and $COO^{\ominus}$ substituent] $\longrightarrow$

[structure: bicyclic benzodioxinone with O, O, and $=O$] $\xrightarrow[\text{H}_2\text{O}]{\times}$ [structure: benzene ring with OH and $COOH$]

$+$ $+$

CH_3OH [structure: H and H with $=O$, formaldehyde]

However, this first mechanism does not occur because synthesis of the cyclic intermediate by a different route showed that it is stable to the reaction conditions.

(b) [structure: benzene ring with $\overset{\oplus}{O}-CH_2-O-CH_3$ bearing H, and $COO^{\ominus}$ substituent] $\longrightarrow$

[structure: benzene ring with OH substituent and $\underset{O}{\overset{\|}{C}}-O-CH_2-OCH_3$ substituent] $\longrightarrow$ products

This corresponds to a $\overset{+}{C}H_2OCH_3$ migration. However, no isosbectic points* in the UV is seen between the spectrum of the reactant and product, as would be expected if this mechanism were correct.

(c) [structure: benzene ring with $\overset{+\delta}{C}H_2-\overset{..}{O}CH_3$ over $O---H$, and $\underset{O}{\overset{\|}{C}}-O^{-\delta}$ substituent] $\longrightarrow$ [structure: benzene ring with OH and $COO^{\ominus}$] $+ \ CH_2^{\oplus}\cdots O-CH_3$
                                                                                                                                      oxocarbonium ion

---

* Isosbectic points are those points of equal amplitude at a given wavelength that occur in superpositions of three or more spectra of a system under different concentrations. It can be observed only for a system containing an equilibrium between *two* spectrally active states in which only the relative populations of the states have been changed.

This last mechanism is the favored one. In a way, it mimics lysozyme because the carboxylate behaves as a general acid (Glu-35) and the CH_3O— group as the glucose ring oxygen.

Finally, mention should be made that the participation of neighboring carboxylic acids in the hydrolysis of amides is also a phenomenon pertinent to the understanding of enzyme-catalyzed amide hydrolysis. One such enzyme is the acid protease pepsin, found in the gastric juice, and it obeys general-acid catalysis. R. Kluger and C. H. Lam, of the University of Toronto, prepared rigid model compounds to study the carboxylic acid participation in amide hydrolysis (112). They found that anilic acid derivatives of endo-*cis*-5-norbornene meet the criteria of rigid geometric proximity of the interacting functional groups.

$$R = C_6H_5$$
$$= C_6H_4—4'—OCH_3$$
$$= C_6H_4—4'—Cl$$
$$= C_6H_4—3'—NO_2$$
$$= C_6H_4—4'—NO_2$$

These substrates are rigid, so that the entropy barrier to formation of an intermediate is minimized. Both general-acid and general-base catalysis are readily observable over a wide pH range for certain of these compounds.

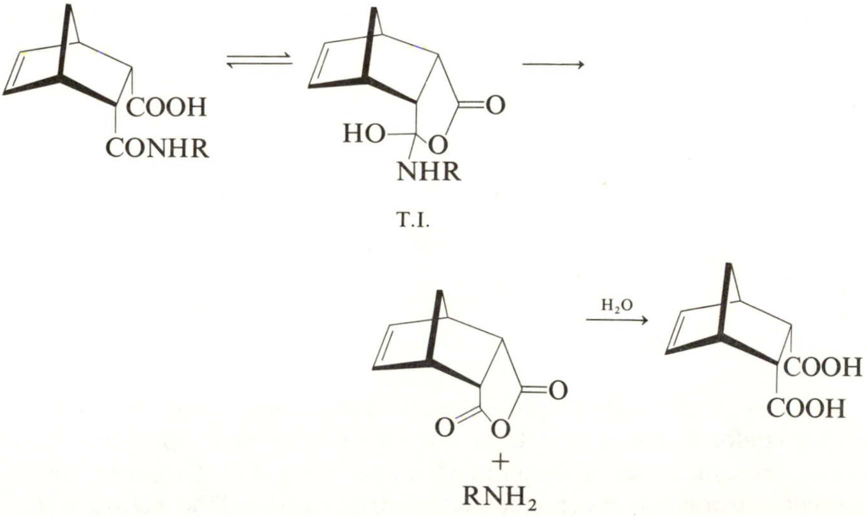

T.I.

The possibility that the acid protease, pepsin, may form a covalent amino enzyme derivative between its active site aspartic acid group and the substrate peptide suggests that a reactive anhydride may form. Since a second carboxylic acid is available at the active site (113), a mechanism analogous to the one above may be relevant.

4.6 Stereoelectronic Control in Hydrolytic Reactions

Now that we have seen examples of hydrolytic reactions and acetal hydrolysis by enzymes, we may wonder how important the stereochemistry of the products and reactants is in these transformations.

This section is devoted to this question through the presentation of a relatively new concept in organic chemistry, *stereoelectronic control*, exploited by P. Deslongchamps from the University of Sherbrooke (114, 115). It uses the properties of proper orbital orientation in the breakdown of tetrahedral intermediates in hydrolytic reactions. This concept is quite different from Koshland's "orbital steering" hypothesis where proper orbital alignment is invoked for the formation of a tetrahedral intermediate. Here we are interested by the process that follows: the cleavage of the tetrahedral intermediate in the hydrolysis of esters and amides.

It is generally accepted that the most common mechanism for the hydrolysis of esters and amides proceeds through the formation of a tetrahedral intermediate. Deslongchamps argued that the conformation of this tetrahedral intermediate (hemiorthoester from ester and hemiorthoamide from amide) is an important parameter in order to obtain a better understanding of the hydrolysis reaction.

$$R-\overset{\overset{\textstyle O}{\|}}{C}-X + H-OR' \longrightarrow \left[R-\overset{\overset{\textstyle OH}{|}}{\underset{\underset{\textstyle OR'}{|}}{C}}-X \right] \longrightarrow R-\overset{\overset{\textstyle O}{\|}}{C}-OR' + H-X$$

$$
\begin{array}{lll}
X = OR'' & R' = H & \text{T.I.} \\
\quad \text{or} & \quad \text{or} & \\
\quad NR_2'' & \quad \text{alkyl} &
\end{array}
$$

Since 1971, he has developed a new stereoelectronic theory in which the precise conformation of the tetrahedral intermediate plays a major role. In other words, the stereochemistry and the ionic state of the tetrahedral intermediate, the orientation of nonbonded electron pairs, and the relative energy barriers for cleavage and for molecular rotation are the key parameters in the stereoelectronically controlled cleavage of the tetrahedral intermediate formed in the hydrolysis of an ester or an amide. He postulated that the precise conformation of the tetrahedral intermediate is transmitted into the product of the reaction and that the specific decomposition of such an intermediate is controlled by the orientation of the nonbonded electron pairs of the heteroatoms.

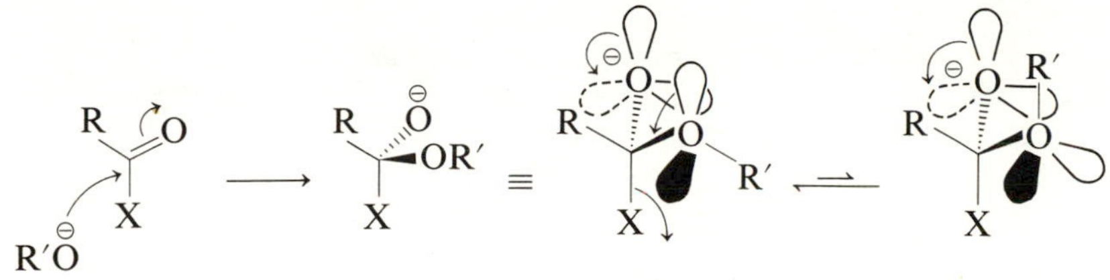

X = OR'', NR₂''   *trans*   *gauche*

The *trans* conformer (the two R groups are away from each other) gives a lower energy pathway than the *gauche* which cannot effectively compete in the cleavage process. *Consequently, the cleavage is stereoelectronically controlled when and only when two heteroatoms of the intermediate, each having one nonbonded electron pair, orient antiperiplanar to the departing O-alkyl or N-alkyl group.*

Experimental evidence to prove this postulate come from four sources. A study of the oxidation of acetals by ozone, acid hydrolysis of cyclic orthoesters, concurrent carbonyl-oxygen exchange and hydrolysis of esters by using oxygen-18 labeling, and basic hydrolysis of N,N-dialkylated imidate salts. We will briefly examine results from the last three approaches. Then we will try to apply this concept to the hydrolysis of ester and amide substrates by serine proteases.

4.6.1 Hydrolysis of Orthoesters

The acid hydrolysis of an orthoester gives a hemiorthoester intermediate that cleaves to an ester and an alcohol.

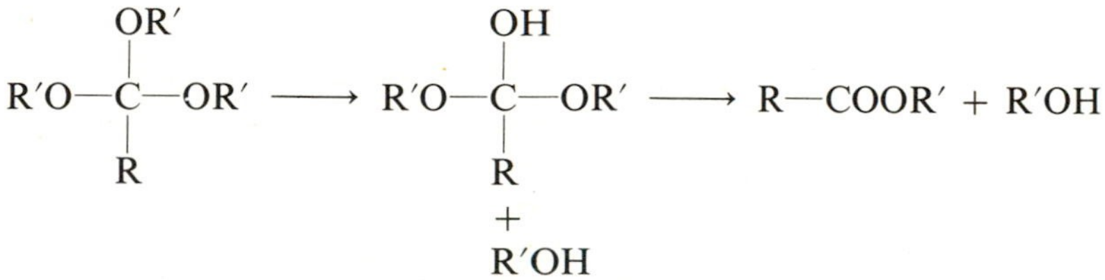

For a cyclic orthoester, a similar equation can be written:

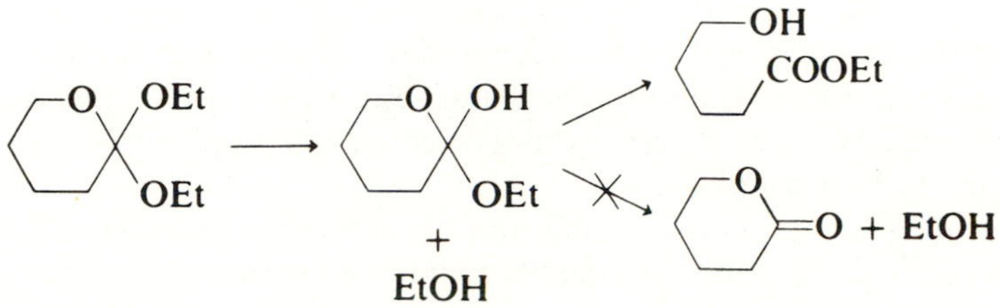

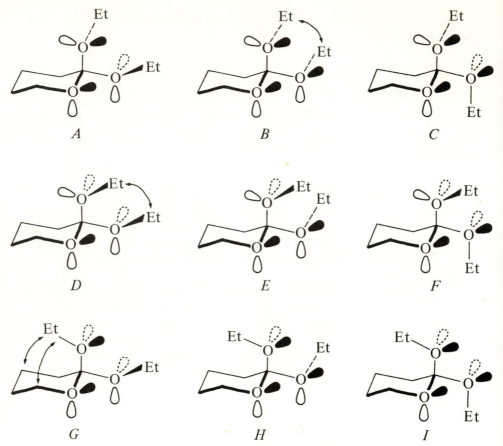

Fig. 4.6. The possible conformers of a cyclic orthoester (114). Reprinted with permission. Copyright © 1975 by Pergamon Press Ltd.

The hemiorthoester intermediate decomposes in a completely specific manner yielding exclusively an hydroxyester and no trace of a lactone is detected. To explain this result we have to examine all the possible conformers of the intermediate. Theoretically, there are nine different *gauche* conformers for an hemiorthoester tetrahedral intermediate. Figure 4.6 represents the nine possibilities for the case of the cyclic orthoester.

The task is then to define which conformer should be taken into consideration to explain the formation of an hydroxyester as the only product of hydrolysis. The stereoelectronic theory predicts that the precise conformation of the tetrahedral intermediate is transposed into the product of the reaction and a cleavage of a C—O alkyl bond is allowed only if the other two oxygens of this intermediate each have an orbital oriented antiperiplanar to the C—O alkyl bond to be broken.

Consequently, a detailed examination of each conformer shows that five of them (B, D, G, H, and I) are readily eliminated because of severe 1,3-*syn* periplanar interactions between the two ethyl groups or with the methylenes of the ring. Since the population of these compounds will be small at equilibrium, they can be neglected. Conformer C must also be eliminated as a reactive conformer simply because it does not have proper orbital orientation on two oxygen atoms to permit the cleavage of the C—O bond of the third oxygen atom. The remaining three conformers A, E, and F do not suffer any strong steric interactions and each have two oxygens with proper orbital orientation to cleave the C—O bond of the third atom.

The cleavage of each of these three conformers will give a corresponding dioxolenium ion according to the proper orbital orientation rule just mentioned. Now if we look at the relative stability of each of these dioxolenium ions for conformers A, E, and F, we see that the ion from E is *cis* and the ones from A and F are *trans*. It is known that *trans*-dioxolenium ions are more stable than *cis*-dioxolenium ions, just as *trans* esters are more stable than *cis* esters. Consequently, the cleavage of conformer E should be a higher energy process and can thus be eliminated on that basis.

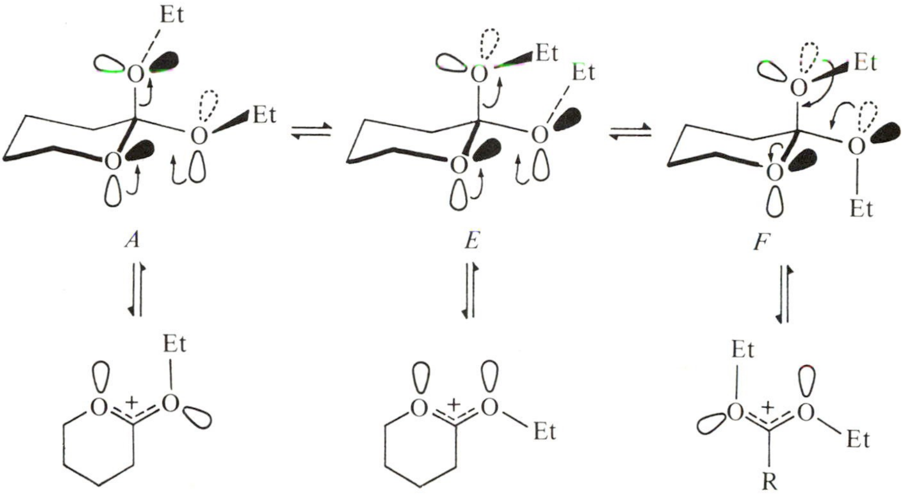

It is more difficult to find arguments to differentiate between conformers A and F. In principle, the formation of a cyclic dioxolenium ion should be favored over an acyclic one when the starting orthoester is cyclic. Furthermore, cleavage of conformer A gives two molecules, while conformer F gives only one molecule. This entropy factor alone should favor conformer A over conformer F as the reactive species. In conclusion, the cyclic orthoester is hydrolyzed preferentially via conformer A only, even if it exists in rapid equilibrium with conformers F and E (114).

A large number of geometrically differently constrained acetals, esters and imidate salts have been examined, with uniform consistency with this theory. For instance, a methyl lactonium salt can be prepared by reaction of a conformationally rigid bicyclic lactone with trimethyloxonium tetrafluoroborate:

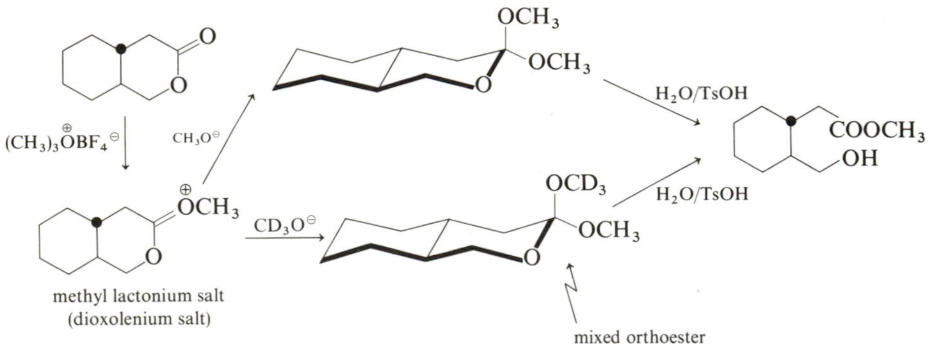

methyl lactonium salt
(dioxolenium salt)

mixed orthoester

Reaction with sodium methoxide in methanol afforded the dimethoxyester. In the presence of deuterated sodium methoxide the mixed cyclic orthoester was obtained in over 90% yield with the deuterated methoxy group exclusively in the axial orientation. The hydrolysis of the orthoester in water containing p-toluenesulfonic acid resulted in the formation of only the non-deuterated hydroxyester. These results confirm that the hydrolysis of cyclic orthoesters proceed by loss of the axial alkoxy group. It is a further proof that conformer F can be eliminated as a reactive conformer and consequently the hydrolysis of the cyclic dialkoxyorthoester takes place through the reaction of conformer A.

A further experimental proof of the stereoelectronic control in ester (and amide) hydrolysis comes from concurrent carbonyl–oxygen exchange and hydrolysis using oxygen-18 labeled esters (116).

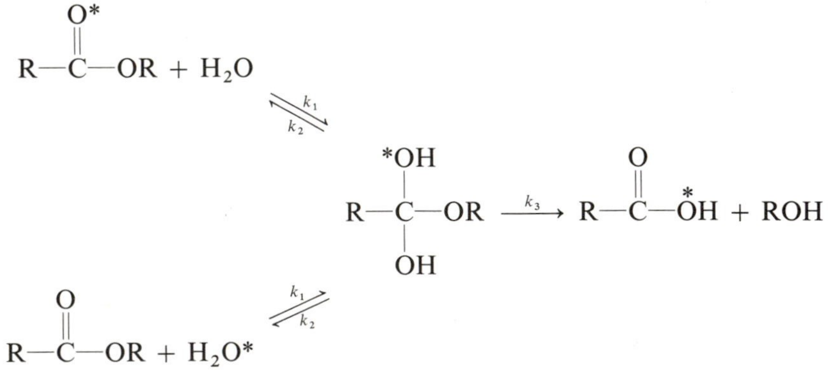

If it is indeed true that there is no conformational change in the tetrahedral intermediate when stereoelectronically controlled cleavage is allowed, then concurrent carbonyl–oxygen exchange during hydrolysis can be used to demonstrate both the lack of conformational change and the stereoelectronic theory. Application of these postulates to the hydrolysis of esters led to the following predictions: (Z)-esters can undergo carbonyl-oxygen exchange but (E)-esters cannot.* An ^{18}O-labeled δ-lactone is an example of a (E)-ester. Reaction with hydroxide ion gives a tetrahedral intermediate which has the required orientation of electron pairs to breakdown in two directions only to give either the starting labeled ester or the product of the reaction.

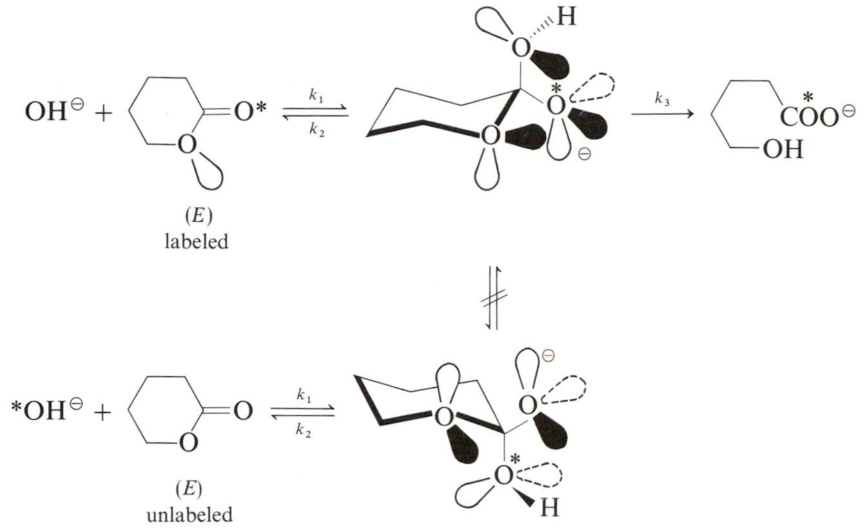

Since the tetrahedral intermediate cannot yield the unlabeled ester, carbonyl–oxygen exchange has not occurred with the (E)-ester. The experimental result is in accord with results previously described in the literature which show that lactones do not undergo carbonyl–oxygen exchange. On the other hand, hydrolysis of (Z)-esters always occur with carbonyl–oxygen exchange with the solvent. The results with lactones can also be explained by using a kinetic argument. Indeed, the extent of carbonyl exchange can

---

* The (Z) for *Zusammen* and (E) for *Entgegen* nomenclature for double bonds and carbonyl functions in molecules is another application of the Cohn–Ingold–Prelog priority rules. If the two groups of highest priority at the ends of the double bond are on the same side of the double bonds (*cis*), the double bond configuration is designated (Z). When the two groups of highest priority are on opposite sides on the double bond (*trans*), the double configuration is (E). These rules can also be applied to ester conformations by looking at the C—O bond. Since the lone pair of electron has the lowest priority, the conformation is said to be (Z) when the R group of the alkoxy is on the same side as the oxygen of the carbonyl. If it is on the other side (such as with lactones) the conformation is the (E).

vary depending on the relative value of k_3 and k_2. If k_3/k_2 is greater than 100, there will be a very low exchange which will be difficult to detect. This argument agrees with experimental results obtained with (E)-esters.

4.6.2 Application to Serine Proteases

It is reasonable to inquire whether recognition of such a concept might apply to mechanistic ideas about enzyme reactions. It now appears that the stereoelectronic theory that Deslongchamps and co-workers have developed is specifically applicable to peptide hydrolyses by serine proteases. Hydrolysis of simple esters will be used first to illustrate the approach.

A hemiorthoester intermediate can be cleaved to yield two different esters, each of which has a (E) or (Z) conformation.

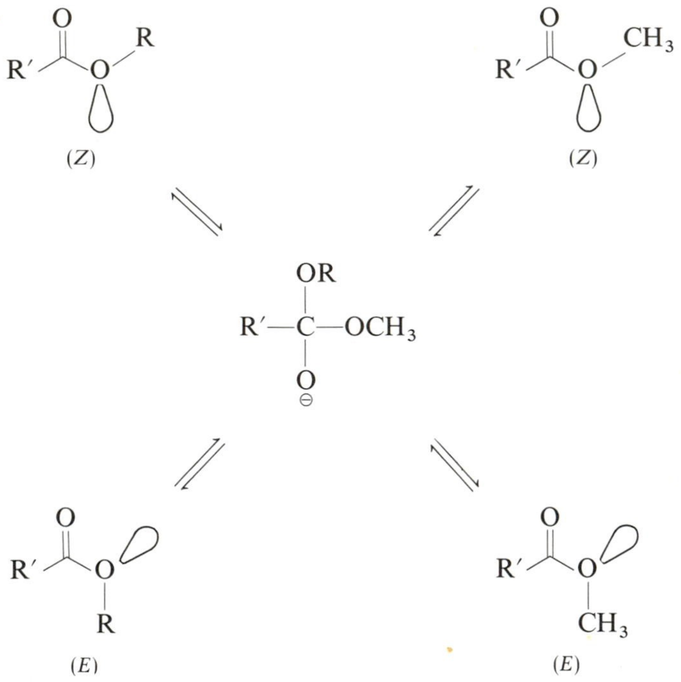

According to the principle of microscopic reversibility, the generation of a hemiorthoester via alkoxide ion attack on an ester should also occur with stereoelectronic control.

Since a *transoid* (Z)-ester is generally more stable than the corresponding *cisoid* (E)-ester, we will use it to analyse the attack by an active site serine-OH. It is expected that the serine residue will approach the ester with an orientation away from the more bulky R group of the ester substrate.

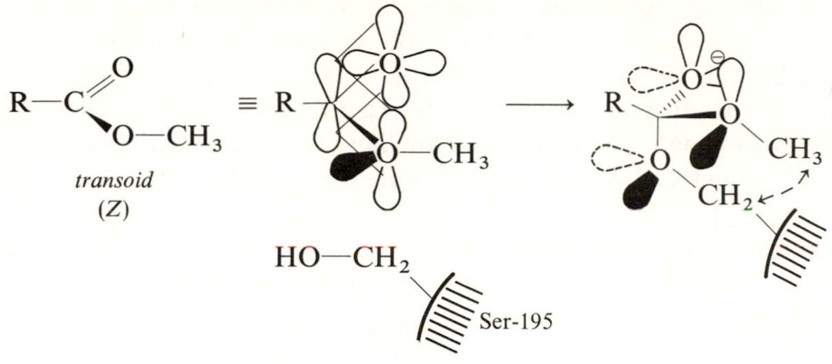

However, the introduction of a serine residue in this tetrahedral intermediate form will cause a large steric hindrance (~ 33.4 kJ or ~ 8 kcal) with the alkoxy chain. To reduce this strain, a conformational change *must* occur at the active site by a rotation of $120°$ around the C—O bond connecting the tetrahedral carbon to the serine-OH group.

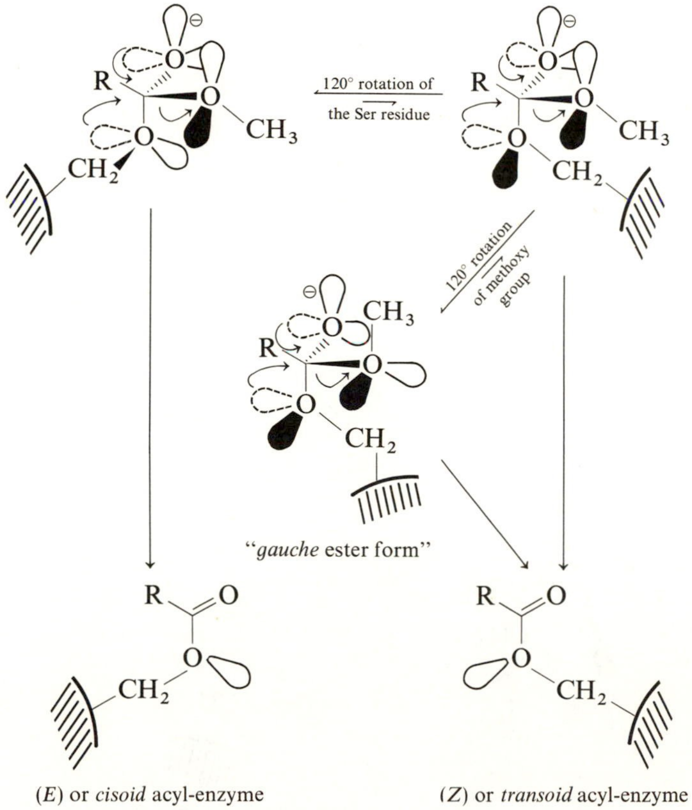

"*gauche* ester form"

(*E*) or *cisoid* acyl-enzyme   (*Z*) or *transoid* acyl-enzyme

This new tetrahedral intermediate will eventually generate a *cisoid* acyl-enzyme intermediate and it should be the preponderant product.

Another possibility is a rotation of 120° of the —OCH$_3$ group. Here the ester group has changed its conformation from *trans* to *gauche* relative to the R group. This new tetrahedral intermediate and the former one will both yield a *transoid* acyl-enzyme intermediate after orbital-assisted cleavage of the —OCH$_3$ group.

The third possible conformer can be excluded since it cannot break down, even after rotation of the —OCH$_3$ group. It is a nonproductive binding intermediate.

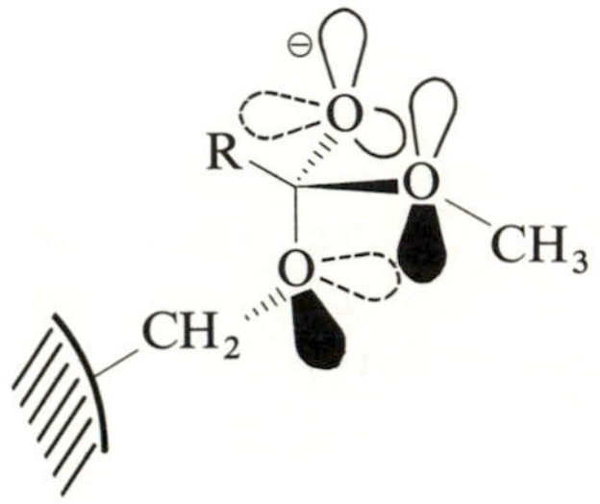

If however, a *transoid* → *cisoid* enzyme mediated isomerization of the substrate ester takes place prior to the attack by the serine residue, a different tetrahedral intermediate will be formed:

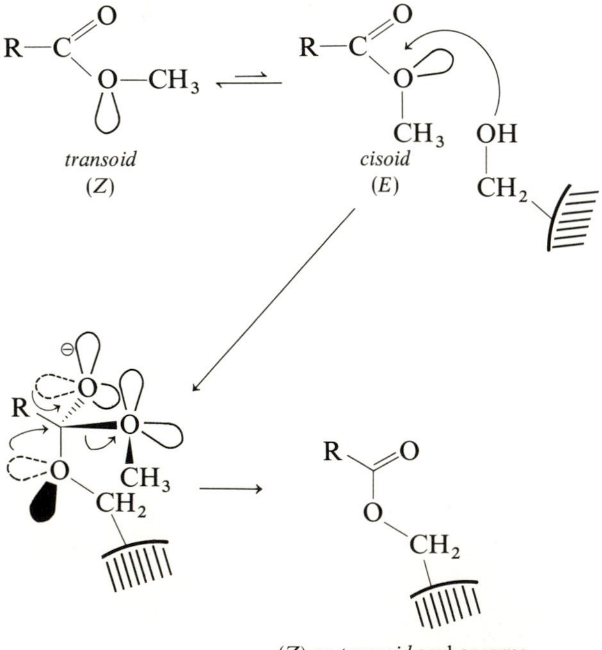

(Z) or *transoid* acyl-enzyme

In this case the steric strain is not as serious as before and this intermediate will break down to give mainly a *transoid* acyl-enzyme.

The hydrolysis of N,N-alkylated imidate salts is also under stereoelectronic control and has been used to understand the mechanism of amide hydrolysis. Two paths are possible: ejection of ROH or R_2NH.

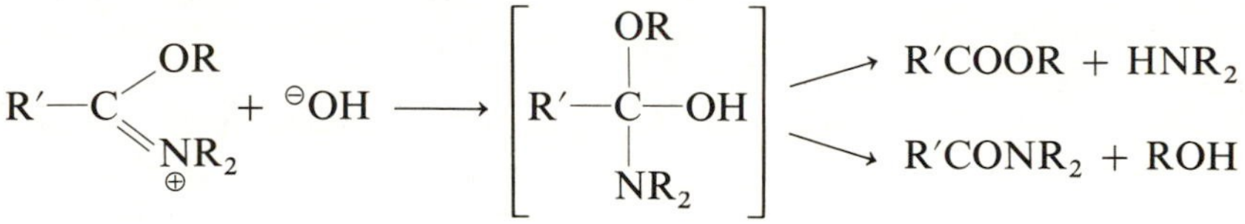

Here again, the ejection of a residue critically requires the assistance of two antiperiplanar lone-pair orbitals on the remaining heteroatoms of the incipient amide. In the event that only one pair is potentially available, facile decomposition will be observed only after bond reorientation (117).

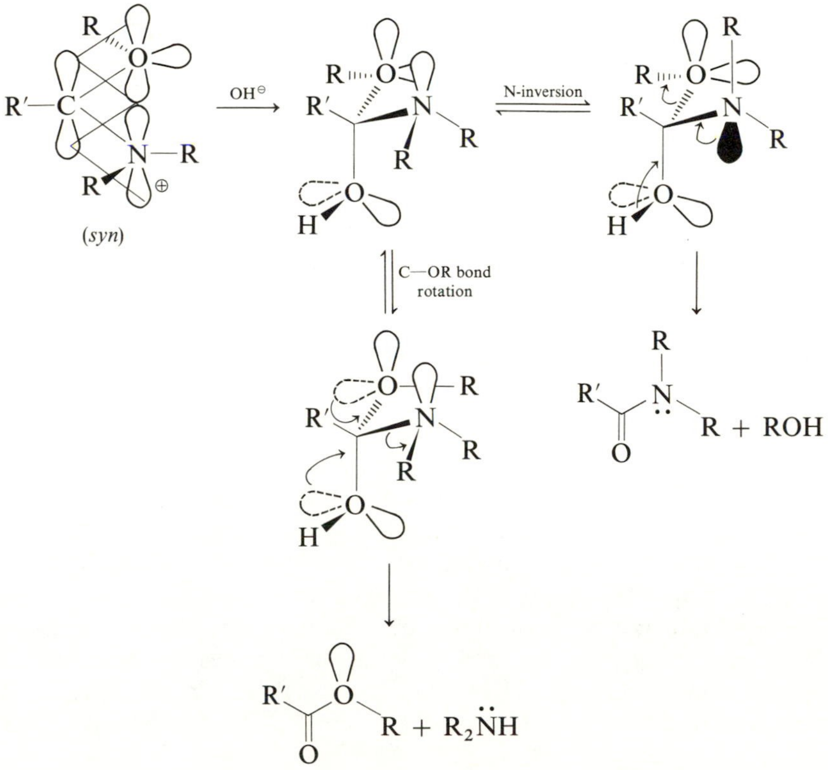

This model is applicable to amide bond hydrolysis by proteases and it would be surprising if enzyme hydrolyses do not obey this principle.

This simple presentation leads to interesting speculation on the pathways preferred by the enzyme. Does the conformational change occurs at the level of the substrate tetrahedral intermediate or at the level of the enzyme's active site residue? Obviously rigid bioorganic models are necessary to solve this dilemma. Such models are yet to come but will offer the capacity of defining the enzyme mechanism of proteases from relatively limited conformations of the tetrahedral intermediate.

Furthermore, we suspect that the concept of stereoelectronic control hydrolysis of esters and amides is broadly applicable to other fields of enzymology, and consequently some well-accepted catalytic principles will require reconceptualization.

One of the closest approaches so far developed is by Bizzozero and Zweifel (118) who tried to explain in 1975 why a proline residue involved in a peptide bond is resistant to α-chymotrypsin cleavage. The objective was to find if the unreactivity of the peptide bond results from an unfavorable interaction of the methylene groups of the proline ring with the enzyme active site or whether the steric hindrance occurs upon formation of the enzyme–substrate complex or during the subsequent bond-change steps, and whether this steric hindrance is related to the ring structure of proline or simply to substitution of the amido nitrogen. In order to answer these questions, the dipeptides N-acetyl-L-phenylalanyl-L-proline amide and N-acetyl-L-phenylalanyl-sarcosine amide were synthesized and their behavior as model substrates of α-chymotrypsin studied.

proline amide   sarcosine amide

Both products were found to be unreactive but proved to be good competitive inhibitors of a specific substrate, N-Ac-L-Phe-OMe. This indicates that they form enzyme–substrate complexes of normal stability and that the reason for their unreactivity has to be sought in the nature of the enzyme–substrate interactions occurring during the subsequent bond-change steps. In other words, their unreactivity can be understood by considering the stereoelectronic course of the transformation leading to the acyl-enzyme intermediate.

The attack of a *trans*-peptide bond by the hydroxyl of Ser-195 of α-chymotrypsin will give the steric situation depicted below:

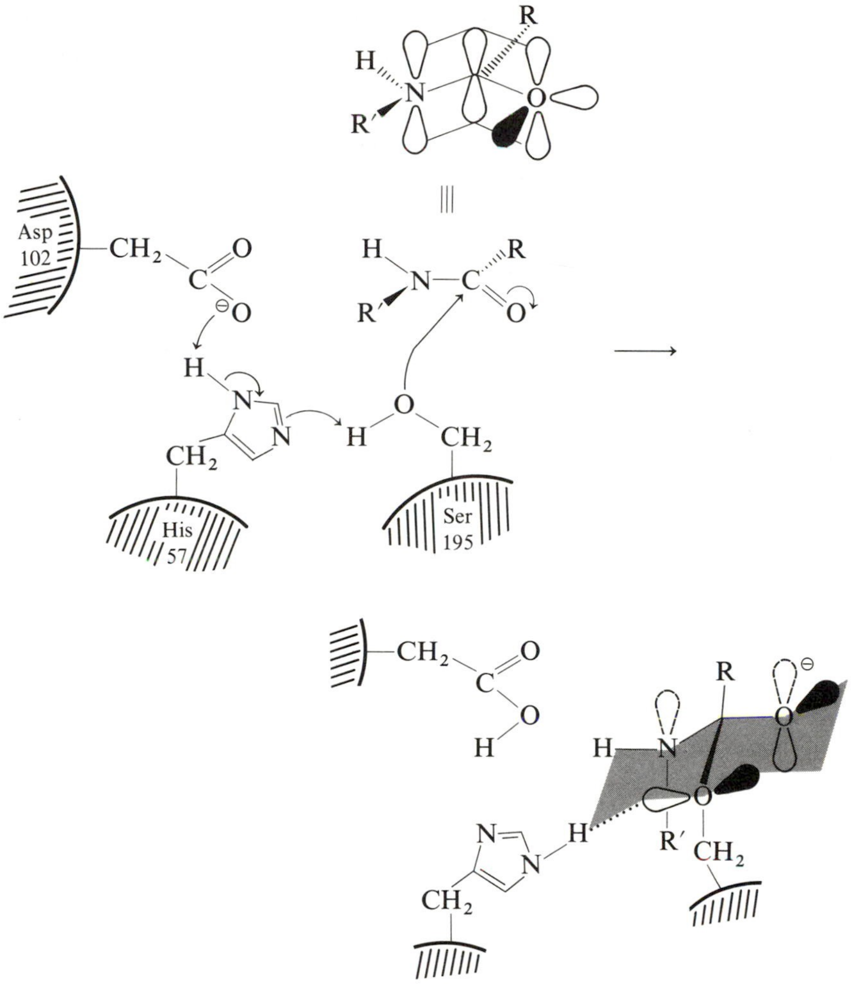

From the principle of microscopic reversibility, the stereoelectronic control theory predicts that the developing lone-pair orbitals on the heteroatoms must be antiperiplanar to the new carbon–oxygen (from Ser-OH and amide carbonyl) or carbon–nitrogen bond. The important point to be considered here is that the nonbonded pair of electrons on the nitrogen atom points toward the solvent and the N—H bond toward the inside of the enzyme active site. To facilitate the spacial view, the relevant atoms are superimposed on a *trans*-decalin frame (shaded area).

When the N—H hydrogen is replaced by an alkyl group, as in the case of a proline residue, this substituent would come too close to the imidazole ring of His-57. Hence, a dipeptide containing a proline residue is inactive toward α-chymotrypsin hydrolysis because the steric hindrance prevents formation of the tetrahedral intermediate.

With regard to the mechanism of the acylation step, the stereoelectronic requirements for cleavage of the carbon–nitrogen bond to form the acyl-enzyme are fulfilled, but the cleavage step still requires protonation of the nitrogen atom. Since it is generally accepted that the imidazole of His-57 is the protonating agent, the situation shown above could not occur since the nonbonded orbital of the leaving nitrogen points in the wrong direction.

Assuming that a protonation–deprotonation mechanism mediated by the solvent does not occur, an inversion mechanism at the leaving nitrogen is needed to produce a new intermediate in which the orientation of the N—H bond and the nonbonded orbital are interchanged.

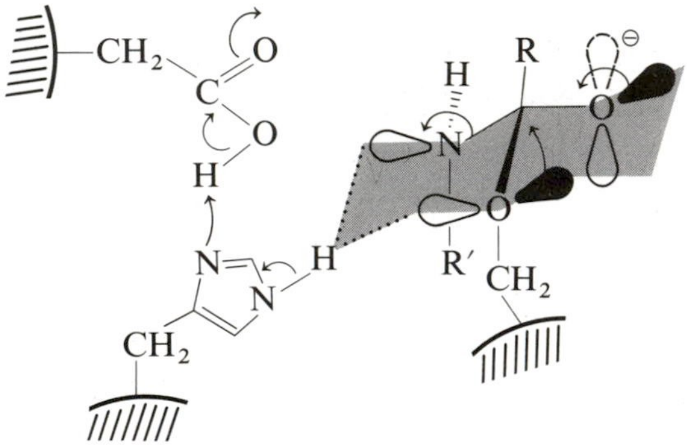

Such an inversion would not only make direct protonation of the leaving nitrogen by His-57 possible; it would also stabilize the C—O (Ser-195) bond since this would now be antiperiplanar to the N—H bond rather than to the nonbonded electron pair. This conformational change would probably not take place in the presence of a proline residue.

Deslongchamps kindly provided us with his own view of the mechanistic path by which α-chymotrypsin and other serine proteases can hydrolyze secondary amides by stereoelectronic control (Fig. 4.7). Petkov et al. (119) also arrived at a similar proposal by studying the influence of the leaving group on the reactivity of specific anilides in α-chymotrypsin-catalyzed hydrolysis. Furthermore, the stereoelectronic control theory has been applied to the mode of action of ribonuclease A, staphylococcal nuclease and lysozyme (120).

Finally, it should be mentioned that a concept of torsional strain of amide bonds was also developed by W. L. Mock in 1976 and applied to the mode of action of carboxypeptidase A (121). It involves a torsion of the amide bond to allow either cis or trans addition of the nucleophile and the proton on the developing orbitals in the transition state. This principle complements Deslongchamps' theory of optimum geometry of the transition state tetra-hedral intermediate.

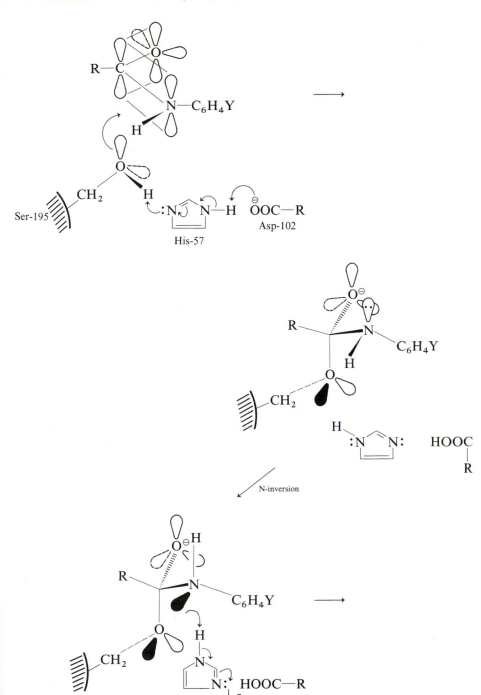

Fig. 4.7. Deslongchamps' proposal of stereoelectronic control in secondary amide hydrolysis by α-chymotrypsin.

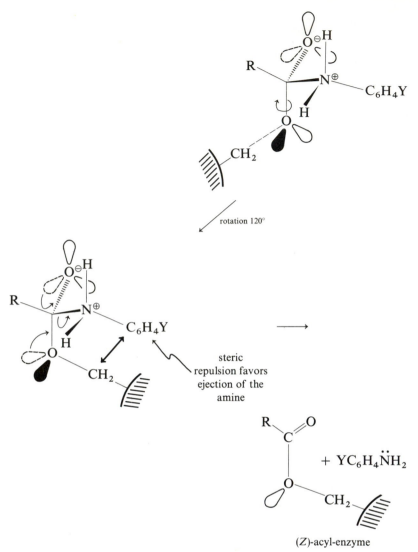

Fig. 4.7. (continued)

4.7 Immobilized Enzymes and Enzyme Technology

We shall end this chapter by giving examples of a growing technology where enzymes are anchored to a solid matrix and used to perform specific trans-formation of biomaterials. The enzyme of interest is covalently attached to a carrier polymer by a "spacer" molecule.

The inert carrier could be a polyurethane or other types of resin or a natural polymer such as collogen, easily available from animal skin. A flexible "spacer-arm" is attached to a functional group on the polymer gel. The length of the arm is an important parameter because the free end must be accessible to a functional group of the enzyme to make a second covalent bond without affecting the enzymatic activity. Such a matrix-bound enzyme is usually referred to as an *immobilized enzyme* (122–125). Contrary to the well-exploited method of *affinity chromatography*, here the enzyme rather than the substrate analogue is covalently fixed on a solid support. But the concept of biospecific recognition is similar.

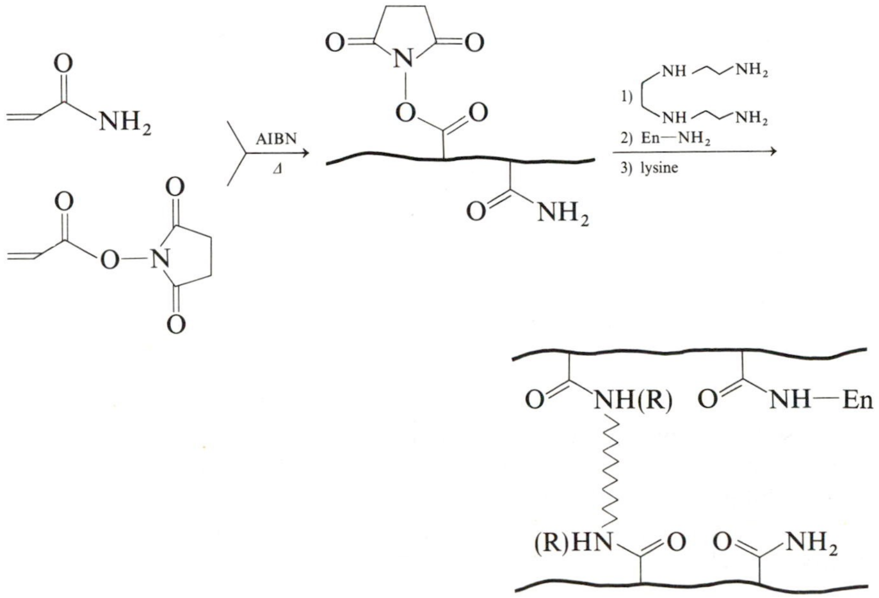

In conventional affinity chromatography, agarose and cross-linked sepharose are used as support to immobilize substrates. Usually BrCN is the activating agent and the spacer-arm is an α,ω-diamine. These poly-saccharide supports are biodegradable and consequently an organic polymer gel is a more useful matrix and is amenable to a wider range of chemical modifications. It is these reasons that prompted the group of G. M. Whitesides from M.I.T. to develop in 1978 a new procedure for the immobilization of enzymes in cross-linked organic polymer gels (126). The procedure surpasses in its operational simplicity and generality the earlier methods. It is also especially valuable in immobilization of relatively delicate enzymes for enzyme-catalyzed organic synthesis for application in large-volume enzymatic reactors.

First, a non-cross-linked water-soluble copolymer bearing active ester groups is prepared by heating acrylamide and N-acrylyloxysuccinimide with azobisisobutyronitrile (AIBN), an initiator of radical polymerization. Reaction of this polymer with an α,ω-diamine as a cross-linking agent and with the enzyme of interest results in enzyme immobilization and gel formation. Operations involving oxygen-sensitive enzymes are carried out under argon. A solution of lysine is added at the end to destroy residual active ester groups.

An important feature of this procedure is that addition of the enzyme to the reaction mixture during the formation of the gel minimizes enzyme deactivation. Furthermore, covalent incorporation of the enzyme into the gel provides some protection against proteases. Second, the procedure is simple and of general use and should be directly applicable to a variety of enzyme systems as well as immobilization of whole cells and organelles. Finally, the gel can be rendered susceptible to magnetic filtration by including a ferrofluid in the gel formation step.

It is important to realize that developing methods for stabilizing the enzyme in its native form is also a major goal in this field (123). Successful progress along these lines have been obtained recently by the use of bifunctional reagents to cross-link the peptide chains within the enzyme tertiary structure.

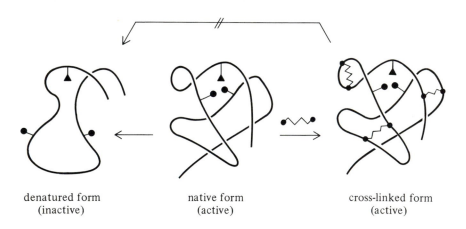

denatured form   native form   cross-linked form
(inactive)   (active)   (active)

Cross-linking prevents denaturation from occurring under the severe conditions of manipulation and keeps the active site region intact. The most used cross-linking agents are imino esters. By reacting with available lysine side chains, amide bonds are eventually formed between two sections of the polypeptide chain.

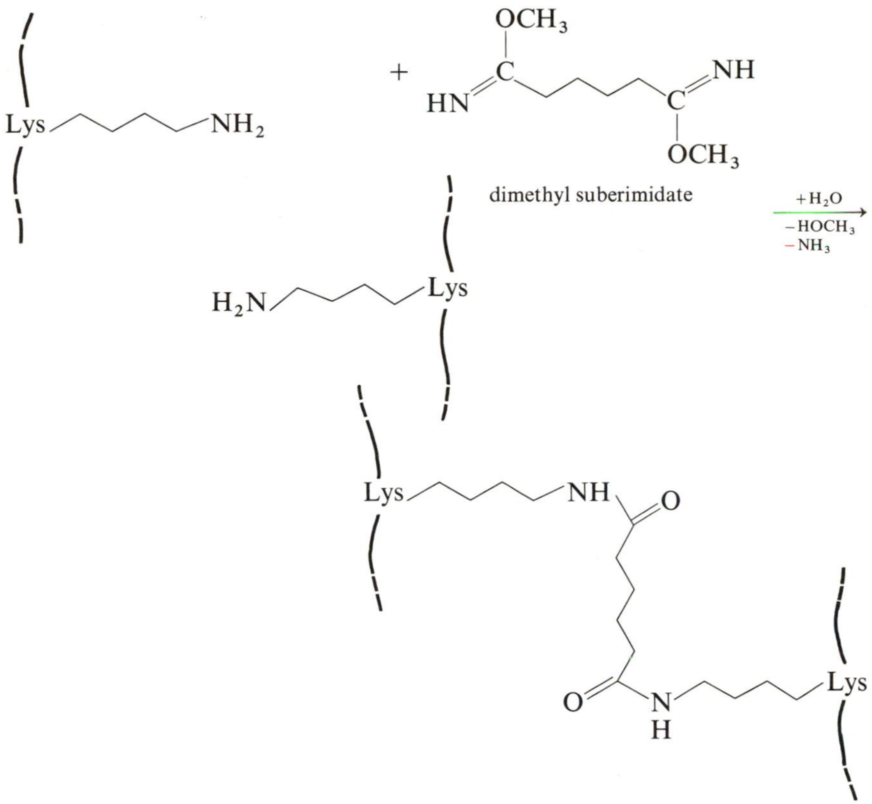

However, the number of carbons separating the reactive imino ester functions must be kept small. Otherwise intermolecular links will also occur.

This technique of immobilized enzymes combines the unique features of both forms of catalysis: the specificity of the enzymes with the stability and ease of handling and storing of supported heterogeneous catalysts. Furthermore, immobilized enzymes can be reused and are applicable to flow systems. Consequently, they find increasing applications in several analytical areas and in medicine.

A useful application of the system is in commercial processes and the term *enzyme engineering* or *enzyme technology* has been attached to it. Enzyme technology has been called a "solution in search of problems" (122). Actually, the technology has yet to fulfill much of the promises that many of its advocates are convinced it holds. Nevertheless, we should be aware of its potential and sensitized to this future technology.

Let us examine few applications specially in the field of food processing. One effort which has been made to provide better quality foodstuff is the following. Immobilized β-galactosidase is now produced by molding a polyisocyanate polymer to a magnetic stirring bar. This fiber-entrapped enzyme is used to reduce the lactose content of milk to overcome the problem of lactose intolerance. Furthermore, by this process milk can be stored frozen for a longer period of time without thickening and coagulating, caused before by lactose crystallization in the milk.

Other efforts in the food industry have ben made to utilize immobilized enzymes. One is to use cellulose from waste paper, wood chips, or sugar cane, degrade it to glucose, and convert this sugar unit back to starch, as edible material. All these processes are enzyme-catalyzed and should be applicable to enzyme technology.

A more spectacular example is the possible conversion of fossil fuel (oil) derivatives to edible carbohydrates. This transformation needs an industrial breakdown of petroleum products to glyceraldehyde. Then glyceraldelyde can be enzymatically converted to fructose, glucose, and starch.

The ultimate application of enzyme technology to carbohydrate synthesis will be the mimicking of nature's method; the fixation of CO_2. This will require in addition to immobilized enzymes, *immobilized coenzymes*. Many efforts have been made in this direction (124).

For instance, an NAD^+ analogue bound to a water-soluble dextran polymer has been applied in the preparation of an *enzyme electrode* (127) and a *model enzyme reactor*. Medicinal applications in this field are obvious. One example will illustrate the principle. The oxidoreduction reaction, $NAD^+ + substrate \rightarrow NADH + H^+ + product$, can be coupled to immobilized enzymes. In this model system, the substrate is pumped into a chamber containing the dextran-bound NAD^+ and two NAD^+-linked dehydrogenases. At the other end the product of the reaction is removed at the same rate by ultrafiltration. Hence, the process can be recycled.

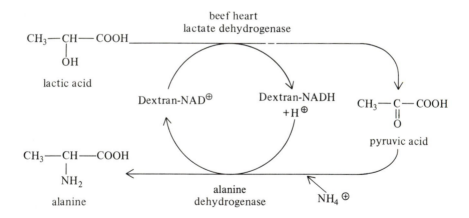

The feasibility of such reactor has been demonstrated by the production of alanine from lactic acid. The unfavorable equilibrium in this lactate dehydrogenase-alanine dehydrogenase reactor is offset by using a high concentration of substrate and by the rapid consumption of pyruvic acid by the second enzyme. It is worthy of mention that such a system also serves as a model for potential therapeutic applications whereby enzymes and coenzymes immobilized together could function as a self-contained unit to correct a metabolic imbalance.

Some of the examples presented may be seen by skeptics as less than reality. However, it is our belief that these projects or a variant of them will be developed on a practical scale in the near future. As K. Mosbach pointed out, the examples discussed above may be looked upon as first steps in the field of *synthetic biochemistry* (124).

In this context, two processes now have useful and practical application in the pharmaceutical industry. The first one is the synthesis of a cortisone analogue, prednisolone, which is used as a drug against rheumatoid arthritis. A steroid precursor, Reichstein's compound, is passed through a series of two different columns, each containing a specific enzyme attached to a polyacrylamide polymer support (128). The hydroxylation reaction at the prochiral C-11 position occurs with retention of configuration.

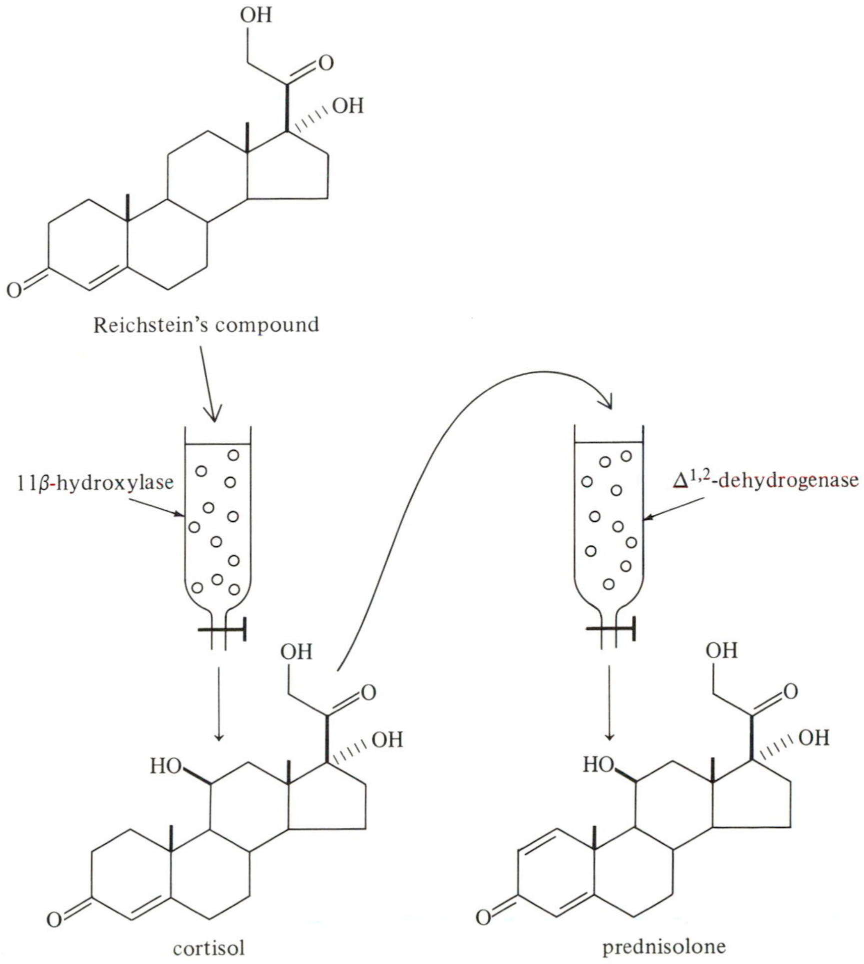

Reichstein's compound

11β-hydroxylase

$\Delta^{1,2}$-dehydrogenase

cortisol

prednisolone

This way the synthesis is rapid, regio- and stereoselective, and economical. These transformations were in part responsible for the lowering in the price of cortisone from about $30.00 per gram in the 1950s to few cents per gram in the 1970s.

The second example, which is also in daily use, is the specific hydrolysis of natural penicillin G by penicillin amidase in preparative scale, again through a chromatographic process (129). It produces a clean hydrolytic product (6-aminopenicillanic acid) free of possible contaminant and it can then be used to prepare chemically all kinds of semisynthetic penicillin derivatives. Unlike the rather unstable soluble enzyme, the insolubilized preparation shows no loss activity after up to 11 weeks of continuous operation at 37°C. Furthermore, the procedure is free from potentially allergy-inducing contaminants.

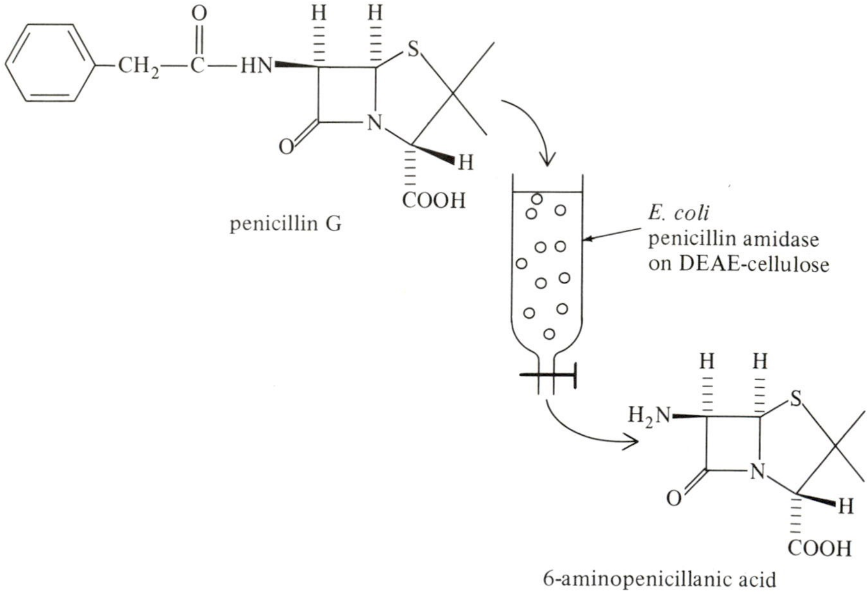

penicillin G

E. coli
penicillin amidase
on DEAE-cellulose

6-aminopenicillanic acid

Since enzymes are valuable not so much for their purely catalytic abilities as for their selectivity during reactions, enzyme technology offers many capabilities in synthetic chemistry applications and will attract more organic chemists to use enzymes to make biological molecules.

Chapter 5

Enzyme Models

*"We dance round in a ring and suppose
But the Secret sits in the middle and knows!"*

R. Frost
From "The Secret Sits"

Enzyme models are generally organic synthetic molecules that contain one or more features present in enzymatic systems. They are smaller and structurally simpler than enzymes. Consequently, an enzyme model attempts to mimic some key parameter of enzyme function on a much simpler level. To dissect out a particular factor responsible for the catalytic efficiency of the enzyme within the biological system would be a tremendous task requiring a knowledge of each of the components that would contribute to the overall catalysis. Instead, with appropriate models, it is possible to estimate the relative importance of each catalytic parameter in absence of those not under consideration. One noticeable advantage of the use of "artificial structures" for modeling enzymatic reactions is that the compounds can be manipulated precisely for the study of a specific property. The state of the art may be further refined by combining those features that contribute most and designing models that actually approach the efficiency of the enzyme. That is, with the tools of synthetic chemistry, it becomes possible to construct a "miniature enzyme" which lacks a macromolecular peptide backbone, but contains reactive chemical groups correctly oriented in the geometry dictated by an enzyme active site. It is often referred to as the *biomimetic chemical approach* to biological systems.* Therefore, biomimetic chemistry represents the field

---

* The word "biomimetic" was introduced first by R. Breslow in 1972 and generally refers to any aspect in which a chemical process imitates a biochemical reaction (197).

which attempts to imitate the acceleration and selectivities characteristic of enzyme-catalyzed reactions (350, 351). It is hoped that such an approach will eventually bridge or at least reduce the gap between the known complex structures of organic biomolecules and their exact functions in life. In order to do this, many factors related to the mechanism of action of a particular enzyme must be known. These include (a) the structure of the active site and the enzyme–substrate complex; (b) the specificity of the enzyme and its ability to bind to the substrate; and (c) the kinetics for the various steps and a knowledge of possible intermediates in the reaction coordinate.

Enzymes are complicated molecules and only a few mechanisms have been definitively established. This is one of the reasons why model systems are necessary. Among the functional groups found on polypeptide chains, those generally involved in catalytic processes are the imidazole ring, aliphatic and aromatic hydroxyl groups, carboxyl groups, sulphydryl groups, and amino groups.

How can such a limited number of functional groups participate in the large variety of known enzymatic reactions and how can the rate of the enzymatic reactions be accounted for in mechanistic terms? These are the fundamental questions that should be asked during the planning of a bioorganic model of an enzyme (130).

In general, an enzyme model should fulfill a twofold purpose: (a) it should provide a reasonable simulation of the enzyme mechanism and (b) it should lead to an explanation of the observed rate enhancement in terms of structure and mechanism. Of course all the informations obtained with enzyme models must ultimately be compared and extended to the *in vivo* enzymatic system under study in order to correlate the bioorganic models to the real natural system.

A model can represent general features for more than one enzyme! Viewed from a different angle, the requirements necessary for the design of a good enzyme model can be summarized in these five criteria:

1. Because noncovalent interactions are the key to biological flexibility and specificity, the model should provide a good (hydrophobic) binding site for the substrate.
2. The model should provide the possibility of forming electrostatic and hydrogen bonds to help the substrate bind in the proper way.
3. Carefully selected catalytic groups have to be properly attached to the model to effect the reaction.
4. The structure of the model should be rigid and well-defined, particularly with respect to substrate orientation and stereochemistry.
5. Of course, the model should preferably be water-soluble, and catalytically active under physiological conditions of pH and temperature.

These criteria are only of limited application but implicit in that summary is the understanding that an efficient catalyst will be constructed by a proper choice of a matrix that can bring catalytic groups and substrates together.

This assumption implies that the matrix does not take an active part in the catalysis other than holding and orienting the substrate and the catalytic group (or groups) rigidly in proximity and correctly with respect to one another. Hopefully, the matrix, like an enzyme, can in the binding process raise the ground state energy of the substrate by rigidification and bond distortion. In addition, proper stereochemistry between the model catalyst and the substrate will result in better specificity and efficiency of reaction. These principles are of fundamental importance in this chapter.

In summary, for an enzyme model to be operative, a certain number of criteria, characteristic of enzyme catalysis, must be fulfilled, among which is substrate specificity—that is, selective differential binding. The enzyme-like catalyst must also obey Michaelis–Menten kinetics (saturation behavior), lead to a rate enhancement, and show bi- and/or multifunctional catalysis (348).

So far, enzyme processes imitated to date have been mostly hydrolytic (131), but stepwise assembly of macromolecules such as proteins and nucleic acids may be possible soon. For instance, structures resembling drug receptors may be incorporated into synthetic membranes, facilitating studies of these receptors without immunological and toxic complications. Furthermore, the ability of membranes to segregate charged species may find commercial use in systems for energy storage or hydrogen generation.

This chapter on enzyme models will start by describing the chemistry and properties of some man-made organic host molecules which possess the capacity of enantiomeric discrimination. Six additional aspects of enzyme models will be developed and other interesting and fascinating enzyme and coenzyme models will be presented in the forthcoming chapters, especially Chapter 7 on coenzyme design and function.

5.1 Host–Guest Complexation Chemistry

The discovery in 1967 by C. J. Pedersen (132, 133) that crown ethers have the unique ability to form stable complexes with metal ions and primary alkyl ammonium cations opened new horizons in organic chemistry (352–354).*

12-crown-4 ether

* Two books have appeared recently on the syntheses and applications of crown ethers: R. M. Izatt and J. J. Christensen, Eds. (1978), *Synthetic Multidentate Macrocyclic Compounds*. Academic Press, New York; and R. M. Izatt and J. J. Christensen, Eds. (1979), *Progress in Macrocyclic Chemistry*, Vol. 1. Wiley, New York.

The abbreviated nomenclature used is simple; the first number represents the total number of atoms in the ring and the second the total number of heteroatoms. It is easy to see an analogy between such complexes having a "cavity" to bind the ligand (L) and the active site of an enzyme which recognizes its specific substrate. The size of the macroring can be varied to allow the binding of ligands of different shapes. The cyclic polyethers of the "crown" types are relatively easy to make and are subject to wide structural modification. To this field of chemistry, D. J. Cram from the University of California coined the name *host–guest complexation chemistry** (134–136). We recall Hans Fisher's lock and key metaphor back in 1894 to describe the match of an enzyme with its substrate in an enzyme–substrate complex. Besides enzyme catalysis and inhibition, complexation plays a central role in biological processes, such as replication, genetic information storage and retrieval, immunological response, and ion transfer. Enough structural information is now known about the complexes involved to inspire organic chemists to design highly structured molecular complexes and to study the chemistry unique to complexation phenomena.

A highly structured molecular complex is composed of at least one host and one guest component. The host component is defined as an organic molecule or ion whose binding sites converge in the complex. The guest component, on the other hand, is defined as any molecule or ion whose binding sites diverge in the complex (136). Guests can be organic compounds or ions, or metal ions, or metal–ligand assemblies. In general, simple guests are abundant, whereas hosts usually have to be designed and synthesized.[†]

A host–guest relationship involves a complementary stereoelectronic arrangement of binding site in host and guest. Therefore, any man-made synthetic host–guest complex must have binding sites (polar and dipolar) and steric barriers located to complement each other's structures. Micelles and cyclodextrins are naturally occurring hosts and their properties will be the object of forthcoming sections in this chapter. The prosthetic groups of hemoglobin, chlorophyll, or vitamin B_{12} are also in this category because they bind selectively iron, magnesium, and cobalt ions.

Figure 5.1 represents some available synthetic host compounds. Can chemists use such organic crown ethers for enantiomeric discrimination (or racemic resolution) as mimic of an enzyme? Cram and others have reported that chiral crown ether complexes have this remarkable property of binding selectively one antipode of amino acid derivatives (134–136). In the design of host molecules, Corey–Pauling–Koltun (CPK) molecular models are invaluable (137, 138). Space-filling scale models provide a possible guide to searching for host structures to bind given amino acid guest compounds.

---

[*] Although this expression is very useful and directly understandable conceptually, it is not easily translated in other languages. We thus suggest that the expression *receptor–substrate complexation* could be of a more universal use.

[†] Association constants between simple organic ammonium salts and 18-crown-6 ether in $CDCl_3$ at 25°C are on the order of $10^6 \ M^{-1}$, corresponding to a $\Delta G°$ of complexation of about -33.5 kJ/mol (-8 kcal/mol).

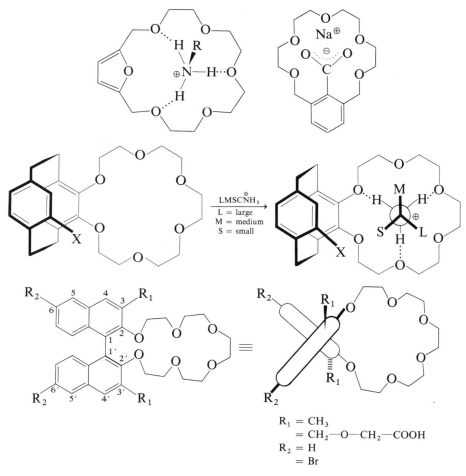

Fig. 5.1. Examples of host molecules.

For instance, a fundamental question in host design is that of the effect of preorganization of binding sites on binding ability. Another problem is that of the placement of substituents in positions that converge on the functional or binding sites of guest compounds (137).

After many trials in the molecular design of chiral crown ethers, a 1,1'-binaphthyl unit incorporated in a macroring by substitution in the 2,2'-positions proves to possess the desirable properties. The naphthalene-containing system, chosen for practical and strategic reasons, imparts rigidity and lipophilicity to conventional cyclic polyethers. The synthesis of such a host is presented in Fig. 5.2.

This host is chiral, possesses a C_2 axis of symmetry, and the dihedral angle between the planes of the two naphthalene rings attached to one another can vary between 60° to 120°. Including a binaphthyl system in a crown ether

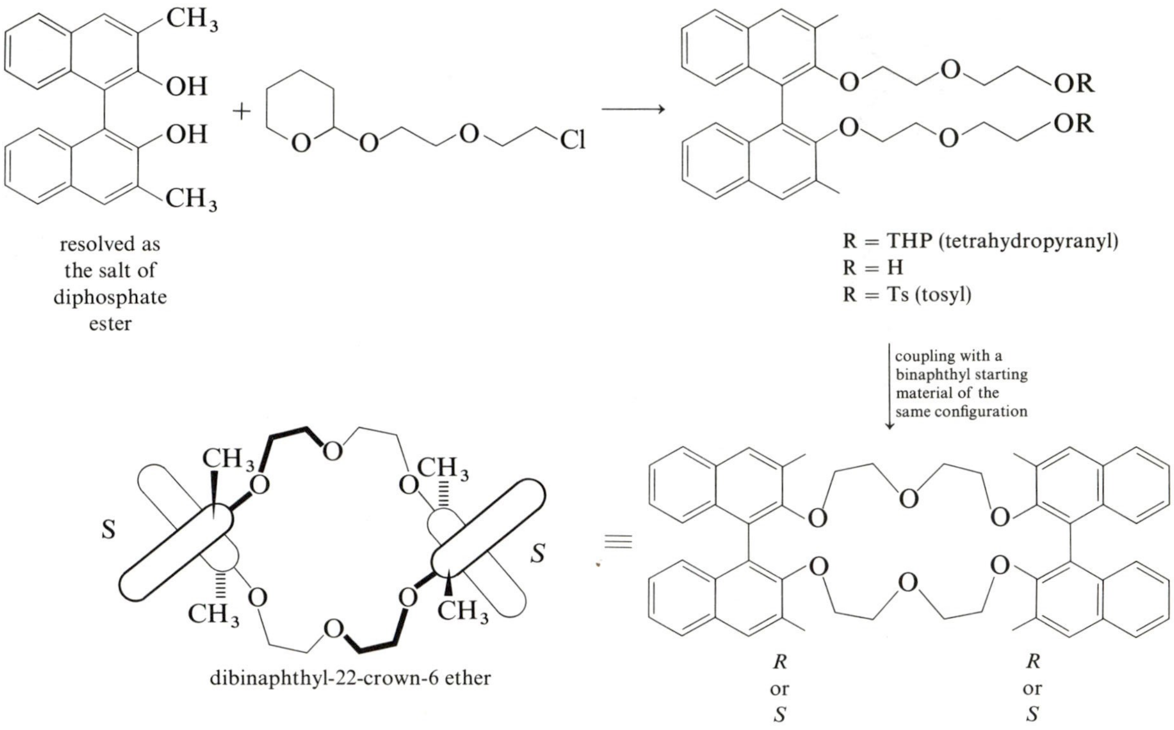

Fig. 5.2. Synthesis of a chiral host molecule.

Fig. 5.3 Chromatographic optical resolution by (*R,R*)-host of methyl phenylglycinate hexafluorophosphate salt (136).

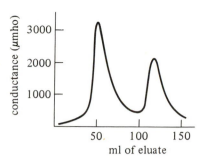

causes the macrocyclic ring to twist like a helix rather than being planar. Both the (*S,S*)- and (*R,R*)-configurations are known, and being optically active they can be used to resolve racemic primary amine salts and amino esters.

The corresponding diastereomeric "activated complexes" formed are highly structured, and their differences are the basis of what Cram calls *chiral recognition*. A liquid–liquid chromatography system is employed for the optical resolution of racemic amine and amino ester salts by the chiral hosts, mimicking enzyme stereospecificities. The ability to resolve optically active compounds by passage of racemic mixtures through columns of such complexing agents could have commercial importance.

Experimentally, a silica gel or celite support is saturated with an aqueous solution of $NaPF_6$ or $LiPF_6$. The mobile phase is a chloroform solution of optically pure host. The racemic mixture is then added to the column and after equilibration a "selective elution" of diastereomeric "activated complexes" is obtained. The appearance of salt in the eluate is monitored by its relative conductance. With racemic methyl phenylglycinate, the elution pattern in Fig. 5.3 is obtained.

Pure material is recovered from each peak after neutralization and optical rotation is taken to determine the conformation by comparison with authentic samples. Finally, NMR investigation was used to deduce the structure of the complexes and two diastereomeric complexes were proposed:

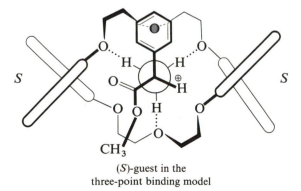

(*S*)-guest in the
three-point binding model

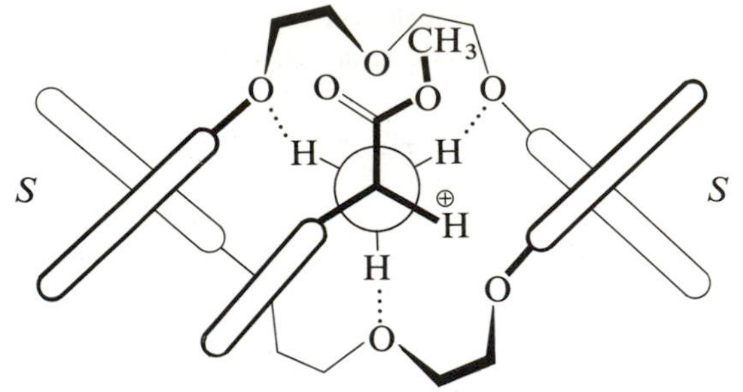

(R)-guest in the
four-point binding model

The terms three-point binding and four-point binding are names given to represent configurational relationship between host and guest. The names are imperfect and incomplete but are convenient and useful (136). At the same time, they identify the more stable diastereomer and point to what is probably the main structural feature responsible for the difference in stabilities of the two diastereomers. These Newman projections show that for both models, the guest molecule binds the (S,S)-host via three N—H hydrogen bindings to the ether oxygens of the macroring. The three substituents of the asymmetric carbon (small, medium, and large) are distributed in space to minimize steric effects. The four-point binding model has an extra dipole–dipole interaction with the ester function because of the stacking imposed by the aromatic rings of the guest and the host.

Nevertheless, the three-point binding model is sterically more stable. The reason is that introduction of substituents at 3- and 3′-positions increases the bulkiness of the complex, causing the system to become more selective and favoring the three-point binding model. In other words, as the complex becomes more crowded by increasing the steric requirements of either the host or the guest, the complexation becomes more stereoselective. As a consequence the (S,S)-host has a preferential selection for the S-isomer of the guest molecule. Diastereomeric association constant ratios as high as 18 has been reported.

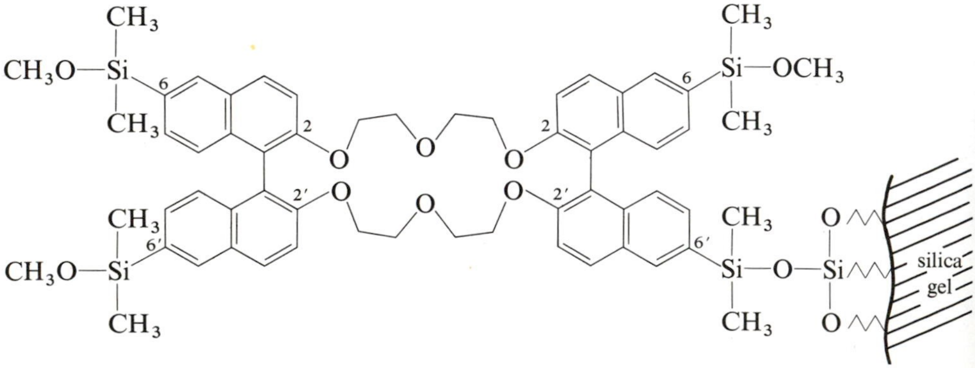

The advantages of a 1,1′-bisbinaphthyl-22-crown-6 ether system can be summarized in three points: (1) The binaphthyl unit is rigid and offers a good chiral barrier. (2) The structure is such that substituents in 2,2′-positions converge where in 3,3′-positions they diverge. This permits a higher degree of stereoselectivity. (3) A functional group at the 6- or 6′-position will not interfere in the complex formation between guest and host, and allows the possibility for the host to be grafted on a solid support.

The naphthalene ring acts as a "spacer" between the solid-phase surface and the host. In this way optically pure (*R,R*)-host was convalently bound to a silica gel solid support, and the resulting material use to resolve primary amine and amino ester salts by solid–liquid chromatography. This technique could be termed *affinity chromatography* specific for enantiomers.

This approach has been extended to the synthesis of macrocyclic ethers (pseudo-crown ethers) incorporated as part of a macromolecular network (styrene–divinylbenzene copolymer) and results in polymers of high coordinating power for various ions (139). The combination of macrocyclic structures with polymer will allow, in the near future, the development of new catalysts containing specific binding properties together with effective catalytic behavior (348).

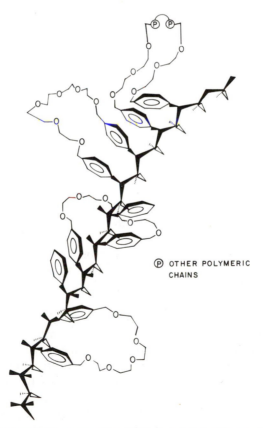

Ⓟ OTHER POLYMERIC CHAINS

Polymeric pseudo-crown ether (139). Reprinted with permission.
Copyright©1979 by the American Chemical Society.

Moreover, some 1,1′-binaphthyl macrocyclic ethers solubilize calcium in hydrocarbon solvents by complexation. Because this behavior resembles transport of calcium across hydrocarbon-like environments of cell membranes, such compounds could be important in studying mechanisms of nerve and muscle function.

In an effort to learn more about the structural parameters involved in complexation, other studies used a pyridinyl unit to form the following host

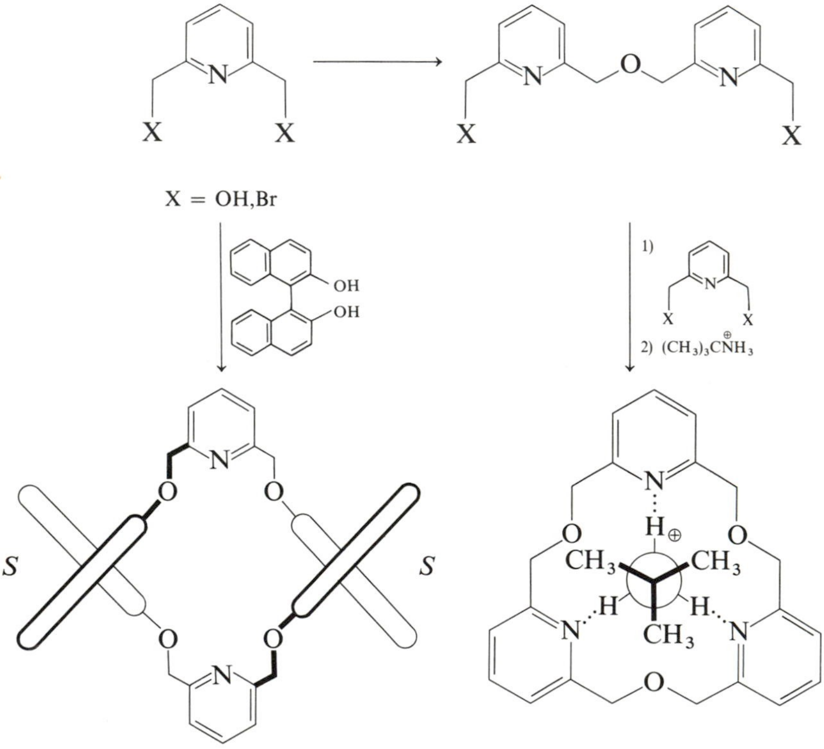

X = OH,Br

compounds. The tripyridine host strongly complexes the *tert*-butyl ammonium thiocyanate salt. The mixed-host, *S,S*-dipyridinyl-dinaphthyl ether, also complexes *S*-phenylglycinate methyl ester preferentially like the polyether analogue. Again, the organization of three hydrogen bonds and three $N^{\oplus}\cdots O$ pole–dipole interactions appear critical for strong and efficient binding (138).

In 1977, a British group under the direction of J. F. Stoddart started to use sugar derivatives for the synthesis of new host compounds (140, 141). Carbohydrates and their derivatives are rich in substituted bismethylendioxy units for incorporation into an 18-crown-6 constitution. They also provide a relatively inexpensive source of chirality and are usually well endowed with functionality.

Starting from L-tartaric acid or D-mannitol they have prepared a series of 18-crown-6 ethers. The D-mannitol precursor permits the association of

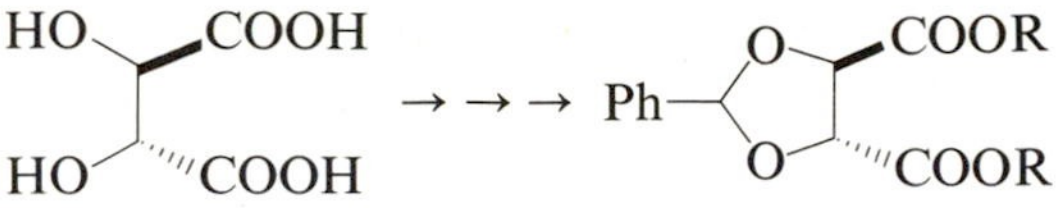

L-tartaric acid

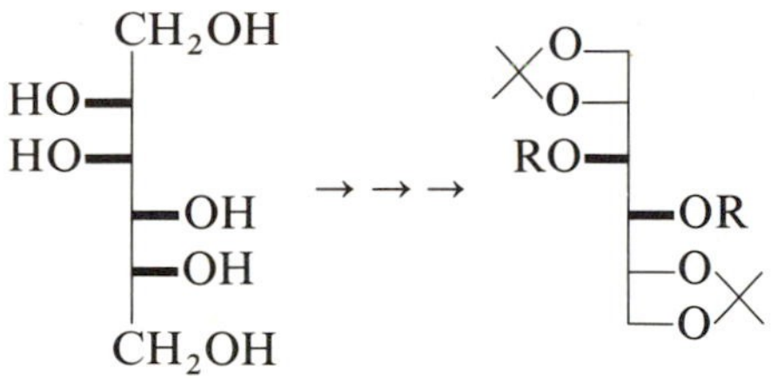

D-mannitol   host precursors

bulky substituents more intimately with the crown ether, and at the same time doubles the number of chiral centers from four to eight.

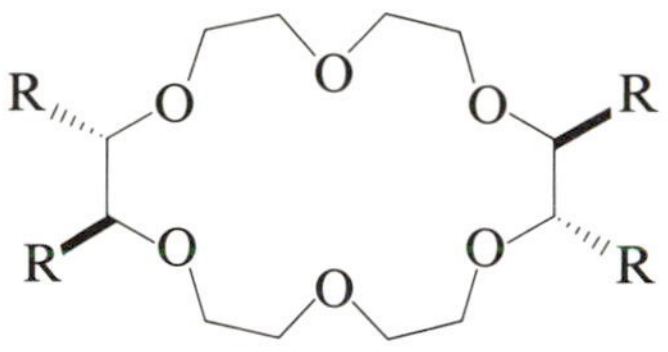

L,L-host from L-tartaric acid
R = CH$_2$OH
R = CH$_2$—OCPh$_3$
R = CONHR$_1$

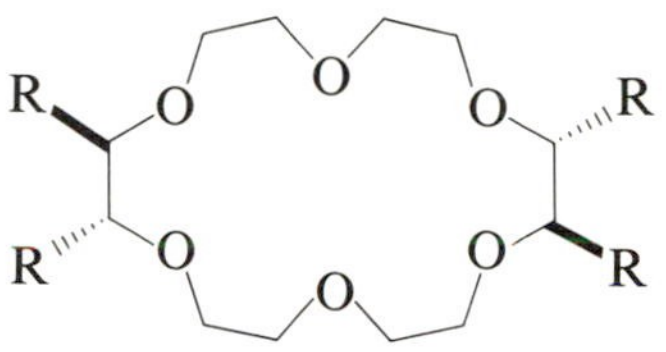

D,D-host from D-mannitol

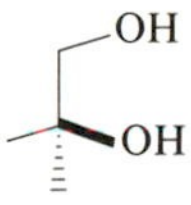

These hosts exhibit enantiomeric differentiation in complexation toward primary alkylammonium salts. Newman projections led to the prediction that the (D,D)-host–(R)-guest complex is more stable. Furthermore, the nature of the anion is important in promoting complexation. *Soft anions** such as SCN$^-$, ClO$_4$$^-$, and PF$_6$$^-$ favor complex formation, whereas *hard anions* such as OH$^-$, Cl$^-$, and Br$^-$ form very stable salts and mitigate against the formation of the host–guest complex.

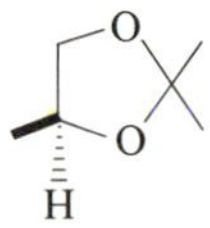

Such an extension in host–guest complexation chemistry is a good example of a long-term objective to build enzyme analogues by lock and key

---

* According to Pearson's nomenclature hardness is associated with high electronegativity and small size of the ion. On the other hand, a soft base is one in which the donor atom is of high polarizability and of low electronegativity and is easily oxidized. The general principle of hard and soft acids and bases is that hard acids prefer to associate with hard bases and soft acids prefer to associate with soft bases (144, 145).

chemistry with crown compounds. Of course, matching of sizes, shapes, and electronic properties of binding portions of hosts and guests is a necessary requisite to strong binding. In this respect, carbohydrates and carbohydrate derivatives are "Nature's gifts to chiral synthesis" because they can be transformed into the structural framework of noncarbohydrate targets (142). This concept will be more and more exploited in the coming years.

A marcrocyclic enzyme model system has also been developed in recent years by Murakami's group in Japan (146, 147). They found that 11-amino-[20] paracyclophan-10-ol catalyzes the deacylation of p-nitrophenylhexadecanoate with a rate 1000-fold greater that 2-aminocyclodecanol.

[20] paracyclophane ring   2-aminocyclodecanol

The greater effectiveness of the paracyclophane ring as compared to the cyclodecanol ring suggests that a sufficient hydrophobicity must be provided by the macrocycle where the cyclodecanol molecule has no ability to incorporate the substrate ester in its cavity. Furthermore, the functional group of the macrocycle must be oriented geometrically in favor of a pseudo-intramolecular reaction with the ester bond of the bound substrate. Substituted paracyclophanes (R ≠ H) are readily amenable to design, and, because of their hydrophobic nature, paracyclophanes remain among the simplest enzyme models.

5.1.1 Chiral Recognition and Catalysis

The previous section demonstrated that chiral macrocyclic polyether hosts discriminate in complexation reactions in chloroform solution between enantiomers of amino ester salt guests. With these results can we go one step further and mimic a catalytic site? We will now describe the design of a host that upon complexation with α-amino ester salts produces a transition state intermediate corresponding to a transacylation (thiolysis) reaction between the chiral host catalytic group (thiol) and the enantiomeric guest salts (143). However, it should immediately be realized that these model systems mimic only the acylation step encountered in serine protease catalysis. So far no acceleration in rate has been observed for the deacylation step.

The following cyclic chiral host has been prepared and the properties for complexation compared to the corresponding opened-chain system.

The hosts are used to study the hydrolysis of L- and D-amino acid p-nitrophenyl esters. The reaction is carried out in 20% EtOH-CH$_2$Cl$_2$

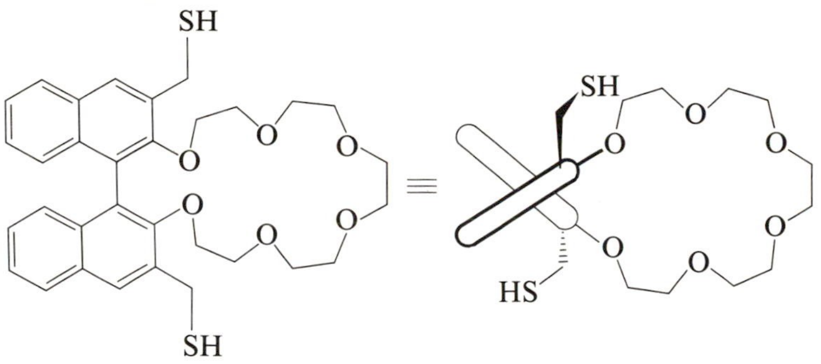

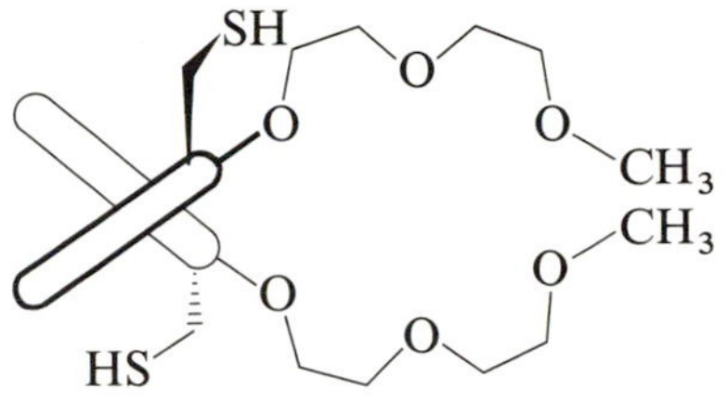

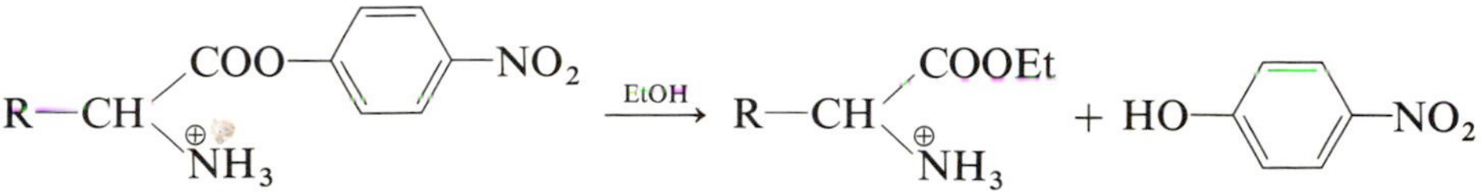

buffered with 0.2 M AcOH and 0.17 M NaOAc at pH 4.8. A burst of p-nitrophenol is followed by ethyl ester formation.

It was observed that the L-amino acid esters react 10^2–10^3 times faster with the cyclic host than with the open host analogue. Clearly, enforced covergence of binding sites enhances complexation by substantial factors. Second, proline esters react at equal rate with both hosts. Hence one needs, as shown before, three protons on the α-nitrogen atom for efficient complexation. In a more polar solvent (40% H_2O-CH_3CN) the rate decreases by a factor of ten. This indicates that the water molecules hydrogen bond to the α-NH_3^+ group and thus compete with the host. In all cases, the S-host reacts faster than the R-host for the natural L-amino acid. The ratio of reactivity depends on the size of the substituent (R) on the glycine moiety.

R	S-host/R-host
CH_3—CH— $\quad\quad\;\;\vert$ $\quad\quad CH_3$	9.2
Ph—CH_2—	8.2
$(CH_3)_2$—$CHCH_2$—	6.0
CH_3—	1.0

These reactions resemble a transacylation where the designed host has some of the properties of trypsin recognition of an NH_3^+ group and papain (a cysteine residue at the active site).

The following structures show the "transition state"-like relationships between S-host and either L- or D-amino ester guest (143).

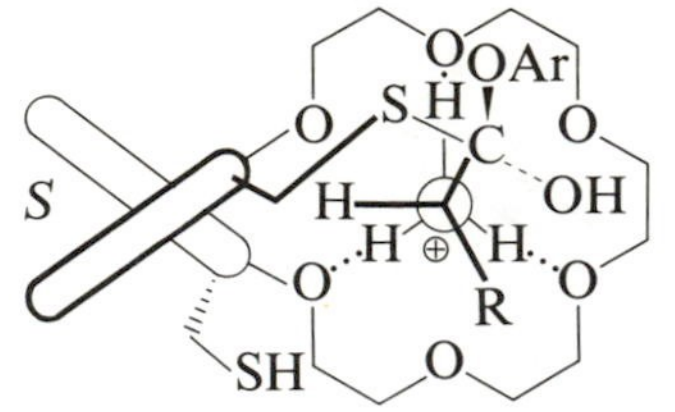

(S) to (L) relationships, more stable 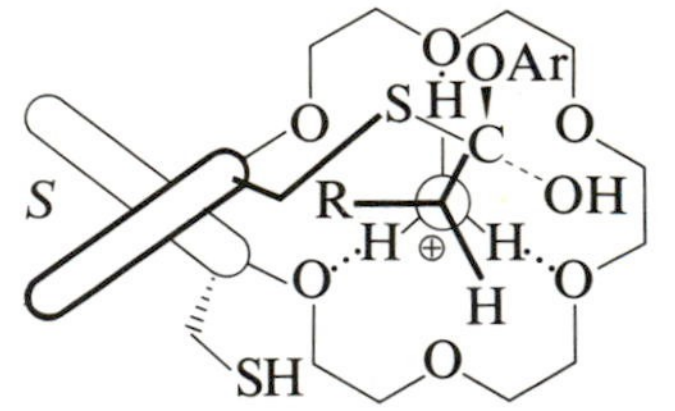   (S) to (D) relationships, less stable

It is rather easy to understand now why the S-isomer was preferentially selected and hydrolyzed. The side chain R being away from the chiral barrier minimizes the steric factors and leads to the formation of a more stable intermediate.

In the same line of thought, Lehn and Sirlin (148) have prepared a chiral macrocyclic molecular catalyst bearing cysteinyl residues. The catalyst complexes primary ammonium salts and displays enhanced rates of intramolecular thiolysis of the bound substrates with structural selectivity for dipeptide esters and high chiral recognition for the L-enantiomer (70 times faster) of a racemic mixture of glycylphenylalanine p-nitrophenyl esters. The representation below shows the complex between the chiral crown ether and the dipeptide glycylglycine p-nitrophenyl ester salt.

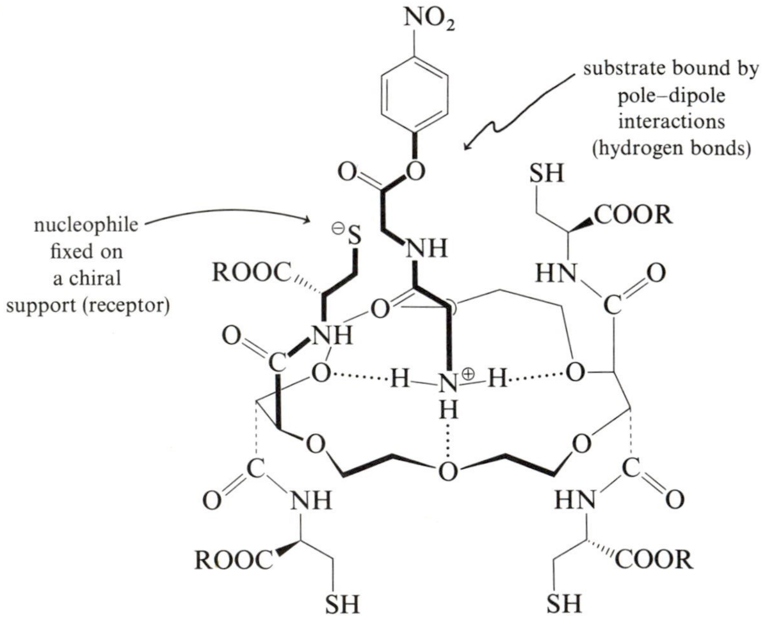

The rate accelerations observed (10^3–10^4 times) are due to complexation of the primary ammonium salt in the crown ether cavity and the participation

of an —SH group of the cysteinyl residues to give an S-acyl intermediate. This "artificial enzyme" model displays molecular complexation, rate acceleration, and structural and chiral discrimination analogous to true biological catalysts.

5.1.2 Stereoelective Transport

Since the discovery of crown ethers it was predicted that chiral carriers would provide a possibility for chiral specificity of guest transport through liquid membranes. This prediction was confirmed in 1974 when Cram and his colleagues found that *enantiomeric differentiation* occurs when designed, neutral, lipophilic, and chiral host compounds carry amino ester salts (guest compounds) from an aqueous solution through bulk chloroform to a second aqueous solution (136).

A binaphthyl dissymmetric (R,R)-host and racemic phenylglycinate methyl ester were used as guest molecules:

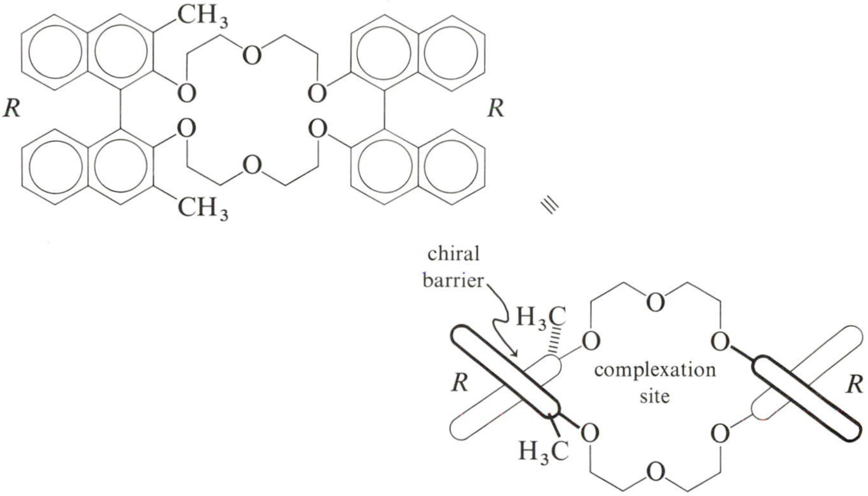

CH₃ / O / O / R / O / O / R / O / O / CH₃

chiral barrier / H₃C / O / O / R / complexation site / R / O / O / H₃C / O

The cell used was a U-shaped tube filled as shown in Fig. 5.4.

A chloroform solution of the host is placed at the bottom of the tube. The guest is present in the α-arm. Ultraviolet absorbance and specific rotation can be measured at different times in the β-arm. In this way, the rate constants for transport were measured for the fast moving enantiomer and the slow moving enantiomer. After 12–19 hr the R-isomer was selectively "pulled" in the β-arm. It has an optical purity of about 80%. The entropy of dilution and the changes in solvation energies associated with inorganic salt, "salting out" the organic salt from its original organic solution, provided the thermodynamic driving force for transport.

By adding an additional bridge to a crown host molecule, a third dimension is added to the host and J. M. Lehn from Strasbourg called these molecules *cryptands* (149). These macropolycycles contain an intramolecular

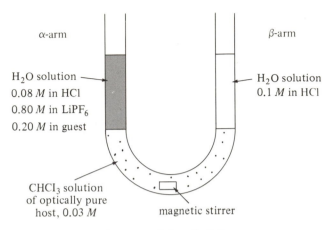

α-arm   β-arm

H$_2$O solution ———   ——— H$_2$O solution
0.08 M in HCl   0.1 M in HCl
0.80 M in LiPF$_6$
0.20 M in guest

CHCl$_3$ solution
of optically pure
host, 0.03 M magnetic stirrer

Average path length, 6.5 cm

Fig. 5.4. Cell for chiral recognition in transport (136). Reproduced with permission from *Techniques of Chemistry* (1976), A. Weissberger, ed., Vol. X, part II, Wiley-Interscience, New York.

cavity (or crypt) and have recently been used by the group of Lehn for selective transport of alkali metal cations though a liquid membrane (150).

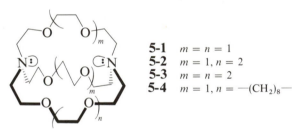

5-1 $m = n = 1$
5-2 $m = 1, n = 2$
5-3 $m = n = 2$
5-4 $m = 1, n = -(CH_2)_8-$

These cryptands form *cryptate*-type inclusion complexes with Na$^+$, K$^+$, or Cs$^+$ picrates. The cryptands function as cation carriers by dissolving the alkali metal picrate into a bulk liquid chloroform membrane as a 1:1 cryptate–picrate ion pair and releasing it from an *in* to an *out* aqueous phase interface (149). Comparison of properties shows, for instance, that **5-4** carries Na$^+$ and K$^+$ much faster than **5-1**. This means that by removal of two oxygen binding sites, the cryptand is transformed from a specific K$^+$ *receptor* (**5-1**) into a specific K$^+$ *carrier* (**5-4**). The work of Lehn on cryptates allowed the design of ligands that can be either a cation receptor or carrier depending on their structure. For instance, **5-1** has the transport sequence K$^+$ < Na$^+$ ≪ Cs$^+$ opposite to the stability sequence of the complexes Cs$^+$ ≪ Na$^+$ < K$^+$. Similarly, **5-2** has opposite complexation Na$^+$ < K$^+$, Cs$^+$ and transport Cs$^+$, K$^+$ < Na$^+$ selectivities, whereas **5-3** has Na$^+$ < K$^+$, Cs$^+$ for both complexation and transport selectivities. The origin of these differences in receptor and carrier behavior lies in the extent of carrier saturation; the most stable cryptates, like **5-1**, have low cation dissociation rates. Thus, the transformation of a cation receptor into a cation carrier may be achieved by simple structural changes and may be accomplished simply by replacing one or two oxygen-binding sites by nonbinding CH_2 groups.

This new coordination chemistry can be extended to anions and other organic ligands. The cryptands have the interesting property of showing a preference for binding one ion over others. At will, the cavities can be constructed either spherical or cylindrical in shape. Such a vast domain of macrocyclic and macropolycyclic metal cation complexes thus has a future in the construction of specific receptors for ions and molecular recognition and the design of molecular catalysts and selective carriers. It allows the design of minimolecules (miniature catalysts) that embody properties of much larger proteins.

These examples all represent *passive* but stereoselective transport where asymmetric recognition can be achieved with organic model systems. However, there is an analogy to be made with these results and the process of mediated transport across biological membranes. All lipidic membranes are practically impermeable to intracellular proteins and to highly charged organic and inorganic ions that surround either side of the membrane. The diffusion of Na^+ out of a cellular membrane and K^+ into the cell occurs along a negative gradient of chemical potential and is called *passive transport*. The passive transport of ions though membranes may be prompted by ionophores (next section for definition). Fortunately, the concentration of cations on either side of the membrane are different and the situation is maintained by *active transport*, which is dependent on metabolic energy. The mechanism for this process is termed the *sodium pump* which functions to keep intracellular K^+ concentrations high and Na^+ concentrations low. Calcium is also thought to be actively pumped out of the cells. For these cases, the energy for the transport is provided by the hydrolysis of ATP. Diffusion of sugars and amino acids to strategic targets in the cell, however, are examples of simple facilitated passive transport.

5.1.3 The Ionophores

It is logical at this point to mention the role and importance of some of nature's chelating agents, the *ionophores*. Figure 5.5 gives representative examples. They contain polyamide, polyester, and polyether functions and most are cyclic. They have the capacity to selectively bind metal ions and act as carriers across membranes (151).

An ion (cation) is too hydrophilic to efficiently cross a thick (~ 10 nm) hydrophobic layer of lipids and lipoproteins such as those found in natural or artificial membranes. However, by binding selectively with the polar functions located inside the macrocyclic ring, the cation is now coated with an hydrophobic shell and can then more readily pass across the membrane.

The naturally occurring ionophores are found mainly in microorganisms and many are used as antibiotics. By complexing with metal ions, they apparently disrupt the control of ion permeability in bacterial membranes. An example is nonactin, an antibiotic that functions by transporting sodium ions into the bacteria until the resulting osmotic pressure causes rupture of the cell wall. Valinomycin coordinates selectively K^+ ions where the permeability of the membrane for potassium is offset. In contrast antamanide, a decapeptide containing only L-amino acids has a binding cavity of a different geometry than valinomycin and shows a strong preference for Na^+ ions over K^+ ions. Nigericin also invokes the opposite effect induced by valinomycin and its analogues and reversible ring closure occurs upon binding to metal ions. Beside carrying ions, ionophores are probably associated with other inhibition processes in the cell such as the control of hormone transport across membranes, regulation of certain metabolisms and the permeation of neurotransmitters in nerve cells.

It is interesting to realize that some ionophores contain amino acids of the D-configuration. Such a situation is not found in higher organisms.

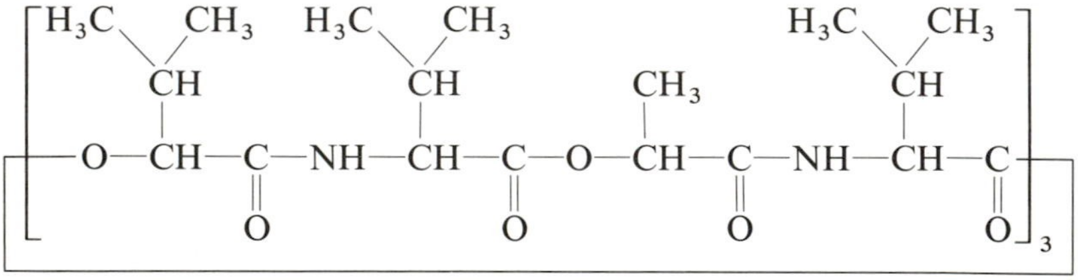

valinomycin

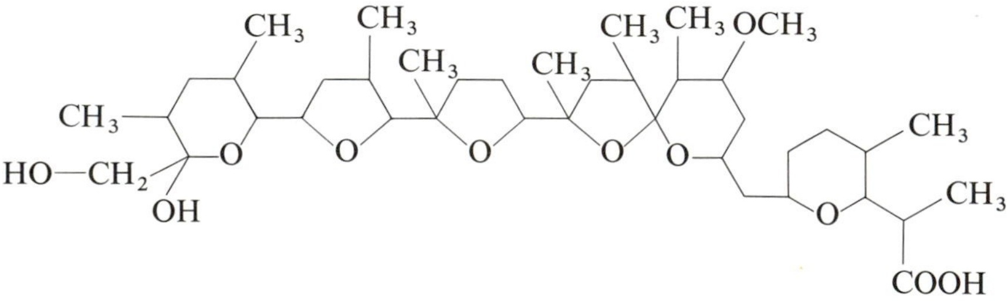

nigericin

$$Pro \rightarrow Ala \rightarrow Phe \rightarrow Phe \rightarrow Pro$$
$$Pro \leftarrow Val \leftarrow Phe \leftarrow Phe \leftarrow Pro$$

antamanide

A cyclic peptide that neutralizes the effect of phalline B, a toxin (also a cyclic peptide) present in the deadly mushroom *Amanita phalloides*.

Fig. 5.5. Structures of some ionophores.

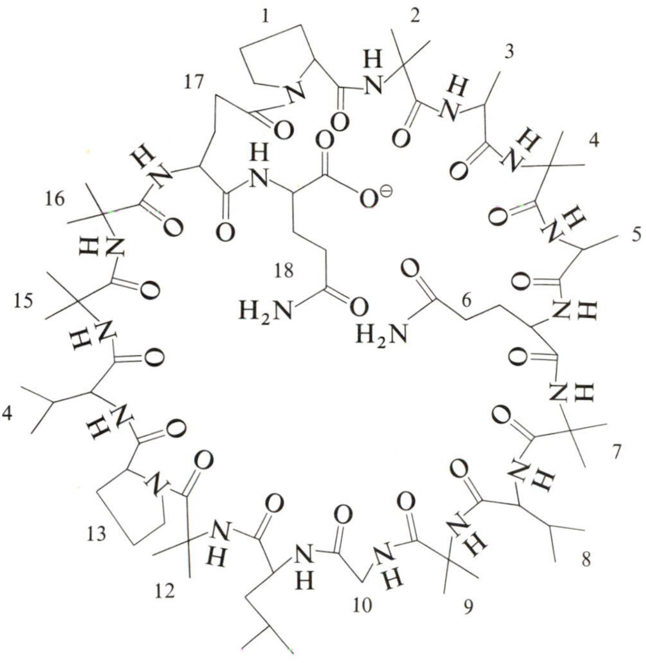

alamethicin

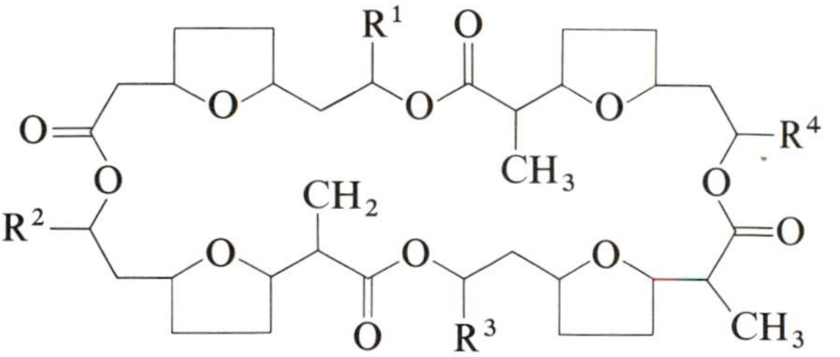

nonactin $R^1=R^2=R^3=R^4=CH_3$
monactin $R^1=R^2=R^3=CH_3$, $R^4=C_2H_5$
dinactin $R^1=R^3=CH_3$, $R^2=R^4=C_2H_5$

Fig. 5.5. (continued)

Stimulated by the discovery of natural ionophores in the 1960s, chemists have successfully synthesized a number of compounds, composed of natural building blocks, that can complex inorganic and organic ions. One example will be presented here. In 1974, E. R. Blout (152) from Harvard University, reported the enantiomeric differentiation between D- and L-amino acid salts in complexes with cyclo-(L-Pro-Gly)$_n$ peptides (n = 3,4) (Fig. 5.6). In chloroform, ($\pm$)-Pro-OBz·HCl, Phe-OMe·HCl or Val-OMe·HCl were mixed with the cyclopeptide and the complexes analyzed by ^{13}C-NMR. Diastereomeric

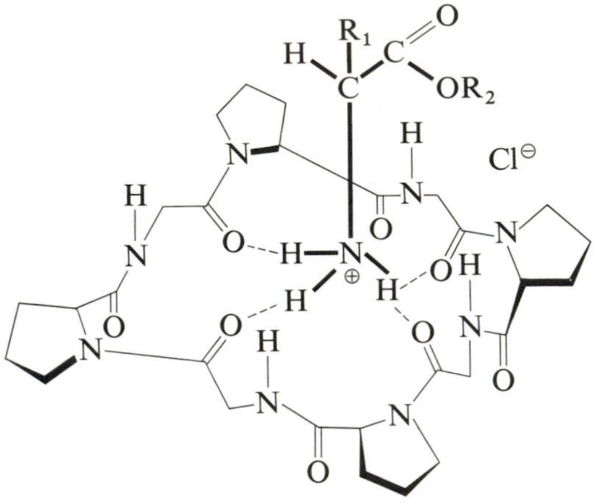

Fig. 5.6. Complex of cyclo-(L-Pro-Gly)$_4$ with an amino acid ester (152). Reprinted with permission. Copyright © 1974 by the American Chemical Society.

pairs of complexes were formed and distinct resonances were observed and attributed to different orientational effects upon complexing D- or L-amino acids.

The binding scheme shows that the four carbonyl functions of the glycine residues are oriented toward the interior of the ring and allow hydrogen bonding with the α-$\overset{+}{N}H_3$ group of the guest molecule in the cavity. Blout's compounds, although structurally resembling ionophores, are closely related to Cram's compounds in function.

In leaving this subject we note that, in the near future, systems like those of Cram and Blout will find application in the synthesis of dissymmetric molecules by an asymmetric induction of prochiral molecules.

5.2 Micelles

Surfactants are amphiphilic molecules, that is, they have both pronounced hydrophobic and hydrophilic properties. A detergent is an excellent example. In solution, such low molecular weight electrolytes form ion pairs with the counterions. By increasing the monomeric concentration, clusters and then low molecular weight aggregates are formed. Finally, larger aggregates called *micelles* are produced (153–158). Therefore, micellization of monomeric surfactants is observed when the surfactant concentration exceeds the so-called *critical micelle concentration* (cmc). In general the cmc varies from 10^{-2} to 10^{-4} M and the conductance of the solution changes sharply above this concentration.

Most often micelles are spherical. In a polar solvent such as water, the hydrophobic hydrocarbon chains of the surfactants are directed toward the interior while the polar or ionic head groups are distributed on the surface

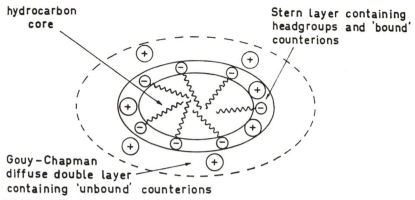

hydrocarbon
core

Stern layer containing
headgroups and 'bound'
counterions

Gouy–Chapman
diffuse double layer
containing 'unbound' counterions

Fig. 5.7. Elliptical cross section of an idealized spherical model of an ionic micelle (156). Reprinted with permission. Copyright©1977 by The Chemical Society.

of the sphere facing the counterions in the aqueous solution. It is the coming together of hydrophobic groups in water that leads to a favorable entropy change because of the liberation of water molecules from the aqueous–apolar interfaces where the hydrophobic groups appear to have considerable freedom of movement in the micelle. It is this gain of entropy that leads to the favorable free energy change on micellization.

Figure 5.7 gives an idealized spherical model of a micelle. Micellization of a surfactant such as dodecyltrimethyl ammonium bromide creates positively charged surfaces composed of cationic "heads." Coulombic attraction gathers the bromide ions into the vicinity of the quaternary nitrogens. This region forms the Stern layer and the most interesting micellar chemistry occurs in this region. Very little water is present in the interior of a micelle which is hydrocarbon-like and it is this difference in polarity between the interior and the surface that makes micelles resemble globular proteins. In effect, the polarities of micelle surfaces are generally similar to those of proteins and intermediate between that of water and ethanol. The fact that the active site of an enzyme is apparently quite apolar, even though the enzyme as a whole is water soluble, makes micelle studies very pertinent (154, 155).

It is well known that organic compounds, particularly nonpolar ones, can absorb onto or in micelles, increasing their solubility relative to that in pure water and often altering their chemical reactivity. At the same time it is the micelles, rather than individual surfactant molecules, which are responsible for altering the rate of organic reactions in aqueous solution of surfactants. Therefore, a proper choice of surfactant can lead to rate increases of 5- to 1000-fold compared to the same reaction in the absence of surfactants. Depending on the type of micelle used, this result in a large concentration of H^+ or OH^- ions which are gathered in the Stern layer and are responsible for the increase in the rate of the reaction. Other basic groups or nucleophiles in the micelle, should also have an effect on catalysis. There is a much weaker interaction between the micelle and the counterions in the wider Gouy–Chapman layer, which extends for several hundred angstroms from the micelle surface and causes a gradual ion gradient.

Rate enhancements have been observed with cationic, anionic, and nonionic micelles.

For instance, bis-2,4-dinitrophenyl phosphate is rapidly hydrolyzed at pH 8 by cationic micelles. Anionic micelles will inhibit the reaction because of repulsions with the negatively charged product and competition with the negatively charged OH⁻ which causes the reaction.

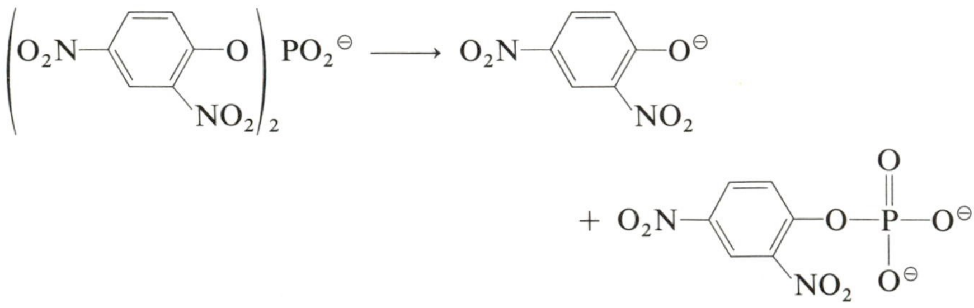

$$\left(O_2N-\!\!\bigcirc\!\!-O\atop {}^{\backslash}NO_2\right)_2 PO_2^{\ominus} \longrightarrow O_2N-\!\!\bigcirc\!\!-O^{\ominus}$$

$$+ \; O_2N-\!\!\bigcirc\!\!-O-\overset{O}{\underset{O^{\ominus}}{P}}-O^{\ominus}$$

On the other hand, zwitterionic micelles which are usually relatively ineffective in catalysis are effective in the following decarboxylation reaction (154):

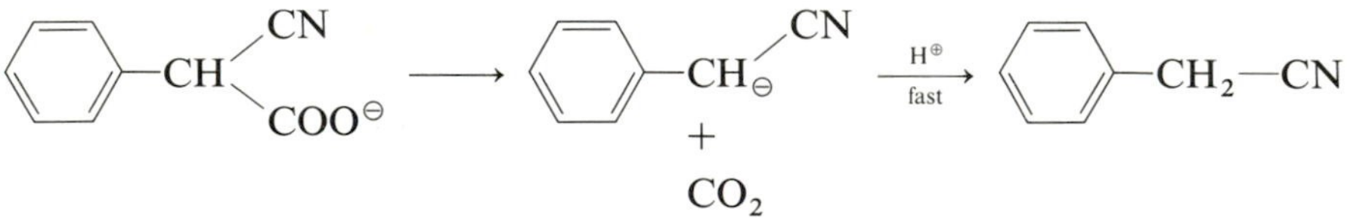

$$\bigcirc-CH{\overset{CN}{\underset{COO^{\ominus}}{}}} \longrightarrow \bigcirc-CH_{\ominus}{\overset{CN}{}} + CO_2 \xrightarrow[fast]{H^{\oplus}} \bigcirc-CH_2-CN$$

The rationale is based on favorable coulombic interactions as follows:

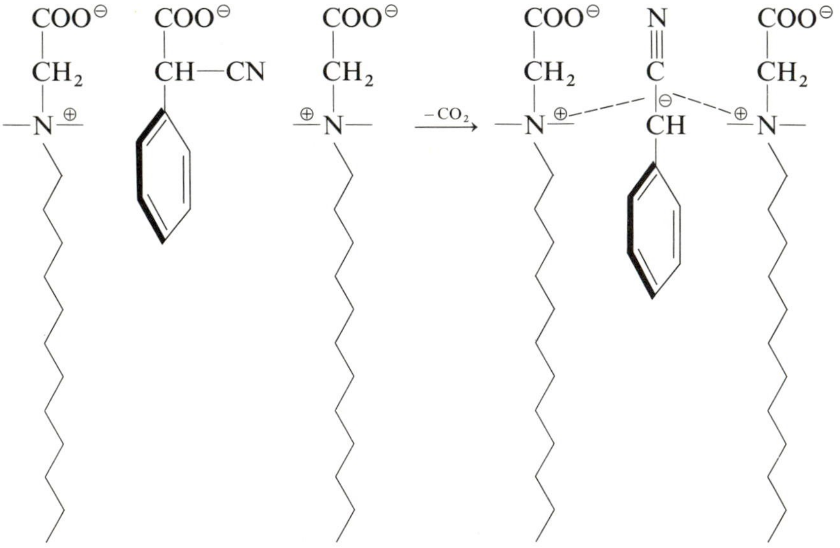

Consequently, many features of kinetics in micellar systems are related to reactions in monolayers and polyelectrolytes surfaces.

Sodium dodecyl sulfate (SDS) $[CH_3(CH_2)_{11}SO_3^-Na^+]$, a well-known surfactant, forms spheres containing 50 to 100 molecules. The potential between bulk and micellar phases is about 50–100 mV and electrostatic and hydrophobic interaction forces are important factors for maintaining the stability of the micelles. SDS is often used to denature proteins where similar forces are present in their tertiary structure.*

Examples of micellar systems are given below where the surfactant provides the medium for catalysis but does not directly participate in the reaction. For instance, N-acetyl histidine can bring about the hydrolysis of p-nitrophenyl esters via an acyl-imidazole intermediate.

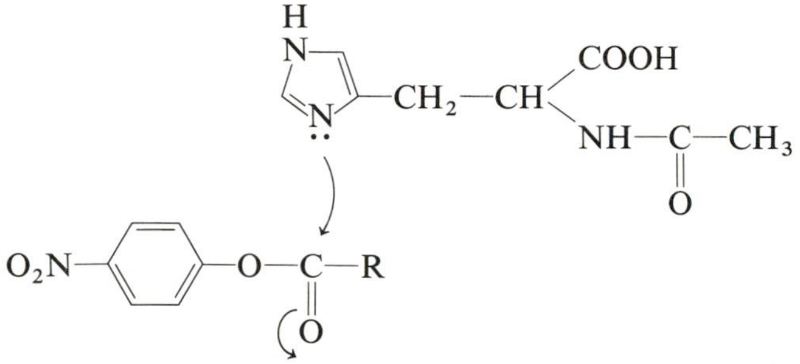

The nucleophilicity of the imidazole ring toward the ester function is enhanced when the system is in a micellar form, i.e., in the presence of SDS, as a result of a favorable high concentration of catalyst and substrate in the micelle.

Cyanide ions are known to react with n-alkyl pyridinium salts:

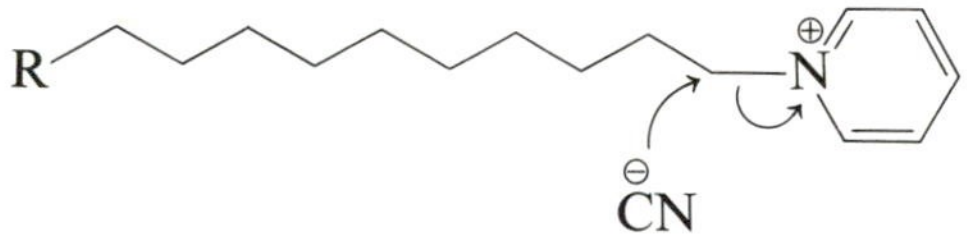

However, the rate is markedly increased by cationic surfactants. The longer the alkyl chain, the faster the rate. This shows that hydrophobic binding is largely responsible for the increase in rate of substitution in this reaction.

Cyanide ions can also be added to 3-carbamoyl pyridinium bromide (NAD$^+$ analogue):

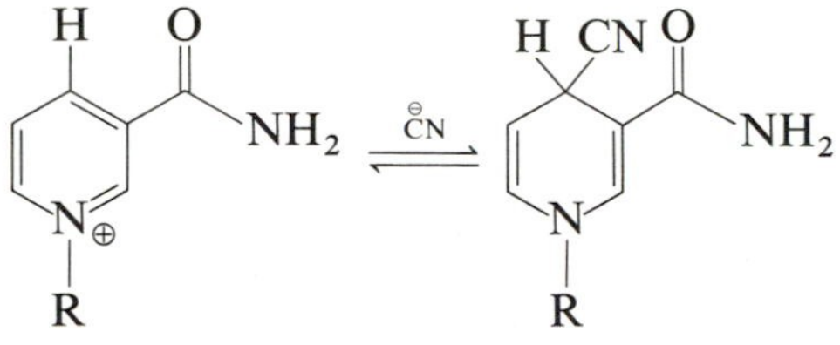

$R = C_{16}H_{33}$

---

* That is, the final folded form of the polypeptide chain or active form of a protein.

The presence of 0.02 M cetyltrimethyl ammonium $[CH_3\text{-}(CH_2)_{15}\text{—}\overset{+}{N}(CH_3)_3]$ salts increases the rate constant by 950-fold for the addition of CN^- to the N-hexadecyl substrate ($R = C_{16}H_{33}$) and increases the corresponding association constant about 25,000-fold (159). Notice that in this transformation, the charged substrate becomes neutral after reaction. So the product of the reaction will be reoriented within the micelles and will be pulled inside. The hydrophobic interactions destabilize the reactant with respect to product and it has been suggested that the presence of the surfactant is responsible for the large acceleration in the rate of attack by the CN^- ion. As a consequence, the proportion of [substrate] vs [product] of the reaction is displaced to the right in the micellar system.

Mention should be made of reversed or inversed micelles (160). Sulfosuccinate surfactants form reversed micelles where a remarkable amount of water (50 moles/mole of solute) can be incorporated inside the micelle in octane solutions.

di-2-ethylhexyl sodium sulfosuccinate

The concept of "water pools" has been introduced by F. M. Menger in 1973 to describe the nature of the cavity inside reversed micelles (160). Addition of p-nitrophenyl acetate in the presence of imidazole to this micellar system results in a 53-fold increase of hydrolysis of the acetate as compared to bulk water. Clearly, imidazole in the micelle is able to come very close to the substrate and to catalyze its hydrolysis. Therefore, remarkable rate enhancements in reversed micelles have been ascribed to favorable substrate orientation in the interior of the reversed micelles, where bond breaking may be assisted by proton transfer.

Dodecylammonium propionate (DAP) also forms reversed micelles at 0.10 M concentration in benzene and is able to entrap 0.55 M of water. This system was used by J. H. Fendler to investigate the protonation of pyrene-1-carboxylic acid using a nanosecond time resolved fluorescence technique (161). The most striking feature of the data obtained is the extraordinarily large rate constant for the protonation of the carboxylate group in the surfactant solubilized water pool; a value of the order of 10^{12} M^{-1} s^{-1} was observed! Figure 5.8 gives a model of this ultrafast proton transfer system in the reversed micelle which is only feasible if the donor and acceptor are in close proximity. A fraction of DAP is hydrolyzed to propionic acid which is pulled into the water pool region and proton transfer occurs within the hydration shell of the surfactant.

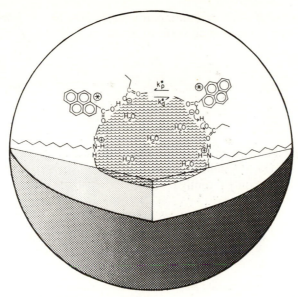

Fig. 5.8. A proposed model for the ultrafast proton transfer at the hydration shell of the surfactant headgroups in reversed micellar DAP in benzene. Since the concentration of surfactant is in a very large excess over the probe, proton transfer must occur from dodecylammonium propionic acid to pyrene-1-carboxylate. For the sake of clarity, two pyrene moieties are drawn in the aggregate shown. In reality, there is much less than one probe per aggregate. The shaded area indicates the extent water which hydrates the surfactant headgroups (161). Reprinted with permission. Copyright © 1978 by the American Chemical Society.

Charge and proton relay through hydrogen bonds have been proposed to contribute to the catalytic efficiency of enzymes, and in this sense reversed micelles provide an appropriate model to delineate the importance of such factors at the enzyme active site. Micellar surfaces also provide a convenient means for the reduction in dimensionality, an important factor in enhancing reaction rates. They also serve as good models to demonstrate the feasibility of ultrafast proton transfer when the reactants are localized in a suitable environment such as membrane surfaces and other complex biomacromolecules.

Micelles with attached catalytic groups are also known and can act as catalysts. For example, esters and carbonates can be hydrolyzed by long-chain *N*-acyl histidine surfactants (162).

substrate (carbonate)

N-Dodecyl-*N'*,*N'*-dimethyl aminoethyl carbonate ion is hydrolyzed 2240 times faster with the following surfactant than with *N*-acetyl histidine.

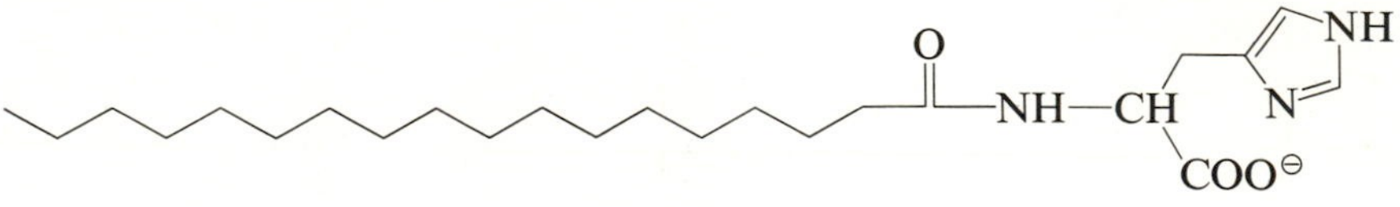

catalyst (*N*-acyl histidine)

In this context, many micellar enzyme analogues of serine proteases have been prepared (131). Cysteine proteases, however, such as papain and ficin, have been modeled only recently (163).

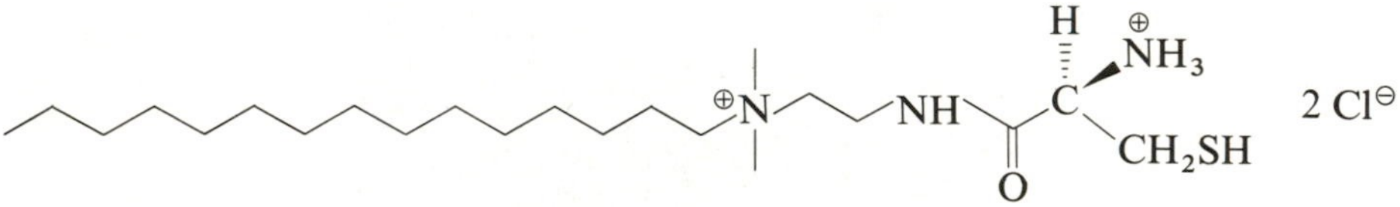

self-contained thiol-functionalized surfactant

The above cysteine-containing long hydrocarbon chain forms micelles between 0.003 and 0.05 *M* and cleaves *p*-nitrophenyl acetate with a pseudo-first-order rate constant. The surfactant is 180 times more reactive than cetyltrimethyl ammonium chloride, a micellar system without a functional group present.

An imaginative example of micellar catalysis (164) is in the acyloin (benzoin) condensation in the presence of *N*-lauryl thiazolium bromide (see Section 7.3).

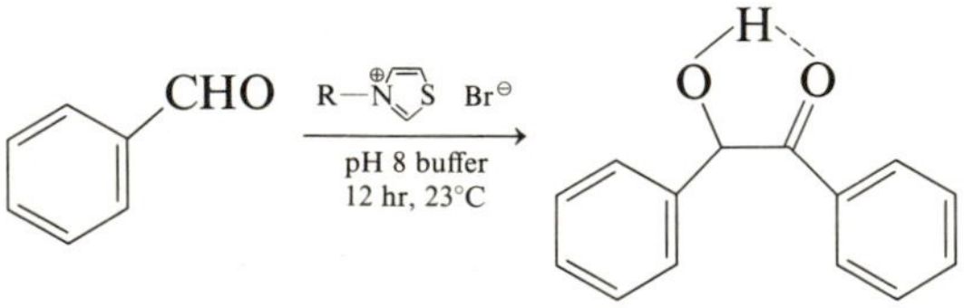

When R = butyl the reaction does not work, but if R = dodecyl, micelles are formed, benzaldehyde molecules intercalate and the yield reaches up to 95% conversion.

In summary, impressive catalytic effects are obtained by incorporating reactants in micelles, thereby increasing their effective concentration and reducing the entropy loss in the transition state by providing an effective medium for the reaction. The other advantages of a micellar system are:

(a) favorable hydrophobic interactions;
(b) model to some degree the behavior of enzymes and membranes (phospholipid vesicules are catalysts);
(c) the forces that hold micelles are similar to those for tertiary structure of proteins;

(d) increases local concentration of ions responsible for the catalysis;
(e) rate enhancement up to 1000-fold have been observed.

However, it is difficult to have a greater increase in rate of reaction with micelles because of the inherent uncertainty as to the structure of the reaction site. Therefore the limitations are:

(a) the structure of the micelle is not well defined;
(b) the structure depends on surfactant and substrate concentrations;
(c) rigid orientational effects are not expected unless immobilized surfactants on a polymer support can be achieved;
(d) nothing is known about the relative orientation of reactive groups.

Thus, a micelle remains a very crude enzyme model.

5.2.1 Stereochemical Recognition

How about stereochemical recognition? So far only few micelles of optically active surfactants have been used as catalysts in a number of reactions with chiral substrates, but in general the effects are small. Two examples will be given here where chiral micelles could stereoselectively catalyze the hydrolysis of chiral esters.

Cationic surfactants derived from D($-$)-ephedrine analogues show different catalytic efficiencies in hydrolyses of p-nitrophenyl esters of D- and L-mandelic acid (165). Hydrolysis of the racemic mixture is slower than its enantiomers with D($-$)-surfactant, suggesting that more than one substrate molecule is incorporated into each micelle. Therefore, an enantiomeric substrate molecule perturbs the micellar structure in such a way that the resulting complex then exhibits markedly different activities toward the two enantiomers.

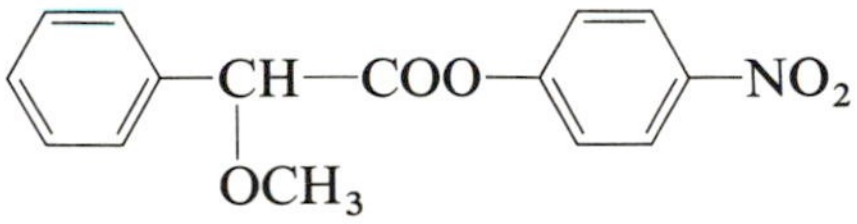

D($-$)-mandelic ester
L($+$)-mandelic ester
substrates

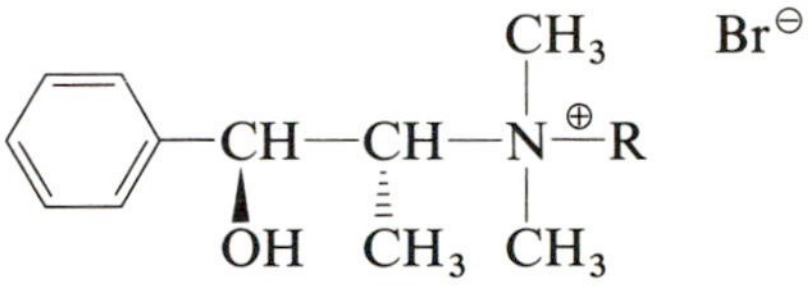

R = C$_{10}$H$_{21}$
= C$_{12}$H$_{25}$

D($-$)-surfactant

Fig. 5.9. Selective hydrolysis of mandelic ester by a chiral micelle (165).

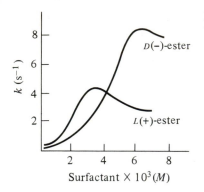

The liberation of *p*-nitrophenol can be followed spectrometrically at 25°C, pH 9.0 in 0.01 *M* borate buffer plus 0.5% dioxane using a substrate concentration of 10^{-3} *M*. The results are presented graphically in Fig. 5.9.

A stereoselective hydrolysis of D(−)-mandelic ester over L(+)-enantiomer is taking place. At high surfactant concentration, some inhibition is observed, probably from the presence of surfactant counterions (salt effects).

A more dramatic example, again from C. A. Burton's work (166), is the use of an optically active L-histidyl-cationic micelle. It shows a larger degree of stereoselective control (approximately threefold) for hydrolysis of amino acid derivatives.

The synthesis of the micelle is outlined below:

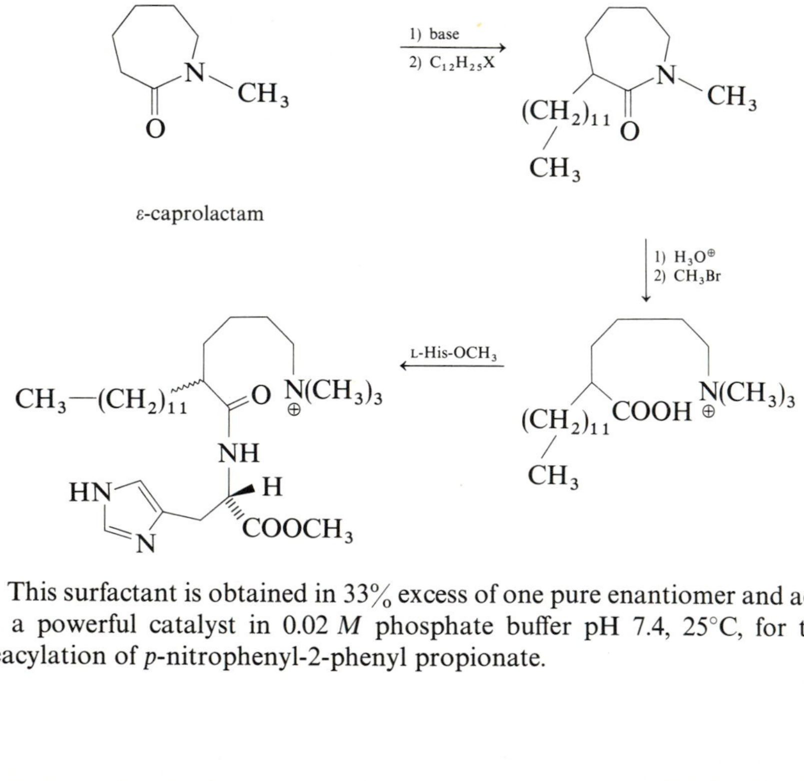

This surfactant is obtained in 33% excess of one pure enantiomer and acts as a powerful catalyst in 0.02 *M* phosphate buffer pH 7.4, 25°C, for the deacylation of *p*-nitrophenyl-2-phenyl propionate.

$$CH_3-\overset{*}{C}H-C\overset{\displaystyle O}{\underset{\displaystyle O-}{\big\langle}}-NO_2$$

The rate of the reaction is pH dependent and shows that the group participating in the catalysis has a pK of 6.4 to 7.5. As compared to bulk buffer, rate enhancements of 260 and 283 for R- and S-isomers are obtained, respectively, but the binding constants are the same for both isomers.

However, a better stereoselectivity is obtained with N-acetyl-phenylalanine esters.

$$\bigcirc\!\!-CH_2-\overset{*}{C}H-C\overset{\displaystyle O}{\underset{\displaystyle O-\bigcirc-NO_2}{}}$$
$$\underset{\underset{CH_3}{\overset{|}{C}=O}}{\overset{|}{NH}}$$

The S-isomer is deacylated faster with an enantiomeric reaction rate ratio of 3 to 1. The binding constants were again found to be the same for both enantiomers so the difference in rates is probably due to ΔG differences in the transition state for the enantiomeric amino acid ester. The results are presented in Fig. 5.10.

The initial rates are similar confirming that the stereospecificity of the reaction depends on the transition state rather than initial state interactions. Addition of a competing surfactant such as cetyl ammonium salts still yields an enantiomeric rate ratio of 2 to 2.5 for the S-isomer depending on the concentration. Consequently, it is the presence of a functional group in the micelle rather than the individual surfactant molecules that is responsible for the selectivity observed, the imidazolyl residue functioning as the nucleophilic catalyst.

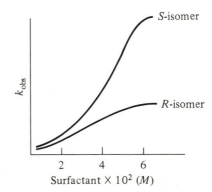

Fig. 5.10. Selective hydrolysis of N-acetyl-Phe-ester by the L-histidyl-cationic micelle (166).

A hypothetical transition state intermediate between the chiral micelle and the *S*-isomer can then be presented:

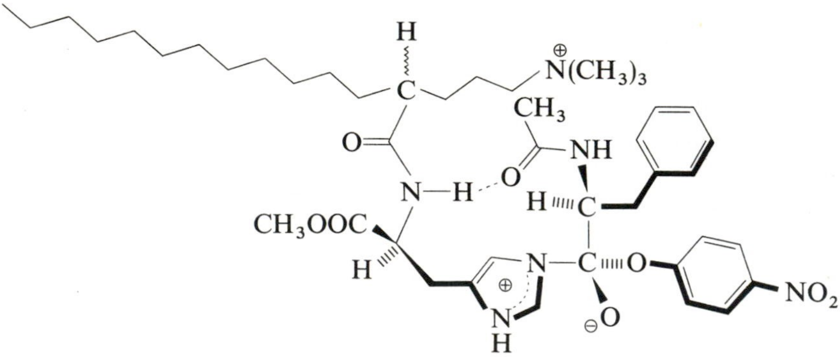

The *R*-isomer, however, would suffer from a severe imidazole-phenyl ring repulsion.

This approach of micellar stereoselectivity by functional surfactants has been extended to cleavage of dipeptide diasteriomeric substrates (167).

In conclusion, solvation changes and loss of some of the translational entropy in forming a transition state are two important factors responsible for catalysis and rate enhancements observed with micellar systems. In this respect they resemble enzymes. Another formal similarity between enzymatic and micellar catalysis is the strong hydrophobic binding with the substrate. However, the fact that micelles are of limited rigidity results in poor specificity in catalysis and only moderate rate enhancements are obtained.

Nevertheless, micellar systems have found applications in pharmacology and in industry particularly in emulsion polymerization. Furthermore, synthetic organic chemists are frequently faced with the problem of reacting a water-insoluble organic compound with a water-soluble reagent (OH^-, MnO_4^-, IO_4^-, OCl^-, etc.). The use of surfactants can now alleviate this problem. Two-phase reactions may be undertaken where the surfactants disperse organic liquid in water, generating higher yields and shorter reaction times (157).

5.3 Polymers

Enzymes are copolymers composed of various amino acid monomers. It is then easy to understand that the utilization of synthetic organic polymers to change the reactivities of low molecular weight substances has received more and more attention lately (168). These reactions can serve as models for more complex enzymatic processes. Although polymeric catalysts are considerably less efficient than enzymes, several analogies between natural and synthetic macromolecular systems have been revealed. In particular, a polymer with charged groups will tend to concentrate and/or repel low molecular weight ionic reactants and products in its vicinity and, consequently, will function as either an inhibitor or an accelerator of the reaction between two species. However, if catalytically active functions are added to a polymer which contains charged groups, the polymer itself, and not its counterions, will take part in the catalysis (169, 170).

We will describe in this section examples of this last type of polymer, in particular imidazole-containing polymers, which have esterolytic properties and in many ways resemble serine proteases (169).

Early investigations with poly(methacrylic acid) showed that this polymer can catalyze the nucleophilic displacement of bromine ion from α-bromoacetamide.

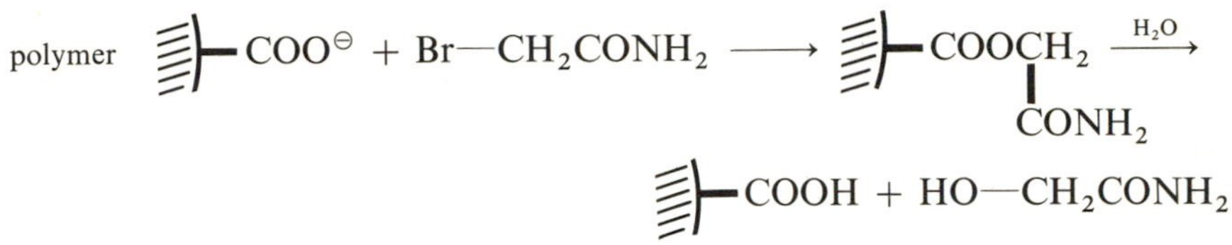

$$\text{polymer} \equiv\!\!-COO^{\ominus} + Br\!\!-\!\!CH_2CONH_2 \longrightarrow \equiv\!\!-COOCH_2 \xrightarrow{H_2O}$$
$$\qquad\qquad\qquad\qquad\qquad\qquad\qquad\qquad\qquad\quad | \atop CONH_2$$

$$\equiv\!\!-COOH + HO\!\!-\!\!CH_2CONH_2$$

However, if the degree of ionization of this polyacid increases, the catalytic power of the polymer decreases markedly.

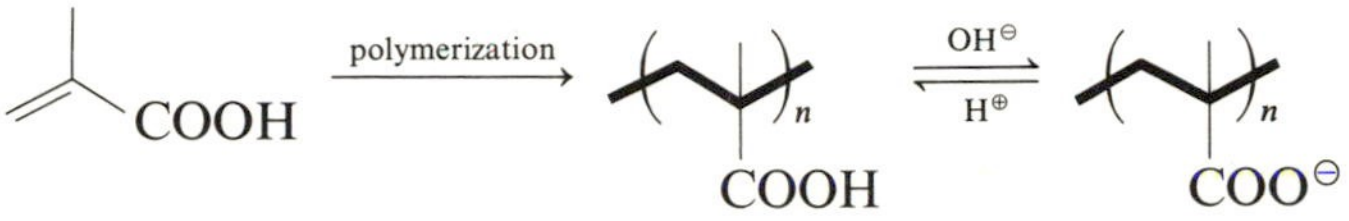

methacrylic acid   poly(methacrylic acid)

Poly(vinyl-4-pyridine) has the catalytic capacity to accelerate the solvolysis of 2,4-dinitrophenyl acetate but the rate constant of the reaction increases as the fraction of neutral pyridine residues increases. In fact, both free and charged species are probably needed for catalysis; the charged residues providing electrostatic binding.

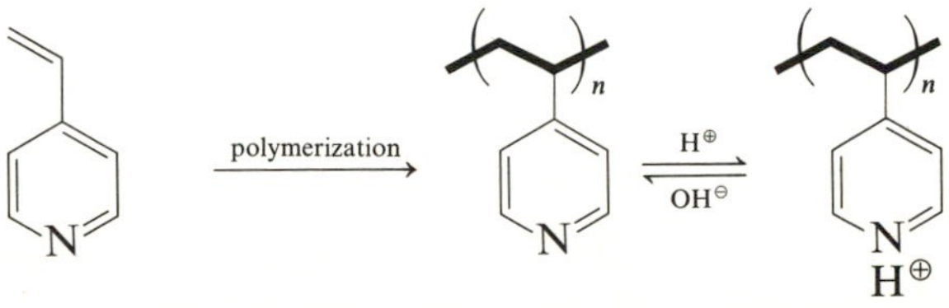

vinyl-4-pyridine   poly(vinyl-4-pyridine)

From these two examples, it can be concluded that a balance of neutral and charged functions must be important for the polymer to act as an efficient catalyst. The above two polymers can be classified respectively as anionic and cationic polymers. Ion-exchange resins are among these categories. R. L. Letsinger was among the first to find applications of substrate binding to a polymeric catalyst using the concept of electrostatic interaction.

In 1965 C. G. Overberger and colleagues showed that vinyl polymers, containing imidazole and benzimidazole, have cooperative multifunctional interactions which also lead to enhanced catalytic action of the polymers in comparison to low molecular weight precursors (170).

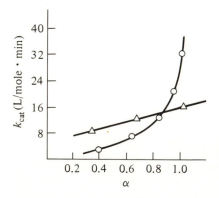

However, these polymers have a peculiar behavior. For example, the rate of solvolysis of p-nitrophenyl acetate (PNPA) in 28% EtOH—H$_2$O by poly(vinylimidazole) is presented in Fig. 5.11 and is compared to imidazole alone. The α value represents the fraction of the functional group in the nonionized (neutral) form. At low pH values, the imidazole ring is protonated and the α value is small.

It is clear that protonated imidazole ring (low α value) does not participate in catalysis. The upward curvature for the polymer shows that it is less effi-

Fig. 5.11. Solvolysis of PNPA catalyzed by poly(vinylimidazole) (○) and imidazole (△) (28.5% ethanol–water, ionic strength 0.02, 26°C)(170).

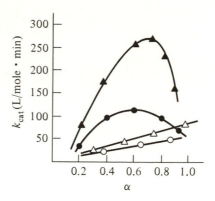

Fig. 5.12. Solvolyses of NABA and NABS catalyzed by poly(vinylimidazole)(●,▲) and imidazole (○,△), respectively (28.5% ethanol–water, ionic strength 0.02, 26°C) (170).

cient than imidazole itself at $\alpha < 0.8$, but more efficient at $\alpha > 0.8$. However, as the pK for the formation of anionic imidazole is ~ 14, it is impossible with this polymer to study the catalytic system as a function of (total) dissociation in a hydroxylic system. However, if one uses a poly(vinylbenzimidazole), which has a smaller pK value of ~ 12.2, better results are obtained. The rate of hydrolysis of the same substrate does increase dramatically as a function of alkaline pH. Interestingly, a polymer of N-vinylimidazole, which cannot form the corresponding anion, is a much less efficient catalyst.

If anionic subtrates such as 4-acetoxy-3-nitrobenzoic acid (NABA) and 4-acetoxy-3-nitrobenzenesulfonate (NABS) are used, a different rate profile is observed (Fig. 5.12).

Bell-shaped curves are obtained for both cases with maximal activity at 75% neutrality ($\alpha = 0.75$) with poly(vinylimidazole). These results are best explained if we assume that enough cationic sites (25%) have to be present on the polymer for substrate binding but also a large portion of the polymer residues must be neutral (75%). It is these neutral imidazole rings that are probably responsible for the hydrolysis of the ester substrates.

If neutral–neutral imidazole interaction or bifunctional catalysis is involved at neutral or near neutral pH, then three mechanisms can be proposed to describe the interactions between the polymer catalyst and the substrate.

The first one is a general-base nucleophilic type catalysis:

But it cannot operate if the pH is too low because the first imidazole ring would be protonated and not act as a nucleophile.

   The second possibility is general-acid nucleophilic type catalysis which
seems unlikely since in general RO^- would not deprotonate an imidazole.

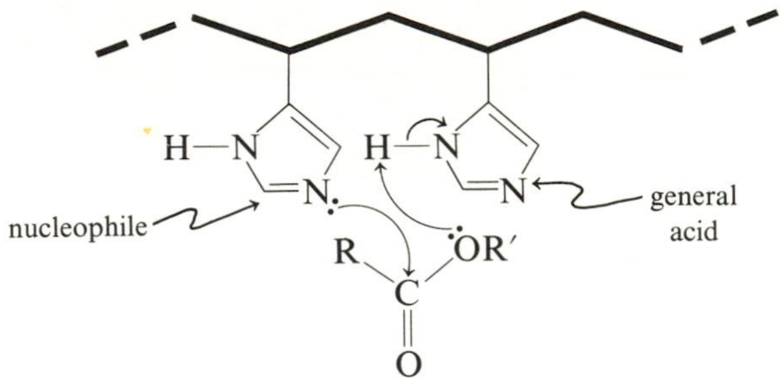

   Finally, the third possible interaction is via stabilization of the tetrahedral
intermediate:

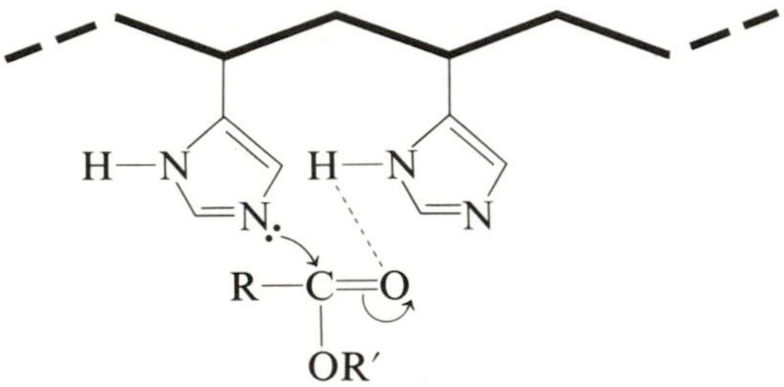

   No evidence has been presented to rule out these proposed mechanisms.
However, to explain the catalytic role of the polymer, two factors must be
taken into consideration. First, increasing protonation of imidazole leads to
a decrease in hydrophobic interactions with neutral and charged substrates,
but an increase in ionic interactions with oppositely charged substrates.
Second, a polyion would be expected to be in a more extended conformation
as the degree of protonation increases because of charge repulsions. This
would render difficult interactions of two imidazole functions. Consequently,
the efficiency of catalysis is believed to take place via the cooperativity of two
imidazole rings and not just one anionic ring. Furthermore, in the solvent
of high polarity used, water molecules are important and may be involved in
some way to help the imidazole rings to work in a cooperative fashion. In
effect, the k_{cat} of the reaction is increased in water as opposed to a medium
containing 30% alcohol.
   Synthetic copolymers have also been shown to have catalytic power comparable to enzymes. The following poly(vinylimidazole)-co-poly(vinyl alcohol) has been prepared to verify if a cooperative interaction of imidazole and
hydroxyl groups is possible. This situation is reminiscent of the enzyme
α-chymotrypsin. However, the polymer is only slightly more active than
poly(vinylimidazole) in esterolytic reactions.

poly(vinylimidazole-*co*-poly(vinylalcohol)

Perhaps a more dramatic indication of a bifunctional participation is in the solvolysis of a positively charged substrate, 3-acetoxy-*N*-trimethylanilinium iodide (ANTI), by poly(vinylimidazole)-*co*-poly(acrylic acid).

ANTI

poly(vinylimidazole)-*co*-poly(acrylic acid)

Figure 5.13 indicates that at high imidazole content in the copolymer, there are insufficient anionic sites to bind the positively charged substrate. On the other hand, at low imidazole content, the polymer begins to behave as a polyanion. As expected, the polymer was much less efficient with neutral substrates.

The efficiency and selectivity of the copolymer for positively charged substrate is then rationalized by the electrostatic attraction of the substrate with the anionic carboxylate groups in the polymer which accumulate the substrate in a high local concentration of imidazole nucleophiles. This type of cooperative effect could serve as a model for the nervous system enzyme acetylcholinesterase. The enzyme catalyzes the hydrolysis of its positively charged substrate, acetylcholine.

Recently, high rate enhancements were observed with a synthetic polymer having long alkyl chains (10 residue mole %) attached to it. The groups

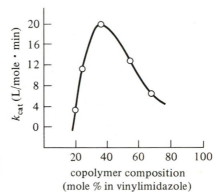

Fig. 5.13. Solvolysis of ANTI catalyzed by copolymers of vinylimidazole with acrylic acid (pH 9.0, 28% ethanol–water, ionic strength 0.02, 26°C) (170).

k_{cat} (L/mole · min)

copolymer composition
(mole % in vinylimidazole)

of I. M. Klotz (171), Northwestern University, developed the system and was able to attach dodecyl chains to a small cross-linked water-soluble poly(ethylenimine) matrix ($\overline{DP} \simeq 600$).*

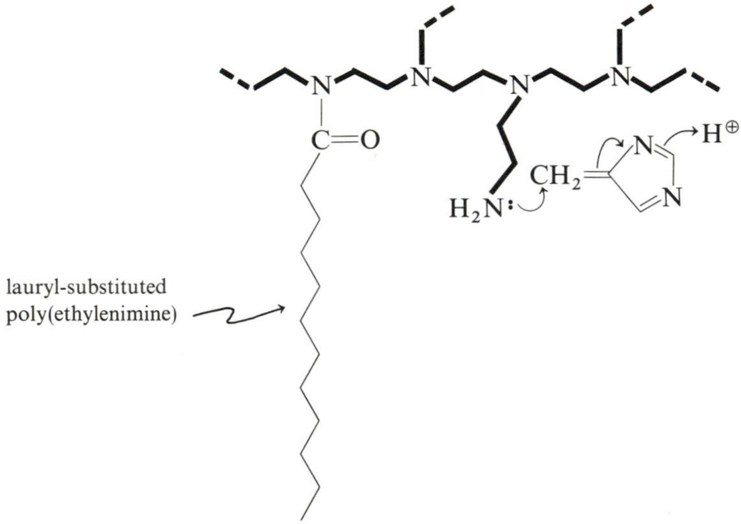

lauryl-substituted
poly(ethylenimine)

Reaction of this polymer with methylene-imidazole (or chloromethyl-imidazole) leads to a 15% incorporation of imidazole ring on the polymer backbone. Hydrolysis of phenolic sulfate esters (catechol sulfate) was studied and accelerations of 10^{12}-fold, compared to unbound imidazole, were obtained! This remarkable macromolecular catalyst, possessing a high local

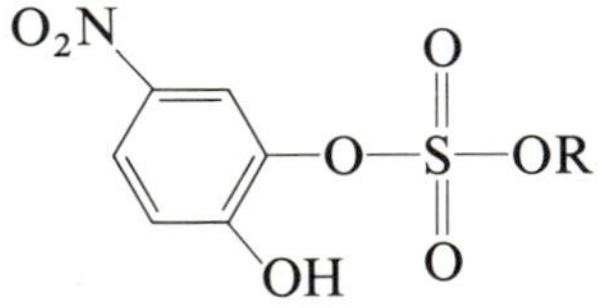

4-nitrocatechol sulfate

concentration of binding and catalytic groups, approaches catalytic constant values observed for the hydrolysis of nitrophenyl esters by α-chymotrypsin. Furthermore, the rates observed with sulfate esters are making this true polymer catalyst 10^2 times more effective than the type IIA aryl-sulfatase enzyme, although the substrates are not physiological.

This rigid macromolecular matrix possessing catalytic imidazole groups and micellar hydrophobic regions is the closest enzyme-like synthetic polymer made to date. It has been called "synzyme" (synthetic enzyme) by Klotz since its reactivity is purported to be of the same order of magnitude as that of an enzyme (171).

* $\overline{DP}$, mean degree of polymerization.

Of course, not all polymeric catalysts have comparable reaction rates but enzyme models have progressed significantly in the last decade. In the years to come, considerable progress will be expected in this field until we have enzyme models that will show both the speed and specificity of enzymes.

Another elegant example of the imitation of the properties of biopolymers by synthetic polymers comes from the school of E. Bayer of Tübingen (172). They have prepared chiral polysiloxane polymers for resolution of optical antipodes. The prochiral polymeric backbone was a copolymer of poly[(2-carboxypropyl)methylsiloxane], octamethylcyclotetrasiloxane, and hexa-methyldisiloxane. Amino acids or small peptides were covalently linked to this polymer in order to introduce a chiral surface. For this, the free carboxyl function of the polymer was reacted with the L-amino acid in the presence of DCC (see Chapter 2). The individual chiral centers (amino acids) on the polymer surface were separated by siloxane chains of specified length in order to achieve optimum interaction with the substrate and polymer viscosity. An example of great value for optical resolution is the polymer designated "chirasil-Val," containing 0.86 mmole of N-*tert*-butyl-L-valin-amide per gram of polymer (Fig. 5.14).

The polymer–substrate interaction has been studied by gas chromatography and has been used for determining the degree racemization of amino

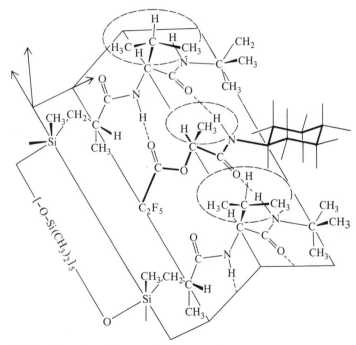

Fig. 5.14. Chirasil-Val, diasteriomeric association complex with N-cyclohexyl-O-pentafluoropropionyl-L-lactamide (172). Reprinted with permission. Copyright© 1978 by Verlag Chemie GMBH.

acids and natural substances. In all cases, the D-enantiomer of a racemic mixture of amino acids is eluted from an L-amino acid phase (polymeric) before the L-form. The illustration above shows the preferential interaction of one enantiomer. No such favorable stacking of the "receptor" polymer surface and the substrate is possible when the substrate has the D-configuration. The importance of the dimethylsiloxane units is also apparent as they keep the L-valinamide units at a distance and prevent formation of intramolecular hydrogen bonds which would give the polymer a quasi-crystalline structure.

Other applications include the separation of optical antipodes of drugs whose enantiomeric composition can be established conveniently and rapidly by gas chromatography. The effort of Bayer and his colleagues bring organic chemists one step closer to the synthesis of "tailor-made" chiral matrices. Like proteins they can undergo selective interaction with enantiomers from a wide variety of substances.

In summary, the potential features of functionalized polymers as catalysts are:

(a) the possibility to achieve high effective concentration of catalytic groups on a polymer backbone;
(b) the possibility to generate a micelle-like binding region by adding hydrophobic side chains to the polymer;
(c) the possibility to have electrostatic binding with high charge density by attaching ionic side chains to the polymer.

On the other hand, the main disadvantages are their limited solubility in water and the random arrangement of the polymer chains. Furthermore, the kinetic profile is more complex because of the multisite nature of the polymer as compared to an enzyme which normally has only one active site region.

5.4 Cyclodextrins

α-Cyclodextrin is a naturally occurring host molecule composed of six D-glucose units linked head to tail in a 1α,4-relationship to form a ring called cyclohexaamylose (Fig. 5.15). It has a relatively inflexible doughnut-shaped structure where the top of the molecule has twelve hydroxyl groups from positions 2 and 3 of the glucose units and the bottom the six primary hydroxyl groups from position 6. So the outside of the α-cyclodextrin molecules has hydrophilic hydroxyl groups while the cavity features mostly C—H, C—C, and C—O bonds, and is rather hydrophobic in nature. This situation is the reverse of that encountered with crown ethers which have rather hydrophilic cavities (173). α-Cyclodextrin can form insoluble, crystalline *inclusion*

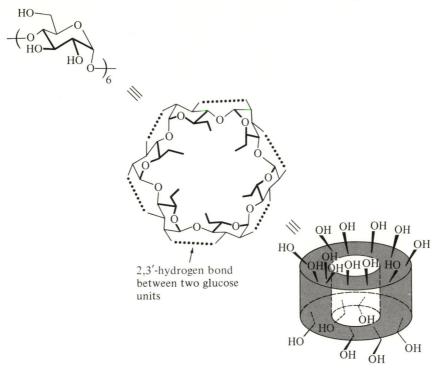

Fig. 5.15. Structural representations of α-cyclodextrin.

complexes with a variety of guest molecules. Usually a 1:1 ratio between host and guest molecules is observed and the size of the guest is the determining factor for the formation of the complex. The inner diameter of the cavity of α-cyclodextrin (6-glucose units) is 0.45 nm. A benzene ring is small enough to penetrate 6-, 7-, and 8-glucose units in cycloamyloses. With anthracene, however, only 8-glucose units can accommodate this aromatic system. Hydrophobic interactions seem to be the most probable driving force for inclusion complex formation. In addition, hydrogen bonding, van der Waals, and London dispersion forces may also play a role.

The application of cyclodextrins to biomimetic chemistry was orignated by F. Cramer in 1965, followed by R. L. Letsinger, H. Morawetz, M. L. Bender, and further extended by R. Breslow and I. Tabushi. R. Breslow, leader of a group at Columbia University, was the first to show that selective aromatic substitution can take place with the α-cyclodextrin system (174). He found that treatment of anisole (10^{-4} M) in water at room temperature with HOCl (10^{-2} M) in the presence of an excess of α-cyclodextrin resulted in 96% chlorination at the *para* position of the anisole ring.

The results indicate not only that the cyclodextrin blocks all but one aromatic ring position to substitution, but also that it actively catalyzes

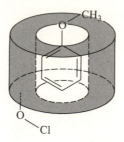

substitution at the unblocked position. The schematic representation above shows an anisole molecule in the cavity of cyclohexaamylose. One or more hydroxyl groups can be converted to a hypochorite to explain the increased rate of chlorination in the complex. This illustrates a noncovalent catalysis (classical Michaelis–Menten binding) in which the host provides the cavity for the reaction without formation of a covalent intermediate.

Similarly, aromatic esters are rapidly hydrolyzed by α-cyclodextrin. The secondary hydroxyl groups are believed to be involved in the catalysis but it is not known which one and how many. An intermediate is formed where

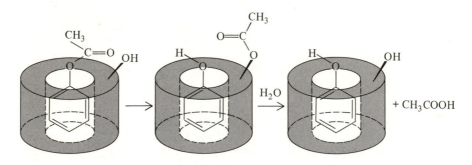

the acyl group is transferred to the cyclodextrin host molecule. This situation is formally analogous to the mechanism for hydrolytic enzymes such as serine proteases and serves as an enzyme model because a complex with the substrate is formed prior to reaction. This transformation is classified in the category of covalent intermediate catalysis. It should be noted, how-ever, that the second step in ester hydrolysis is very slow for cyclodextrins so they are not true catalysts for these reactions.

M. L. Bender's group at Northwestern University studied the hydrolysis of *m-tert*-butylphenyl acetate in the prescence of 2-benzimidazoleacetic acid (175) with α-cyclodextrin in order to probe the charge-relay system of the enzymatic mechanism in serine proteases. The results showed that in the presence of catalyst a 12-fold acceleration of ester cleavage takes place after complex formation between the substrate and α-cyclodextrin. This observa-tion is consistent with the formation of the substrate–enzyme complex in enzymatic reactions. However, the spacial disposition of the imidazole, carboxyl, and alkoxyl groups is apparently different from those shown for serine proteases. Serine proteases function by a nucleophilic attack of an alkoxide ion, whereas here the model shows nucleophilic participation by the imidazole group. Improvement in this system has been possible by selective modification of one of the secondary hydroxyl groups to a hist-amine residue. Unfortunately, one serious inconvenience in using cyclo-dextrin molecules is that they are not very active at neutral pH but only under basic conditions so that the kinetic data at pH 13 cannot be compared directly to those obtained, for instance, with α-chymotrypsin at pH 8. The host molecule is stable to alkaline solution but such condition can be detri-mental to the structure of the substrate. On the other hand, one should keep in mind the host's susceptibility to acidic media.

Note also that one of the chief problems is the binding step; the binding cavity is not apolar enough, is open at both ends and the dissociation constants are larger than with enzymes. Immobilization of the substrate is not always adequate, high concentration of cyclodextrin is usually required, and the conformation of lowest energy of the substrate may not be the one for optimal catalysis. To overcome this difficulty, Breslow (176) attached a bulky group to the more reactive primary hydroxyl groups at one end of β-cyclodextrin (cycloheptaamylose with an inner diameter of 0.7 nm) (Fig. 5.16).

In this way the bottom of the host is "capped" and the cavity is more hydrophobic and shallow. It corresponds to the construction of acylclo-dextrin with a "floor" across one end of the doughnut's hole that forces shallower binding of the substrate. The following transformation was then examined:

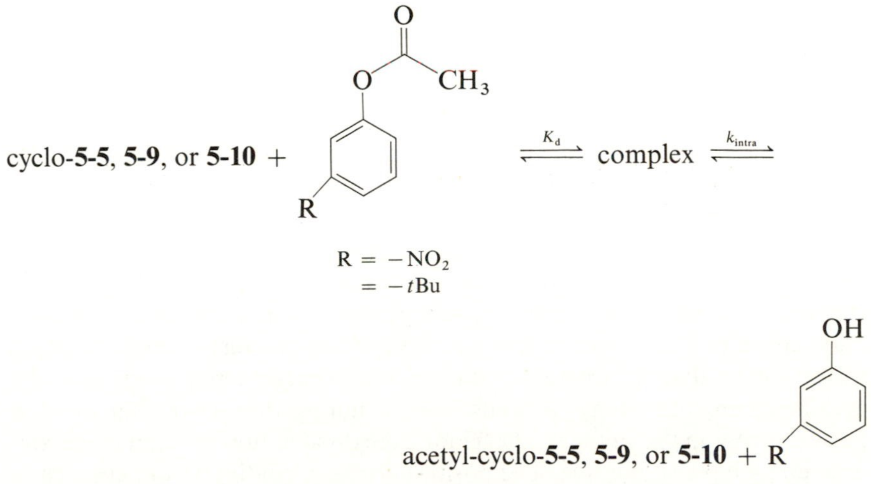

cyclo-**5-5**, **5-9**, or **5-10** + [acetyl-substituted aromatic, R = $-NO_2$, $= -t$Bu] $\underset{K_d}{\rightleftharpoons}$ complex $\xrightarrow{k_{intra}}$

acetyl-cyclo-**5-5**, **5-9**, or **5-10** + R [hydroxyl aromatic]

The results of this acyl-transfer process are presented in Table 5.1.

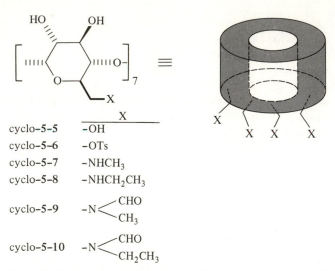

Fig. 5.16. Preparation of "capped" cyclodextrin molecules (176).

Table 5.1. Rate and Dissociation Constants for Reactions of Cycloheptaamylose Derivatives with m-Nitro and m-$tert$-Butylphenyl Acetate (176)

Substrate R	cyclo	$10^3 k_{intra}$ (sec^{-1})	k_{intra}/k_{OH^-}	$10^4 K_d (M)$
—NO$_2$	5-5	11.9 ± 0.05	64	57 ± 7
—NO$_2$	5-9	123 ± 5	660	51 ± 7
—NO$_2$	5-10	210 ± 40	1140	260 ± 50
—tBu	5-5	4.13 ± 0.25	365	1.9 ± 0.2
—tBu	5-10	37 ± 5	3300	4.6 ± 0.9

The reactions show Michaelis–Menten kinetics. cyclo-**5-9** (capped-Me) gives a 10-fold increase in rate and cyclo-**5-10** (capped-Et) an even larger increase. The rates are compared with the reaction (k_{OH^-}) in absence of host. Formation of a Michaelis–Menten complex is one of the reasons to justify the utilization of cyclodextrin-catalyzed reactions as models of hydrolytic enzyme reactions. It is interesting to observe that the presence of a *tert*-butyl group on the substrate dramatically reduces the dissociation constant which means that it is better bound; the *tert*-butylphenyl group fills the cycloheptaamylose cavity. It is also worth noting that as the substrate is pushed higher in the cavity by the "floor," the dissociation constant generally goes up along with the rate. The position for best binding in the absence of the "floor" is too low for good catalysis. Adamantane carboxylic acid, which properly sits on the surface of the cavity of β-cyclodextrin is an excellent competitive inhibitor of complex formation.

adamantane carboxylic acid

In the studies of the hydrolyses of substituted phenyl acetates by α- or β-cyclodextrins, it was observed that *meta*-substituted phenyl esters were more rapidly hydrolyzed than the corresponding *para*-isomers, a phenomenon termed "*meta*-selectivity." This observation indicates that the binding mode is probably asymmetric. This effect is apparently dependent on the depth of the cavity, and Fujita *et al.* (177) showed that appropriate simple modifications of β-cyclodextrin such as "capping" the host, for instance, can alter this selectivity and leads to conversion of the well-established *meta*-selectivity to *para*-selectivity.

In 1978, Komiyama and Bender gave further experimental evidence for the importance of hydrophobic bonding in complex formation of α- and β-cyclodextrins with 1-adamantanecarboxylate (178). Hydrophobic (nonpolar) bonding is characterized by a favorable entropy change which is attributed to a transfer of the guest molecule from aqueous medium to a more apolar medium such as the cavity of a cyclodextrin molecule. This transfer requires breakdown of structural water around the guest, resulting in a large favorable ΔS and a small unfavorable ΔH change. The importance of hydrophobic bonding in the complexation of cyclodextrins is also consistent with the finding of a stronger binding of "capped" cyclodextrin relative to native cyclodextrin with guests. In conclusion, Bender and co-workers showed that a favorable entropy change due to hydrophobic bonding is largely responsible for the stabilization of complexes of cyclodextrins with apolar guest compounds (178).

Further, complex formation of cyclodextrins with guest compounds such as drugs and insecticides introduced new physicochemical features to these compounds. This leads to interesting practical usages and reinforces the view that cyclodextrins are suitable models of enzymatic binding as well as enzymatic reactions.

In this regard, Breslow's group (179) synthesized a β-cyclodextrin-bisimidazole molecule to model ribonuclease A (RNase A) (see Chapter 3). The approach is based on the preparation of a "capped" disulfonate derivative made earlier by I. Tabushi and co-workers of Kyoto University (180, 181).

The model hydrolyzes a cyclic phosphate substrate derived from 4-*tert*-butylcatechol in a selective manner with cooperative catalysis by a neutral imidazole and an imidazolium ion.

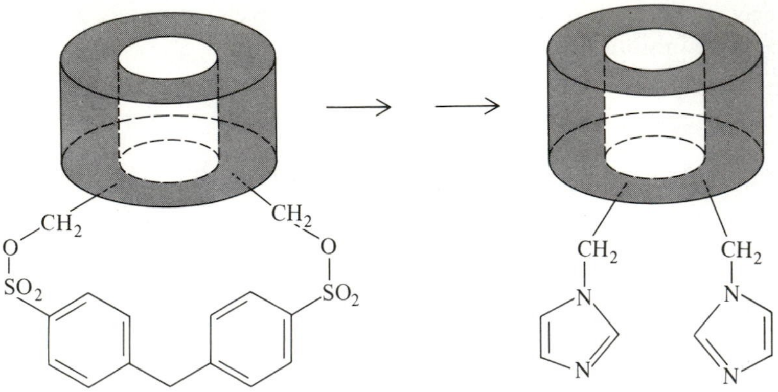

A normal chemical hydrolysis would produce a random mixture of two products whereas hydrolysis by the "artificial enzyme" leads to the production of only the *m*-phosphate isomer.

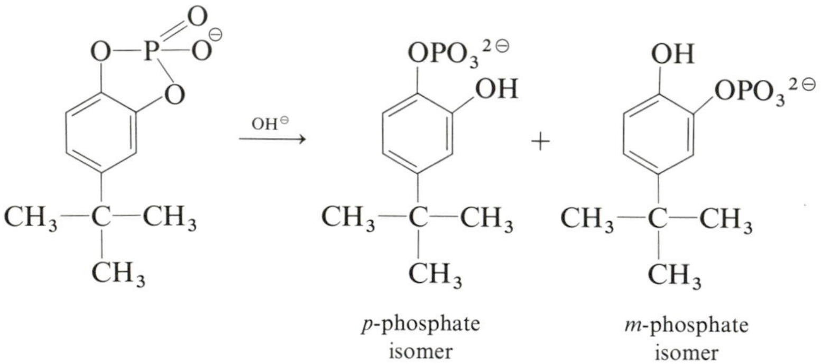

The reaction is much slower than with RNase (17-fold) but the selectivity is in accordance with an in-line mechanism without *pseudo-rotation* as is observed with the enzyme (refer to Section 3.3 for details). As in the case of *para*-chlorination of anisole (p. 291), this example of cyclodextrin reaction gives only one of two possible products.

On treatment with potassium iodide, the "capped" disulfonate β-cyclodextrin discussed above could easily be converted to the corresponding diiodide β-cyclodextrin. With appropriate nucleophiles (imidazole, histamine) a new route to bis(*N*-imidazolyl)-β-cyclodextrin and bis(*N*-histamino)-β-cyclodextrin was developed by Tabushi's team (182). In the presence of Zn(II) ion, both regiospecifically bifunctionalized cyclodextrins hydrate CO_2 and are the first successful carbonic anhydrase models. The Zn(II) ion binds to the imidazole rings located in the edge of the cyclodextrin pocket and the presence of an additional basic group, as with bis(histamino)-cyclodextrin-Zn(II), enhances the activity. Therefore, the present models show that all three factors, Zn(II)-imidazole, hydrophobic environment, and a base seem to help to generate the carbonic anhydrase activity (182). The chemistry of this enzyme is further discussed in Section 6.2, p. 331.

Also of interest is the model developed by Breslow and Overman (183) where a metal ion is introduced into a cyclodextrin–subtrate complex. With p-nitrophenyl acetate as substrate the presence of Ni(II) in the cyclodextrin–substrate complex results in a further increase in rate of hydrolysis by a 1000-fold.

This novel complex can be prepared in the following way:

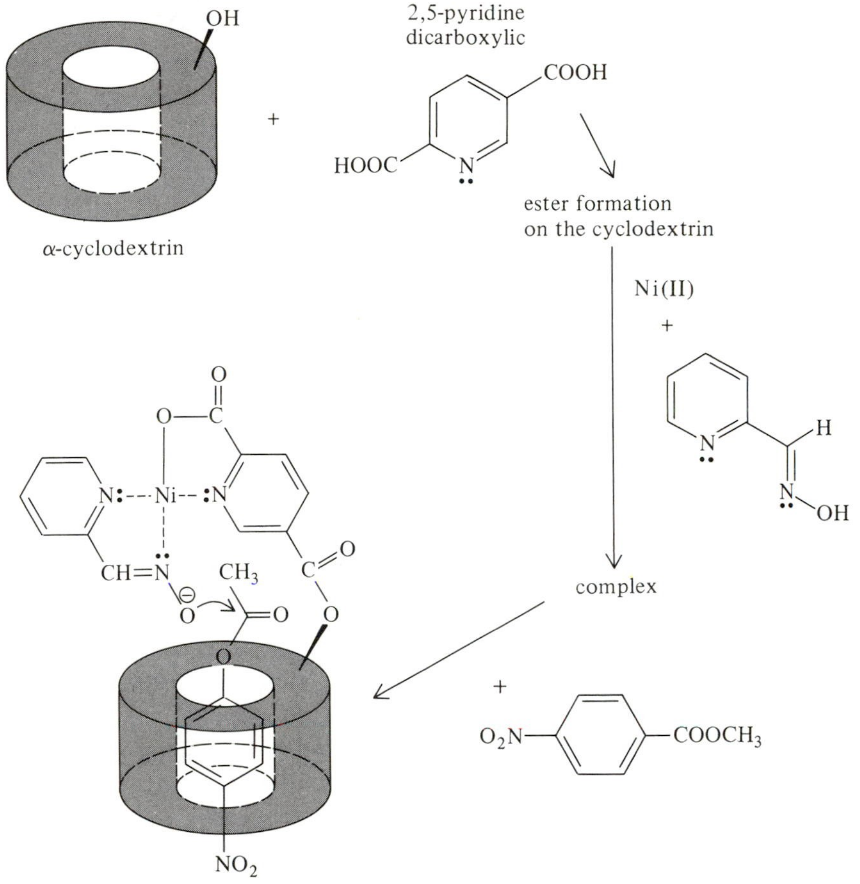

The cyclodextrin "cage" holds the ester while the metal ion positions other groups for attack. If cyclohexanol is added in the solution, it competes with the substrate and the efficiency of the system falls to 60%. Most of the catalytic power of the system is attributed to the binding of the substrate by the cyclodextrin moiety of the complex. In the absence of cyclodextrin, the rate enhancement is 350-fold. Thus, by combining the properties of cyclodextrin with those of a metal ion, a much more efficient catalytic system can be obtained that mimics enzyme features.

In a search for well-characterized polyfunctionalized α-cyclodextrins, Knowles' group (184, 185) developed in 1979 the elegant approach shown below:

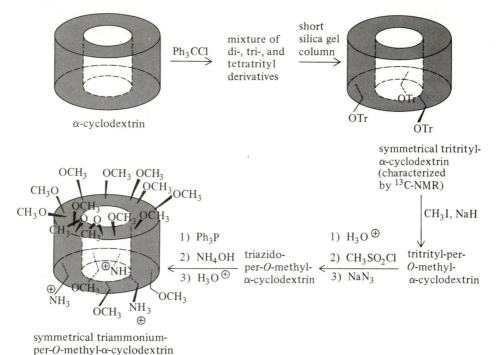

α-cyclodextrin

Ph_3CCl → mixture of di-, tri-, and tetratrityl derivatives

short silica gel column →

symmetrical trityl-α-cyclodextrin (characterized by ^{13}C-NMR)

OTr   OTr   OTr

CH_3I, NaH

1) $H_3O^{\oplus}$
2) CH_3SO_2Cl
3) NaN_3

tritrityl-per-O-methyl-α-cyclodextrin

triazido-per-O-methyl-α-cyclodextrin

1) Ph_3P
2) NH_4OH
3) $H_3O^{\oplus}$

symmetrical triammonium-per-O-methyl-α-cyclodextrin (characterized by ^{13}C-NMR)

This procedure provides access to a wide variety of cyclodextrin derivatives and the rational synthesis of even more sophisticated model system employing the regiospecifically disposed functionality on one side of the cyclodextrin cavity and possible additional functionalities on the other side.

This symmetrical multifunctional guest has the ability to bind complementary hosts. Figure 5.17 shows the binding of benzyl phosphate in the cavity.

Potentiometric titration of the ammonium groups of the host gave apparent pK_a values of 7.42, 8.02, and 8.79. These pK_a values are rather low for primary ammonium groups but could reflect their nonpolar environment. At pH 7, the host is primarily in the fully protonated species and shows a dissociation constant for benzyl phosphate (0.031 mM) which is almost three orders of magnitude smaller than for benzyl alcohol (24.3 mM). It is thus a properly designed host for benzyl phosphate. Furthermore, the dissociation constant for the complexation of benzyl phosphate is pH dependent showing that the guest binds in the cyclodextrin cavity with the phosphate oxygen interacting (at pH 7) strongly with the three symmetrically disposed ammonium ions. Interestingly, inorganic phosphate itself does not bind to this cyclodextrin derivative and does not expel benzyl phosphate from the cavity, showing again the importance of hydrophobic binding between host and guest.

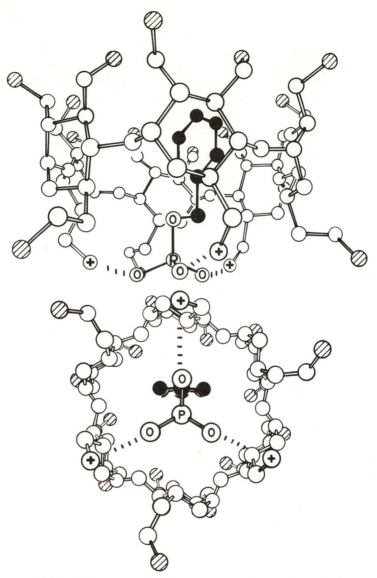

Fig. 5.17. Representation of the symmetrical triammonio-per-*O*-methyl-α-cyclodextrin complexed with benzyl phosphate. Each of the three ammonium ions is shown by the symbol ⊕. The 15 methyl groups are shaded. All hydrogen atoms are omitted and the size of the atoms is arbitrary (185). Reprinted with permission. Copyright©1979 by the American Chemical Society.

Therefore, both hydrophobic and electrostatic interactions in aqueous solution are exploited in this specific host–guest complexation, and this provides a model for multiple recognition sites common in biological systems.

Being formed of D-glucose units, cyclodextrins are chiral and chiral induction on substrates have been observed with cyclodextrin reactions (173).

A recent application of this is the preparation by Breslow's team (186) of a covalently linked coenzyme-cyclodextrin "artificial enzyme." It consists of a β-cyclodextrin-pyridoxamine (refer to Section 7.2 for the chemistry of this coenzyme) that can selectively transaminate phenylpyruvic acid to phenyl-alanine with a 52% excess of the natural L-enantiomer.

W. Saenger's group (187) from the Max Planck Institute has data to show that cyclodextrins can be even better models for enzymes than was hitherto assumed. They have demonstrated that a conformational change takes place in the host when substrates are included in it. This situation is analogous to enzyme which exhibits "induced fit" in their interaction with substrates (see Section 6.2 for an example).

Other interesting systems involving cyclodextrin chemistry have been applied as enzyme models by I. Tabushi. They are specific allylation-oxidation of hydroquinone (188), cyclodextrin having an amino group attached to retinal as a rhodopsin model (189) and specific inclusion catalysis by β-cyclodextrin in the one-step preparation of vitamin K_1 and K_2 analogues (190).

For all the enzyme models that we have examined so far, the catalytic efficiency appears to be understandable in terms of three main effects:

(a) hydrophobic interactions to provide an efficient binding;
(b) a medium (or cavity) effect where proper polarity provides a large driving force for the reaction;
(c) an orientation effect that leads to restriction of conformation and provides a large rate acceleration.

5.5 Enzyme Design Using Steroid Template

Long alkyl chains with a catalytic group at one end can provide enough hydrophobic binding sites with fatty ester substrates to produce a rate enhancement. In the following case,

a 10-fold increase in rate is observed at pH 8 in Tris buffer, 25°C. Micelles are formed and there is no case yet of a large rate enhancement for a 1:1 association of alkyl chains with monomeric binding site. Therefore, a serious problem is the fact that these molecules are rather insoluble in water and thus have a tendency to aggregate or form micelles. Work has to be done at concentration well below the cmc level which is about 10^{-4} M for this system. Furthermore, the conformation of the chain varies even when extended.

Since the alkyl chains are very flexible and a strong hydrophobic area is never produced, a search for a more oriented network was needed.

To overcome this difficulty J. P. Guthrie, from the University of Western Ontario, suggested the use of a planar, large, and rigid hydrophobic backbone, a steroid skeleton (191).

For this he synthesized the following steroid molecule, androstane $3\beta, 11\beta$-diamino-$17\beta(4)$-imidazole.

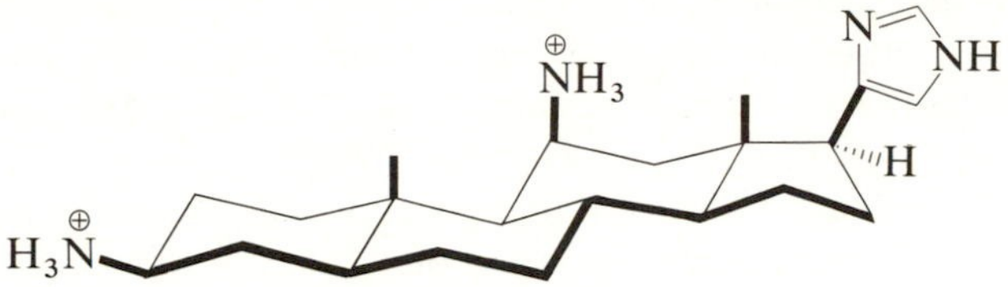

The presence of two ammonium ions makes the molecule water soluble enough to inhibit micelle formation under the experimental conditions. It should be recalled that bile acids are water soluble but form micelles.

As compared to cyclodextrin which is a three-dimensional system, this one is a two-dimensional hydrophobic surface. In this regard it is a useful model to evaluate hydrophobic and electrostatic interactions of well-defined geometries. The presence of an imidazole group at position 17β is responsible for the catalytic effect observed during the hydrolysis of esters.

To illustrate this, a series of aromatic esters, with increasing degree of hydrophobicity have been tested. The relative rate is doubled for $n = 2$ or 3 as compared to $n = 1$:

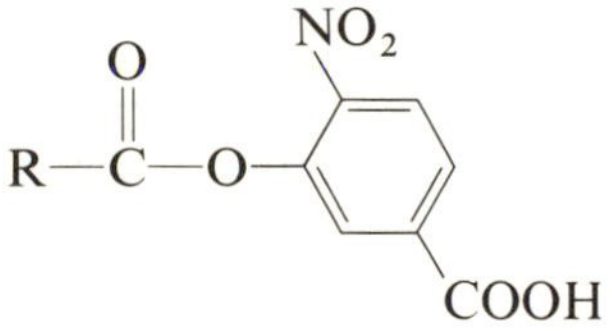

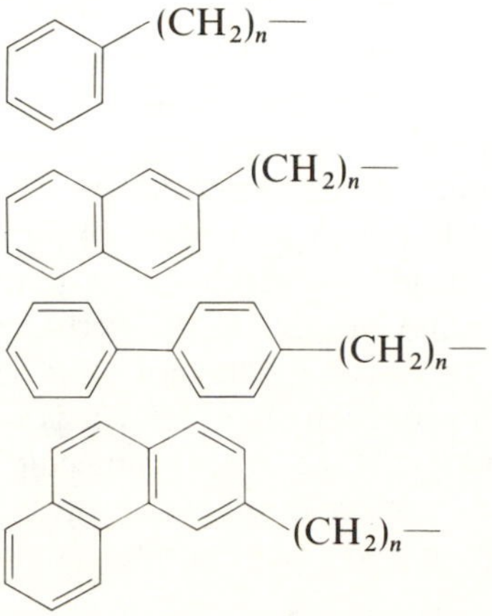

It was found that as the size of the hydrophobic group in the ester increased the rate of catalyzed hydrolysis also increased. However, these esters showed essentially identical reactivity toward imidazole alone as nucleophile. Space-filling models then suggested that two CH_2 groups are required between the aromatic ring system and the ester group in order to permit enough hydrophobic contact with the α-surface of the steroid molecule. It is concluded that imidazole-catalyzed hydrolysis of the ester is facilitated by hydrophobic interactions between substrate and catalyst.

A true second-order rate is observed, indicating a 1:1 stoichiometry. Because of solubility problems, no saturation kinetics are measured, only the second-order rate constants are evaluated but not k_{cat} nor the K_m binding constants. A productive binding representation of a tetrahedral intermediate formed indicates that efficient hydrophobic binding is possible between the substrate and the steroid rings. The advantage of this system is that very little flexibility in either substrate or catalyst is allowed.

Another example is the general-base catalyzed β-elimination of β-acetoxy-ketones which has been studied with this steroid catalyst (192).

The rate-limiting step in this reaction is the base-catalyzed enolization of the ketone. With Ar = phenyl, naphthyl, and phenanthryl, the hydrophobic rate enhancement is 9.0, 27, and 110, respectively.

Since the pK of imidazole is 7.1 and the pK of the corresponding steroid–imidazole is 7.2, the catalyst basicity is not likely to be the factor responsible for the rate enhancement. The results show that the three substrates have comparable kinetic behaviors. In fact, the steroid–imidazole complex should, if anything, retard the reaction. Therefore, the size of the aryl group must make the difference for proper fit. In other words, some favorable noncovalent interactions with the rest of the steroid molecule and the substrate must be involved. In effect, the conformation of the transition state must be such that the rings of the steroid and the substrate interact in such a way that the bottom face of the steroid skeleton is shielded from water. Consequently, strong and favorable hydrophobic bindings are the main reason for the rate enhancement observed with the phenanthryl derivative. This is illustrated below:

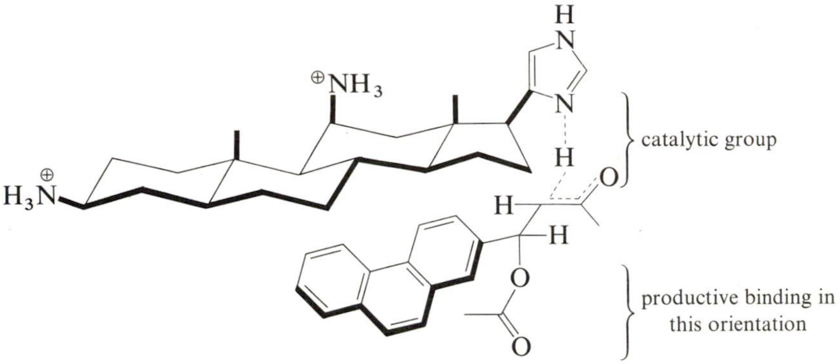

Some studies have also been carried out with anionic and cationic acetyl substrates (193). The general picture that emerges is that rate retardation is observed with cationic species while rate enhancement are obtained with anionic substrates. However, a *meta* anionic group shows a larger rate enhancement than a *para* anionic orientation.

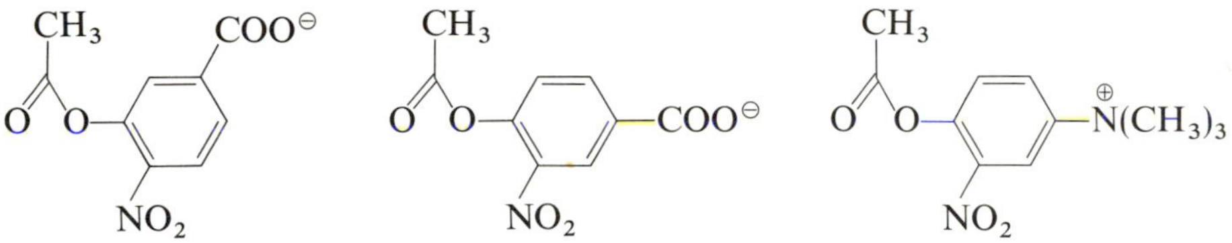

This is the first clear demonstration of catalysis resulting from electrostatic interaction of charged groups remote from the reaction center with reasonably well-specified geometry.

The following structures show that the m-COO$^-$ and $-$NH$_3^+$ groups in the tetrahedral intermediate can come into close contact without strain. The contact between the 11-ammonium ion and p-carboxylate of the substrate is possible but only at the expense of desolvating the ionic groups without formation of a compensating hydrogen bond, whereas the 11-ammonium and m-carboxylate can come into contact, without intervening water, and form a good hydrogen bond. In other words, close contact is necessary for a large electrostatic affect, but is only possible if loss of solvating water is compensated for by formation of a new hydrogen bond. The presence of the C-18 methyl group is not a steric barrier.

In conclusion, the steroid system is a useful enzyme model because there is binding of catalyst and substrate and some selectivity for some substrates over others. Stronger binding affinities within the steroid–imidazole system is needed and should be developed to constrain the substrate to bind in only one mode.

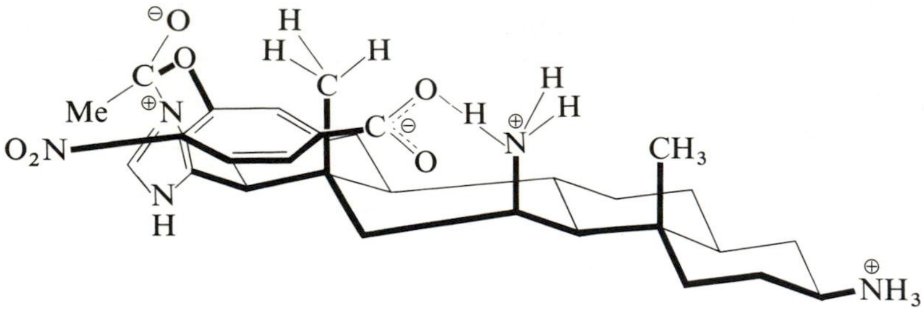

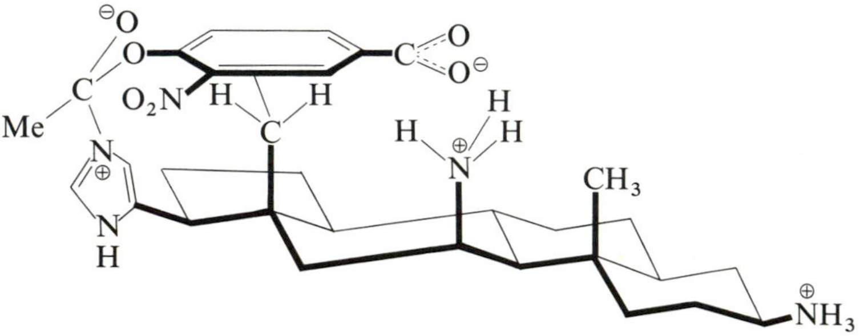

For this, Guthrie's group is planning to synthesize a steroid–imidazole dimer by combining two ketone analogues with an aromatic diamine.

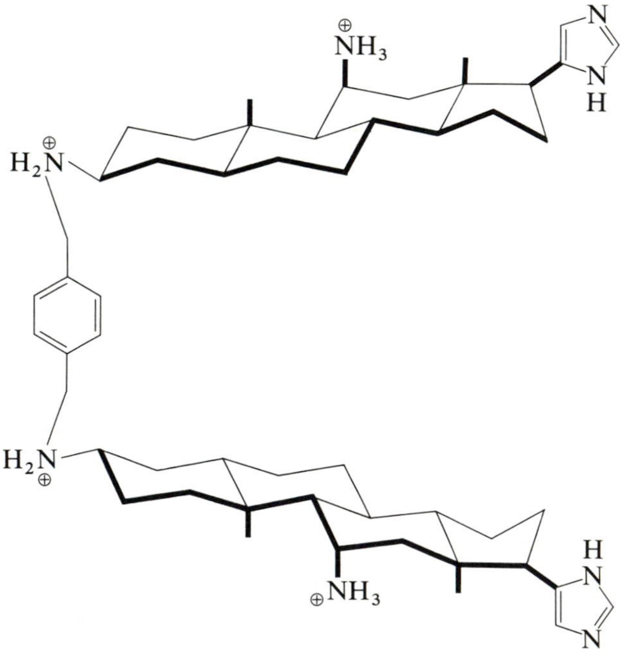

This way, a three-dimensional frame is built and the substrate is expected to be trapped in the "jaw" of the dimeric catalyst where hydrophobic binding now comes from both faces of the substrate reaction center. So far, the bis(11)-keto derivative has been prepared and increased catalysis by 1000-fold over that of simple imidazole has been observed.

5.6 Remote Functionalization Reactions

Nature's ability to carry out selective functionalization of simple substrates utilizes a principle of great power which has not been applied by chemists until recently (194). For example, enzymatic systems such as desaturases can oxidize a single unactivated carbon–hydrogen bond at a specific region on the alkyl chain of stearic acid and convert it to oleic acid, possessing only a *cis* geometry.

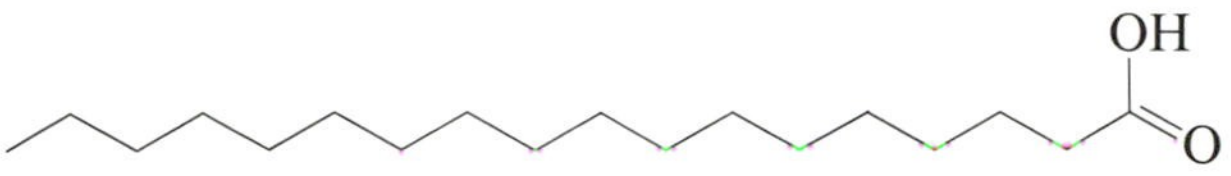

stearic acid

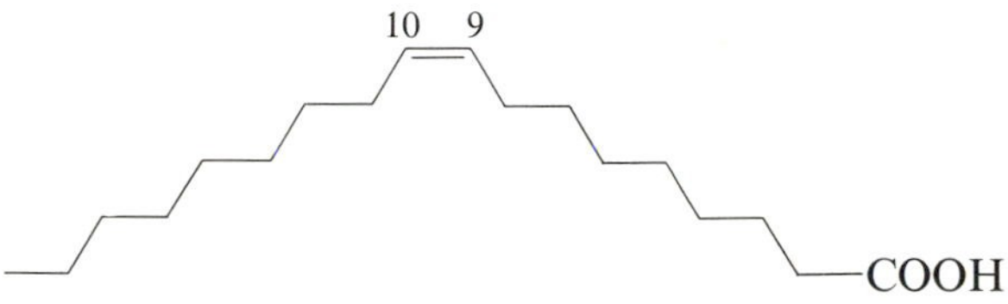

oleic acid

Therefore, high rate enhancement and the production of only one product from a reaction where many are possible are important characteristics of enzymatic transformations. In this context, the team of R. Breslow has produced the first chemical example of the application of this principle of *remote functionalization* by undertaking selective oxidation of alkyl chains (194, 195). As the reagent to attack an unactivated CH_2 group they selected the photochemical activation of benzophenone triplet (Yang triplet reaction) and studied the intramolecular insertion of benzophenone carboxylic ester of alkanols (Fig. 5.18).

The radical reaction proceeds with the formation of a covalent bond intermediate. Benzophenone is regenerated after ozonolysis and the covalently attached alkyl chain (substrate) is selectively oxidized with yields up to 66% (Table 5.2).

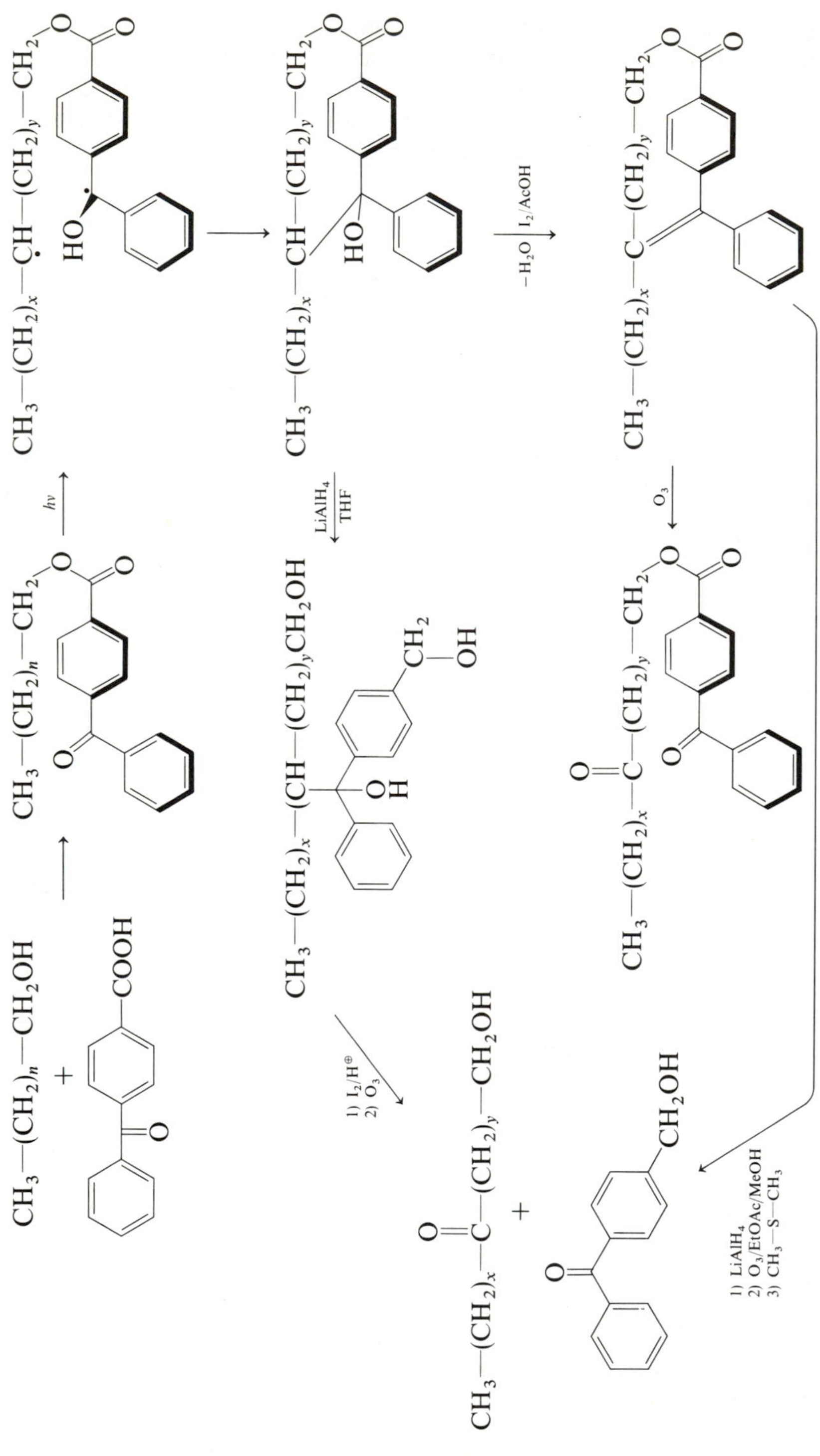

Fig. 5.18. Breslow's strategies to remote functionalization reactions.

Table 5.2. Observed Degree of Oxidation of the Alkyl Chain Given in Percent of Oxidation (194)

Oxidation site	Number of carbons in ester alkyl chain[a]			
	C-14	C-16	C-18	C-20
C-8	—	—	—	—
C-9	—	—	—	—
C-10	—	8	8	5
C-11	11	10 to 12	17	15
C-12	49	8 to 13	21	20
C-13	22	3 to 10	18	19
C-14	—	56 to 66	12	19
C-15	—	7 to 10	5	13
C-16	—	—	13	8
C-17	—	—	6	—

[a] A dash means that no product was detected

Fatty acid esters of benzophenone-4-carboxylate can be incorporated in a micelle such as that from sodium dodecyl sulfate (SDS) or lecithin and can be used as photochemical probes for model membrane structures (196).

So the possibility of developing a variety of specific functionalization reactions utilizing this concept is at hand. In particular, properly designed rigid models such as steroid templates were used to develop new approaches toward simpler syntheses of hormones where greater selectivity could be achieved.

For example, by attaching the benzophenone chromophore to position 3α of a steroid molecule (197), a series of selectively functionalized derivatives have been obtained under various experimental conditions (Fig. 5.19).

It was also demonstrated that a covalent bond between substrate and "catalyst" is not always necessary. In the example below, electrostatic interactions between the two substances are sufficient to bring the two molecules close enough together for bond insertion and hydrogen abstraction.

steroid products with bond insertion and hydrogen abstraction

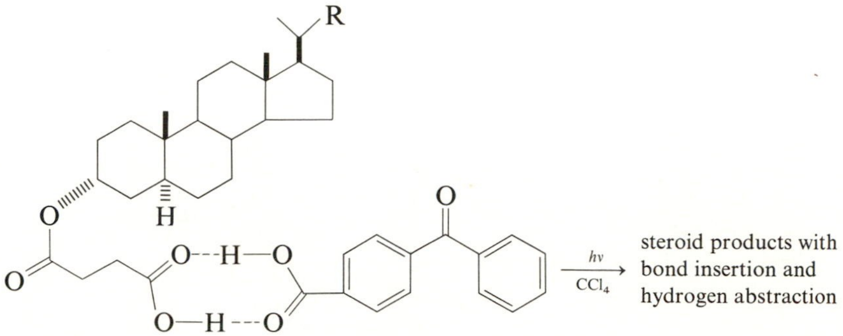

R
HO
H
O
OH
hν
CCl4
1) −H2O
2) O3
R′O
hν
BrCCl3
hν
C6H6
hν
CH3CN
good
yield
R′ = benzophenone
side chain

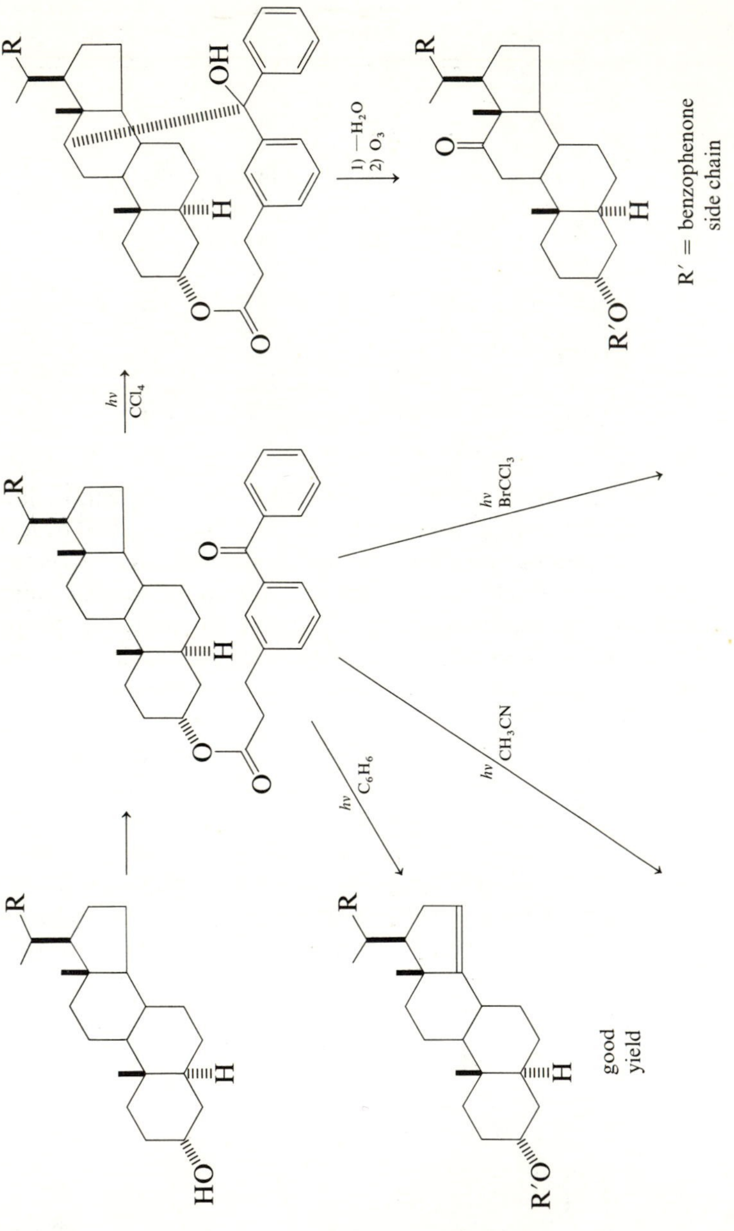

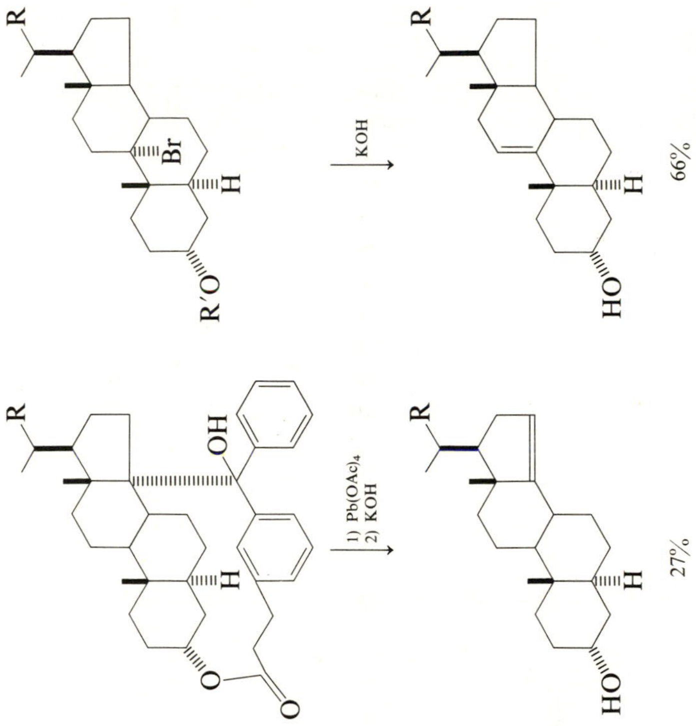

Fig. 5.19. Remote functionalization of steroids.

A variant of this approach of remote functionalization is the use of a *p*-dichloroiodobenzoate ester to provoke intramolecular halogenation and elimination reactions (198):

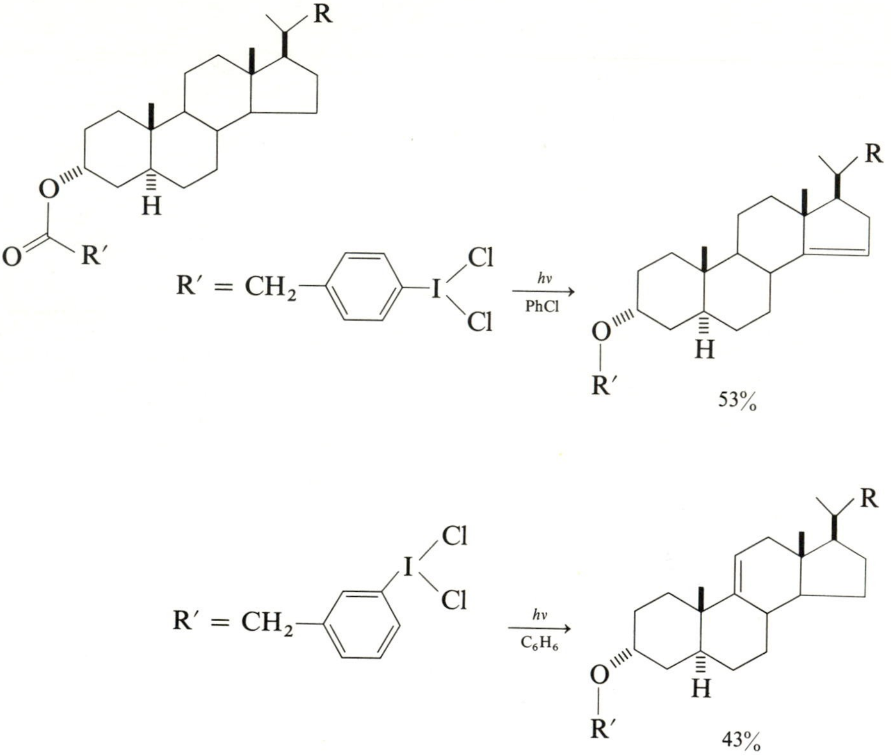

$$R' = CH_2 \overset{Cl}{\underset{Cl}{-\!\!\!\!\bigcirc\!\!\!\!-\ I}} \qquad \xrightarrow[\text{PhCl}]{hv}$$

53%

$$R' = CH_2 \overset{Cl}{\underset{Cl}{-\!\!\!\!\bigcirc\!\!\!\!-\ I}} \qquad \xrightarrow[\text{C}_6\text{H}_6]{hv}$$

43%

This method has been applied to a new synthesis of the adrenal hormone cortisone (199):

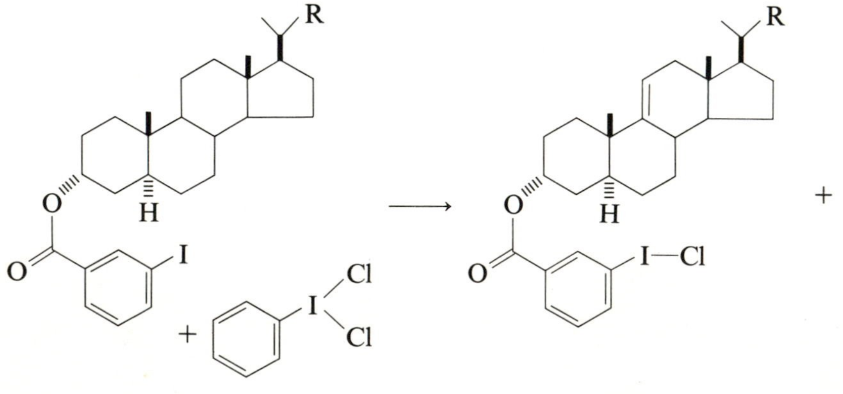

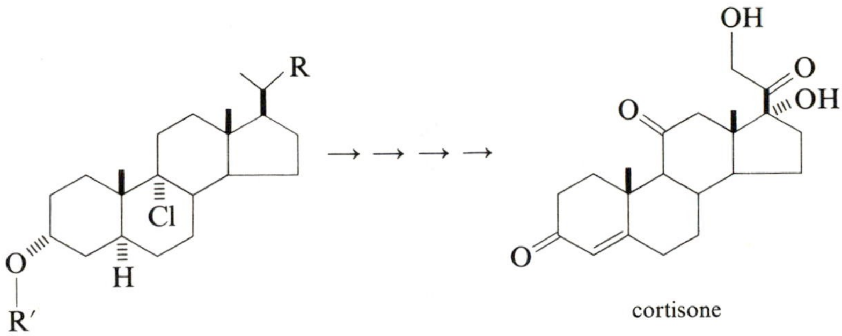

cortisone

By using different templates, different positions on the steroid are attacked. This is a case of enzyme-like geometric control of the site of functionalization rather than the usual control by reactivity seen in chemical reactions. These template-directed halogenation reactions show both catalysis and specificity and are indeed biomimetic (350).

Recently, the group of R. Chênevert of Laval University made an interesting model of the enzyme desaturase by combining the concepts of host–guest complexation and remote functionalization. They synthesized a crown ether possessing lateral chains with a benzophenone unit. The starting material was the optically active form of tartaric acid.

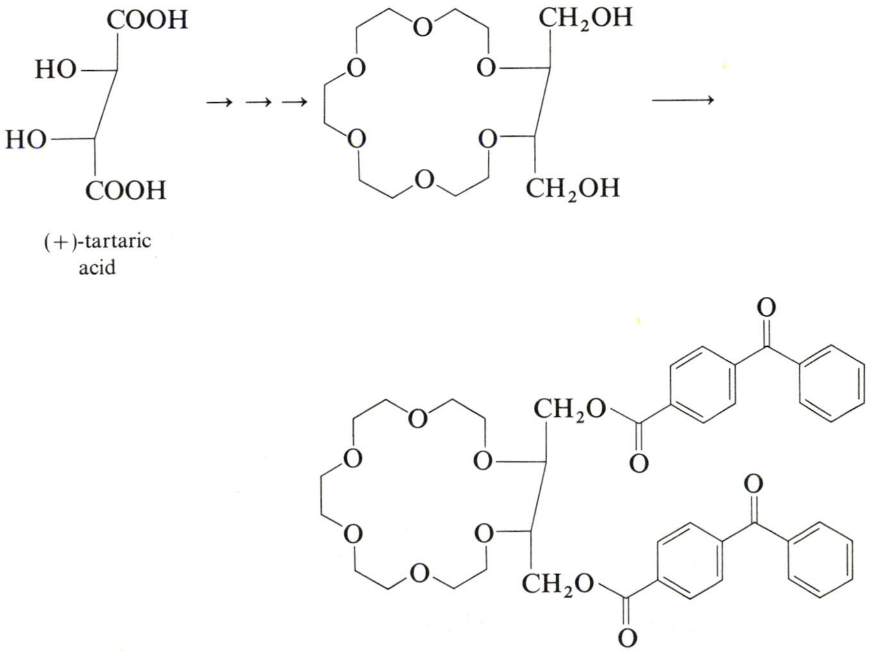

(+)-tartaric
acid

The hydrophilic cavity of the crown ether can bind alkyl ammonium salts, and upon irradiation, the system should regiospecifically introduce a double bond by hydrogen abstraction of the complexed alkyl chain. Such molecular association resembles an enzyme–substrate complex where the geometry of the ensemble allows the regiospecific introduction of a double bond on the alkyl chain of the amine. In a simpler way it does mimic the enzyme desaturase which converts stearic acid to oleic acid by introduction of a *cis* double bond at position 9–10 of the alkyl chain.

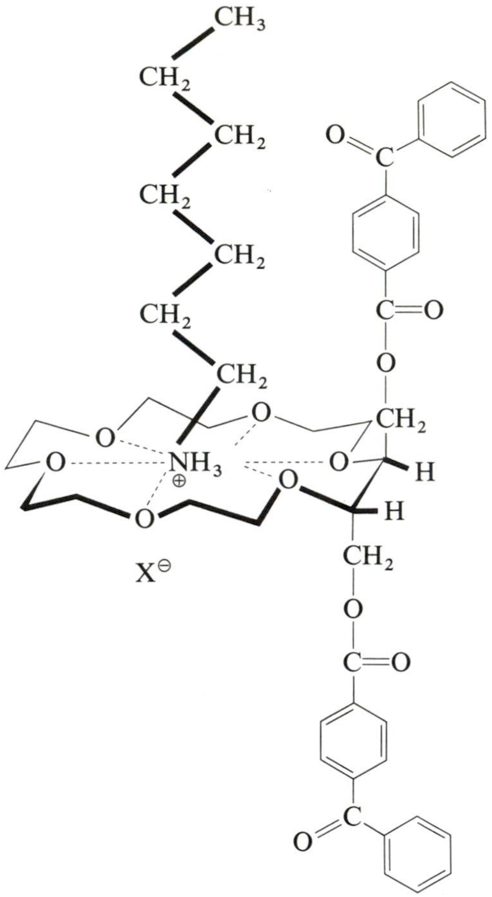

5.6.1 Other Examples Related to Biosynthesis

Examples of biosynthesis include the biosynthetic transformation of many alkaloids where nature couples phenol rings either *ortho-ortho, ortho-para,* or *para-para* by a one electron oxidation mechanism using metal-independent enzymes. Oxidative coupling of phenols has long been recognized as an important biosynthetic pathway, not only for alkaloids but also examples can be found among the tannins, lignins, and plant pigments. Chemically, various oxidants can bring about the coupling of phenols

[Fe^{3+}, $Fe(CN)_6{}^{3-}$, PbO_2, Ag_2O] and a recent chemical model (200) for morphine using thallium(III) ion has been developed to induce such an oxidative phenolic coupling reaction:

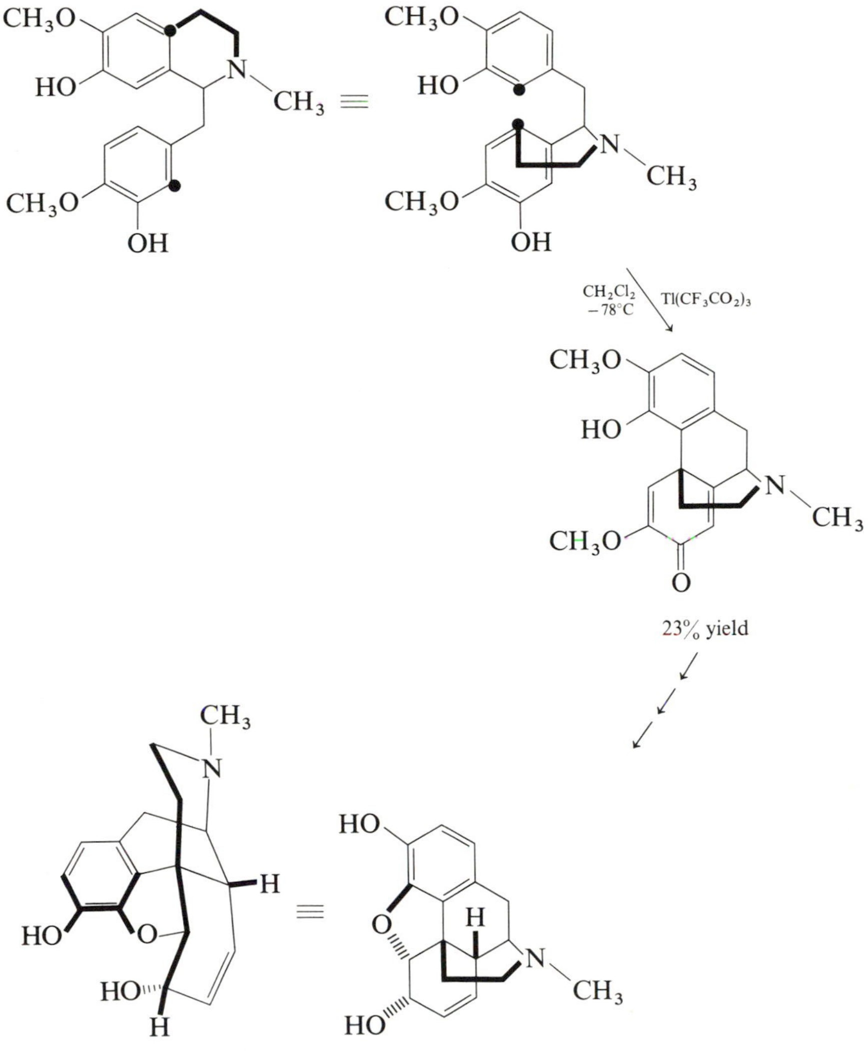

morphine

The mechanism involves the formation and dimerization of phenoxyradicals.

Many stable radicals are known and the commercial use of hindered phenols such as 2,4,6-tri-*tert*-butylphenol or 2,5-di-*tert*-butylhydroxytoluene (BHT) as antioxidants in food and the similar function of vitamin E (α-tocopherol) in cell tissues is based on their ability to form stable free radicals of the type found in oxidative coupling reactions.

OH

2,4,6-tri-*tert*-butylphenol   vitamin E

HO

Another practical application of biological oxidation is to be found in the biosynthesis of the prostaglandins (201, 202). In nature, they are synthesized by selective oxidation of a C_{20} fatty acid precursor that contains three or four double bonds. The polyunsaturated fatty acid is enzymatically oxidized by a cyclooxygenase in two consecutive radical cyclizations by molecular oxygen to give a bicyclic endoperoxide intermediate. Upon decomposition this leads to a variety of prostaglandin molecules among which are PGE_2 and $PGF_{2\alpha}$ as well as thromboxane A_2 and prostacyclin (Fig. 5.20).

The prostaglandins were discovered in the 1930s and were observed to affect a wide variety of physiological processes. They modulate the activities of many of the cells in which they are synthesized. The molecular basis of many of the major actions of prostaglandins is not yet known but they seem to control the action of hormones rather than act as hormones.

Lately, special interest in the prostaglandins has focused on the processes of inflammation and allergic responses because they stimulate the synthesis of histamines. A commonly used medication is Aspirin, an anti-inflammatory and anti-pyretic drug (relief of fever). Some evidence now suggests that the mode of action of this drug is to stop the synthesis of prostaglandins by inhibiting the cyclooxygenase which is responsible for the formation of the endoperoxide intermediate. The endoperoxide and thromboxane A_2, a non-prostaglandin product, are highly active physiological agents known to

portion
modified

R_1
R_2

the natural
endoperoxide

some
synthetic
analogues

HN CH O N O CH_2 S

HN , CH , CH_3- O , N , CH_2 or O , S

relative
activity: ~ 0.1, ~ 0.1,   2 to 3,   3 to 4,   4 to 7,   24

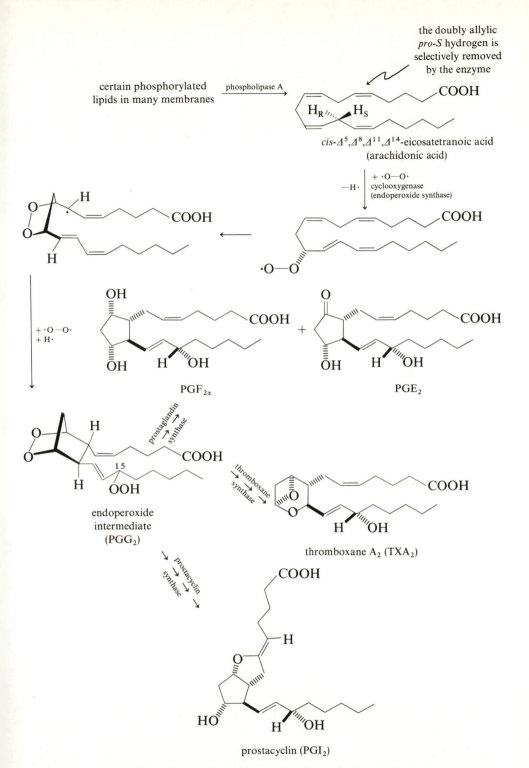

the doubly allylic
pro-S hydrogen is
selectively removed
by the enzyme

certain phosphorylated
lipids in many membranes

phospholipase A

cis-$\Delta^5,\Delta^8,\Delta^{11},\Delta^{14}$-eicosatetranoic acid
(arachidonic acid)

$-H\cdot$ $+ \cdot O—O\cdot$
cyclooxygenase
(endoperoxide synthase)

$+ \cdot O—O\cdot$
$+ H\cdot$

PGF$_{2\alpha}$

PGE$_2$

prostaglandin
synthase

thromboxane
synthase

endoperoxide
intermediate
(PGG$_2$)

thromboxane A$_2$ (TXA$_2$)

prostacyclin
synthase

prostacyclin (PGI$_2$)

Fig. 5.20. Biosynthesis of prostaglandins and other derivatives.

induce aggregation of blood platelets and vasoconstriction. Synthetic analogues of the endoperoxide, where the peroxy function has been replaced,
have been found to possess similar properties. Some are even more potent.

Prostacyclin, another derivative, has the inverse effect of dissolving blood
platelet aggregates. It is the most powerful anti-aggregatory agent known.
The discovery of these two new classes of compounds, the thromboxanes and
prostaglandins, their opposing effects on the cardiocascular system, and the
inhibition of their synthesis by certain anti-inflammatory drugs are providing
promising new insights into the prevention and treatment of arteriosclerotic
heart disease (203). Hopefully, more synthetic analogues of these new compounds will have interesting therapeutic applications in the control of
thrombosis and cardiac disorders in general.

Also, synthetically modified prostaglandin analogues have been evaluated
for potential use as contraceptives. In small amounts they induce abortion.
B. Samuelsson and E. J. Corey have made major contributions in the field of
characterization and synthesis of prostaglandin derivatives (201).

Because of the importance of peroxy radical cyclization to biological
oxidations, simple organic models have been used to elucidate the mechanism. Such model studies also help chemists to understand general principles
of the reactivity of radicals.

The group of A. Porter reported a method of generating specific unsaturated peroxy radicals (204). Of fundamental importance in this method
is the fact that hydroxy peroxide hydrogens (ROOH) are abstracted with
relative ease by *tert*-butoxy radicals compared to the ease of abstraction of
hydrogens attached to carbon. Di-*tert*-butyl peroxy oxalate (DBPO) generates peroxy radicals to give two consecutive radical cyclizations.

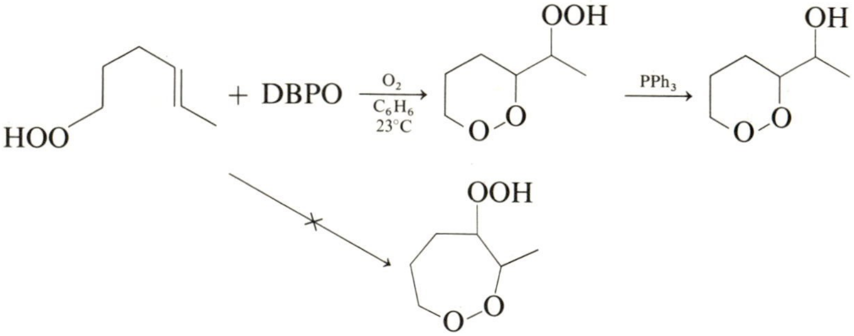

similarly:

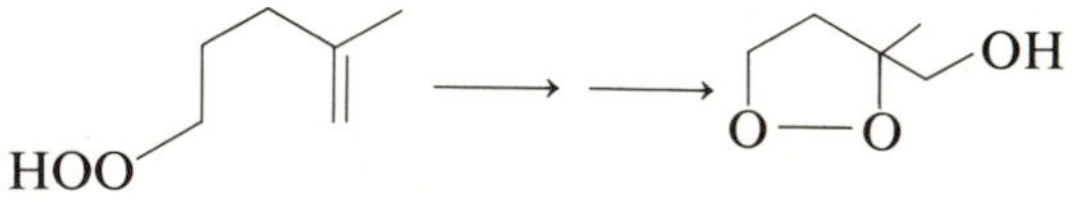

Therefore, these model systems are applicable to a systematic study of
radical cyclization leading to a prostaglandin endoperoxide intermediate.

A reasonable mechanism consistent with the observations is as follows:

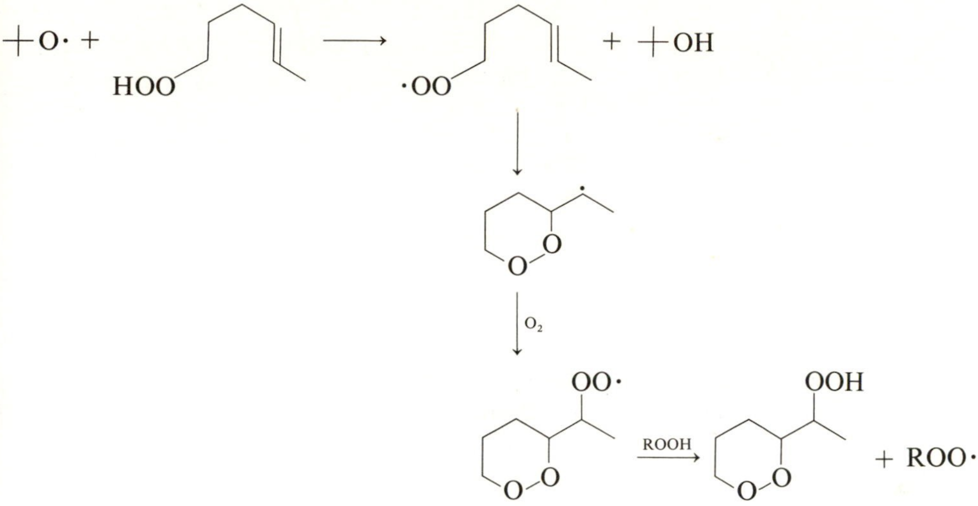

Alternatively, other prostanoid endoperoxide model compounds have been prepared. Of importance is the hydroperoxybromination of cyclopropanes and the cyclization of the γ-hydroperoxy bromides into 1,2-dioxolanes (205). This synthetic strategy is outlined below and shows that cyclopropane rings can also serve as efficient and convenient *synthons* for the preparation of endoperoxide model compounds.

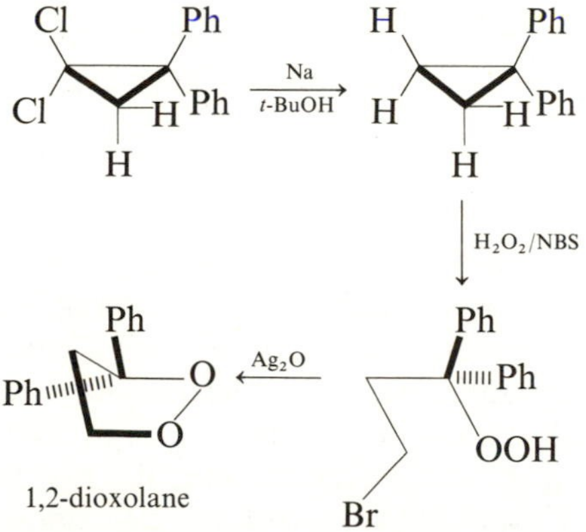

1,2-dioxolane

Finally, mention must be made of the efforts of R. G. Salomon's group to synthesize several derivatives of 2,3-dioxabicyclo[2.2.1]heptane, the strained bicyclic peroxide nucleus of prostaglandin endoperoxide (206). They are prepared from 2,3-dioxabicyclo[2.2.1]hept-5-ene by selective reduction with diimide or chlorination.

It should be realized that the work of Porter and others on intramolecular catalysis was done to sort out various aspects of enzyme function and not to try to reproduce enzyme catalysis. Nevertheless, it is sometimes necessary to look into these problems before designing "artificial enzymes."

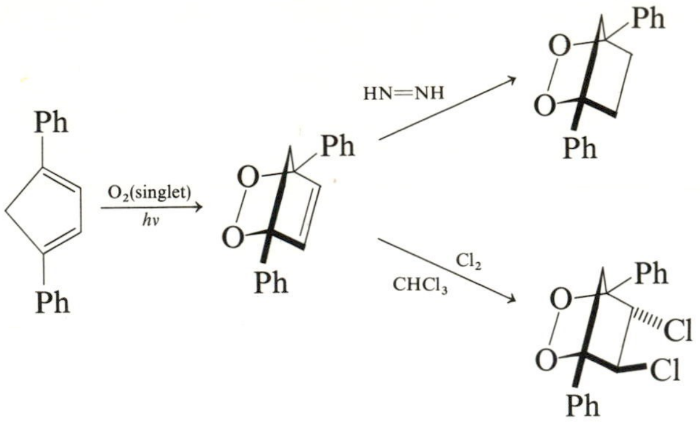

5.7 Biomimetic Polyene Cyclizations

Biomimetic-type synthesis may be defined as the design and execution of laboratory reactions based upon established or presumed biochemical transformations. This implies the development of chemical transformations new to the nonbiological area and the elaboration of elegant total synthesis of various natural product precursors. Efforts in this direction have been conducted by two schools: one by E. E. van Tamelen (207) and the other by W. S. Johnson (208, 209), both at Stanford University.

5.7.1 From Squalene to Lanosterol

Squalene is the precursor of sterols and polycyclic triterpenes. In the 1950s, G. Stork (Columbia University) and A. Eschenmoser (ETH, Zurich) proposed that the biogenetic conversion of squalene to lanosterol involved a synchronous oxidative cyclization pathway. The transformation is acid catalyzed and proceeds through a series of carbonium ions to allow the closure of all four rings. There is now ample evidence that the first step is the selective epoxidation of the $\Delta^{2,3}$-double bond to form 2,3-oxidosqualene (Fig. 5.21).

The earlier work of van Tamelen showed that squalene can be converted chemically to the 2,3-bromohydrin with hypobromous acid in aqueous glyme (210). In addition, N-bromosuccinimide (NBS) treatment of squalene in water selectively gives the desired terminal bromohydrin. Treatment with ethanolic base results in the racemic 2,3-oxidosqualene. Using rat liver homogenate under standardized aerobic conditions, racemic 2,3-oxidosqualene finally gives rise to sterol fractions which can be purified by chromatography.

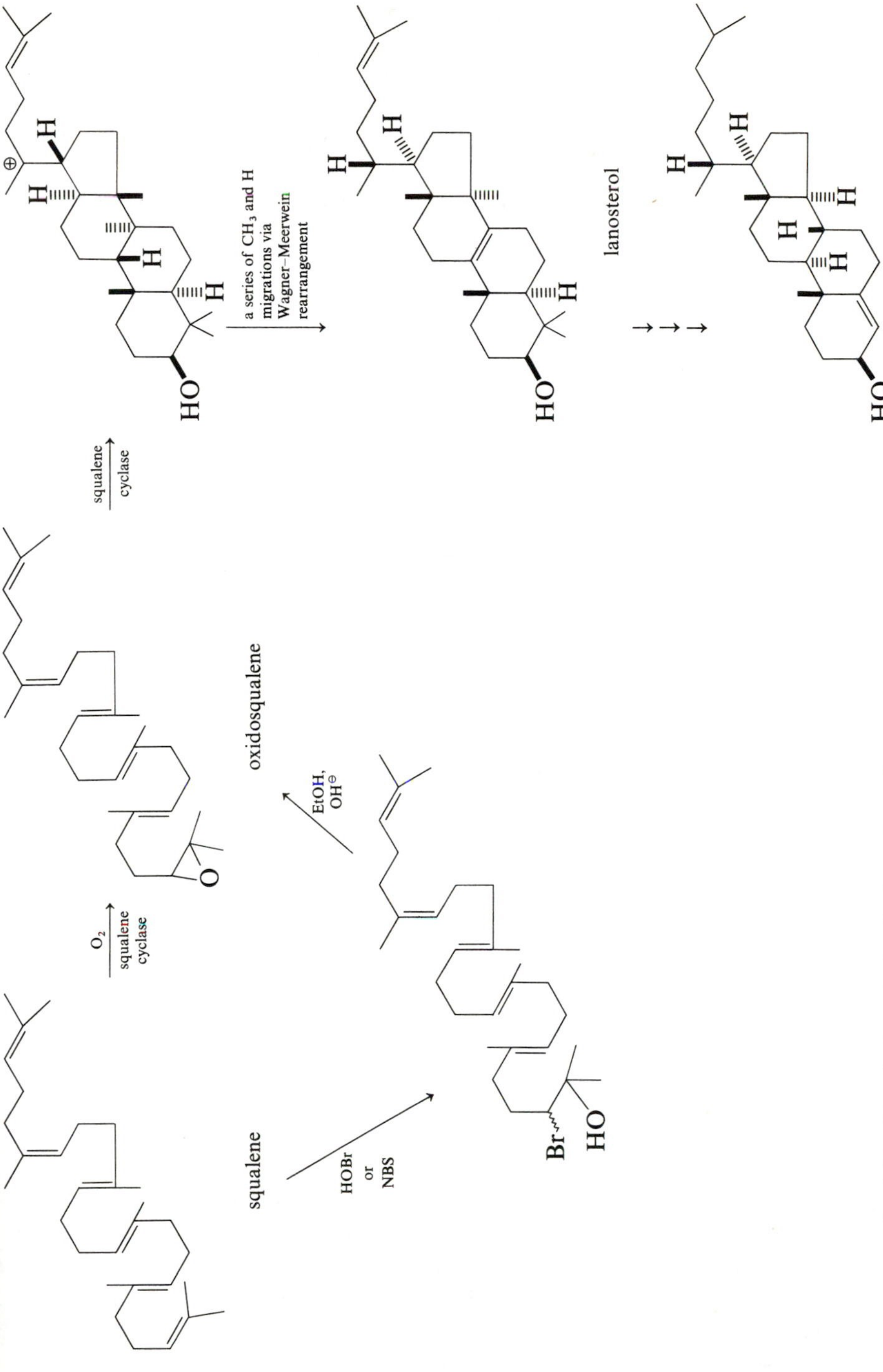

Fig. 5.21. Chemical synthesis of 2,3-oxidosqualene and its biochemical transformation to cholesterol.

Furthermore, it was demonstrated that 2,3-oxidosqualene is synthesized directly from squalene in the sterol-forming rat liver system. The former is a precursor of sterols and is far more efficiently incorporated than squalene under anaerobic conditions. Therefore, it seems very likely that the intermediate 2,3-oxidosqualene is cyclized by an enzymatic mechanism that leads to lanosterol, the precursor of cholesterol and other steroid hormones. Of course, in the enzymatic process only one isomer, the 3-S-isomer, of the epoxide is formed.

Closer examination of the 2,3-oxidosqualene skeleton reveals the presence of three distinct π-electron systems designated α, β, and γ-regions (210, 211).

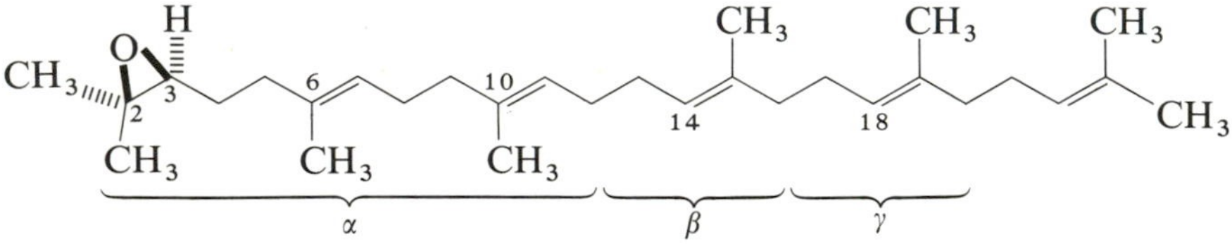

For the cyclization to occur, the α-region is essential for the enzyme cyclase. A second enzymatic control is involved in the β-region for the Δ^{14}-double bond to orient properly on the incoming carbonium ion. The reaction is therefore believed to proceed nonstop until a tetracyclic system is obtained. The role of the squalene oxide cyclase is to maintain the carbon skeleton in place to maximize orbital overlap which allows generation of the σ-bonds in the sterol molecule. This is represented below:

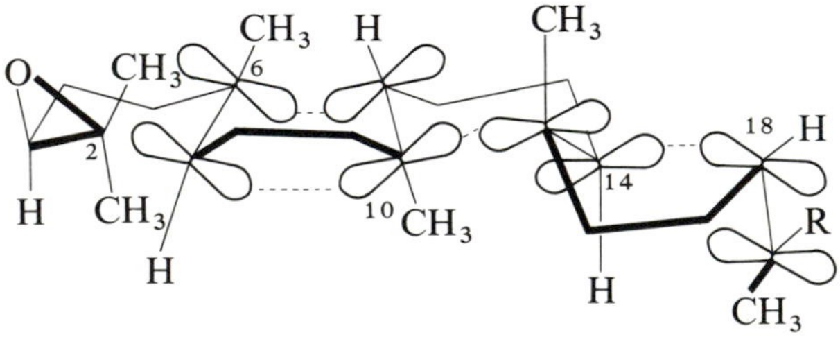

This three-dimensional representation of the skeleton shows the three important orbital overlaps. First the epoxy-Δ^6 bond system where a S_N2 reaction occur at C_2, then the Δ^6–Δ^{10} overlap which is maximized in a boat form, and finally the Δ^{10}–Δ^{14} and Δ^{14}–Δ^{18} systems where the orientation of the π-planes are perpendicular. Note the boat conformation of the B ring. Obviously, the biological cyclization of squalene can be rationalized on stereoelectronic grounds. Such a "helix-like" conformation in the transition state very likely favors the concerted cyclization process. This process of cyclization is quite complex and the group of E. J. Corey was among the first to be concerned with the question of how the cyclization of squalene is initiated (212). In this process only one of six possible double bonds is selectively activated.

In animals, squalene adopts the chair–boat–chair folding shown above to generate a chair–boat–chair tricyclic structure, which then undergoes a series of methyl-hydrogen migrations and finally a proton loss. In plant sterols, however, an all-chair folding mechanism takes place:

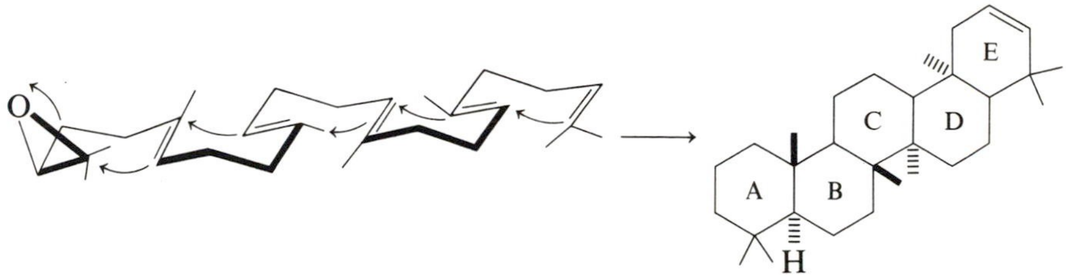

Particularly interesting is the mechanism by which hydrogen and methyl shifts occur in the Wagner–Meerwein rearrangement to lanosterol. A priori two types of methyl migration are possible: a single 1,3-shift or two 1,2-shift of methyls.

Bloch and Woodward, in 1954, suggested a very imaginative experiment to solve this ambiguity (213). Starting from two selectively enriched ^{14}C-labeled ($\bullet$) ψ-ionones, they obtained four differently labeled squalene molecules by condensation with a double Wittig reagent.

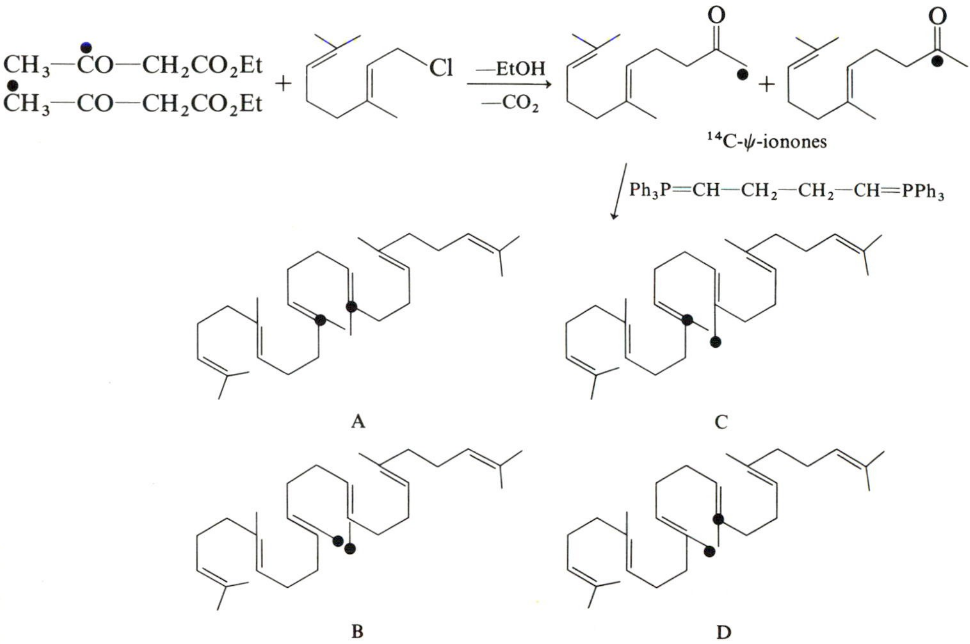

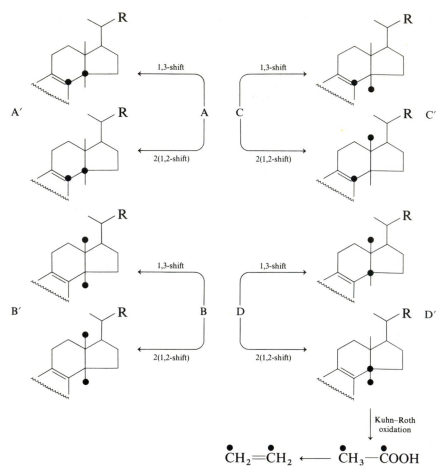

Fig. 5.22. Bloch and Woodward's proof of two 1,2-methyl migrations in the biosynthesis of lanosterol.

Enzymatic cyclization of these labeled squalenes A, B, C, and D led to A′, B′, C′, and D′ lanosterol partial structures shown in Fig. 5.22.

Lanosterol prepared from the labeled squalene mixture, was degraded to acetate by Kuhn–Roth oxidation (sulfochromic treatment) and in turn, the acetate was degraded to ethylene for mass spectrometric analysis. Consideration of the labeling patterns reveals that only structure D can give rise to ethylene containing ^{14}C at *both* carbon atoms. Since such doubly labeled ethylene was actually detected, it was concluded that the process of two 1,2-methyl migrations actually occurs.

With this brief overview of the enzymatic conversion of squalene to lanosterol, we can now envisage strategies for the total synthesis of polycyclic natural products in the laboratory. This field of bioorganic chemistry has been called *biomimetic polyene cyclizations* by Johnson.

5.7.2 The Biomimetic Approach

How much of the biosynthetic operation can one expect to simulate in the organic laboratory? The first studies directed toward nonenzymatic biogenetic-like polyene cyclization were not promising. However, van Tamelen found that if the D-ring is preformed, the isoeuphenol system can be obtained in good yield by treatment of the epoxide precursor with a Lewis acid (207).

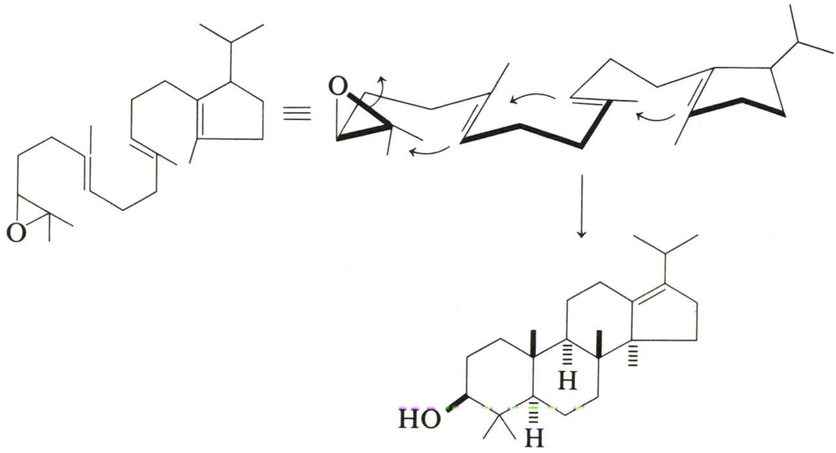

isoeuphenol skeleton
35%

In this cyclization, five asymmetric carbons are formed at one time, based on an all-chair cyclization mechanism. Of course the precursor had to be prepared and its synthesis is outlined in Fig. 5.23.

Van Tamelen argues that by selecting such a key intermediate, the preformed D-ring acts as an "insulator" to deter involvement of the side-chain π-bond with any carbonium ion developing during cyclization (207). It also prevents the C-ring from becoming five-membered, which was one of the previous difficulties in such cyclization.

The biomimetic approach of Johnson, like van Tamelen, also envisages the production of a number of rings stereospecifically in a single step by the ring closure of an acyclic chain having oppositely placed *trans* olefinic bonds (208). The approach was from the beginning very systematic. *trans* olefinic bond precursors of growing complexity were used to study the annelation of one, two, and three ring products with the natural all-*trans* configuration.

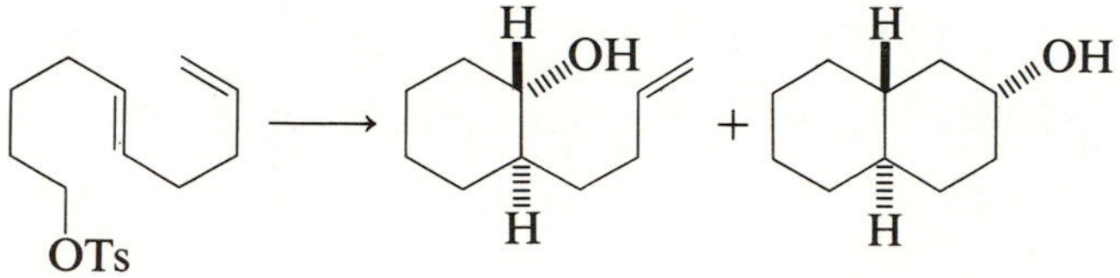

trans-decalin system

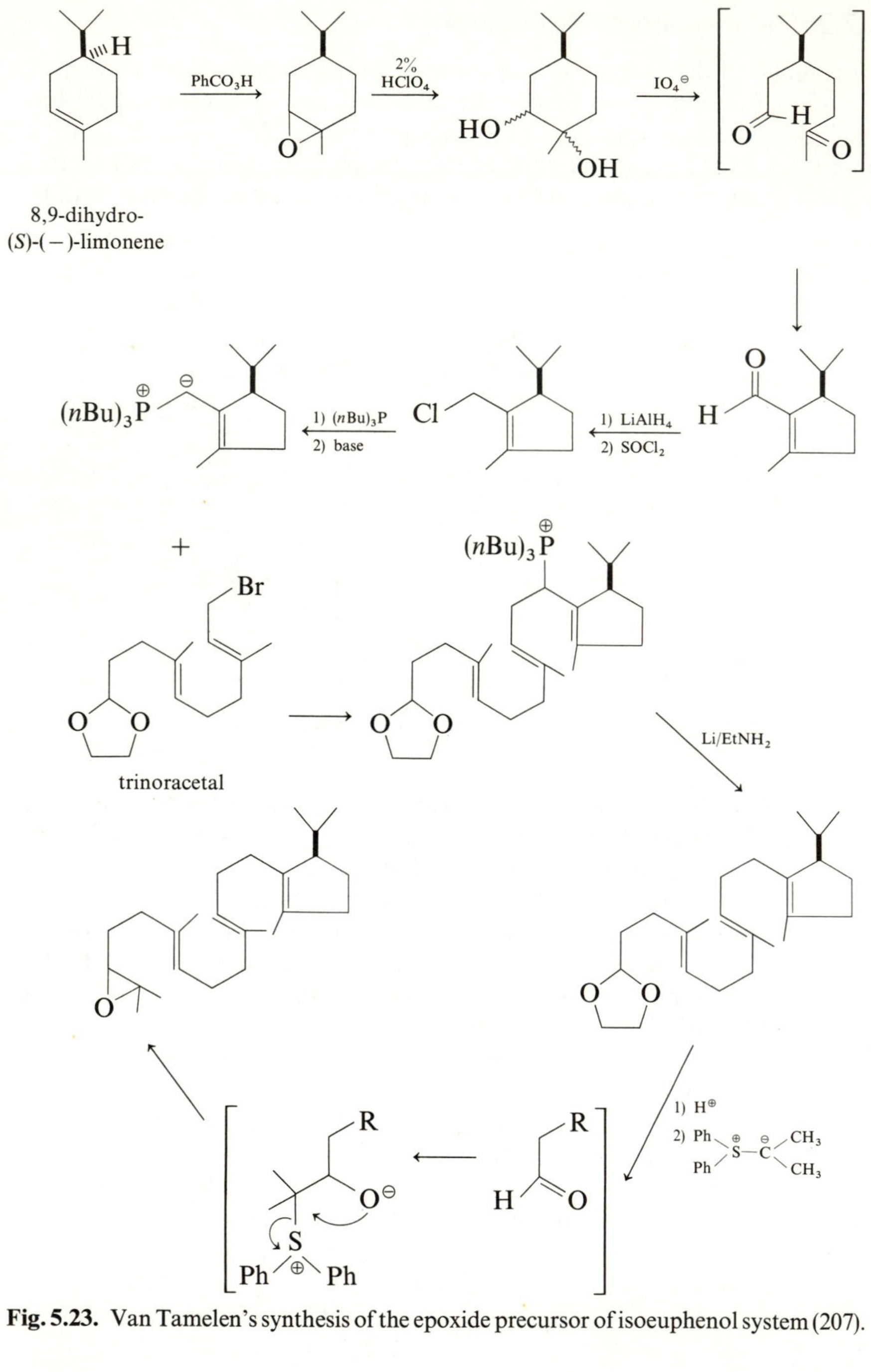

8,9-dihydro-
(S)-(−)-limonene

trinoracetal

Fig. 5.23. Van Tamelen's synthesis of the epoxide precursor of isoeuphenol system (207).

For example, sulfonated diene esters in 1,5-relationship undergo cyclization–solvolysis to yield a *trans*-decalin system. The major bicyclic component is *trans-syn*-2-decalol.

Although the yield is low, the stereospecificity of the reaction represented the first simple example of a system that followed the theoretical predictions of a stereoelectronically controlled synchronous cyclization.

Much better yields are obtained when acetal as well as allylic alcohol functions are used as initiators for these cyclizations. For instance, the following *trans*-dienic acetal gives quantitative conversion to a *trans*-bicyclic material:

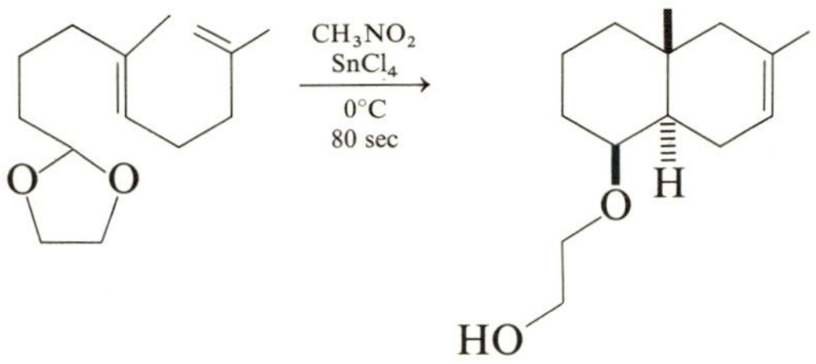

The isomeric *cis*-internal olefinic precursor yielded only a *cis*-decalin derivative, in accordance with the predictions of orbital orientation and overlap. It is important to note that the products of all these biomimetic cyclizations are racemic. However, it is interesting to observe that when an optically active dienic acetal is used, a very high degree of asymmetric induction is realized:

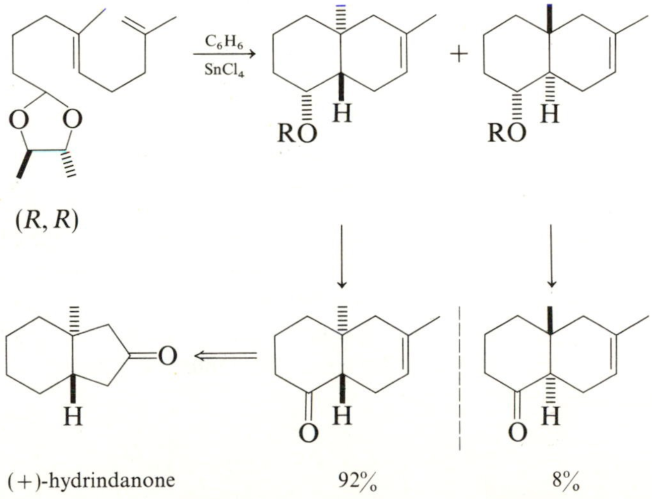

(R, R)

(+)-hydrindanone   92%   8%

For this case, an enantiomeric ratio of 92:8 is observed and the major product was converted to (+)-hydrindanone and its optical activity correlated by optical rotatory dispersion (ORD). The reason for the high selectivity is not clear.

The next objective was to examine the possibility of forming three rings from trienic acetal precursors. Some successful examples are given below:

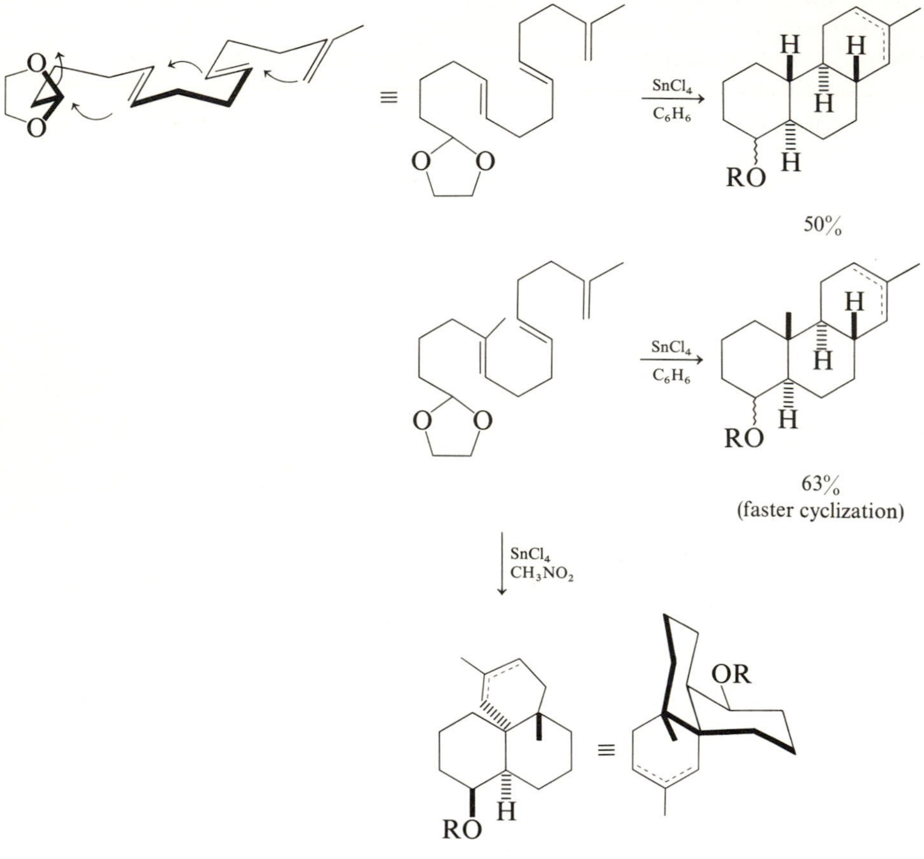

50%

63%
(faster cyclization)

44%
abnormal course

Surprisingly, when the acetal is cyclized in the presence of stannic chloride in nitromethane (conditions which had been so successful with the dienic acetal) the reaction takes a completely abnormal course involving rearrangement. The main product is a tricyclic substance. This product could have arrived through a consecutive 1,2-hydride and methyl shifts of a bicyclic cationic intermediate:

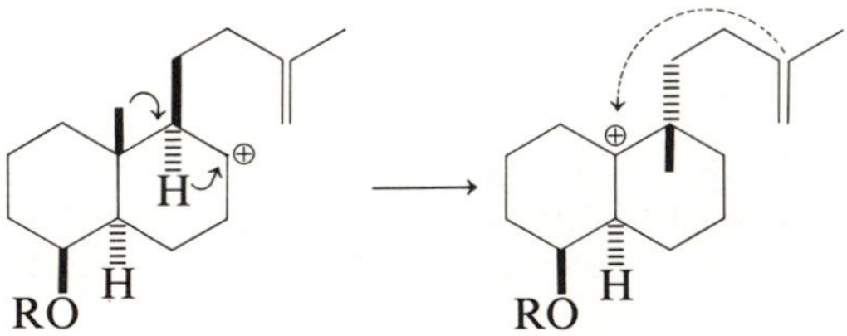

It seems that for this particular case the cyclization is not concerted and a stepwise process is involved leading to mixtures.

Finally, extension to a four-ring system gives D-homosteroidal epimers in the all-*trans* stereochemical series in 30% yield.

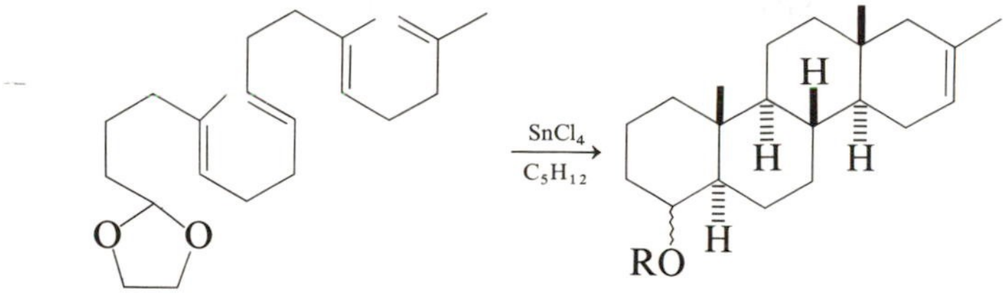

Note that seven asymmetric centers are produced in this cyclization with a remarkable stereoselectivity, yielding only two out of 64 possible racemates. This concerted conversion is the closest nonenzymatic approach realized thus far with all-*trans* configuration. Undoubtedly, the conversion of an open-chain tetraolefinic acetal having no chiral centers into a tetracyclic compound having seven such centers and producing only two out of a possible 64 racemates is a striking tribute to Johnson's method of biomimetic cyclization.

Because the five-membered ring widely occurs in natural products, particularly in the D-ring of steroids, it was of special interest to search for systems which would give five-membered ring closure in biomimetic processes. After many trials, Johnson and co-workers found that the introduction of a methylacetylenic end group is a useful method to induce five-membered ring closure:

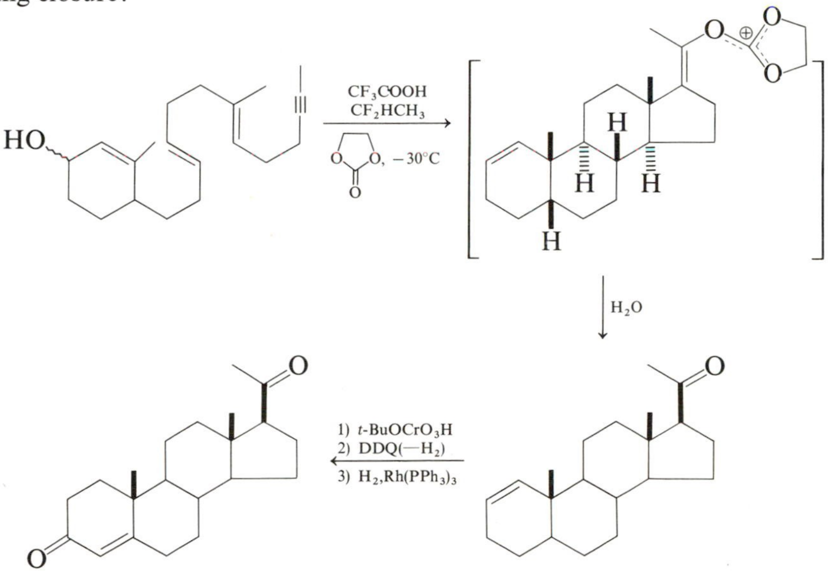

progesterone

DDQ = 2,3-dichloro-5,6-dicyano-1,4-benzoquinone, an oxidizing agent

In this example, the A ring is preformed and the cyclization gives an A/B *cis* junction. The product was then readily transformed to the naturally occurring hormone, progesterone. Thus, acetylenic groups are particularly useful terminators of polycyclization.

In a different approach, a styryl terminating group was also found to favor five-membered ring formation by virtue of the relative stability of the allylic carbonium ion formed.

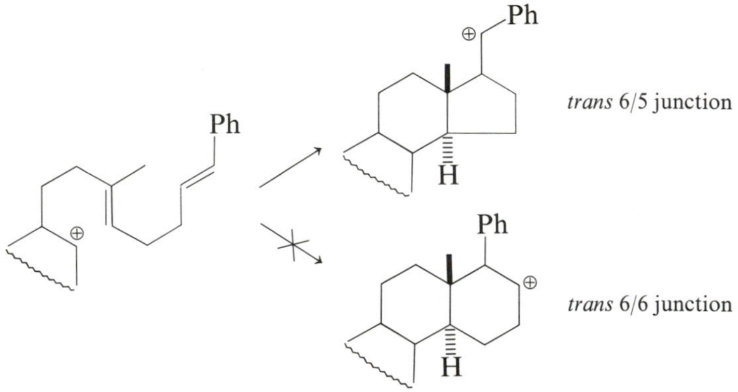

trans 6/5 junction

trans 6/6 junction

This reaction proceeds in 70% yield in CF_3COOH/CH_2Cl_2 at $-50°C$ with exclusive formation of a *trans* 6/5 ring junction.

In summary, certain polyenic substances having *trans* olefinic bonds in a 1,5-relationship and possessing effective initator and terminator functions, can be induced to undergo stereospecific, nonenzymatic, cationic cyclizations. These give polycyclic products of natural interest with an all-*trans* configuration (208, 209).

The mechanism of these biomimetic cyclizations and that of their enzymatic counterparts is still unknown, but the majority of the evidence favors a synchronous process over a stepwise one. The study of these nonenzymatic simulations of sterol biosynthesis is of fundamental importance because it could clarify and explain processes that could have been operative since the beginning of life forms on Earth.

Chapter 6
Metal Ions

"Les belles Actions cachées sont les plus estimables."

Pascal

This chapter will provide some basic concepts concerning the reactivity of biological systems utilizing metal ions. Although N, S, O, P, C, and H are the basic elements used to construct the building blocks of biological compounds, certain metal ions are essential to the organisms. It will be seen that the interactions of metal ions with biological molecules are generally of a co-ordinate nature and are used primarily for maintaining charge neutrality. Also, they are often involved in catalytic processes. Thus, the subjects developed in this chapter interphase organic and inorganic chemical principles.

6.1 Metal Ions in Proteins and Biological Molecules

Many metal ions are essential to living cells. They are Na, K, Mg, Ca, Mn, Fe, Co, Cu, Mo, Zn and constitute about 3% of the human body weight. Na(I), K(I), and Ca(II) are particularly important in the so-called "ion pump" mechanism where active transport of metabolites and energetic processes are taking place. Transitions metals such as Zn(II) and Co(II) are found in various metalloenzymes where they coordinate with amino acids and enhance catalysis at the active site (214). They act as *super acid* catalysts having a directional or template effect. On the other hand, Fe(II) and Cu(II) ions prefer to bind to porphyrin-type prosthetic groups and are involved in many electron transport systems.

In contrast, others metal ions such as Hg, Cd, and the metalloid As are strong chelating agents for thiol groups. Consequently they are efficient enzyme inhibitors. Toxic metals differ from organic poisons in that they cannot be of transformed to harmless metabolites. For example, a very potent poisonous gas is the organo-arsenic compound named *Lewisite*. It is capable of acting in the lungs, the skin, or any other of the inner or outer surfaces of the body (214). Within the last 30 years, the development of specific competitive chelating agents has been responsible for significant advances in

$$Cl-CH{=}CH-As \Big\langle{\begin{array}{c} Cl \\ Cl \end{array}} \qquad\qquad \begin{array}{ccc} CH_2 & CH & CH_2 \\ | & | & | \\ SH & SH & OH \end{array}$$

Lewisite   BAL

the treatment of most types of heavy metal poisoning. One of these antidotes for arsenic poisoning is 2,3-dimercapto-1-propanol or BAL (British anti-Lewisite), developed by Sir Rudolph Peters during World War II. EDTA and some of its analogues are also important and are used to remove, for instance, plutonium in victims of nuclear accidents.

Hopefully, the development of new chelating agents, through a better understanding of the role and mode of action of metal ions in biosystems, will lead in the near future to far more selective and effective agents for the therapeutic control of both toxic and essential metal ions (215, 216).

In recent years there has been a rapid development in the understanding of how metalloenzymes function. Specifically, the position and the sequence of reactions at the reactive site, and some clues as to the mechanism, have been forthcoming. Important in these processes is the hydrolysis (or hydration) of carbonyl- and phosphoryl-type substrate such as CO_2, carbon esters, phosphate esters and anhydrides, and peptides. It is perhaps not so surprising that a significant number of these systems require a divalent metal ion to function. More surprising, however, is the finding that this usually is Zn(II) or Mg(II) (e.g., in enzymes which act on DNA, RNA, cAMP, or cGMP). Thus, the high concentration of zinc in mammals (2.4 g per 70 kg in humans) in exceeded only by that of iron (5.4 g per 70 kg), and most of it is necessary for enzyme function (215).

Zinc ions are important for a number of diverse biological functions. For example, zinc ions serve as a stabilizer for insulin, the hormone that facilitates the passage of glucose into the cell; they are vital components of spermatozoa; they are important in speeding up the healing of wounds; they are involved in the breakdown of protein food in the intestinal tract. Zinc ions are also involved in one of the key reactions in vision.

6.2 Carboxypeptidase A and the Role of Zinc

Enzymes that hydrolyze amide and ester bonds may be divided into three classes: (1) those requiring a thiol group for activity, such as papain, ficin, and other plant enzymes; (2) those inhibited by diisopropylphosphorofluoridate (DFP), such as α-chymotrypin, trypsin, subtilisin, cholinesterase, and thrombin; (3) those that require a metal ion for activity. This last class includes dipeptidases, and exopeptidases such as carboxypeptidase and leucine aminopeptidase. The metal ion is involved in the stabilization of the tetrahedral intermediate (refer to Section 4.4.1).

Another important metalloenzyme is carbonic anhydrase which catalyzes the conversion of carbon dioxide into bicarbonate ion. It is a zinc-dependent

$$CO_2 + H_2O \rightleftharpoons HCO_3^{\ominus} + H^{\oplus}$$

enzyme that is necessary for the absorption of CO_2 into the blood and subsequent discharge at the lungs. It is believed that a coordinated hydroxide ion is the nucleophile in the hydration of CO_2 and it was suggested that an ordered icelike water structure at the active site facilitates ionization (217).

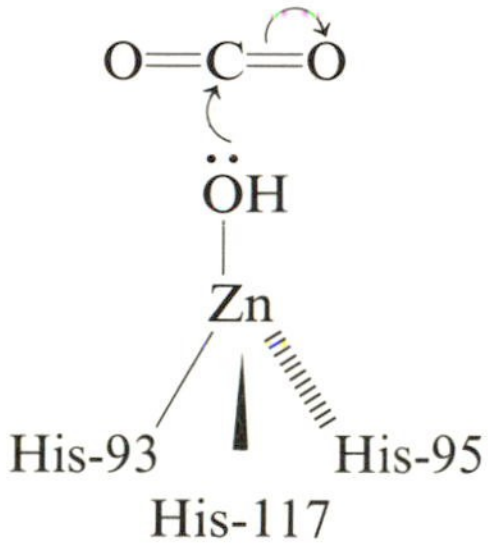

bound to
three histidine
residues

X-Ray studies of carbonic anhydrase show that the active center is composed of three imidazole ligands which have distorted tetrahedral coordination to the Zn(II) ion. Molecular models suggested that a similar geometry could be attained with a tris(imidazolyl) methane derivation. For this reason, R. Breslow's team (218) in 1978 synthesized tris[4(5)imidazolyl] carbinol (4-TIC) and tris(2-imidazolyl) carbinol (2-TIC) as models for the zinc binding site of carbonic anhydrase and alkaline phosphatase. Similarly, bis[4(5)-

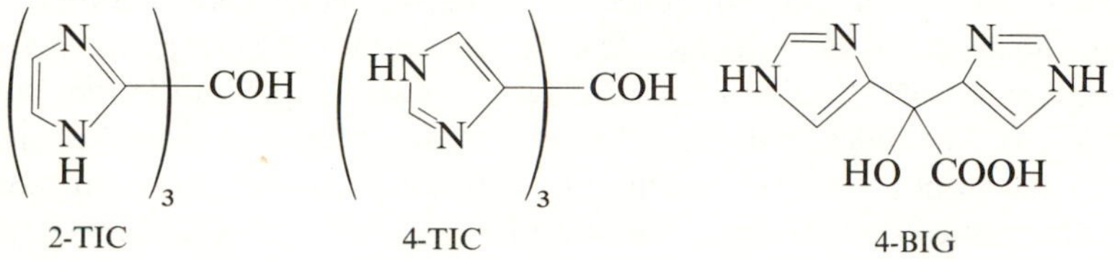

2-TIC   4-TIC   4-BIG

imidazolyl] glycolic acid (4-BIG) has been synthesized to mimic the zinc binding site of carboxypeptidases and thermolysin.

Comparison between the different models clearly showed that 4-TIC is a tridentate ligand capable of using all three imidazole rings for coordination to Zn(II), Co(II), or Ni(II).

4-TIC metal complex (M = Zn, Co, or Ni ions)

However, the Co(II) complex with 4-TIC or with 4-BIG in nonbasic aqueous solution does not have the blue color characteristic of tetraco-ordinate Co(II), as in carbonic anhydrase or carboxypeptidase complexes of Co(II). The models are probably too small so that octahedral complexing is favored. Consequently, spectral and binding studies suggest that the geometry of 4-TIC is not quite right for a good mimic of carbonic anhydrase. It is hoped that a somewhat larger ligand related to 4-TIC might mimic better the spectroscopic behavior of carbonic anhydrase and its extraordinary Zn(II) affinity.

In this section we will orient the rest of the discussion on carboxypeptidase A. This is a zinc-dependent enzyme which catalyzes the hydrolysis of C-terminal amino acid residues of peptides and proteins and the hydrolysis of the corresponding esters. This exopeptidase has a molecular weight of 34,600 and a specificity for aromatic amino acid in the L-configuration.

Carboxypeptidase A is one of those enzymes for which most of the de-tailed structural informations has been obtained by X-ray crystallographic studies, through the efforts of W. N. Lipscomb's group at Harvard University (219). The only zinc atom is situated near the center of the molecule and is coordinated by His-69, Glu-72, His-196 and a water molecule. X-Ray crystallographic studies have also been undertaken in the presence of the substrate Gly-Tyr, bound to the active site. The main features of the binding mode show that the C-terminal carboxylate group of the substrate interacts electrostatically with the guanidium group of Arg-145. The carbonyl oxygen of the substrate amide bond displaces the water molecule coordinated to the Zn(II) ion. There is also an electrostatic interaction between the NH_3^+-terminal group of the substrate and Glu-245. Most important is the move-ment of Tyr-248 which is displaced about 1.2 nm from its original position upon addition of the substrate, thus placing the phenolic OH group about 0.3 nm from the NH of the peptide bond to be cleaved. This is an indication of the *induced-fit* effect, proposed by Koshland, upon enzyme–substrate complex formation.

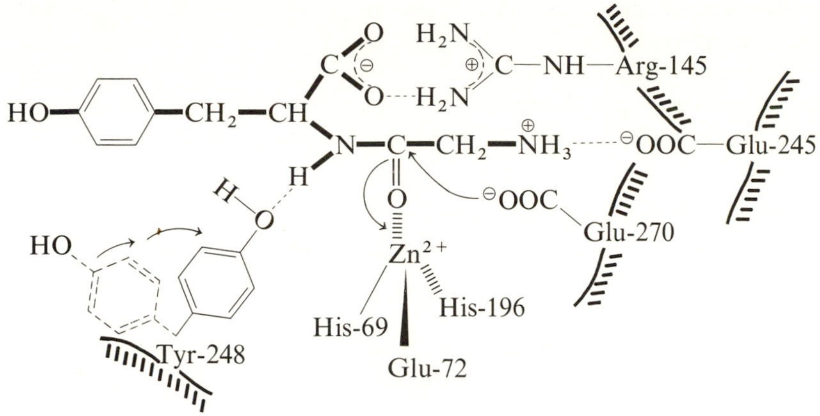

Finally, the carboxylate group of Glu-270 is believed to be implicated in the catalysis by acting as a nucleophile, forming an anhydride intermediate:

anhydride intermediate

$2H_2O$ or a peptide (resynthesis)

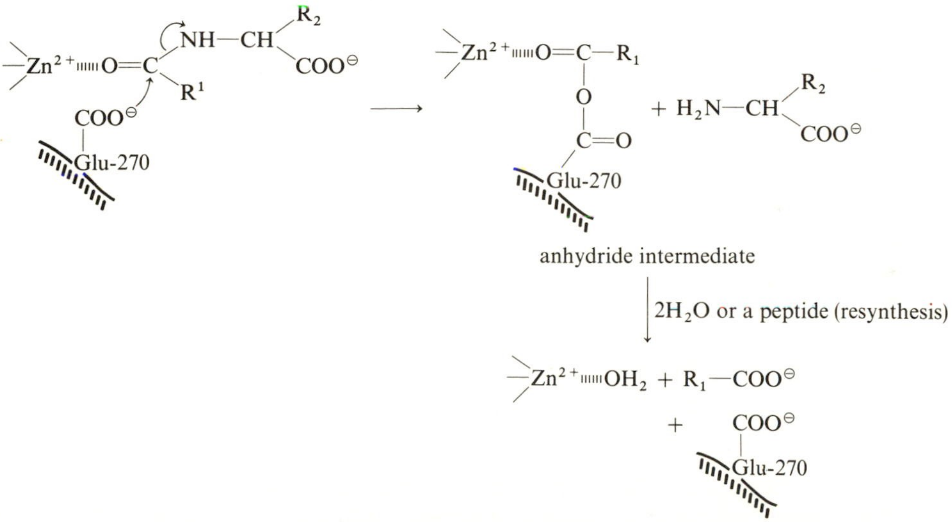

The mechanism shown above involves two steps and an anhydride (acyl-enzyme) intermediate. In the first step Zn(II) of the enzyme electrophically activates the substrate carbonyl towards nucleophilic attack by a glutamate residue. Departure of an alkoxyl group (with ester substrates) or an amino group (with peptide substrates) results in the production of an anhydride between the enzyme glutamate residue and the scissile carboxyl group. In the second step the hydrolysis of this anhydride can be catalyzed by the

Zn(II) ions, the only remaining catalytic group. Although the reaction in $H_2^{18}O$ results in the incorporation of ^{18}O in the carboxylate group of the glutamate residue, this suggested mechanism is ambiguous because the anhydride intermediate has never been trapped by a nucleophile such as hydroxylamine.

Therefore, one could wonder if this mechanism is correct. To approach such a problem three general questions concerning the mechanism of action of carboxypeptidase and metalloenzymes in the broad sense must be examined.

(a) How do metal ions affect ester and amide hydrolysis?
(b) Under what conditions will a neighboring carboxylate group participate in ester and amide hydrolysis and what is the mechanism of such participation?
(c) How will a metal ion affect intramolecular carboxylate group participation in ester and amide hydrolysis?

These questions should be answerable from model studies. Considerable information concerning questions (a) and (b) are available but there are experimental difficulties in attempts to answer (c) because of the difficulty of finding a system where carboxylate group participation is possible but where the metal ion does not bind to it.

To evaluate these points Breslow and his group developed in 1975 and 1976 interesting and more precise models for carboxypeptidase (220, 221). They examined the hydrolysis of the following anhydride models with and

X = COOH, H

without Zn(II) ions. The reaction is pH-dependent only in the presence of Zn(II) ions and pseudo-first-order hydrolysis rate constants at pH 7.5 have been measured (Table 6.1). Therefore, in the presence of Zn(II), the rate falls in the range of carboxypeptidase A and cleavage of anhydride must be a reasonable step. As with carboxypeptidase A, the Zn(II) ion coordinates to two nitrogen atoms and one carboxylate group. Studies of the opening of the anhydride (with X = COOH) by hydroxylamine showed that in the absence of Zn(II), hydroxylamine is an effective nucleophile for attack on the anhydride function, but the attack is not catalyzed by Zn(II). Rather, the pH-dependence indicates that Zn(II) catalyzes the attack of an OH^- ion

Table 6.1. Pseudo-first-order Rate Constants at pH 7.5 for the Hydrolysis of the Anhydride Model Compounds (220)

Model compound		k_{obsd} (sec^{-1})	k_{rel}
with X = COOH	$-$ Zn(II)	2.7×10^{-3}	1.0
	$+$ Zn(II)	3.0	10^3
with X = H	$-$ Zn(II)	5.5×10^{-3}	2.0
	$+$ Zn(II)	1.5	5×10^2

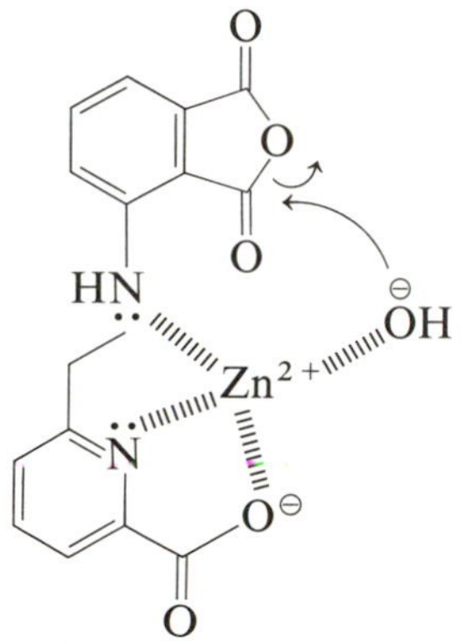

on the anhydride instead of a water molecule and this process is much faster than the uncatalyzed attack by hydroxylamine. Consequently, binding of Zn(II) and OH$^-$ (from H$_2$O) allows OH$^-$ ion to be an extremely efficient nucleophile so that a good nucleophile such as hydroxylamine cannot compete. Remember that in the first step of carboxypeptidase action, the Zn(II) ion catalyzes the anhydride formation between Glu-270 and the substrate by acting as a Lewis acid. The present model shows that it can also catalyze the second step by delivering a specific nucleophile to the anhydride intermediate. Nonetheless, the two-step mechanism proposed for carboxypeptidase A remains one of the most attractive explanations of all the data.

Zinc ions and other ions in metalloproteins could be consider as having a *dual role*. The first is an orientation or template effect while the second is a concentration (of the nucleophile) effect at the site of reaction.

Based on his study on model compounds, Breslow suggested a second mechanism for peptide hydrolysis by carboxypeptidase A which does not involve the formation of an acyl-enzyme intermediate (221, 222). Essentially, in the hydrolysis of a peptide bond there is participation of a zinc ion, a carboxylate ion, and a tyrosine hydroxyl group. The Zn(II) ion still plays the role of a Lewis acid to coordinate the carbonyl oxygen but the carboxylate group rather acts as a general base. The argument is based on the fact that

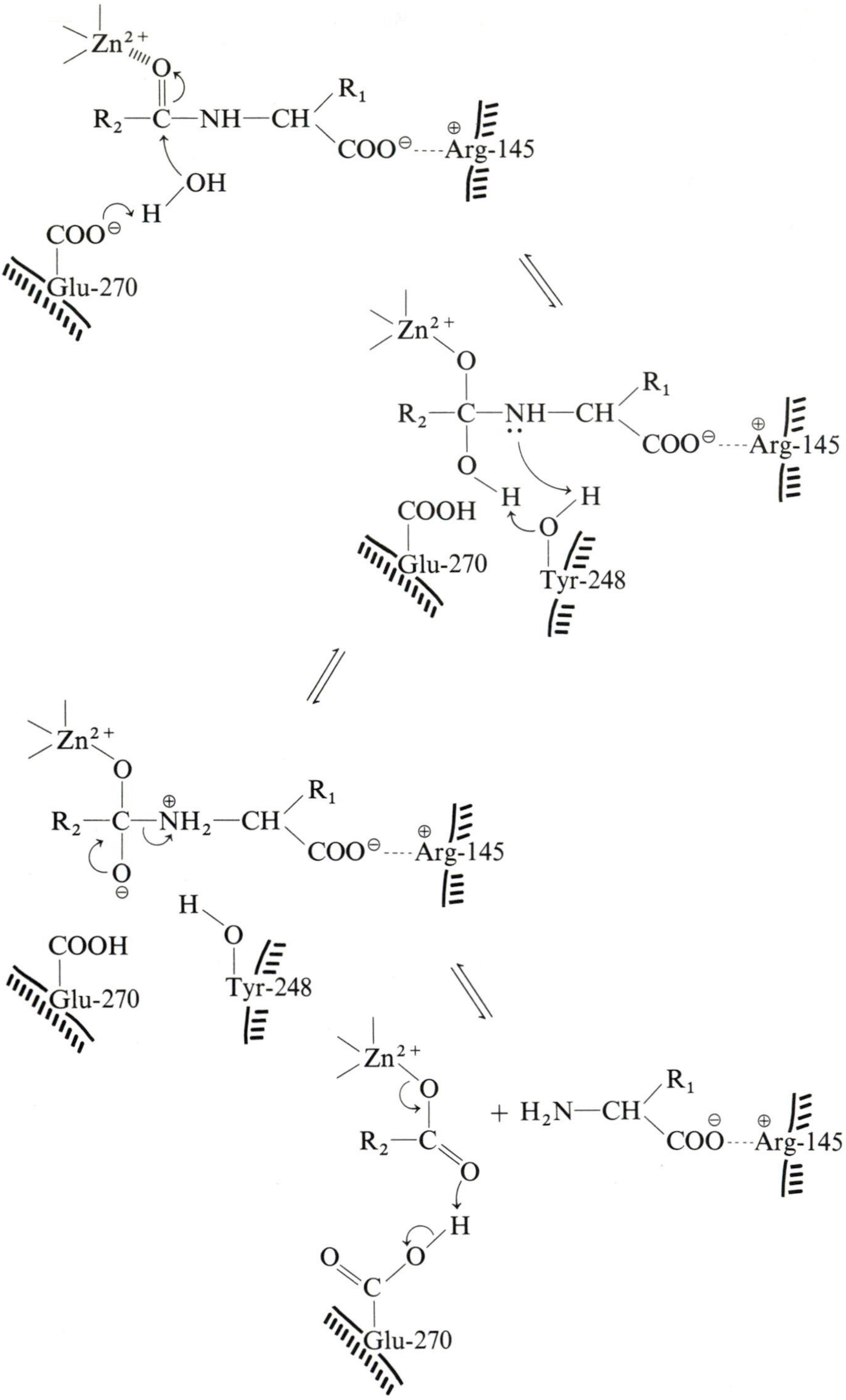
Zn2+
O
R1
R2—C—NH—CH
COO⊖----Arg-145
OH
H
COO⊖
Glu-270
Zn2+
O
R1
R2—C—NH—CH
COO⊖----Arg-145
O
H
H
COOH
O
Glu-270
Tyr-248
Zn2+
O
R1
R2—C—NH2—CH
COO⊖----Arg-145
O
H
O
COOH
Glu-270
Tyr-248
Zn2+
O
R1
R2—C
O
+ H2N—CH
COO----Arg-145
H
O
O
C
Glu-270

in the presence of CH_3OH (to substitute for water), methanolysis of a peptide substrate cannot be directly observed because the equilibrium constant is unfavorable. Thus, the enzyme cannot incorporate methanol in the transition state of the reaction (catalyzed in either direction) for either ester or peptide substrates. This suggests that removal of *both* protons of water is required in the transition state for hydrolysis.

In brief, with this mechanism the carboxylate group of Glu-270 acts as a general base to deliver nucleophilic water to the carbonyl. In the presence of methanol the first step would simply reverse. Consequently, only a second deprotonation could drive the reaction in the forward direction, and this proton transfer might well involve the hydroxyl group of Tyr-248 as a bridge between the OH and the N of the amide bond to be cleaved. This proposal has the advantage of giving a justification for the "induced-fit" effect mentioned earlier in this section.

Finally, much controversy has surrounded the question of whether Arg-145 or Zn(II) is the binding site for the C-terminal carboxyl group of the substrate. Breslow's mechanism suggests that both are true (222). Indeed, the Zn(II) ion binds the carboxylate group of the hydrolyzed product which becomes the substrate for the reverse or next reaction. One would thus expect that exopeptidases should have two alternate binding sites, separated by a distance corresponding to one amino acid residue in the substrate, as in this proposed mechanism.

From these studies of the mode of action of carboxypeptidase A, the following two points can be extracted: (a) the Zn(II) ion presumably complexes the carbonyl of ester and amide substrates, and (b) Glu-270 is also implicated as a participant, and both general-base and nucleophilic mechanisms have been proposed. There is also strong evidnece that the mechanisms are different with esters and amides. However, another mechanistic possibility to be envisaged for carboxypeptidase A is to consider a nucleophilic attack of the substrate ester or amide bond by Zn(II) coordinated hydroxide ion. Such a possibility was scrutinized by T. H. Fife and V. L. Squillacote (223). In particular, they examined the hydrolysis of the carboxyl substituted ester, 8-quinolyl hydrogen glutarate in the presence of Zn(II) ion.

Extensive kinetic studies showed that at pH values above 6, Zn(II) ion catalyzes the hydrolysis of the ester function through an intramolecular attack of the metal bound hydroxide ion or by a metal ion promoted attack of an external OH^- ion. However, at pH below 6, the metal ion facilitated OH^- catalysis, although capable of large rate enhancements, cannot compete with the intramolecular attack of the carboxylate ion to eject the 8-hydroxyquinoline bound to the zinc ion (223).

With regard to the mechanism of carboxypeptidase A, the question is whether a neighboring carboxyl group (Glu-270) functioning as a nucleophile can compete with metal ion promoted OH^- catalysis. The present model seems to indicate that with ester substrates, both mechanisms could be operative at appropriate pH values.

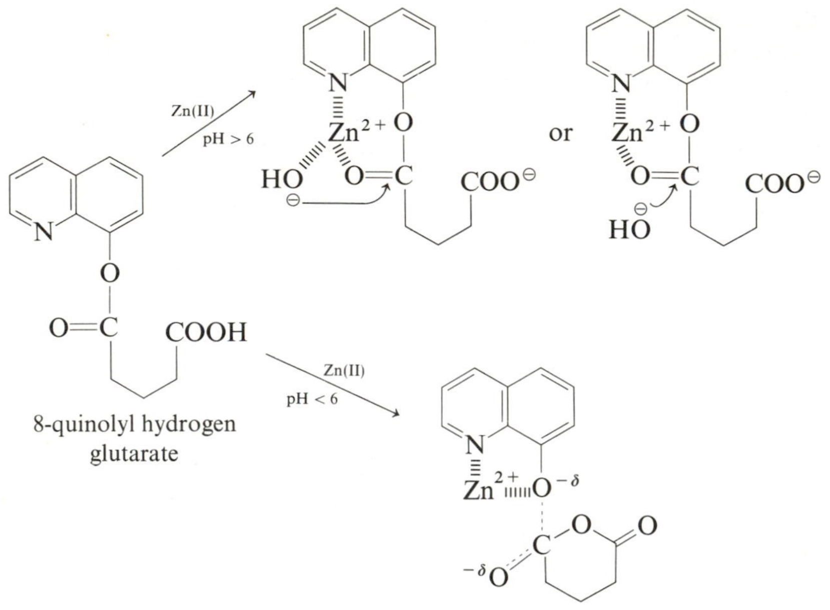

8-quinolyl hydrogen
glutarate

Therefore, a key argument in regard to the proposed enzymatic mechanism of carboxypeptidase A is whether the carboxylate group of Glu-270 is sterically capable of participating efficiently in a nucleophilic reaction. In fact, such evidence has now been obtained by spectral characterization in the subzero temperature range ($-60°C$) study of a covalent acyl-enzyme intermediate obtained in the hydrolysis of the specific substrate *O*-(*trans-p*-chlorocinnamoyl)-L-β-phenyllactate by carboxypeptidase A (224). Furthermore, the results indicate that deacylation of the mixed anhydride intermediate is catalyzed by a Zn-bound hydroxide group.

6.3 Hydrolysis of Amino Acid Esters and Amides and Peptides

Commonly, transition metals have d orbitals which are only partially filled with electrons. In solution, these positively charged metal ions can readily combine with negative ions or other small electron-donating chemical functions called *ligands* to form complex ions. The geometry of the ligand–metal complex depends on the nature of the metal ion and can be as varied as tetrahedral, square planar, trigonal bipyramidal, or octahedral. Two aspects must be considered in the evaluation of transition metal ion complexes with ligands: first, the nature of the ligand–metal bond involved, and second, the geometry of the complex formed. These factors will contribute to the stability of the ionic complexes.

In many metal ion catalytic reactions, the role of the metal ion is similar to that of a proton, but in a more efficient way because more than a single positive charge can be involved and the ion concentration may be high in aqueous solution. Experimentally, concentrations up to 0.1 M are possible whereas with H^+ only $10^{-7}\ M$ is obtained in a neutral medium. Furthermore, the process is energy-favored because the ion forms a complex with a substrate which is more stable than the free form since the complex has a lower ΔG, and a lower $\Delta G^{\ddagger}$ as well in the transition state.

For example, if a molecule such as oxaloacetic acid binds to Cu(II) ion, decarboxylation readily occur:

$$\text{(complex)} \longrightarrow \text{(complex)} + CO_2$$

The metal ion can form many bonds to the substrate and draw electron density from it. It thus behaves as a *super acid*. Furthermore, ligands such as AcO^- or OH^- relative to H_2O should enhance the positive character of the metal and increase its effectiveness. Also, Fe(III) is a better catalyst than Fe(II) because of the additional positive charge.

An enzyme, because of its well-defined three-dimensional structure, will form a catalytic site where catalysis is involved. A small peptide, however, has very little rigidity and will not possess a catalytic capacity. But it is interesting to realize that if a metal ion is bound to the peptide, amide bond hydrolysis can take place, analogous to that observed with hydrolytic enzymes. Thus, hydrolysis of amides (and esters) is susceptible to the catalytic action of a variety of metal ions because the α-amino and the carbonyl oxygen groups are two good potential ligands for complex ion formation. In other words, coordinated ligands (peptide) acquire remarkable reactivity due to the electron-withdrawing effect of positively charged metal ions.

Co(II) and Cu(II) ions can promote the hydrolysis of glycine ethyl esters at pH 7 to 8, 25°C, conditions under which they are otherwise stable. Complexation takes place between the metal ion (M^{2+}) and the amino acid ester to form a five-membered metal chelate. Subsequently, catalysis occurs as a result of the coordination of the metal ion with the amino and ester functions of the amino acid. In either case the metal ion can polarize the carbonyl group, thereby promoting attack of OH^-. The rate of hydrolysis increases with pH showing that OH^- ion participates in the mechanism. Thermodynamically the hydrolysis occurs presumably because the carboxylate anion formed coordinates more strongly to the metal cation than the starting ester.

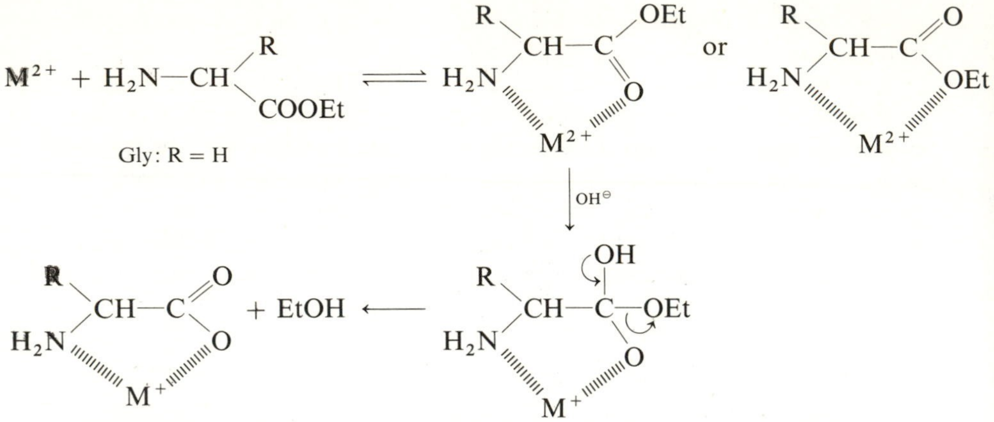

$$M^{2+} + H_2N—CH\underset{COOEt}{\overset{R}{<}} \rightleftharpoons H_2N\underset{M^{2+}}{\overset{R}{\underset{}{}}}\;\;\; or$$

Experimentally it is found that metal ions do not hydrolyze simple esters unless there is a second coordination site in the molecule in addition to the carbonyl group. Hydrolysis of the usual types of ester is not catalyzed by metal ions, but hydrolysis of amino acid esters is subject to catalysis.

In a thermodynamic sense, the transition state energy of a metal–amino acid complex is lowered relative to the free transition state for amino acid hydrolysis principally because of charge stabilization. There is likely a lesser solvent reordering term in $\Delta S^{\ddagger}$ in the metal catalyzed step as well. Consequently, a good template effect of the metal ion in binding the substrate is important. Metal ions will also catalyze the hydrolysis of a number of amides but the effect is not as large as with properly bound esters. The reason lies in the difference in the nature of the leaving group. The poorer amide leaving group makes breakdown of the tetrahedral intermediate rate-determining.

Let us examine few models. The rate of hydrolysis of the ester function of the phenyl and salicyl esters of pyridine-2,6-dicarboxylic acid is significantly enhanced upon complexation with Ni(II) or Zn(II) ions (225).

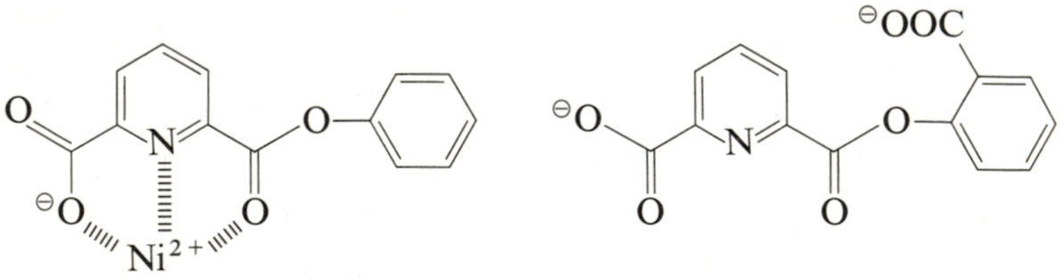

The relative rate with Ni(II) is 9300-fold with the phenyl ester and 3100-fold with the salicyl ester. In the latter case the rate is slower presumably because two anionic sites are in competition for the metal cation. But the reaction rate is almost twice as fast if the second carboxyl group is ionized, suggesting that carboxylate exerts a weak catalytic effect.

Pyridine carboxaldoxime-Zn(II) complex is an excellent catalyst for the deacylation of 8-acetoxyquinoline-5-sulfonate. Again the dual role of Zn(II) ion comes into play (226, 227).

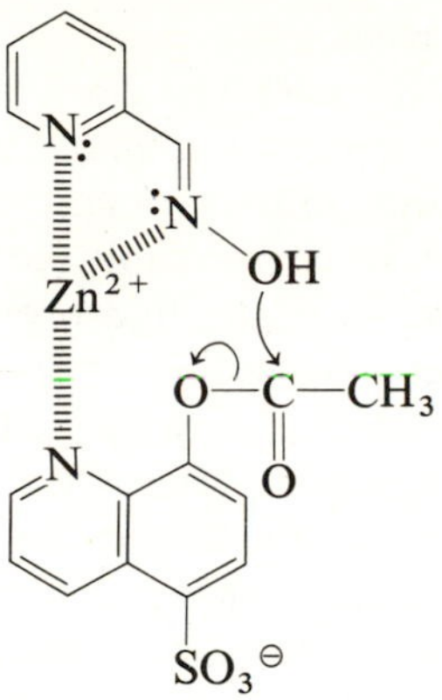

Similarly, phosphorylation of 2-hydroxymethyl phenanthroline by ATP is facilitated by Zn(II) complexation.

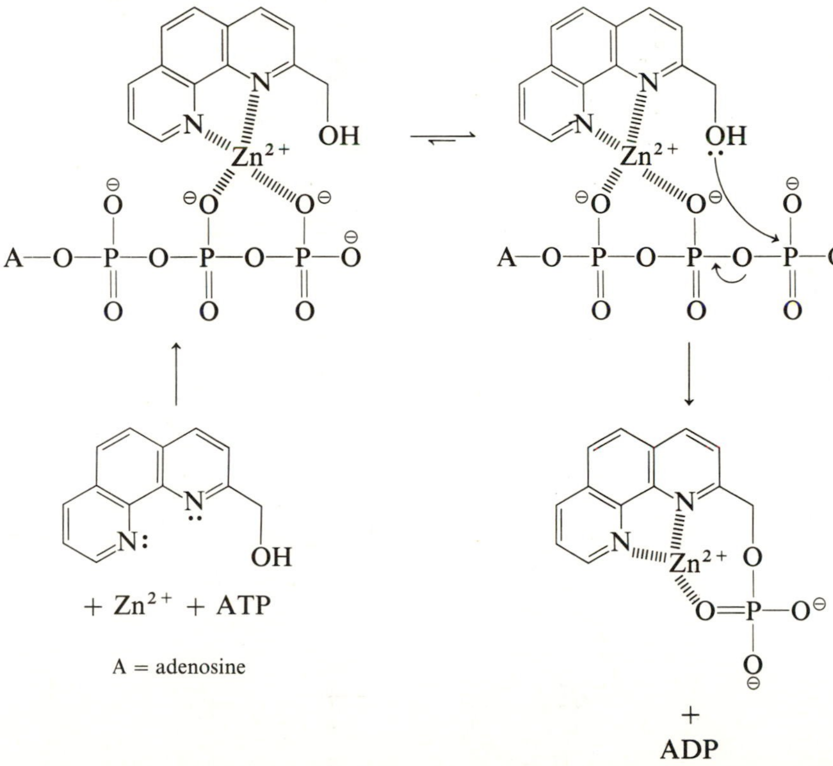

A = adenosine

This example illustrates the catalytic influence of a chelated metal ion. The metal has a *directional effect* or *template effect* as is seen by the binding together of ATP and phenanthroline. In addition it serves a *neutralization* function, thus reducing the electron repulsion between the incoming hydroxymethyl group and the phosphate of the ATP. Of course, the nature of the metal also is important to mediate the position of bond cleavage. The complex also lowers the pK_a of the alcohol function of the phenanthroline moiety to 7.5 so that an appreciable fraction is ionized at the experimental pH value. Further electron delocalization in the aromatic ring system upon complexation with the metal ion increases considerably the acidity of the alcohol group.

In summary, a metal ion facilitates hydrolysis or nucleophilic displacement by some other species in essentially two ways: direct polarization of the substrate and external attack, or by the generation (by ionization) of a particularly reactive (basic) reagent. One of the prime functions of a metal ion in any biological system is thus to provide a useful concentration of a potent nucleophile at a biologically acceptable pH.

Cobalt ions can also exert interesting catalytic properties on ester and amide hydrolysis. For instance Co(III) is much more effective than Zn(II) at polarizing the carbonyl group of a peptide bond. However, it has been estimated that carboxypeptidase A catalyzes the hydrolysis of benzoylglycyl-L-phenylalanine about 10^4 faster than a Co(III) activation could afford. Therefore, the enzyme must also be exerting an additional effect.

The group of D. A. Buckingham (228, 229), then at the Australian National University, has prepared a number stable Co(III) ion complexes with amino acids and peptides, Ethylene diamine and triethylene tetramine are usually used as ligands. A first example is given here where Hg(II) ion drives the reaction.

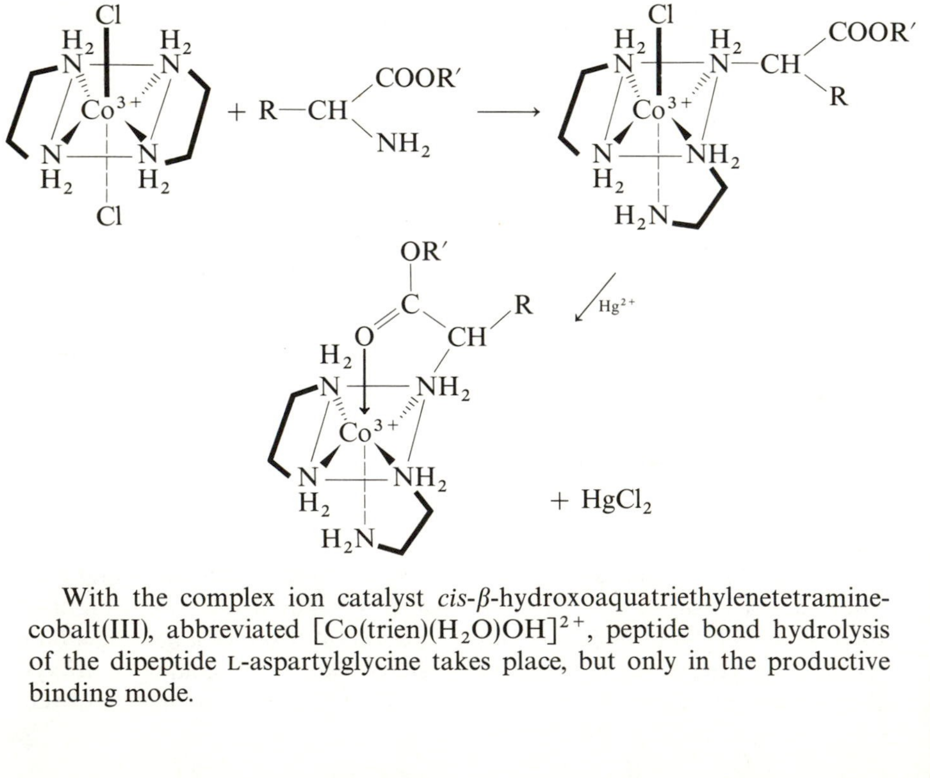

With the complex ion catalyst cis-β-hydroxoaquatriethylenetetramine-cobalt(III), abbreviated $[Co(trien)(H_2O)OH]^{2+}$, peptide bond hydrolysis of the dipeptide L-aspartylglycine takes place, but only in the productive binding mode.

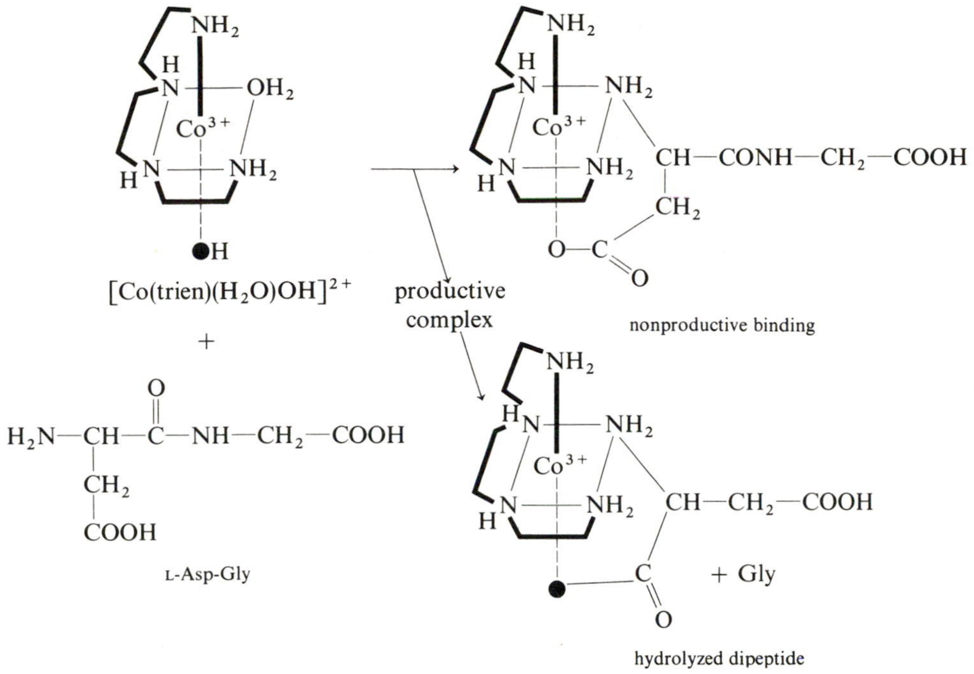

$[Co(trien)(H_2O)OH]^{2+}$

+

$H_2N-CH-C(=O)-NH-CH_2-COOH$
$\quad\quad CH_2$
$\quad\quad COOH$

L-Asp-Gly

productive complex

nonproductive binding

$CH-CONH-CH_2-COOH$
CH_2
$O-C(=O)$

$CH-CH_2-COOH$
$C(=O)$
+ Gly

hydrolyzed dipeptide

The rate-determining initial step involves the replacement of a coordinated water molecule by the terminal NH_2-group of the peptide. The remaining OH-ligand then acts as a nucleophile to promote the hydrolysis. Such intramolecular attack of a coordinated water or hydroxo group is very common in coordination chemistry.

The beauty of the Co(III)-trien systems is that water exchange and substitution in the coordination sphere of the metal ion is in general a very slow process ($t_{1/2} \simeq$ minutes to hours), which means that kinetic parameters can be easily evaluated. The slow exchange of ligands in aqueous solution has thus the advantage of using ^{18}O tracers (indicated as ●) to follow the path of the coordinated aqua or hydroxo group and thereby allowing the possibility of distinguishing between the direct nucleophilic and general-base paths for hydrolysis. These advantages bring with them obvious problems when one considers their parallel (or lack of it) in enzymatic processes. For instance, the Co(III)-trien complexes promoted reactions are stoichiometric rather than catalytic with the product of hydrolysis or hydration remaining firmly bound to the metal center. For this reason Co(III) complexes are not as useful as they could be for enzyme mimicking. However, because of a favorable decrease in $\Delta S^{\ddagger}$ ($\Delta H^{\ddagger}$ remains essentially unchanged) upon complexation with appropriate ligands, rates of $\sim 10^4$ have nevertheless been observed. In spite of that the system still imitates reasonably well the metals in their ability to polarize adjacent substrate molecules and to activate coordinated nucleophilic groups.

Let us look at another example; the hydrolysis of glycyl-L-aspartic acid by $[Co(trien)(H_2O)OH]^{2+}$. The situation is slightly more complicated than the preceding one because three structures can be postulated:

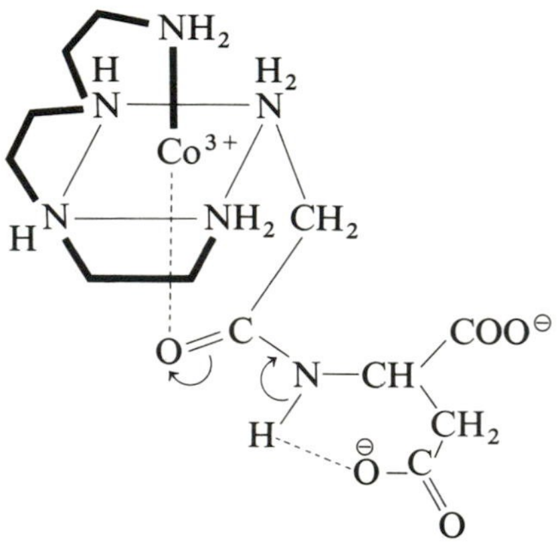

6-1

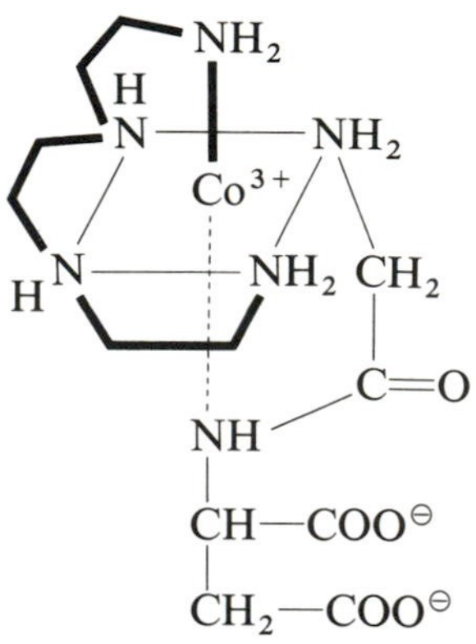

6-2

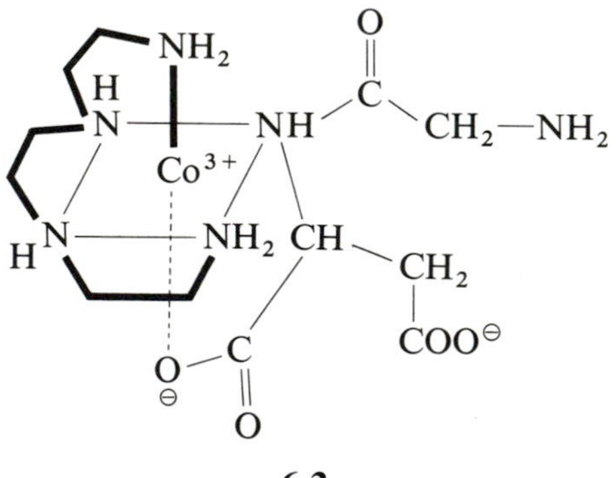

6-3

Structure **6-1** has been favored on the basis of NMR and proton–deuteron exchange studies and hydrolysis occurs in about 20% yield. Again the N-terminal first binds to the cobalt ion followed by the carbonyl of the amide bond. The γ-carboxylate group of aspartic acid might be implicated in the hydrolysis by contributing to the stability of the complex.

In general with a peptide not having polar side chains, two pathways are possible for the hydrolysis of the metal–substrate complex. An external water molecule activates the amide carbonyl or the adjacent coordinated OH group acts as a nucleophile on the amide function. This process corresponds to an intramolecular hydrolysis by a metal–ligand. In other words, process A involves a carbonyl coordination intermediate whereas in process B a carbonyl attack by the coordinated hydroxide takes place. These pathways can be followed with ^{18}O-enriched water.

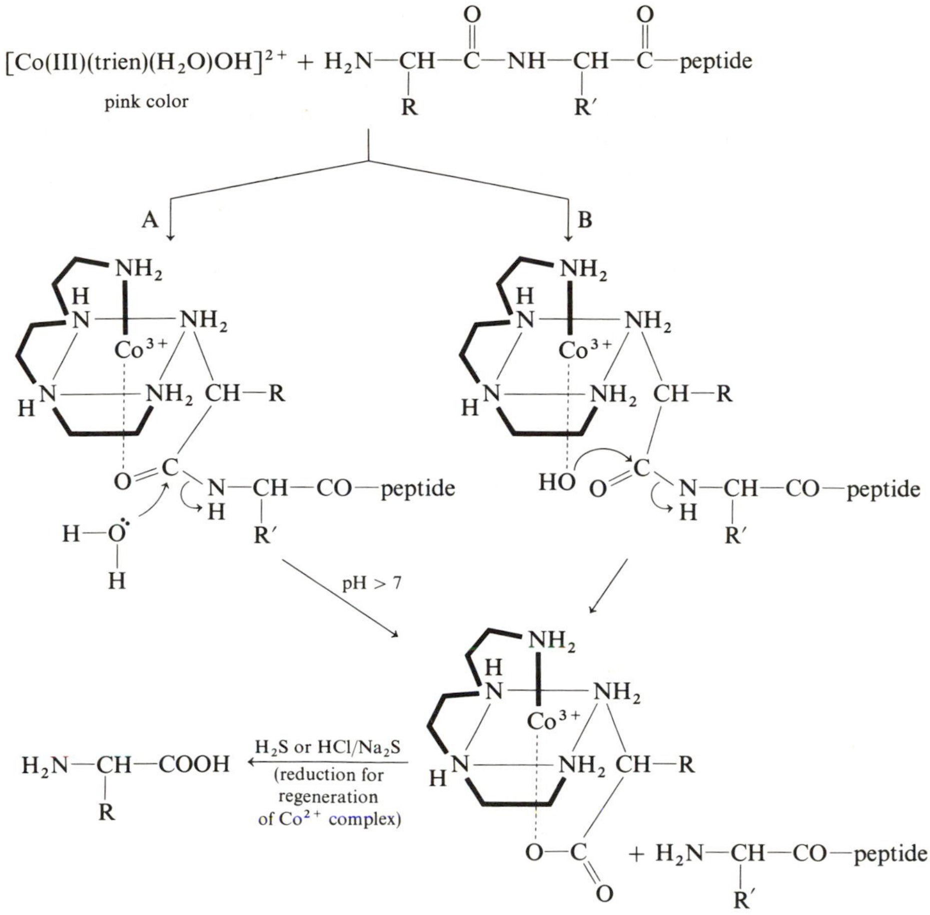

$$[Co(III)(trien)(H_2O)OH]^{2+} + H_2N-CH-\underset{\underset{R}{|}}{C}-NH-\underset{\underset{R'}{|}}{C}-peptide$$

pink color

A ↓   ↓ B

(reaction scheme showing Co(III) trien complexes with process A and process B, leading via pH > 7 to the intermediate complex and regeneration)

$$H_2N-\underset{\underset{R}{|}}{CH}-COOH \xleftarrow[\substack{\text{(reduction for} \\ \text{regeneration} \\ \text{of Co}^{2+} \text{ complex)}}]{H_2S \text{ or } HCl/Na_2S}$$

$$+ H_2N-\underset{\underset{R'}{|}}{CH}-CO-peptide$$

Therefore, water and OH are the two ligands replaced by the amino acid terminal group in process A. In process B, the coordinated OH or H_2O is a potent nucleophile and has the proper geometry to attack the amide bond via a tetrahedral intermediate followed by the liberation of the peptide minus the last N-terminal amino acid residue. It is a particularly interesting case because of its selectivity for the hydrolysis of only the N-terminal amino acid of the peptide chain. The resulting Co(III)–amino acid complex has to be decomposed by a reducing agent. It should be realized that the metal ion is not a true catalyst in this hydrolysis since it must be regenerated. Its role is more of a *promoter* of the hydrolytic reaction.

With such Co(III) complexes, rate enhancement of 10^8 have been observed relative to alkaline hydrolysis of the amide function. Such acceleration of hydrolysis reactions is comparable to rates obtained with carboxypeptidase A for its substrates. Notice that in process A, which is believed to be favored over process B with Co(III) ion, the carbonyl group becomes polarized and susceptible to external water attack much more so than in the noncomplex form. Thus the metal ion plays again the role of a *super acid*. In other words, the direct polarization of a carbonyl function by a metal ion generates a more electrophilic center at the carbon atom. Of course, different metal ions have different abilities in this respect, depending largely on their overall charge, size, coordination number, and ease of displacement of (usually) a coordinated water molecule.

These findings have been applied to a stepwise sequential hydrolysis and analysis of peptides from the N-terminal end. This is comparable to the action of an aminopeptidase. With most amino acids the yield of hydrolysis varies between 30 to 50%. To improve the yield, the approach has been altered to include a solid phase support. In alkaline buffer at 60°C only the N-terminal residue remains bound to the resin support and unreactive materials can be washed and recycled (230).

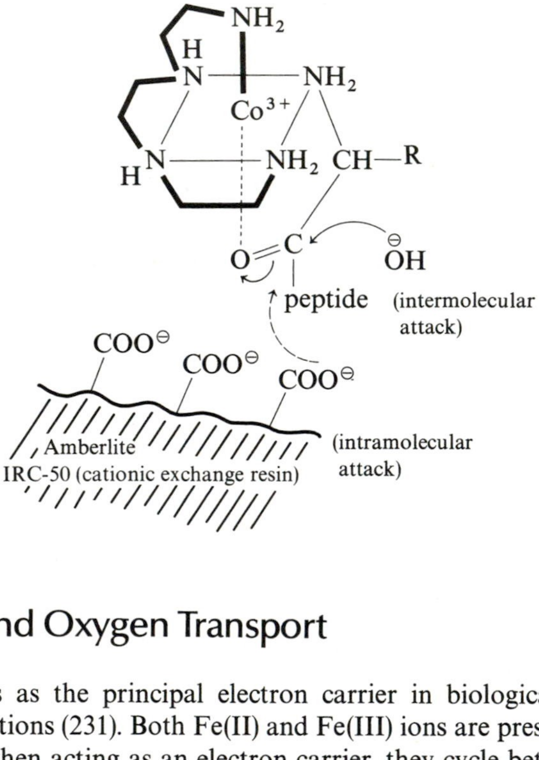

6.4 Iron and Oxygen Transport

Iron functions as the principal electron carrier in biological oxidation–reduction reactions (231). Both Fe(II) and Fe(III) ions are present in human systems and when acting as an electron carrier, they cycle between the two oxidation states. This is illustrated by the cytochromes.* Iron ions also serve to transport and store molecular oxygen, a function that is essential to the life of all vertebrates. In this system only the Fe(II) form exists [Fe(III)-hemoglobin does not carry oxygen]. In order to satisfy the metabolic requirement for oxygen, most animals have developed a circulating body fluid that transports oxygen from an external source to the mitochondria of tissues. Here it is needed in the respiratory chain to permit oxidative phos-

---

* A cytochrome is an electron-transporting protein that contains a heme prosthetic group covalently attached to the peptide chain.

phorylation and production of ATP. However, the solubility of oxygen in water is too low to support respiration in active animals. To overcome this problem, nearly all bloods contain supplies of oxygen-carrying protein molecules. The muscles of many animals also contain proteins that reversibly bind oxygen. These molecules facilitate the uptake of oxygen through the muscle and may also provide a reservoir for the storage of oxygen.

The molecular architecture of oxygen-carrying proteins is fascinating and in the course of biological evolution, nature has developed several alternative molecular devices to serve as oxygen carriers. Each is strikingly colored and these proteins fall into three major families: *hemoglobin*, the familiar red substance in the blood of humans and many other animals; *hemocyanin*, the blue pigment in the blood of many molluscs and arthropods; and *hemery-thrin** the burgundy colored protein in the body fluids of a few minor invertebrates. All are metalloproteins. Hemoglobins contain iron by virtue of the heme group; hemocyanins possess dimeric copper clusters (see Section 6.5) and hemerythrins have dimeric iron centers. Hemoglobin is the red protein of red blood cells which carries oxygen from the lungs to the tissues and corresponds to about three-fourths of all the iron in the human body (232).

The hemoglobin molecule is a tetramer composed of two similar globins (polypeptide chain) of unequal length. In the center of the protein lies a prosthetic group made of a porphyrin nucleus. This acts as a tetradentate ligand for the iron ion. The porphyrin–iron complex is called a *heme* group. The association of a protein molecule with a heme is referred to as a *hemo-protein*.

Hemoproteins, ubiquitous among plants and animals, perform at least four important functions related to oxygen and the production of energy: (1) oxygen transport to tissue; (2) catalytic oxidation of organic compounds; (3) decomposition of hydrogen peroxide; and (4) electron transfer.

Hemoglobin reversibly binds oxygen so that under conditions of high oxygen pressure, such as prevails in the lungs, oxygen will associate preferentially with the protein. Conversely, in the tissues where oxygen is required, the oxygen–hemoglobin complex will dissociate. The oxygen is then transferred to another oxygen-binding hemoprotein, myoglobin. This is a monomeric polypeptide chain. Hence, myoglobin facilitates transfer of oxygen from blood to muscle cells, which then stores the oxygen as an energy source (233).

It should be realized that the ability to bind oxygen reversibly is a unique property found in nature only in iron–porphyrin proteins, iron proteins, and copper proteins. However, other small molecules such as CO, CO_2, or CN^- can also interact with these metalloproteins. In fact, CO binds to hemoglobin even more avidly than oxygen, producing a cellular oxygen deficiency, which is sometimes identified as "carbon monoxide poisoning."

---

* The common prefix *hem*-, in these names does not derive from the presence of heme (iron protoporphyrin IX), which occurs only in hemoglobin, but from the Greek word for blood.

In both proteins the heme is tightly bound to the protein part (globin) through about 80 hydrophobic interactions and a single coordinate bond between an imidazole ring of the so-called "proximal histidine" and the iron atom. In spite of numerous differences in their amino acid sequences, all myoglobin and hemoglobin-heme units have very similar tertiary structure consisting of eight helical regions. The heme is wedged in a crevice between two helical regions; oxygen binds on one side of the porphyrin while the histidine residue is coordinating the other side. It is believed that the unique oxygen binding properties of hemoglobin depends on structural features of the entire molecule of hemoglobin and myoglobin.

In deoxyhemoglobin the iron atom is in a high-spin Fe(II) state and lies slightly out of the porphyrin plane. On binding O_2, however, the iron becomes low-spin Fe(II) and moves into the plane. This apparently results in a motion of the proximal imidazole ring by 0.06 nm which causes conformational changes in the protein backbone, producing a higher O_2 affinity quaternary form of the protein molecule. This structural change is the basis of M. F. Perutz's molecular model to explain *cooperativity* in hemoglobin.

Indeed, the most significant property of hemoglobin is cooperative oxygen binding. That is, the oxygen affinity of the tetramer rises with increasing saturation. The concept of *allostery*, developed by J. Monod and co-workers at Pasteur Institute, took origin with this protein. Cooperativity is required for the transfer of oxygen from the carrier hemoglobin to the receptor myoglobin as well as for responses to other physiological requirements.

Finally it should be recalled that the iron bound to porphyrin (the heme) is in the ferrous state. Upon oxygen binding hemoglobin exhibits a reversible behavior with 1:1 stoichiometry *per* iron atom without oxidation of Fe(II) to Fe(III). Scientists have been searching for ways to study this reversible binding of molecular oxygen by Fe(II) in heme. In fact, heme becomes capable of reversibly binding oxygen when it is incorporated into a large protein structure, However, when the heme is removed from any of the proteins and placed in solution at room temperature, molecular oxygen irreversibly oxidizes the iron to the ferric state.

Model studies have been undertaken in an attempt to understand the manner by which hemoglobin and myoglobin regulate oxygen affinity (237). The goal in making heme-model compounds is to answer the following questions: (1) How is oxidation to Fe(III) retarded during physical measurements of hemoglobin such as X-ray crystallography? (2) What is the detailed molecular geometry of the heme, heme–CO, and heme–O_2 complexes? (3) How are oxygen affinity and oxidation rate controlled in the hemoprotein?

In 1970, it was first shown that an organic iron complex absorbs oxygen from the air, taking a single oxygen atom into its molecular structure (234). The oxygen pickup occurs in a reaction of triphenylphosphine with $(Bu_4N)_2$ $\{Fe_2[S_2C_2(CF_3)_2]_4\}$. X-Ray works showed that molecular oxygen slips in between iron and phosphorus to give an oxide adduct, $(Bu_4N)\{(C_6H_5)_3P OFe[S_2C_2(CF_3)_2]_2\}$. In preparing this complex, E. F. Epstein and I. Bernal,

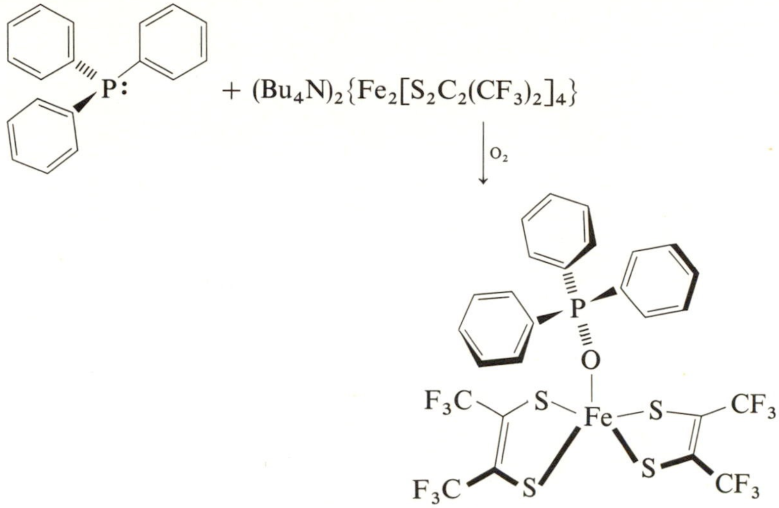

$+ (Bu_4N)_2\{Fe_2[S_2C_2(CF_3)_2]_4\}$

at Brookhaven National Laboratory and A. L. Balch at University of California, Los Angeles, found the compound much more stable than similar cobalt complexes previously made.

Of course, the current direction is to synthesize complexes which will more closely mimic hemoglobin and myoglobin and to develop oxygen-activating species in general. In 1973 and 1974, several model systems were devised. We will thus be concerned here with the interaction of molecular oxygen (dioxygen) and carbon monoxide with metalloporphyrin complexes as models of biologically important hemoproteins.

J. Baldwin and co-workers (235) at M.I.T. (now at Oxford) first synthesized a porphyrin-like structure surrounding an Fe(II) atom that reversibly bound oxygen in solution. However, this occurred only at $-85°C$. Later he developed a simple method to prepare a cagelike or a "capped" porphyrin molecule. This was done in the hope that upon O_2 binding the irreversible Fe(II) → Fe(III) process will not occur and O_2 will bind reversibly to the model molecule at room temperature (236).

For this they have examined the direct condensation of suitable tetra-aldehydes with pyrrole, as a route to capped porphyrins. This is schematically represented as follows:

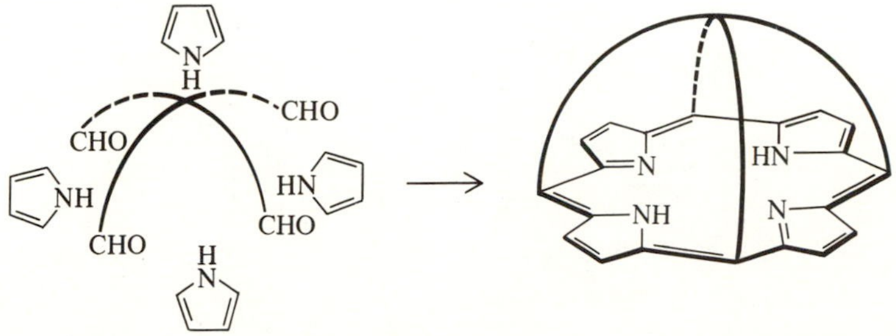

The suitable tetraaldehyde was prepared from salicylaldehyde, and condensed with pyrrole (Fig. 6.1).

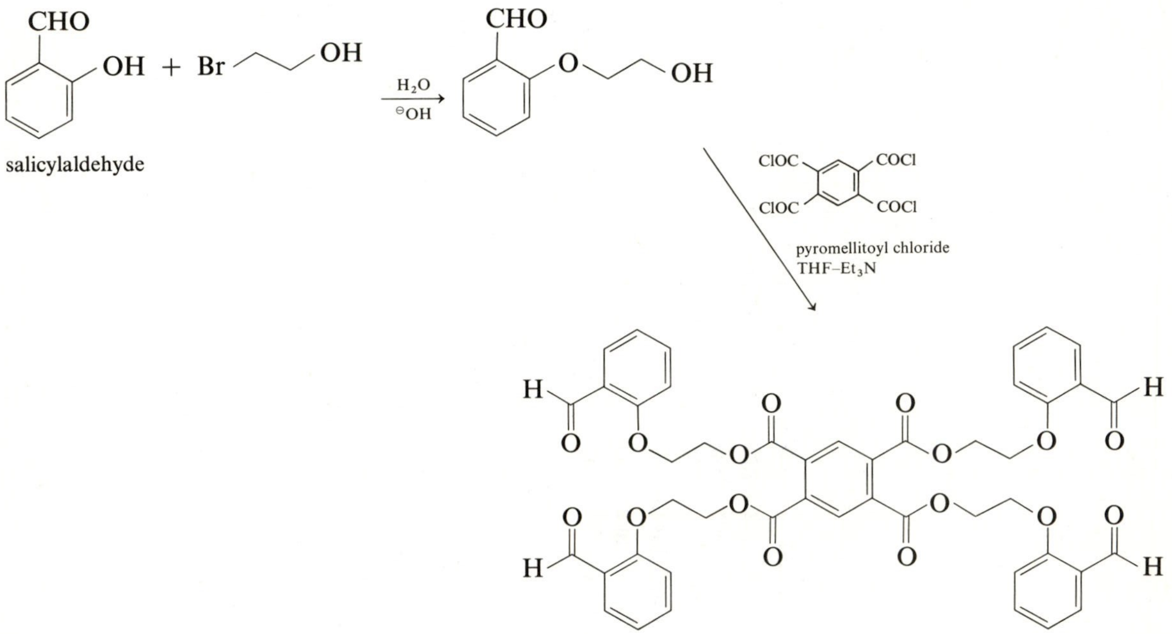

CHO

OH + Br ⌒ OH

salicylaldehyde

$\xrightarrow[\ominus OH]{H_2O}$

CHO

O ⌒ OH

ClOC — COCl
ClOC — COCl

pyromellitoyl chloride
THF–Et$_3$N

→

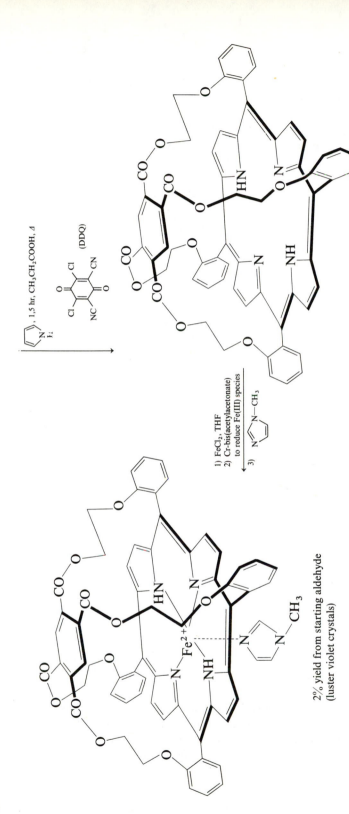

Figure 6.1. Baldwin's synthesis of "capped" porphyrin (236). Reprinted with permission. Copyright © 1975 by the American Chemical Society.

The corresponding "capped" porphyrin–Fe(III) complex was reduced with chromous bis(acetylacetonate) in benzene to a crystalline ferrous porphyrin which reversibly oxygenates in solution at room temperature. The kinetic stability of solutions containing this oxygen complex depends on many factors such as the temperature, the nature and coordination of the porphyrin and of the axial base (1-methylimidazole is much superior to pyridine), the partial pressure of oxygen, and the solvent polarity.

In a different approach, T. Traylor and his group (237–239) at the University of California, San Diego, introduced an Fe(II) atom within a modified porphyrin ring having the desired imidazole side chain properly designed to mimic the neighboring histidine groups in hemoproteins. This man-made active site of myoglobin now has the proximal histidine residue covalently attached to the porphyrin nucleus. It was obtained by the following sequence of chemical transformations, from natural chlorophyll a (Fig. 6.2). The strategically placed imidazole group creates both physical and chemical conditions that induce the synthetic molecule (protoheme) to bind oxygen reversibly (but only at $-45°C$) in approximately the same way as the natural molecule does. In CH_2Cl_2 at $-45°C$, the model shows almost the same binding constant with oxygen as does sperm whale myoglobin at room temperature. Interestingly, this result shows that the protein might not be as important to the active site as was formerly believed although current theories explain hemoglobin's efficient binding of oxygen through some type of beneficial interaction among the four subunits. Notice that this ion selectivity by the porphyrin ring is analogous with host–guest designed chemistry, discussed in Chapter 5.

If the imidazole group is replaced by a pyridine molecule, the compound does not bind oxygen but is oxidized to the ferric state. So the basic nature of the imidazole ring and its geometry relationship to the Fe(II) ion in the heme results in a unique binding power for oxygen. The model binds CO much more strongly than it binds O_2.

The binding efficiency for oxygen is also present in the frozen solid state, and when the model compound is dissolved in a film of polystyrene. Practically, it is now possible to make materials similar to a synthetic active site that can be built into plastic film to facilitate oxygen transfer in artificial lung machines. Of course, such technology can become really practical only with the development of new models that can bind oxygen more efficiently at room temperature without being oxidized. Before 1975, reversible oxygenation of model complexes containing heme has been observed only at low temperatures or in nonphysiological media.

The close proximity of the porphyrin system to certain residues of the peptide chain in hemoglobin can create steric hindrance toward CO or O_2 ligation to the Fe(II) ion. In 1979, Traylor's group developed two variants of the previous model to evaluate this steric effect (240). Figure 6.3 outlines the synthesis of these molecules that are referred to as cyclophane porphyrins.

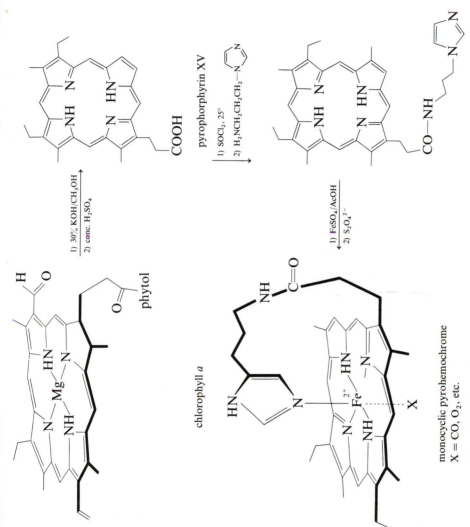

Figure 6.2. Traylor's approach to man-made active site of myoglobin (237).

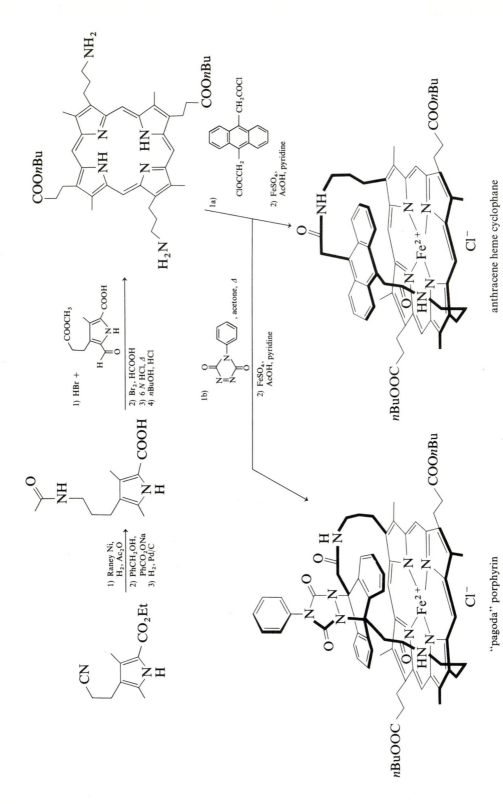

Figure 6.3. Synthesis of anthracene-heme [6,6]cyclophane and of "pagoda" porphyrin (240).

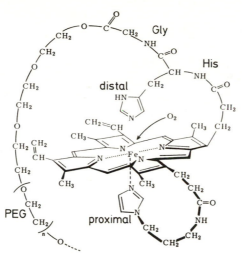

Figure 6.4. Bayer's hemopolymer (241). Reproduced with permission. Copyright © 1977 by Verlag Chemie GMBH.

With these two new and more sophisticated heme models a large distal side steric effect is introduced because the distance between the anthracene ring and the porphyrin ring is smaller than 0.5 nm. By converting the anthracene system to the "pagoda" system via a Diels–Alder reaction, further restriction and a tighter pocket is achieved. Addition of 1-methylimidazole in methylene chloride to the models give molecules showing a five-coordinated visible spectra as previously observed with "capped" heme. The corresponding CO complex has also been prepared in dry benzene and corresponds to pure monocarbonyl heme. Therefore, the anthracene ring above the porphyrin ring has greatly reduced the affinity of the second CO molecule. The binding is reduced by 10^3-fold as compared with chelated hemes without the steric effect. The presence of an additional steric factor in the "pagoda" molecule results in a further decrease by 10-fold the rate of CO association. These dynamic models thus illustrate that the dramatic reduction in CO association rates can be attributed to the steric hindrance in the synthetic heme pocket (240).

In 1977, E. Bayer's group in Tübingen synthesized for the first time a heme group embedded in a polymer network (241). The structure is shown in Fig. 6.4 and was obtained from poly[ethylene glycol bis(glycine ester)] and polyurethanes from polyethylene glycols and diisocyanates as the basic polymers, using the procedure of *liquid-phase peptide synthesis*.* At the end

---

* Liquid-phase peptide synthesis resembles the solid-phase method (see Chapter 2) in that the synthesis takes place on a polymer backbone (i.e., polyethylene glycol). However, the polymer is soluble in the solvent system used, so that a homogenous reaction results. This methodology has found far less use than the analogous solid-phase synthesis.

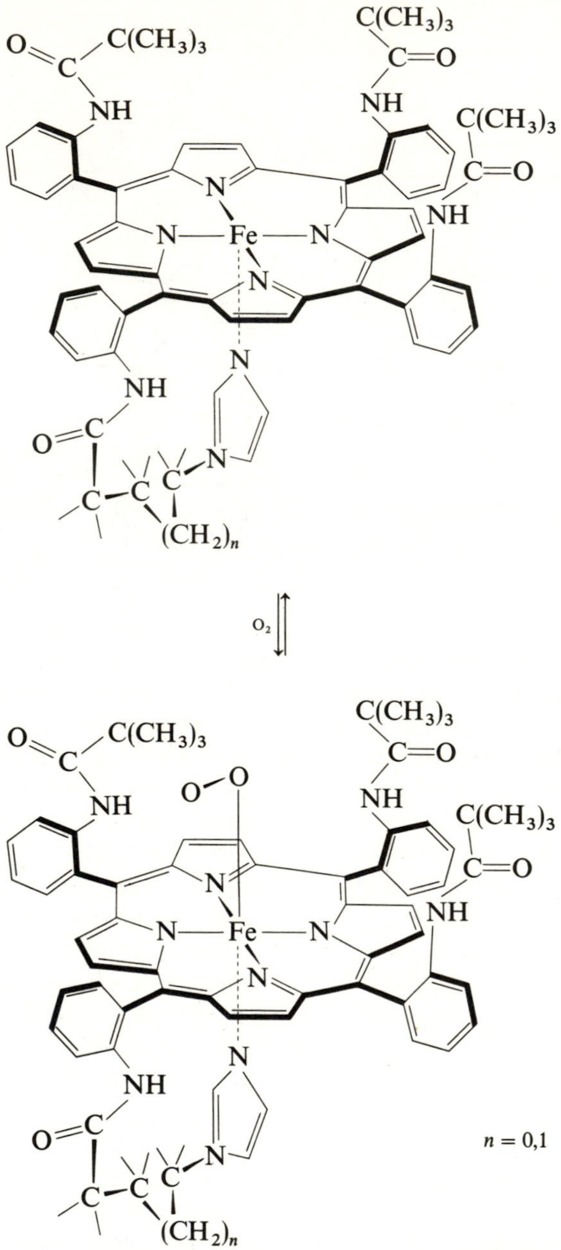

Figure 6.5. Collman's "picket fence" ferrous porphyrin model (243). Reprinted with permission. Copyright © 1977 by the American Chemical Society.

of the synthesis the iron is present in the Fe(III) state and must be reduced with sodium dithionite prior to oxygenation.

The polymer shares the following properties with hemoglobin and myoglobin: (a) good solubility in water in order to achieve high concentration of O_2; (b) hindrance of irreversible oxidation of the oxygen complex by the functionalized polymer; and (c) imitation of a distal imidazole.

Furthermore, a second imidazole ligand corresponding to the proximal histidine of natural oxygen carriers was introduced by modification of the propionic acid side chain of the porphyrin ring. Interestingly, the structural units of the active site of myoglobin or hemoglobin which are essential for binding oxygen are all present, albeit in a different arrangement (241). In particular, the oxygen coordination site is shielded both by histidine and by the synthetic polymer. The oxygen absorption curve for the complex at 25°C is in good agreement with the values for myoglobin (0.2 to 1.2 torr). Furthermore, the sigmoid shape of the curve is indicative of a cooperative effect in oxygen binding, as in the case of hemoglobin.

Consequently, the synthetic hemopolymer has properties closely resembling those of natural oxygen carriers, although the basic polymer is completely different. This demonstrates that specificity and reactivity are by no means privileges of naturally occurring polymers. Such findings could have been obtained only from model studies.

Another leader in this field of bioorganic models of oxygen-binding site in hemoproteins is J. P. Collman (242, 243) of Stanford University. He used an approach that modified the molecular geometry of the porphyrin such that the oxidation process of iron is sterically hindered and diminished. The effort consisted in constructing a porphyrin ring with four bulky o-pivalamidophenyl groups projecting from one side of the ring. The other side is left unencumbered with the hope that it could be protected by a bulky axial imidazole group (Fig. 6.5). This new compound, although less of a model for myoglobin than protoheme itself, prevents heme–heme approach and Fe—OO—Fe formation (237). Collman called his model "picket-fence" porphyrin. It does bind molecular oxygen reversibly at room temperature and the dioxygen complex is crystalline. This model system shows significant cooperativity in its binding of O_2 but only in the solid state so far. At low O_2 pressures a low O_2 affinity form exists, and at high O_2 pressures a higher O_2 affinity form develops. This ability to show high affinity at high partial pressures of O_2 (as in the lungs) and low affinity at low O_2 pressures (as in the muscles and tissues) is critical to efficient O_2 transport. In this regard, the model mimics hemoglobin quantitatively (243).

By making precise measurements of the physical properties of the oxygen–iron bond in their model, Collman's group found that the oxygen molecule binds in an end-on fashion (Pauling's model) to the iron of the protoheme, rather then in the sideways manner proposed by some theories (Griffith's model). In fact, the way in which molecular oxygen binds to certain iron-containing biological molecules has been the subject of great theoretical

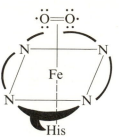

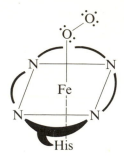

Griffith's model   Pauling's model

debate for the past 30 years. Finally, X-ray crystallographic studies of Coll-man's model prove that oxygen binds in the end-on fashion, a suggestion first proposed by L. Pauling in 1948!

In 1978, a new approach by the team of D. Dolphin (244) at the University of British Columbia has been taken in devising molecular models that mimic the reversible binding of molecular oxygen by heme. They have successfully incorporated ruthenium(II) into *meso*-tetraphenylporphyrin and octaethyl-porphyrin systems. They found that these porphyrin systems complexed with two molecules of the solvent but only *one* molecule of oxygen per ruthenium atom is absorbed reversibly at 25°C and at 1 atm in DMF. The group also has made monocarbonyl and dinitrogen (one N_2 molecule) species reversibly in polar aprotic solvents. Therefore, these new systems parallel that of corresponding Fe(II) systems, but at more convenient temperatures.

Artificial hemoglobins containing zinc, manganese, copper, and nickel have also been reconstituted, and their properties compared with those of the native iron proteins. These artificial systems, however, are incapable of reversible oxygenation. In contrast, cobalt substituted hemoglobin and myoglobin are functional, although their oxygen affinities are 10–100 times less than those of the native oxygen carriers. In fact, it has been known since the 1930s that several synthesized Co(II) complexes bind molecular oxygen reversibly. Cobalt is adjacent to iron in the periodic table, and it was hoped that the elucidation of the behavior of cobalt complexes might shed some light on that of hemoglobin and the other naturally occurring oxygen carriers. In other words, the Co(II) complexes were recognized as possible models for hemoglobin and other natural O_2 carriers. A second reason for selecting cobalt is that more extensive thermodynamic data are available for related cobalt systems.

A recent synthetic model of cobalt-substituted hemoglobin is shown in Scheme 6.1 (245). The long side chain facilitates the coordination of the pyridine ring to the central cobalt atom. The Co(II) complex of this so-called "looping-over porphyrin" reacts reversibly with molecular oxygen at low temperatures (−30°C to −60°C) but the side chain has a negligible en-

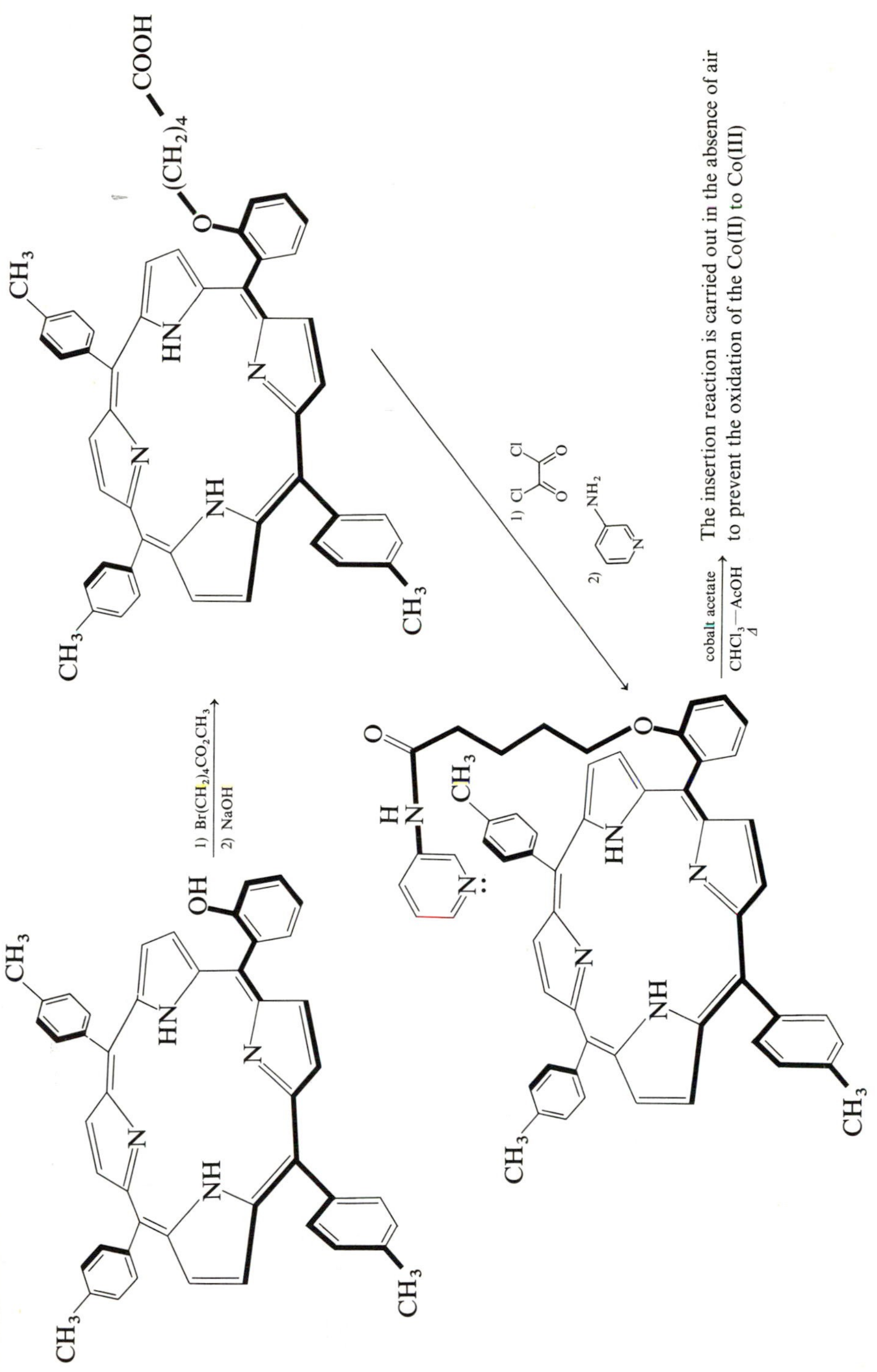

COOH

(CH$_2$)$_4$

O

CH$_3$

HN

N

NH

N

CH$_3$

CH$_3$

CH$_3$

1) Cl—C(=O)—C(=O)—Cl

2) NH$_2$ (pyridine)

The insertion reaction is carried out in the absence of air to prevent the oxidation of the Co(II) to Co(III)

$\xrightarrow[\text{CHCl}_3\text{—AcOH},\ \Delta]{\text{cobalt acetate}}$

1) Br(CH$_2$)$_4$CO$_2$CH$_3$

2) NaOH

OH

CH$_3$

HN

N

NH

N

CH$_3$

CH$_3$

CH$_3$

O

H
N

CH$_3$

O

N:

HN

N

NH

N

CH$_3$

CH$_3$

CH$_3$

Scheme 6.1

hancement effect on the oxygen affinity for such models, as compared to iron–porphyrin systems.

Collman's group has also examined oxygen binding to "picket fence" porphyrin–cobalt complexes of *meso*-tetra-($\alpha,\alpha,\alpha,\alpha$-*o*-pivalamidophenyl) porphyrinato-cobalt(II)-1-methylimidazole and 1,2-dimethylimidazole (246).

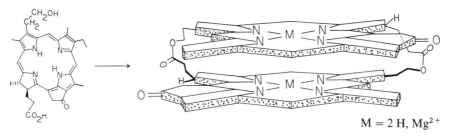

B, a base such as *N*-methyl-
    imidazole or
    1,2-dimethylimidazole

B′, the position where
    O_2 can bind

Interestingly, these complexes bind oxygen with the same affinity as cobalt substituted myoglobin and hemoglobin in the solid state and in toluene. More elaborated models await the synthesis of "picket fence" porphyrins with different "pickets."

Although not related to iron and oxygen transport, it should be mentioned that synthetic biomimetic models of special pair bacteriochlorophyll *a* have also been prepared (247). The reason is that, in the molecular organization of chlorophyll in the photoreaction centers of both green plants and photosynthetic bacteria, it is believed that special pairs of chlorophyll molecules are oxidized in the primary light conversion event in photosynthesis. Dimeric chlorophyll derivatives such as the one in Fig. 6.6 in which the porphyrin

$M = 2\,H,\ Mg^{2+}$

Figure 6.6. Biomimetic model of pair bacteriochlorophyll *a* (247). Reprinted with permission. Copyright © 1978 by the American Chemical Society.

macrocycles are bound by a simple covalent link, exhibit several photochemical properties which mimic *in vivo* special pair chlorophyll.

In addition to the iron–porphyrin compounds, there are a large number of nonheme proteins implicated in many oxidation–reduction processes. They contain relatively large amounts of sulfur and iron and have thus been designated *iron–sulfur proteins* (248). They are widely distributed in nature, occurring in all living organisms, and their physiological function is that of electron transfer rather than catalysis of chemical transformations. Among these proteins are the *rubredoxins, adrenodoxins,* and *ferredoxins.* They are the most ubiquitous electron carriers in biology and have been directly associated with, or implicated in, a larger area of metabolic reactions, especially in photosynthetic processes and CO_2 and N_2 fixation in bacteria and plants. The first ferredoxin protein was isolated and purified in 1962 from *Clostridium pasteurianum* and spinach. This nonphotosynthetic anaerobic bacterium is one of the oldest bacteria, present on Earth more than 3.1 billion years ago.

The presence of a Fe—S spectroscopic chromophore and their relatively low molecular weight (about 6,000 to 12,000) make these proteins fascinating molecules for investigation. Little can be said at this time about their mechanism of action but they are known to function as strong reducing agents. They are also characterized by evolution of H_2S in acidic solution (acid-labile sulfur). It is believed that part of the primitive and reductive atmosphere prevailing at the origin of organic molecules on Earth could have come from such transformations. Hence ferredoxins might become an essential link in the comprehension of the origin of life.

These iron–sulfur redox proteins contain a cagelike cluster of iron and sulfur linked to cysteine residues of the peptide chain.

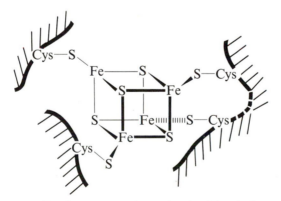

cagelike cluster proposed as active site of ferredoxin

It is these metal–sulfur bridges which are responsible for the biological properties of these sites. The sulfur ligands make the Fe(II) state very unstable and account for their powerful reducing properties in various biological processes. Such cagelike clusters behave as one-electron redox catalysts where the Fe(II) $\rightleftharpoons$ Fe(III) oxidoreduction reactions occur within the cluster.

Only few synthetic analogues of the Fe—S clusters are known today and this particular chemistry is still in the development stage. In this line of thought, mention must be made of R. H. Holm (249) at Stanford University who has made a series of synthetic analogues in an approach toward elucidation of the active sites of certain iron–sulfur proteins and enzymes (355).

6.5 Copper Ion

Copper ions in the (I) and (II) oxidation state are biologically important. These ions appear to stabilize walls of certain blood vessels including the aorta and the sheath around the spinal cord. They are involved in the body's production of the color pigments of the skin, hair, and eyes, and in the *in vivo* synthesis of hemoglobin (232, 250).

In certain copper-containing proteins the copper appears to serve principally in electron transport with no evidence of Cu–O_2 interaction, such as in cytochrome oxidase. Of importance, however, is that many copper proteins and enzymes participate in reactions in which the oxygen molecule is directly or indirectly involved. An example is hemocyanin, the oxygen carrier in the blood of certain sea animals such as snails, octopus, and crustacea. Oxygenated hemocyanin is blue and the cephalopods (crabs and lobsters) are literally the blue bloods of the animal kingdom. Hemocyanins are giant molecules of $MW > 10^6$ that occur free in solution.

Both hemocyanin and tyrosinase, the enzyme that activates molecular oxygen for the oxidation of tyrosine, rely on direct covalent interaction between $Cu(I)$ and O_2, forming an observable dioxygen adduct. The mushroom *Gyroporus cyanescens* (Bluing Boletus), which turns blue instantly when bruised, also contains a copper protein or a "blue protein," as it is called.

Basically, three different type of copper centers are known (251). "Blue" or type I copper occurs in the blue electron-carrying proteins such as stellacyanin, plastocyanin, and azurin. The spectroscopic display of blue copper is very useful for characterizing the different ligand environments and coordination numbers around the copper atom. There are also a "nonblue" or type II copper and an electron paramagnetic resonance (EPR) "nondetectable" or type III copper center, apparently containing a pair of contiguous Cu atoms. Type II is usually found in combination with types I and III but it occurs alone in galactose oxidase. Type I is also found accompanied with type II and III, as in the blue oxidases, laccase, ceruloplasmin, and ascorbic oxidase. For instance, laccase is known to be a four-copper atom protein which binds covalently with molecular oxygen.

Amino acid analyses of a variety of hemocyanins indicate that a large amount of histidine and methionine per copper pair is present as well as cysteine, although the number involved in disulfide bridges has not been determined. Intuitively, three types of donor atoms are likely to be involved in these protein complexes, namely, oxygen (carboxylate, phenolate, and water), nitrogen (amine, amide anion, and imidazole), and sulfur (thioether and thiolate). Furthermore, copper (II) can adopt square-planar, square-pyramidal, trigonal-bipyramidal, octahedral, and tetrahedral geometries.

So far only a few biomodels of copper proteins have been made where the structure and ligand environments of the copper sites are based on the analysis of the electronic spectra.

B. Bosnich's group of the University of Toronto designed a variety of ligands based on the above considerations (251). Some are presented in Fig. 6.7.

This extended series of ligands and their Cu(II) complexes have been prepared as spectroscopic models (observable in visible and near-IR regions) for determining the geometries and ligand coordination of copper proteins. From their results it was possible to propose the following structures for the copper coordination and geometry in (blue) type I copper proteins and for the copper site in oxyhemocyanin (251).

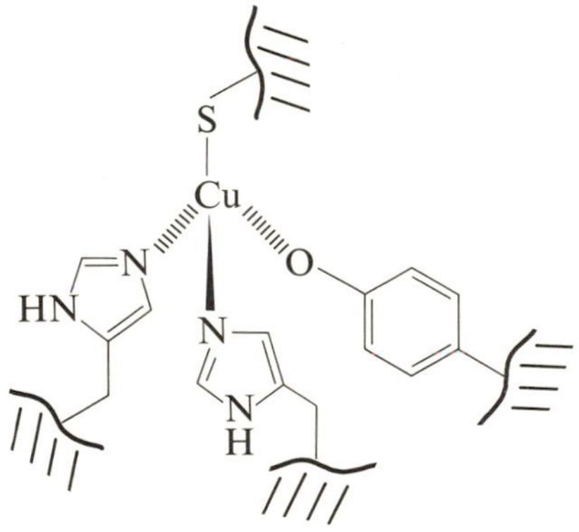

proposed structure and ligand coordination in the blue type I Cu proteins; the copper is bound at the active site via imidazole, phenolate and thioether ligands

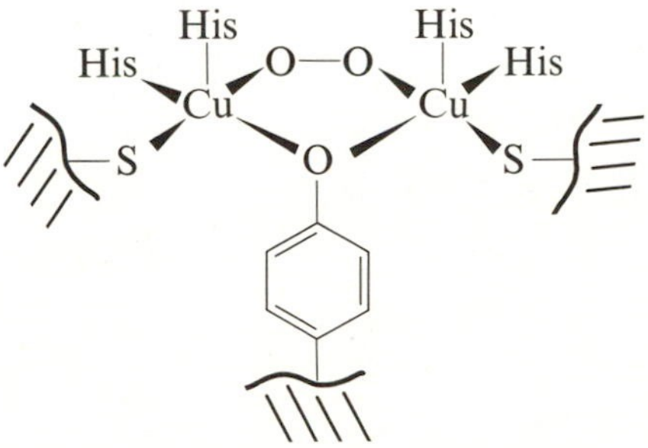

proposed environment of the Cu site in oxyhemocyanin; type III copper may also have a similar molecular architecture

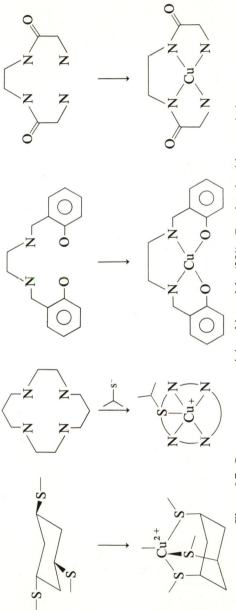

Figure 6.7. Some copper-containing biomodels (251). Reprinted with permission. Copyright © 1977 by the American Chemical Society.

With the same objective of elucidating the active site geometry and mechanism of copper proteins, the following model has also been made by other workers (252).

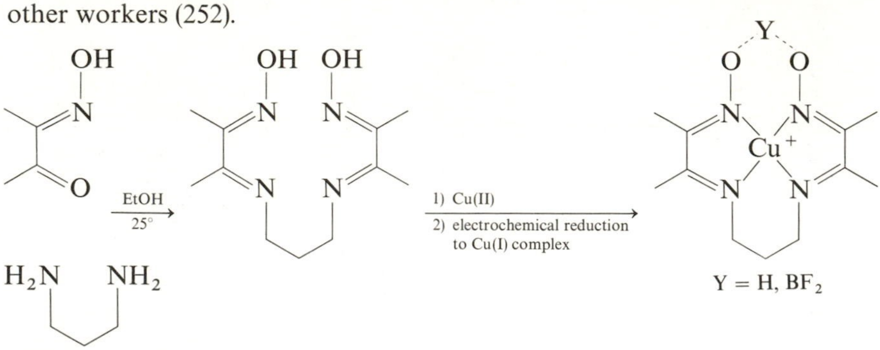

To inhibit binding of a second copper atom and to prevent possible hydrogen atom transfer reactions, the bridging oxime hydrogen was replaced with BF_2 via treatment with BF_3–etherate in dioxane. The Cu(I) complex reacts with monodentrate (i.e., CO, 1-methylimidazole, acetonitrile) yielding five-coordinate adducts. Of course such a model is not an accurate mimic of the active sites of any copper protein but suggests that consideration of Cu(I) active site structures must include the possibility of five-coordination.

Since many enzymes have two metals ions in their active site it would be interesting to mention the synthesis, in 1978, by Buckingham of a model (Fig. 6.8) for bismetallo active sites based on the "capped" or "picket-fence" porphyrin concept (253). The structure of the model, shown below, consists of a tetrapyridine coordination arrangement joined via nonflexible links to a tetraphenylporphyrin. This arrangement, which is capable of considerable

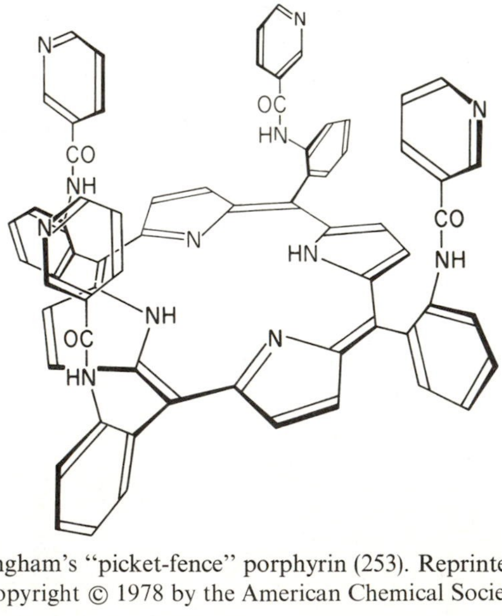

Figure 6.8. Buckingham's "picket-fence" porphyrin (253). Reprinted with permission. Copyright © 1978 by the American Chemical Society.

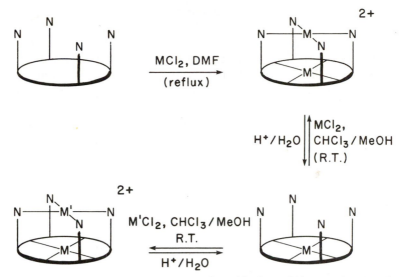

Figure 6.9. The bismetallo active site concept of Buckingham (253). Reprinted with permission. Copyright © 1978 by the American Chemical Society.

modification, allows the insertion of the same or different metal ions into the two different coordination sites.

Figure 6.9 illustrates this arrangement where M and M' can be iron, copper, or nickel ions. Interconversions have been repeated several times without displacement of the metal ion from the porphyrin nucleus.

Electron paramagnetic resonance (EPR) spectra show strong interactions between the two metal centers and for the bis-Cu system, the Cu $\cdots$ Cu separation was estimated to be 0.59 nm. This very imaginative synthetic model appears capable of considerable extension.

The group of R. Weiss (254) and the group of J.-M. Lehn (255) from Strasbourg in collaboration with S. J. Lippard from Columbia University have developed in 1979 new bimetallic copper models based on macrocyclic complexes. The first group synthesized the following elongated macrocyclic ligand that can bind two Cu(I) or Cu(II) ions.

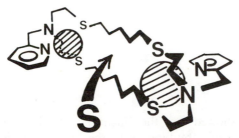

The spheres represent the space occupied by the copper ions and S the region where substrates can bind (254). Reprinted with permission. Copyright©1979 by the American Chemical Society.

Insertion of linear diatomic and triatomic substrate molecules such as CO, NO, O_2, or N_3^- are possible. The metal ions are located inside the macrocyclic ligand, each linked to a NS_2 ligand donor set as well as to the substrate (4 molecules for the case of N_3^-) trapped in the middle. Since some metalloproteins use binuclear metal centers to perform catalytic function, such a model represents a mimic for type III copper pairs in copper enzymes. In the binuclear Cu(II) complex, the Cu $\cdots$ Cu distance was estimated to be 0.52 nm. Interestingly, this complex exhibits antiferromagnetism and is diamagnetic at room temperature. The groups of Lehn and Lippard (255) tackled the problem in a different way although conceptually the approach is quite similar. They synthesized an imidazolate bridged dicopper (II) ion incorporated into a circular cryptate macrocycle (see Section 5.1.3).

The present approach is to prepare an imidazolate bridged bimetallic center that would prevent the two metal ions from migrating away from each other when the bridge is broken. The preparation involves the synthesis of a dinuclear cryptate macrocycle containing two Cu(II) ions and where a substrate such as sodium imidazolate is able to bind.

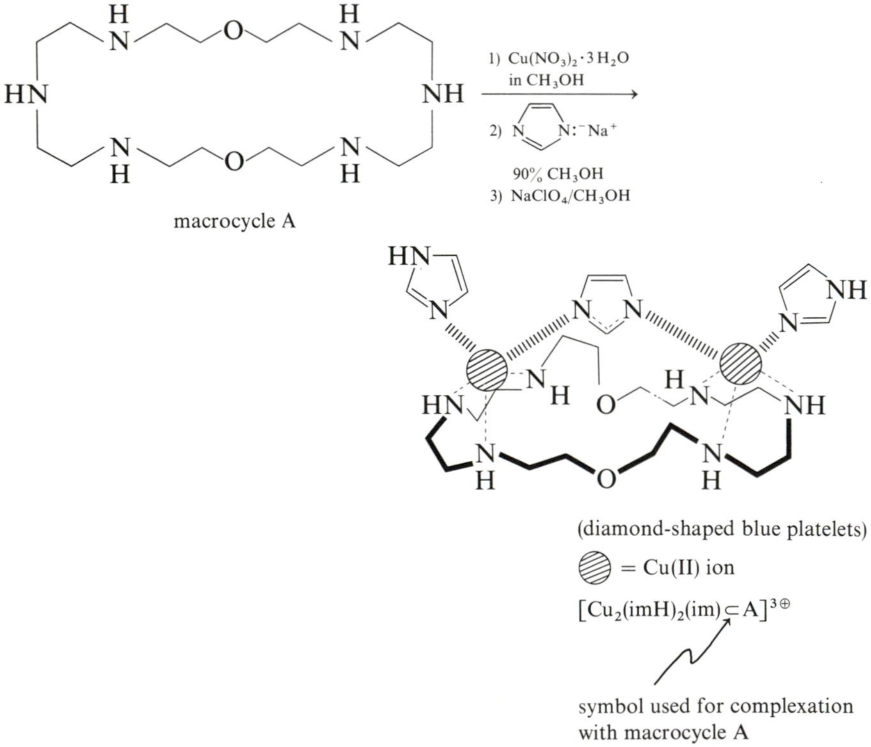

macrocycle A

1) $Cu(NO_3)_2 \cdot 3 H_2O$
   in CH_3OH
   $\longrightarrow$

2) imidazolate: N_2^- Na^+

90% CH_3OH

3) $NaClO_4/CH_3OH$

(diamond-shaped blue platelets)

= Cu(II) ion

$[Cu_2(imH)_2(im) \subset A]^{3\oplus}$

symbol used for complexation
with macrocycle A

X-Ray work showed that the $Cu_2(im)^{3+}$ ion is incorporated into the circular cryptate cavity by binding to the two diethylenetriamine units at each end of the macrocycle. Each Cu(II) ion is further coordinated to a neutral imidazole (imH) ligand, achieving overall pentacoordination. The synthesis of such a bimetallic complex offers the possibility of mimicking the 4-Cu(II) form of bovine erythrocyte superoxide dimutase and opens the way for the preparation of selective coreceptors as well as the development of bioinorganic models of metalloproteins in general.

6.6 Cobalt and Vitamin B$_{12}$ Action

Although cobalt ions are found in both the (II) and (III) oxidation states, the most important biological compound of cobalt is vitamin B$_{12}$ or cobalamin where the Co(III) form is present (256) (Fig. 6.10). Cobalamin or related substances are important biological compounds that are involved in a great variety of activities, particularly in bacteria. Vitamin B$_{12}$ is also necessary in the nutrition of humans and probably of most animal and plant species. It is of critical importance in the reactions by which residues from carbohydrates, fats and proteins, are used to produce energy in living cells. Pernicious anemia is a severe disease in elderly people. This disease is usually accompanied in mammals by the increased excretion of methylmalonic acid in the urine. Today it is effectively controlled by a 100 μg injection of vitamin B$_{12}$.

Vitamin B$_{12}$ (cyanocobalamin, R = CN) is one of the most complicated naturally occurring coordination compounds and certainly the most complex nonpolymeric compound found in nature. In the 1950s, H. A. Barker demonstrated the involvement of a derivative of vitamin B$_{12}$ in the following reaction:

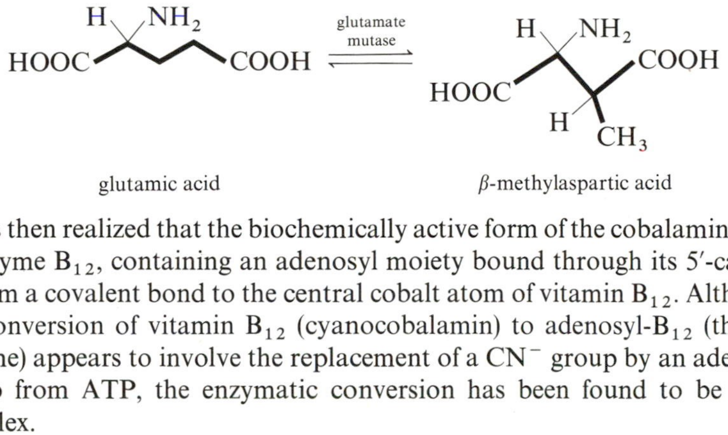

| glutamic acid | β-methylaspartic acid |

It was then realized that the biochemically active form of the cobalamin is the coenzyme B$_{12}$, containing an adenosyl moiety bound through its 5′-carbon to form a covalent bond to the central cobalt atom of vitamin B$_{12}$. Although the conversion of vitamin B$_{12}$ (cyanocobalamin) to adenosyl-B$_{12}$ (the coenzyme) appears to involve the replacement of a CN$^-$ group by an adenosyl group from ATP, the enzymatic conversion has been found to be fairly complex.

One of the remarkable properties of B$_{12}$ is its ability to form alkyl derivatives (256). Before the discovery of vitamin B$_{12}$ by Barker it was thought that a Co—C$_\sigma$ bond should be unstable if at all capable of existence. This is the first and only example of a water-stable organometallic compound occurring in nature. Its complete structure was revealed in 1956 by the crystallographic work of D. C. Hodgkin, aided by the earlier chemical studies of Lord A. R. Todd and A. W. Johnson. Its full synthesis was completed in the early

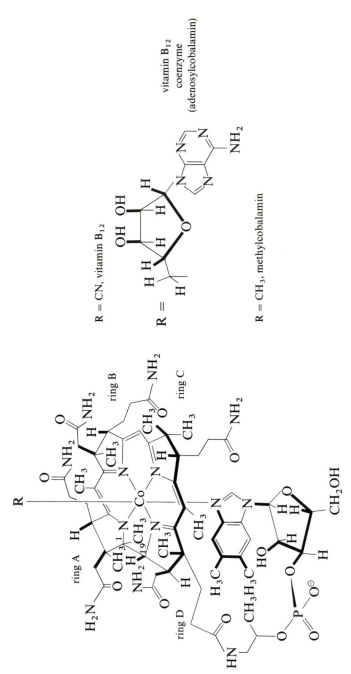

Figure 6.10. Structure of vitamin B_{12}.

1970s by a team effort of R. B. Woodward (Harvard) and A. Eschenmoser (ETH, Zurich).

The core of the molecule resembles an iron–porphyrin system but here the ring is called a *corrin* and has two of the four modified pyrrole rings directly bonded (C$_1$–C$_{19}$ link). All the side chains are made of acetamide and/or propionamide groups. One of them is linked to an isopropanol phosphate residue attached to a ribose and finally to a dimethylbenzimidazole ring to the Co(III) ion.*

Surprisingly, the cobalt ion can have the $+1$(B$_{12s}$), $+2$(B$_{12r}$), or $+3$(B$_{12a}$) oxidation state. Indeed, one of the remarkable properties of alkyl cobalamins is that three routes for cleavage of the Co—C bond are possible:

$$R—Co(III) \longrightarrow R^{\cdot} + \overset{\cdot}{Co}(II) \text{ (called B}_{12r}\text{ form)}$$

$$R—Co(III) \longrightarrow R^{\oplus} + \overset{\cdot\cdot}{Co}(I) \text{ (called B}_{12s}\text{ form)}$$

$$R—Co(III) \longrightarrow R^{\ominus} + Co(III) \text{ (called B}_{12a}\text{ form)}$$

The B$_{12s}$ form is a potent nucleophile (also a reducing agent) and coenzyme B$_{12}$ is believed to be formed via a nucleophilic attack on ATP. In the presence of diazomethane, the B$_{12s}$ form is converted to methylcobalamin (R = CH$_3$ in the general structure). Therefore, an alkyl radical, a carbocation, and a carbanion can all be produced in B$_{12}$-chemistry. In both bacteria and in liver, the 5′-deoxyadenosyl coenzyme is the most abundant form of vitamin B$_{12}$, but lesser amounts of methylcobalamin also exist.

The rearrangement described above by Barker using the coenzyme B$_{12}$-dependent enzyme glutamate mutase is a most remarkable reaction. Until very recently, no analogous chemical reaction was known. In fact, elucidation of the structure of the coenzyme form of vitamin B$_{12}$ did not clarify its mechanism. Beside this transformation, nine distinct enzymatic reactions requiring coenzyme B$_{12}$ as cofactor are known. Most of which are without precedent in terms of organic reactions. They are listed in Fig. 6.11. In choosing vitamin B$_{12}$ derivatives as coenzymes, enzymes appear to have reached a peak of chemical sophistication which would be difficult to mimic by the chemist.

Interestingly, all the reactions mentioned above can be generalized as the migration of a hydrogen atom from one carbon to an adjacent with concomitant migration of a group X from the adjacent carbon to the one where the hydrogen was originally bound.

$$-\overset{\displaystyle H^*}{\underset{\displaystyle X}{C_1}}-\overset{\displaystyle |}{\underset{\displaystyle H}{C_2}}-Y \;\rightleftharpoons\; -\overset{\displaystyle |}{\underset{\displaystyle H^*}{C_1}}-\overset{\displaystyle H}{\underset{\displaystyle X}{C_2}}-Y$$

---

* Naturally occurring vitamin B$_{12}$ analogues have been found where the benzimidazole ring is replaced with purines (adenine, guanine, etc.).

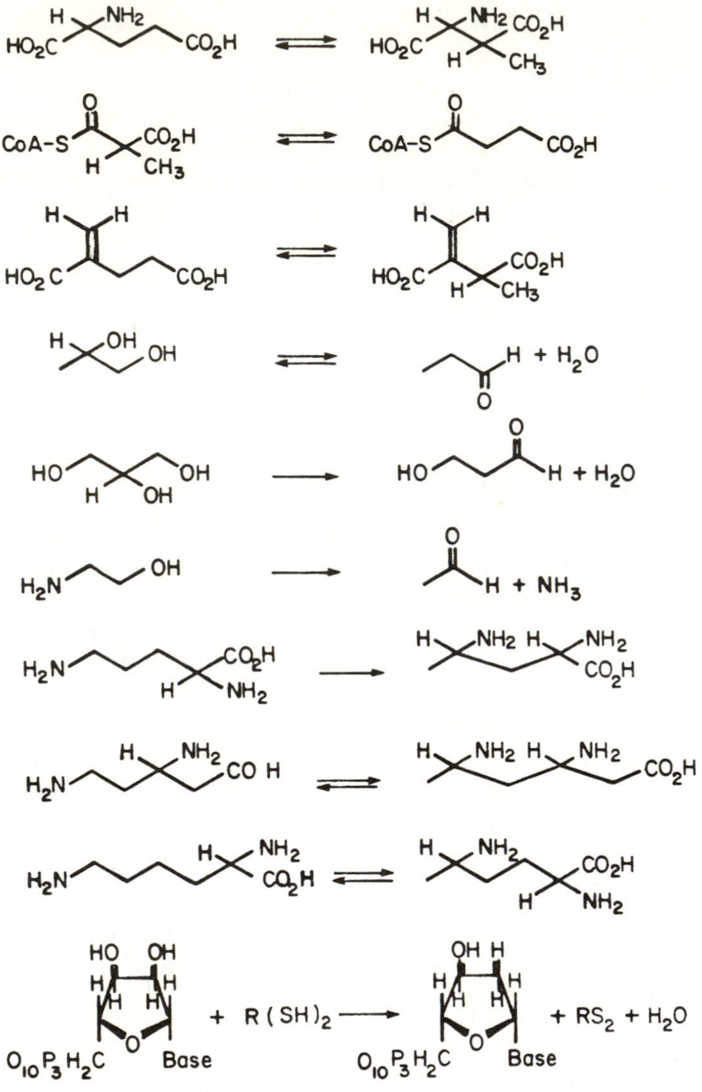

$$(6\text{-}1)$$

$$(6\text{-}2)$$

$$(6\text{-}3)$$

$$(6\text{-}4)$$

$$(6\text{-}5)$$

$$(6\text{-}6)$$

$$(6\text{-}7)$$

$$(6\text{-}8)$$

$$(6\text{-}9)$$

$$(6\text{-}10)$$

Figure 6.11. The ten different reactions that require coenzyme B_{12}.

This unique carbon skeleton rearrangement has been studied in some detail with most of the B_{12}-dependent enzymes. Of course, the crucial question concerning the mechanism is: *How does the 1,2-shift occur?* Before examining the details of the mechanism, let us examine some useful experimental observations.

It was observed that for methylmalonyl-CoA (reaction 6-2), the X group (—COSCoA) is transferred intramolecularly and the hydrogen which is transferred does not exchange with water during the process. The same observation applies for the dehydration of propanediol (reaction 6-4) and the conversion of glutamic acid to methylaspartic acid (reaction 6-1). However, for these last two cases, inversion of configuration takes place at the carbon atom to which the hydrogen migrates. But for the methylmalonyl-CoA, the hydrogen that migrates from the methyl group to the C$_3$ position of succinyl-CoA occupies the same steric position as previously occupied by the —COSCoA group. The migration occurs with retention of configuration.

$$\text{HOOC}-\underset{\underset{\text{H}}{|}}{\text{C}}-\underset{\underset{\text{H*}}{\nearrow}}{\overset{\overset{\text{CO}-\text{S}-\text{CoA}}{|}}{\underset{}{\text{C}}}}\!\!-\text{H*} \longrightarrow \text{HOOC}-\underset{\underset{\text{H}}{|}}{\overset{\overset{\text{H* CO}-\text{S}-\text{CoA}}{|}}{\text{C}}}-\text{CH}_2 \qquad (6\text{-}11)$$

prochiral center
(retention of configuration)

$$\underset{\underset{\text{H*}}{|}}{\overset{\overset{\text{HOOC}\quad\text{H}}{|\quad|}}{\text{CH}_2-\text{C}-\text{CH}}}\!\!\overset{\text{COO}^\ominus}{\underset{\text{NH}_3^\oplus}{}} \longrightarrow \underset{\text{H*}\;\;\text{H}}{\overset{\overset{\text{COOH}}{|}}{\text{CH}_2-\text{C}-\text{CH}}}\!\!\overset{\text{COO}^\ominus}{\underset{\text{NH}_3^\oplus}{}} \qquad (6\text{-}12)$$

(inversion of configuration at this center)

$$\underset{\underset{\text{H}}{|}}{\overset{\overset{\text{OH}}{|}}{\text{CH}_3-\text{C}-\text{CH}}}\!\!\overset{\text{OH}}{\underset{\text{H*}}{}} \longrightarrow \left[\underset{\underset{\text{H*}}{|}}{\overset{\overset{\text{H}}{|}}{\text{CH}_3-\text{C}-\text{CH}}}\!\!\overset{\text{OH}}{\underset{\text{OH}}{}}\right] \longrightarrow \underset{\underset{\text{H*}}{|}}{\overset{\overset{\text{H}}{|}}{\text{CH}_3-\text{C}-\text{CHO}}}$$

prochiral center
(inversion of configuration)

$$(6\text{-}13)$$

In this last example, it has been shown that the migrating hydrogen atom does not exchange with the solvent but exchanges with similar hydrogen atoms of other substrate molecules. The transformation corresponds to an internal redox reaction, but is not exclusively intramolecular. Hence:

$$\underset{\text{HO}\quad\text{OH}}{\text{CH}_3-\text{CH}-\text{CD}_2} \longrightarrow$$

$$\text{CH}_3-\text{CH}_2-\overset{\overset{\text{O}}{\diagup}}{\underset{\diagdown\text{D}}{\text{C}}}\;, \;\text{CH}_3-\text{CH}_2-\overset{\overset{\text{O}}{\diagup}}{\underset{\diagdown\text{H}}{\text{C}}}\;,\;\text{CH}_3-\underset{\underset{\text{D}}{|}}{\text{CH}}-\overset{\overset{\text{O}}{\diagup}}{\underset{\diagdown\text{D}}{\text{C}}} \qquad (6\text{-}14)$$

$$\underset{\text{HO}\;\;\text{T}}{\overset{\overset{\text{T}\;\;\text{T}}{|\;\;|}}{\text{CH}_3-\text{C}-\text{C}-\text{OH}}} + \underset{\text{HO}\quad\text{OH}}{\text{CH}_2-\text{CH}_2} \longrightarrow \text{CH}_3-\overset{\overset{\text{O}}{\diagup}}{\underset{\diagdown\text{T}}{\text{C}}} + \text{CH}_3-\text{CH}_2-\overset{\overset{\text{O}}{\diagup}}{\underset{\diagdown\text{T}}{\text{C}}}$$

$$(6\text{-}15)$$

Thus, the stereochemical course of vitamin B$_{12}$-dependent enzymes is a fascinating subject for the chemist to study (256). For instance, the enzyme propanediol dehydrase catalyzes also the dehydration of deuterated analogues of 1,2-propanediol (reactions 6-16 and 6-17). Migration of the deuterium atom is observed only with one isomer. These initial experiments showed that the dehydration proceeds by way of a 1,2-shift.

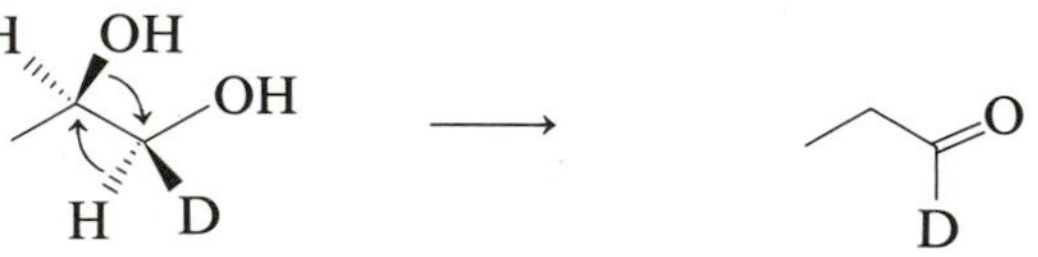

(6-16)

(1*R*,2*S*)-[1-^{2}H]propanediol — H migrates in this isomer to produce [1-^{2}H]propionaldehyde

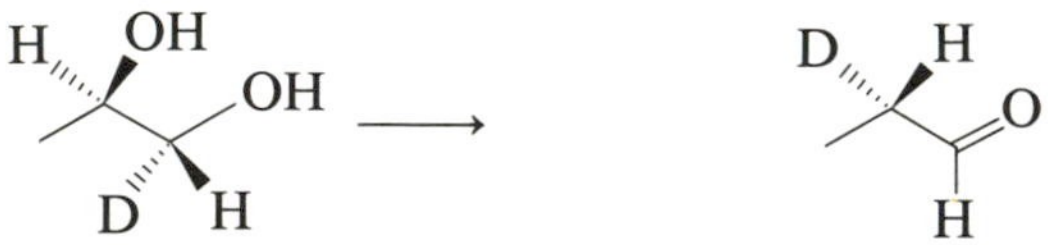

(6-17)

(1*R*,2*R*)-isomer — ^{2}H migrates in this isomer to produce [2-^{2}H]propionaldehyde of 2*S*-configuration

Arigoni and colleagues (257) used 2*S*- and 2*R*-1-[^{18}O]-1,2-propanediol to bring an elegant and additional proof for the concomitant 1,2-shift of an OH from C-2 to C-1 as the hydrogen migrates from C-1 to C-2 in this process.

More interesting is the finding by R. H. Abeles of Brandeis University that the C-5′ hydrogen of the coenzyme is also replaced by tritium when 1,2-[1-^{3}H] propanediol is used as substrate and the tritium could be transferred back to the product. In his view, the three nonequivalent hydrogen atoms of the 5′-deoxyadenosine moiety have equally a one-third probability of transfer back to the substrate. Consequently, the enzyme does not distinguish between the two prochiral hydrogens at C-5′ and shows that the role of the coenzyme B_{12} is basically to act as carrier. This apparent lack of stereoselectivity involved an homolytic cleavage of a cobalt–carbon bond (see below).

The experimental evidence suggests that a coenzyme-bound intermediate must exist in the initial event of some B_{12}-catalyzed rearrangements. It can be pictured in the following manner, using ethylene glycol as substrate:

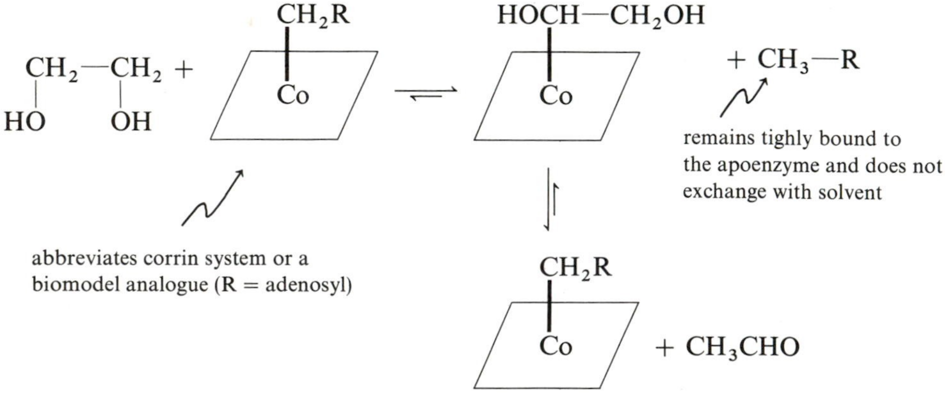

CH₂—CH₂ +
HO OH
CH₂R
Co
HOCH—CH₂OH
Co
+ CH₃—R
remains tighly bound to the apoenzyme and does not exchange with solvent
abbreviates corrin system or a biomodel analogue (R = adenosyl)
CH₂R
Co
+ CH₃CHO

The detailed mechanism of this fascinating biochemical transformation was elucidated mainly through the efforts of the groups of R. H. Abeles (256), D. Arigoni (257), D. Dolphin (261, 262), G. N. Schrauzer (258–260) and P. Dowd (264). For simplification, various biomodels of the corrin system were used. Some are presented in Fig. 6.12. The most commonly used is the bis-(dimethylglyoximato)–Co complex (called cobaloxime) which shows many properties of the cobalt atom in the corrin ring.

If a cobaloxime is coordinated with a suitable basic group (B) in one of the axial positions, it behaves in many ways like vitamin B$_{12}$. When reduced to the Co(II) state it reacts with alkyl halides to form alkyl cobaloxime, analogous in many properties and chemical reactivity to vitamin B$_{12}$ coenzyme.

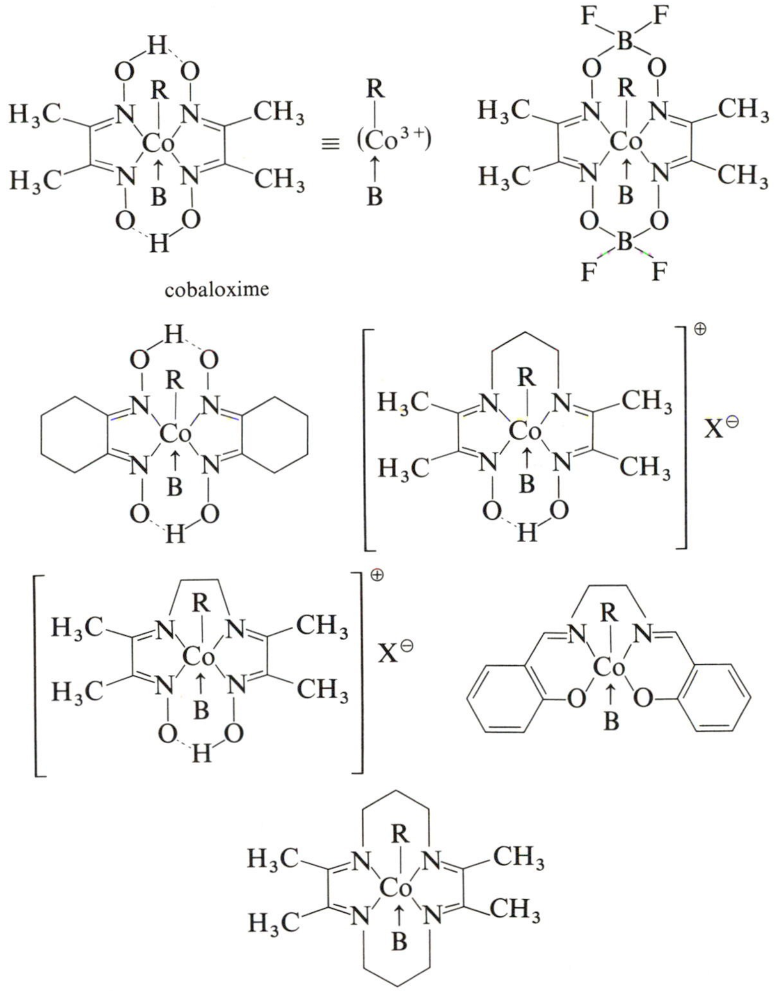

Figure 6.12. Some biomodels of the corrin system (259). Reproduced with permission. Copyright © 1976 by Verlag Chemie GMBH.

As seen previously, breaking the Co—C bond of the coenzyme B_{12} is a necessary step in the catalytic cycle. The evidence also suggests that the cleavage is *homolytic*, producing a B_{12r} radical [Co(II) species and a C-5′-methylene radical of deoxyadenosine]. This cleavage can also be induced nonenzymatically both thermally and photochemically with chemically related cobaloximes.

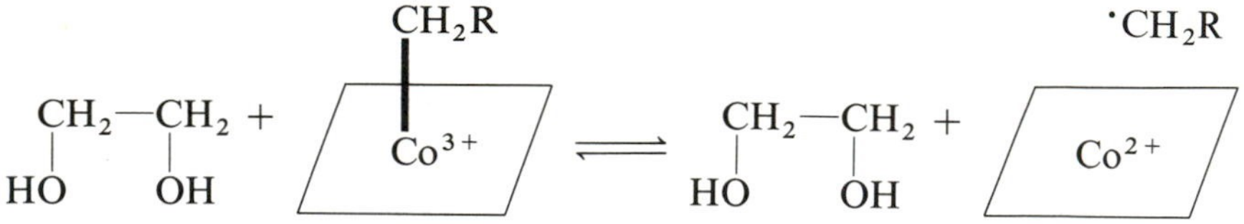

This homolytic cleavage results in the formation of a reactive intermediate which can abstract a hydrogen atom from the substrate to give the CH_3—C-5′-adenosyl intermediate (CH_3—R) and a substrate radical:

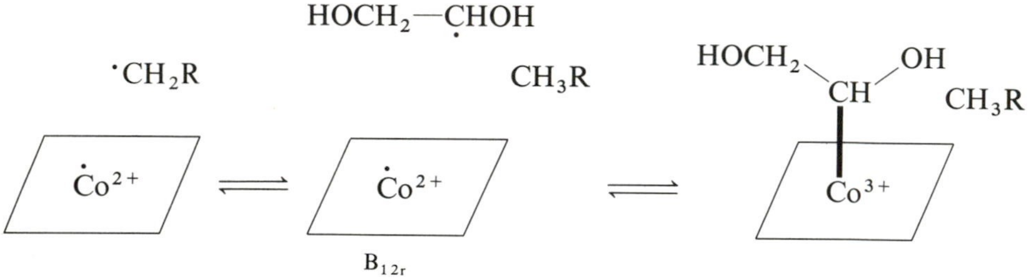

The combination of this radical with the B_{12r} species generates a new alkyl cobalamin with the substrate as ligand. We have thus accomplished a trans-alkylation of the cobalt atom. How does the Co—C bond become activated toward homolytic cleavage? It is believed that the presence and proper orientation of the propionamide side chains on the corrin ring are responsible for the ease of the enzymatic system, possibly by some distortion of the corrin (269). Support for this hypothesis comes from the fact that hydrolysis of a side chain to the corresponding acid results in an inactive coenzyme B_{12} molecule.* It is not known why nature proceeds by homolytic fission, a unique situation in coenzyme chemistry!

Now the crucial point. How does the following rearrangement occur?

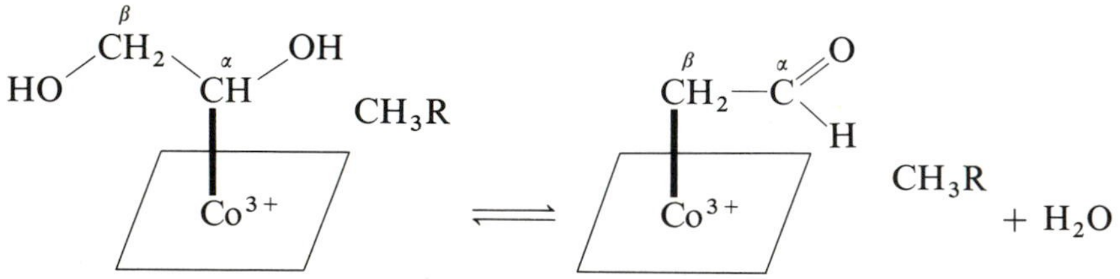

---

* In a personal communication, N. T. Anh of Orsay, France, indicated that X-ray work on coenzyme B_{12} showed that the angle of the Co—CH_2—CH—bond of the methylene 5′-adenoxyl moiety is unusually wide. The peculiar hybridization of this carbon could account for the ease of breaking of the Co—C bond.

Since OH$^-$ has to be a leaving group, three extreme electronic forms of the resulting intermediate can be envisioned.

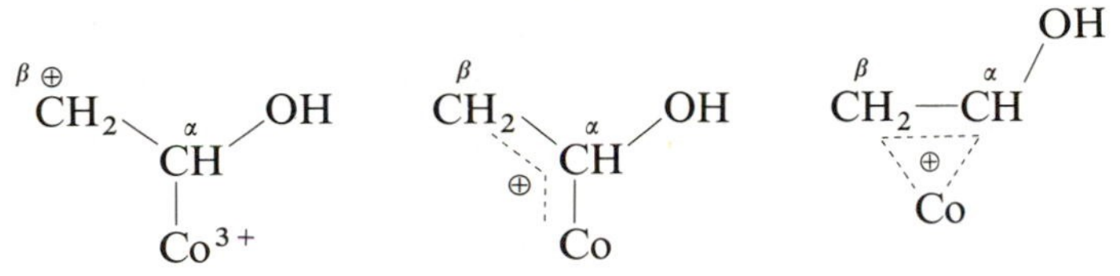

primary   delocalized   π-complex
carbonium ion   carbonium ion

Model studies indicated that solvolysis of ^{13}C-labeled 2-acetoxy-ethyl(pyridine) cobaloxime in methanol gives equal amounts of the two ^{13}C-labeled products.

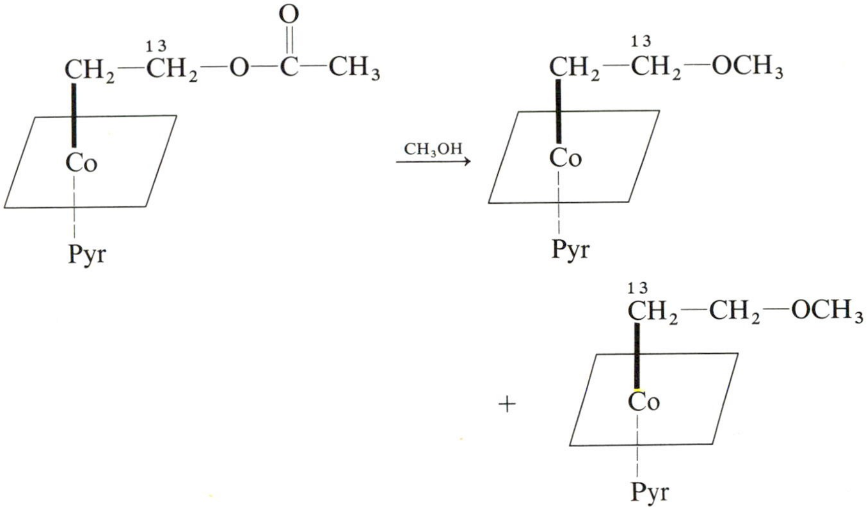

This suggests that a π-complex intermediate is the simplest pathway for the rearrangement. Consequently:

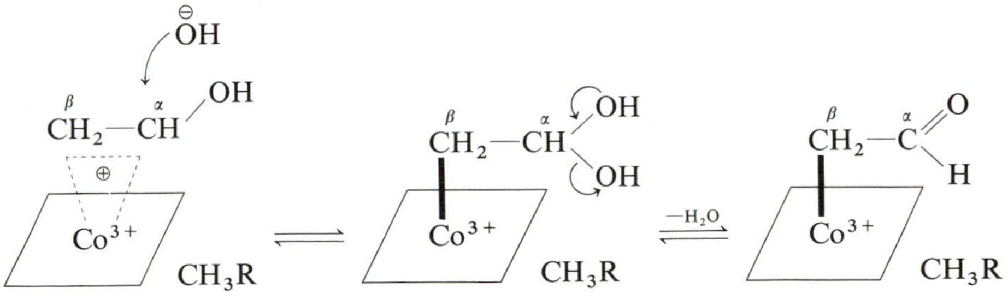

enol   diol
intermediate   intermediate

Since the hydroxyl group on the α-carbon stabilizes the developing positive charge, a nucleophile (H$_2$O, or OH$^-$) will more likely attack this position to generate a new σ-complex on the cobalt ion with the β-carbon of the substrate.

In summary, the sequence of events taking place are first loss of the β-substituent from the cobalt σ-complex, followed by formation of a π-complex and finally readdition of the leaving group on the π-complex to generate a new cobalt σ-complex with the substrate.

The transformation is completed by a second transalkylation that will regenerate the original coenzyme B_{12} and the dehydrated product.

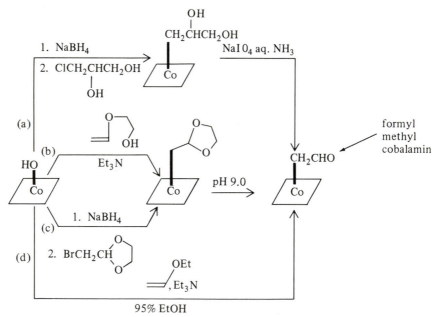

Therefore, the critical intermediate in B_{12}-chemistry is a Co(III)-olefin π-complex. The formation of this intermediate is chemically reasonable because an enol is an electron-rich species whereas a metal (trivalent Co) is electron deficient.

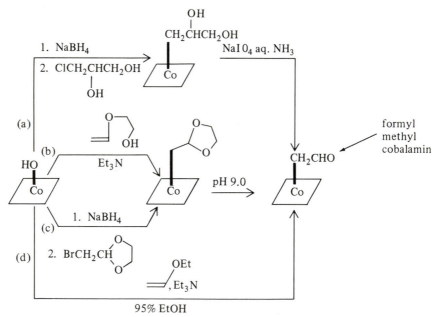

Figure 6.13. Four different routes for the synthesis of formyl methyl cobalamin (263). Reprinted with permission. Copyright©1974 by American Chemical Society.

An analogy to such a process is found in the heterogeneous polymerization of olefins (Section 4.1) where the existence of a $\sigma-\pi$ complex between a metal (TiCl$_3$) and a double bond is well documented.

Further evidence for the $\sigma-\pi$ rearrangement with the model system cobaloxime comes from Dolphin's work with electron-rich olefins, such as vinyl ethers (261, 262). As expected, the quenching of the π-complex by a nucleophile occurs on the most positive carbon (Fig. 6.13).

In these sequences, formyl methyl cobalamin, the intermediate in the enzymatic conversion of ethylene glycol to acetaldehyde, is synthesized in four different ways.

With the more elaborate systems, β-methylaspartate to glutamate (reaction 6-18) and methylmalonyl-CoA to succinyl-CoA (reaction 6-19), it has been proved that the largest group is the one to migrate as illustrated below:

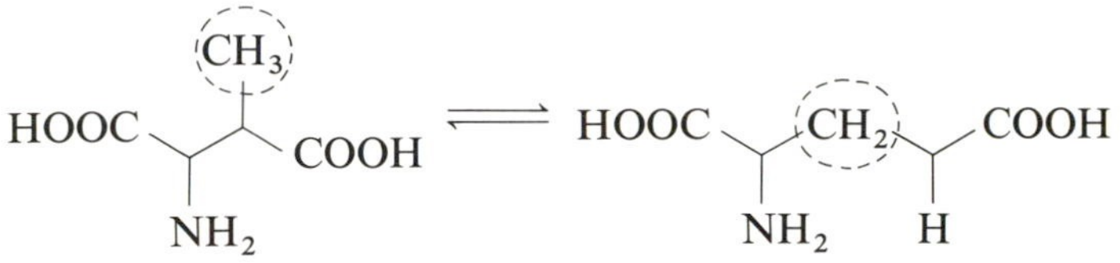

$$
\begin{array}{ccc}
\text{HOOC} \underset{\text{NH}_2}{\overset{(\text{CH}_3)}{\diagup\!\!\diagdown}} \text{COOH}
& \rightleftharpoons &
\text{HOOC} \underset{\text{NH}_2 \quad \text{H}}{\diagup\!\!\diagdown} (\text{CH}_2) \diagup\!\!\diagdown \text{COOH}
\end{array}
\qquad (6\text{-}18)
$$

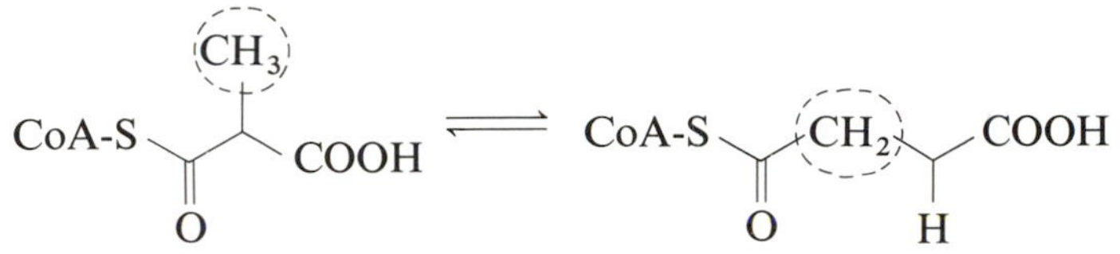

$$
\begin{array}{ccc}
\text{CoA-S} \underset{\text{O}}{\overset{(\text{CH}_3)}{\diagup\!\!\diagdown}} \text{COOH}
& \rightleftharpoons &
\text{CoA-S} \underset{\text{O} \quad \text{H}}{\diagup\!\!\diagdown} (\text{CH}_2) \diagup\!\!\diagdown \text{COOH}
\end{array}
\qquad (6\text{-}19)
$$

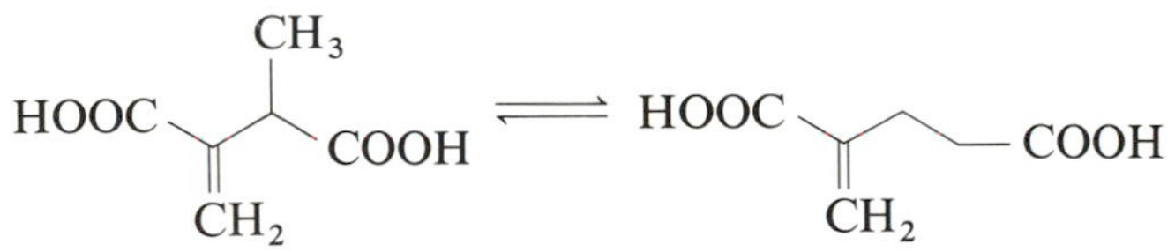

$$
\begin{array}{ccc}
\text{HOOC} \underset{\text{CH}_2}{\overset{\text{CH}_3}{\diagup\!\!\diagdown}} \text{COOH}
& \rightleftharpoons &
\text{HOOC} \underset{\text{CH}_2}{\diagup\!\!\diagdown} \diagup\!\!\diagdown \text{COOH}
\end{array}
\qquad (6\text{-}20)
$$

In the first case the migration occurs with inversion whereas in the second case it occurs with retention of configuration.

For the system methylitaconate to α-methyleneglutarate (reaction 6-20), P. Dowd's group at the University of Pittsburgh made an interesting model to demonstrate the migration of the acrylate fragment in the carbon-skeleton rearrangement leading to α-methyleneglutaric acid (264).

The coenzyme model intermediate was synthesized by the reaction of vitamin B$_{12s}$ [Co(I)] with bis(tetrahydropyranyl) bromomethylitaconate. On standing in aqueous solution the model yields rearranged α-methyleneglutaric acid together with unrearranged methylitaconic acid and butadiene-2,3-dicarboxylic acid.

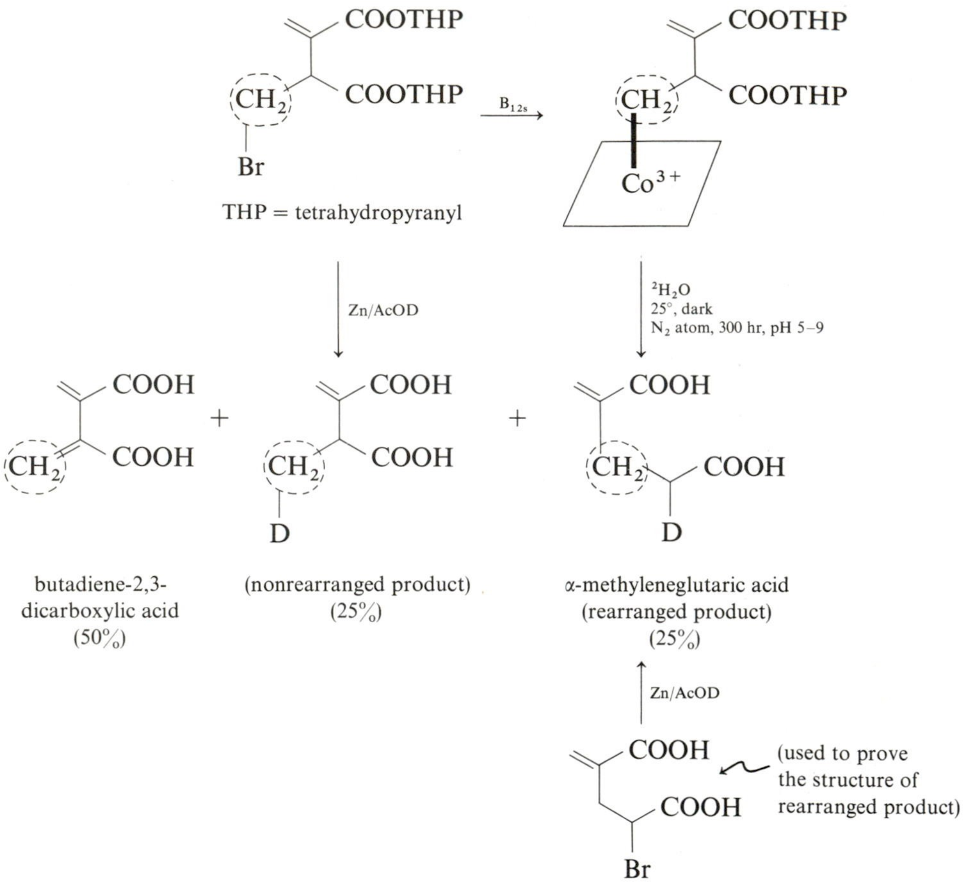

THP = tetrahydropyranyl

butadiene-2,3-
dicarboxylic acid
(50%)

(nonrearranged product)
(25%)

α-methyleneglutaric acid
(rearranged product)
(25%)

(used to prove
the structure of
rearranged product)

Since the model reaction provides no role for deoxyadenosine, its 5'-methy-
lene being the instrument of hydrogen transfer in all the coenzyme B_{12}-
dependent carbon-skeleton rearrangement reaction, it was important to
learn the source of the hydrogen introduced into the products. This was
accomplished by performing the reaction in ^2H_2O and analyzing the product
by NMR and mass spectrometry. Both rearranged and nonrearranged products contained deuterium but not the butadiene-2,3-dicarboxylic acid. The
position of labeling was further proved by Zn/AcOD reduction of the
starting material.

It thus establishes that the C—Co bond in the model series is hydrolyzed
by proton transfer from the solvent water. It should be recalled that in the
corresponding enzymatic process, exchange with solvent water does not
occur in the course of the rearrangement because of the presence of the 5'-
deoxyadenosine moiety. However, if the assumption is made that the position
of the deuterium is indicative of the position of the cobalt, prior to hydrolysis,
then the presence of deuterium exclusively at the γ-carbon in this model

requires that the acrylic acid moiety be the migrating group in the rearrangement reaction (264). Thus, this model system leads one to expect that in the enzyme catalyzed rearrangement, it is the more complex group which migrates.

Whether the real coenzyme actually takes opportunity of this mechanism remains to be proved. The group of L. Salem (265) of the University of Paris-Sud proposed, on theoretical grounds, that in coenzyme B$_{12}$ rearrangements the more electronegative group of the substrate should migrate preferentially. This is in accordance with Dowd's finding (264).

It is also known that a malonic ester substituted in the β-position with cobalt can rearrange into the corresponding succinic ester. To test the possible role of the metal and to better simulate the situation of the enzyme, the group of J. Rétey (266) synthesized the following cobalt complex in which the "substrate" is anchored covalently by two methylene bridges to planar cobaloxime.

Irradiation in methanolic solution, followed by acid workup, gave the rearranged product in good yield. This observation strongly suggests that the homolytic rearrangement of the bound-substrate to the product is promoted by the cobalt atom.

These extensive efforts on the biochemical reaction catalyzed by coenzyme B$_{12}$ led to the following unified picture. First, coenzyme B$_{12}$ inserts into an unactivated C—H bond of the substrate. The result is that the adenyl group of the coenzyme now carries a hydrogen atom of the substrate, while the substrate methylene group (in the case of methylmalonyl-CoA to succinyl-CoA) becomes bonded to the cobalt of B$_{12}$, replacing the original adenyl-B$_{12}$ C—Co bond. Second, a rearrangement occurs within the B$_{12}$-substrate molecule to produce a B$_{12}$-product molecule. Finally, this undergoes the equivalent of a reverse of the first reaction in which the adenyl-cobalt bond of coenzyme B$_{12}$ is again formed, and the reaction product is released carrying one of the hydrogen atoms of the 5'-deoxyadenosine methyl group.

In 1976, Breslow (267) examined various potential model systems of B_{12}-homolytic cleavage and found simple chemical transformations which furnish good precedent for the step involving insertion into an unactivated carbon. He made a biphenyl-cobaloxime compound which by irradiation generates the expected benzylic radical.

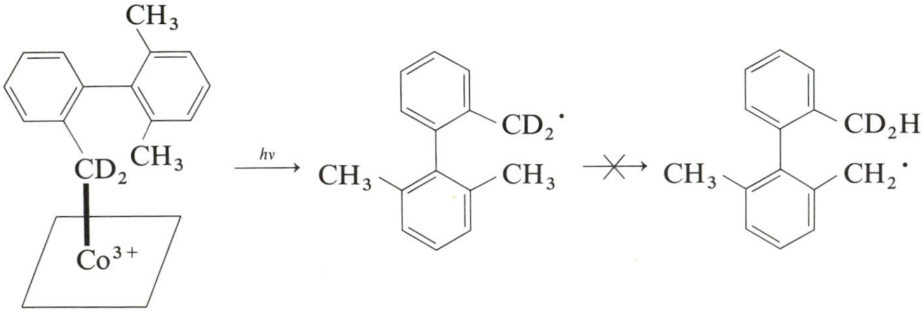

However, the resulting radical dimerized without hydrogen transfer from the nearby methyl group. Breslow thus turned to a cyclododecyl radical, a system in which intramolecular hydrogen transfer to a carbon free radical has already been demonstrated. Again deuterium was used as a marker.

Since B_{12s} [Co(I)] is a powerful nucleophile it reacts rapidly with the alkyl iodide to give a cyclododecyl-cobalamin. Upon bromination this affords cyclododecyl bromide with the deuterium marker distributed over several ring carbons.

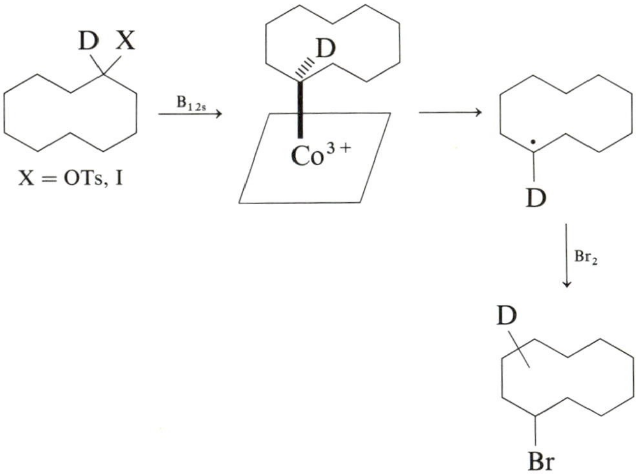

Transannular hydrogen transfer thus occurs in the overall sequence. Since the reaction of cyclododecyl iodide with B_{12s} is certainly not a simple displacement at carbon but also involves transannular hydrogen transfer via homolytic fragmentation, the insertion reactions common to all coenzyme B$_{12}$ catalyses now seem to have a good chemical precedent (267).

However, convincing direct evidence for the intermediacy of carbon radicals in natural coenzyme B$_{12}$ rearrangements is still lacking and for this reason E. J. Corey proposed in 1977 an interesting and different mechanism, consistent with current knowledge of organometallic reactions (268). A key feature of the proposal is an electrocyclic opening of the corrin ring of the coenzyme, cleaving the unique direct chemical bond that joins ring A and D (see Fig. 6.10) and thus allows a role for the side chain of the corrin system. Corey argues that nature's construction of such direct linkage in a corrin system is both nonaccidental and essential. A rational stepwise mechanistic interpretation of the rearrangement taking place involves a cobalt–carbene complex with the substrate.

This interesting proposal is beyond the scope of the present discussion but puts into reevaluation the existence of a Co π-complex in coenzyme B$_{12}$ rearrangements.

Because of the complexity of the enzymatic systems involved in coenzyme B$_{12}$ chemistry there are several reports on the purification of B$_{12}$-dependent enzymes or B$_{12}$-binding proteins by vitamin B$_{12}$ affinity adsorbents. In fact, for purification of enzymes or proteins, *affinity chomatography* has been widely used as one of the most attractive methods (270). For that purpose, the synthesis of a cobalamin–Sepharose insoluble support has been prepared and applied to the purification of N^5-methyltetrahydrofolate-homocysteine cobalamin methyltransferase from *E. coli*. The scheme for the synthesis of the solid support is summarized in Fig. 6.14.

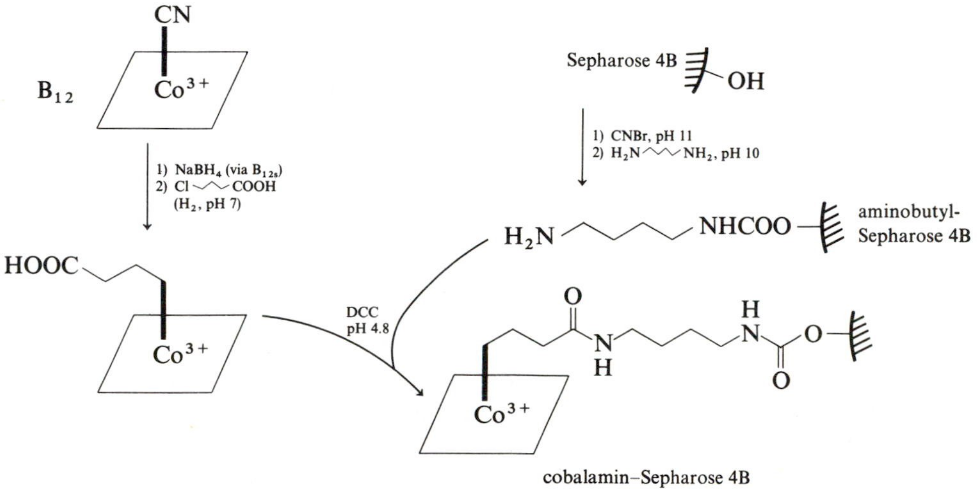

Figure 6.14. Preparation of a cobalamin—Sepharose insoluble support for purification of B$_{12}$-dependent enzymes by affinity chromatography (270).

Here the apoenzyme* of methionine synthetase binds to the adsorbent and can be obtained as the holoenzyme by elution after the cleavage of the carbon–cobalt bond by light irradiation:

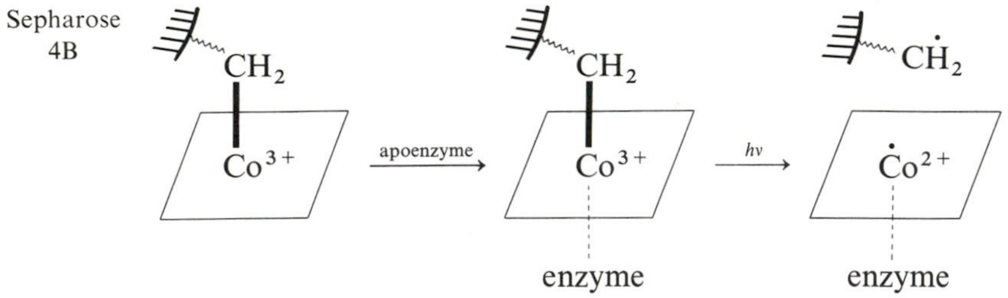

So far the affinity adsorbent is not reusable, but if a method is developed of eluting the protein without photolysis so that the adsorbent would be reusable, it might open the way for the wide application of this type of adsorption in the purification of B_{12}-dependent enzymes.

Finally, because of the growing importance of mercury in the environment, a few words should be said about its methylation and its relationship with vitamin B_{12}. J. M. Wood, a biochemist then at the University of Illinois, was one of the first to study in detail the mechanism for methylation of mercury (271). The research of Wood proved that the biosynthesis of methylmercury could occur under anaerobic conditions. Historically, it was the discovery that inorganic mercury could be alkylated to methylmercuric ion in natural systems that pointed out the present dimension of the mercury pollution problem. Mercuric ion is an extremely strong methyl acceptor and the monomethyl derivative is stable in aqueous solution. Methylmercuric ion in aqueous solution can then be absorbed by unicellular organisms, thereby entering an aquatic food chain. The concentration builds up along the chain, reaching highly toxic levels in fish, and most strongly affecting those species that eat fish. Wood reasoned that B_{12}-containing enzyme systems should be capable of methylmercury synthesis in biological systems via *trans-methylation*. This refers to the transfer of a methyl group from one substrate to another, usually as part of biochemical reaction. This process is also referred to as *biological methylation*. Three major coenzymes are available in nature to carry such methyl transfer reactions in biological system: S-adenosylmethionine (SAM), N^5-methyltetrahydrofolate derivatives (N^5-CH_3-THF), and methylcorrinoid derivatives (methylcobalamin). The first two involve methyl group transfer as a carbonium ion. For the third one, Wood argued that methylcorrinoids are particularly efficient methyl

---

* In the classical terminology of enzymology, the complete enzyme–cofactor complex is termed *holoenzyme*, and the protein component minus its cofactor is termed an *apoenzyme*. The apoenzyme is generally inactive as catalyst (see Chapter 7).

transfer agents since they are capable of transferring methyl group as carbanions, carbonium ions, or radicals (see page 371):

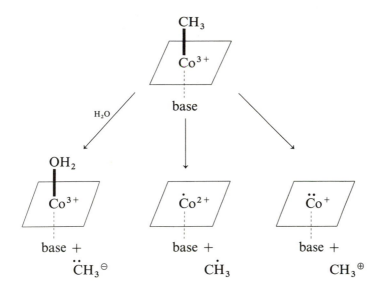

Among the enzymes implicated in mercury methylation are methionine synthetase, acetate synthetase, and methane synthetase. The last one is a very common enzyme in anaerobic ecosystems, such as lake and river sediments. These enzymatic processes can generate CH$_3^+$ and CH$_3^-$ species where vitamin B$_{12}$ derivatives act as mediator to methylate Hg(II) ions.

Furthermore, it has also been shown on models that when methylcorrinoids are photolyzed under anaerobic conditions, homolytic cleavage of the Co—C bond can also occur to give Co(II) and methyl radicals:

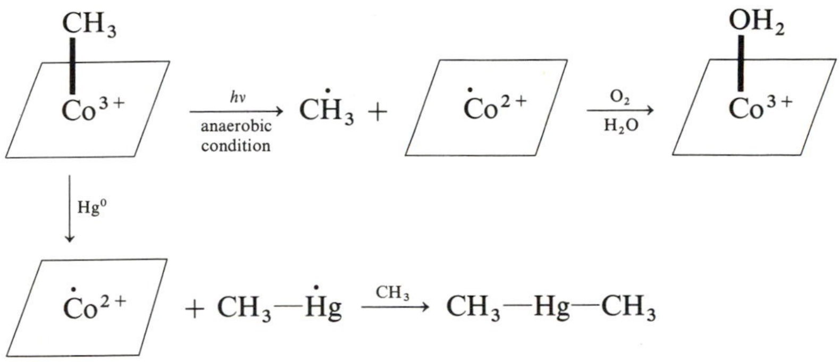

Dimethylmercury is then believed to be synthesized also by methyl radical addition to metallic mercury. Indeed it seems feasible in certain organisms that Hg(II) is transported across cell membranes, reduced to metallic mercury, and then methylated. Dimethylmercury, being volatile, would readily diffuse out of the microbial cells and be released to the water. If the pH is acidic, it would be converted to monomethylmercury and methane:

$$CH_3-Hg-CH_3 + H^\oplus \longrightarrow CH_3\overset{\oplus}{H}g + CH_4$$

volatile   soluble

$$\downarrow h\nu$$

$$C_2H_6 + CH_4 + Hg(0)$$

Apparently, volatile methylmercury in the atmosphere is degraded to ethane, methane, and Hg(0) by irradiation. The metallic mercury then goes back to the soil and lakes and the "environmental mercury cycle" just mentioned is repeated again (272).

Vitamin B_{12} and its derivatives are thus remarkable catalysts of versatile functions and are implicated in numerous unusual biochemical transformations in various organisms. The transformations of mercury in the environment is only one of the many facets of vitamin B_{12} action.

In concluding this chapter, it should be noted that we have presented various models of enzyme mechanisms in which metal ions participate. We have seen that reactions catalyzed by metalloenzymes or enzymes activated by metal ions span a remarkably broad spectrum of reaction types. Of course many facets such as the remarkable speed and specificity of enzyme catalysis have not received complete explanation on the basis of model system chemistry. However, the missing link may be found at the point where biological molecules deviate from the model systems. Here, perhaps the most interesting chemical features will be found (258, 259).

Chapter 7

Coenzyme Chemistry

*"... it is by logic that we prove, but by intuition
that we discover."*

J. H. Poincaré

Cellular metabolism is under enzymatic control and often the enzymes
involved need a substance or cofactor in order to express their catalytic
activity. In these systems the protein portion of the enzyme is designated
the *apoenzyme* and is usually catalytically inactive. The cofactor is a metal
ion or a nonprotein organic substance. Many enzymes even require both
cofactors. A firmly bound cofactor is called a *prosthetic group*. If, however,
the organic cofactor is brought into play during the catalytic mechanism,
it is referred to as a *coenzyme*. The complex formed by the addition of the
coenzyme to the apoenzyme is referred to as a *holoenzyme* (or enzyme, for
short).

Some coenzymes serve as carriers of chemical groups, hydrogen atoms, or
electrons. Others such as ATP function in energy coupling reactions within
the cell and are often regarded as a substrate rather than a coenzyme. Other
coenzymes have more complex structures and are derivatives of *vitamins*.
They act at the active site of the enzyme by combining with the substrate in
a way that permits the reaction to proceed more readily. By definition,
vitamins cannot by synthesized by the host, and must therefore be provided
by the diet. Their presence are thus required for normal growth and health
and their absence causes specific or "vitamin deficiency" diseases.

As is often the case, coenzyme biochemistry leads to unconventional organic chemistry. In this regard, coenzymes are nature's special reagents and their well-defined structures make them ideal molecules to use for developing the concept of structure–function relationship by bioorganic approaches (273). This chapter will thus be concerned with this aspect and special attention will be devoted to model design of coenzyme action.

7.1 Oxidoreduction

Enzymatic oxidoreduction sequences lie at the heart of cellular energy metabolism. The energy released in oxidation of reduced organic or inorganic compounds is captured with varying efficiencies in useful forms such as ATP, membrane potentials, or reduced coenzymes. Because of their physiological role, the mechanism of action of enzymes catalyzing electron transfer processes has been actively studied.

It should be recalled that redox enzymes or *oxidoreductases* are divided into four main classes.

(a) *Dehydrogenases* are enzymes that catalyze the transfer of a hydrogen from one substrate to another. This reversible system normally needs a nicotinamide coenzyme such as NAD^+. Chiral alcohols are made from the corresponding ketones or aldehydes this way. Some dehydrogenases contain an essential atom of Zn(II) (280).

(b) *Oxidases* are enzymes that catalyze the transfer of a hydrogen from a substrate to molecular oxygen. A flavin coenzyme, FAD, is usually required. Water or hydrogen peroxide may be the end product, depending on the enzyme specificity.

(c) *Oxygenases* catalyze the incorporation of oxygen, from molecular oxygen, into a substrate. *Monooxygenases* incorporate one atom of molecular oxygen into the substrate; the other appears as H_2O_2 or H_2O. *Dioxygenases* insert both oxygen atoms into the substrate.

(d) *Peroxidases and catalases*, both catalyze reactions involving hydrogen peroxide as oxidant. Water, the reduced form, is produced. These enzymes require ferric and/or cuperic ions for catalysis.

7.1.1 NAD⁺, NADP⁺

Nicotinamide adenine dinucleotide (NAD^+) and nicotinamide adenine dinucleotide phosphate ($NADP^+$) are pyridine nucleotides, first identified by O. Warburg in 1935. NAD^+ is the coenzyme involved in dehydrogenase reactions and is reduced to NADH during the process.

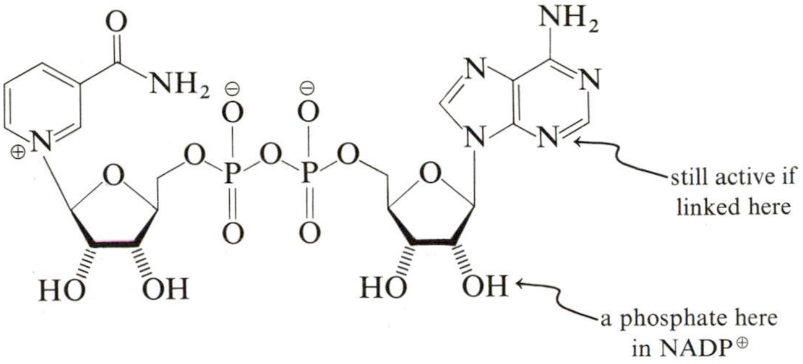

NAD⊕

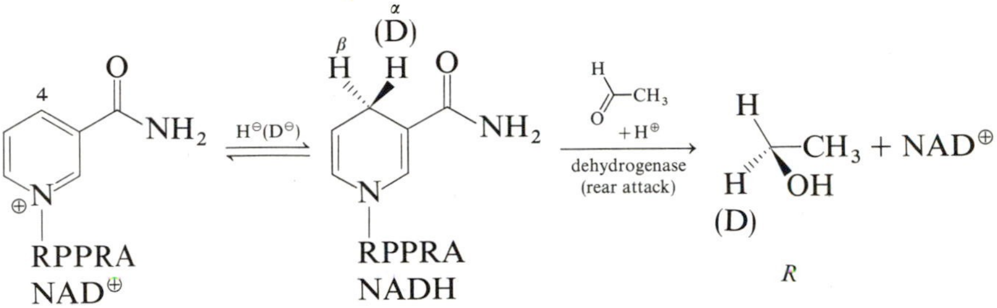

still active if
linked here

a phosphate here
in NADP⊕

RPPRA = ribose-phosphate-phosphate-ribose-adenine

The reaction is stereospecific and only one isomer (the α- or R-isomer in the above example) of NADH is produced. Other dehydrogenases with a different specificity are know to yield the β-isomer stereospecifically rather than the α-isomer.

Theoretical calculations of charge distribution on the nicotinamide ring by B. Pullman in Paris agree with a hydride attack at position C-4. Experimental evidence comes from reaction with a deuterated alcohol and isolation of the pyridinium salt. Methylation and oxidation take place without loss of deuterium, proving that the labeling occurs only at C-4.

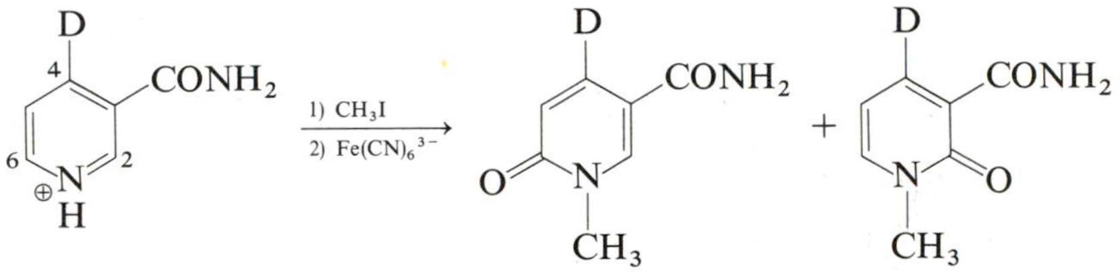

Additional reactions show that a good nucleophile such as cyanide ion can add to NAD$^+$. It can then be lost by dilution to give the rearomatized product. Basic treatment in ^2H_2O affords the 4-d analogue of NAD$^+$.

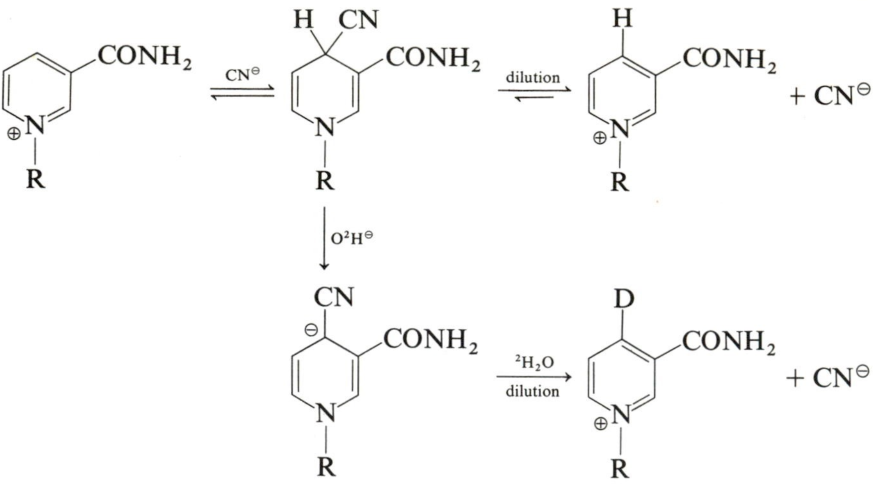

The specific preparation of the deuterated analogue demonstrates the electrophilic character of position C-4. In this regard, bisulfite addition at pH 7 is easily obtained and a dithionite treatment of NAD^+ leads to a reduction reaction at C-4.

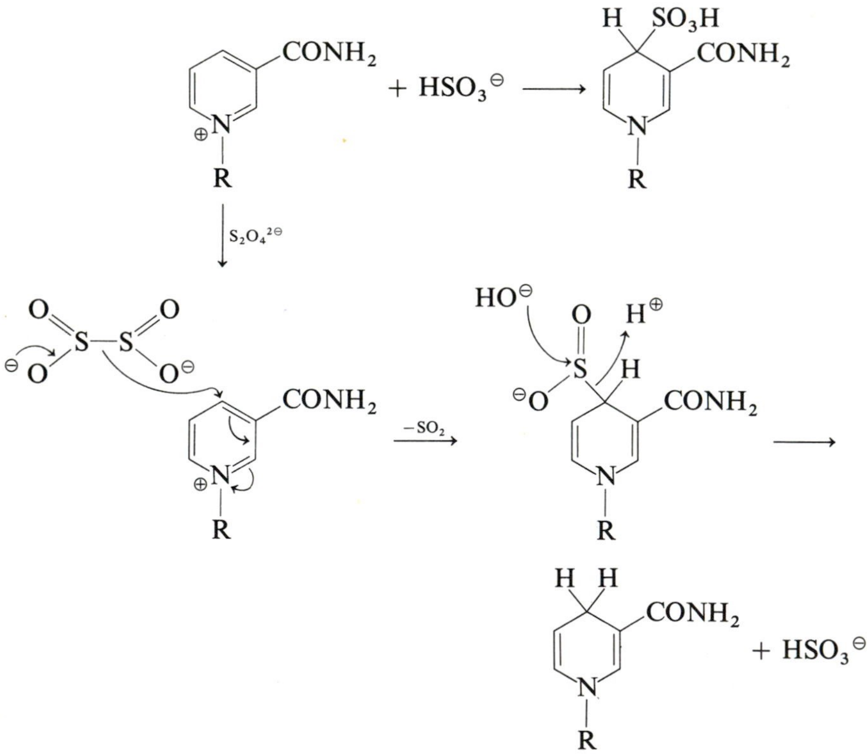

Nucleophilic addition to NAD^+ was also demonstrated in the case of pyruvate at alkaline pH. Addition presumably occurs via the enol (or the corresponding enolate) to give the reduced and oxidized adducts (274).

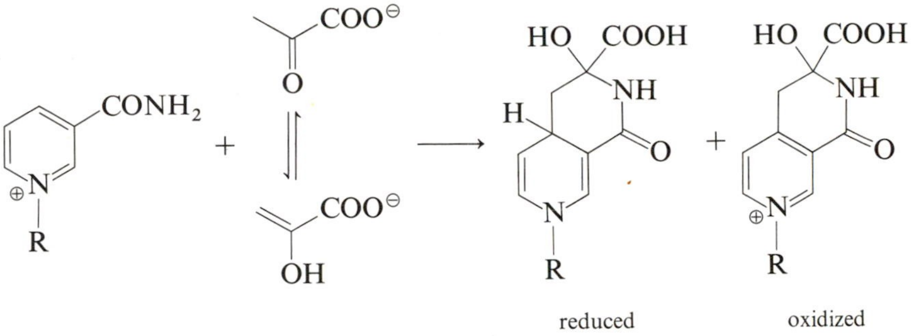

reduced   oxidized

These data suggest the following equilibrium:

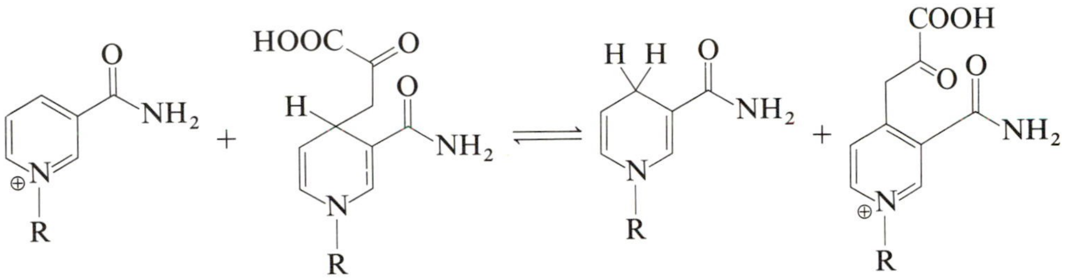

In the $NADH \rightleftharpoons NAD^+ + H^-$ reaction, what is the evidence for the direct transfer of a hydride ion?

O. J. Creighton and D. S. Sigman (275) made, in 1971, the first model for an alcohol dehydrogenase. They used a zinc ion to catalyze the reduction of 1,10-phenanthroline-2-carboxaldehyde by N-propyl-1,4-dihydronicotin-amide, a simple model compound of NADH.

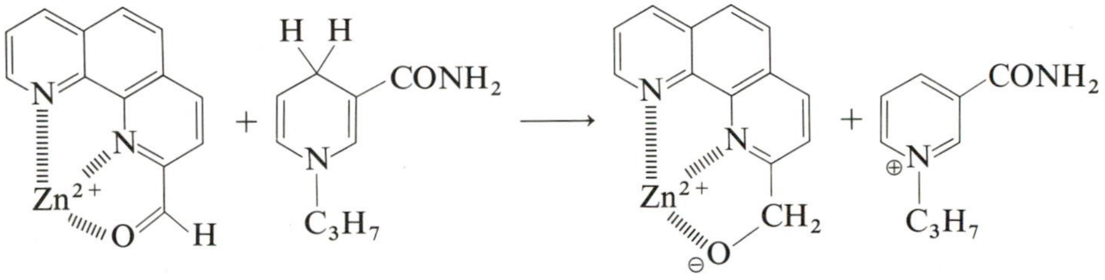

Interestingly, in acetonitrile at 25°C the presence of Zn(II) was found essential for efficient catalysis of this reaction. Coordination of the metal ion to the substrate is thus a driving force for the reduction to take place.

This was the first example of a reduction of an aldehyde by an NADH analogue in a nonenzymatic system. If 1-propyl-4,4-dideuterionicotinamide is used, monodeuterated 1,10-phenanthroline-2-carbinol is produced. This result demonstrates that the product is formed by direct hydrogen transfer from the reduced coenzyme analogue. It strengthens the view that coordination or proximity of the carbonyl to the zinc ion is probably important in the enzymatic catalysis. Zn(II) ion is known to be essential for the catalytic activity of horse liver alcohol dehydrogenase (280).

Surprisingly, if the proximity model below is used to mimic NAD$^+$ reduction, no transfer occurs.

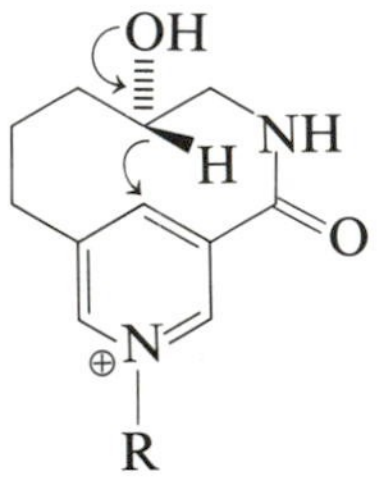

Although, favorable factors are present, the system prefers to remain aromatic. Hence, the formation of NADH in the enzymatic system could be driven by conformational changes that shift the equilibrium toward the nonaromatic species. However, in 1978, a German group (276) observed an *intra*molecular hydride transfer in the presence of pig heart lactate dehydrogenase using a coenzyme–substrate covalent analogue composed of lactate and NAD$^+$.

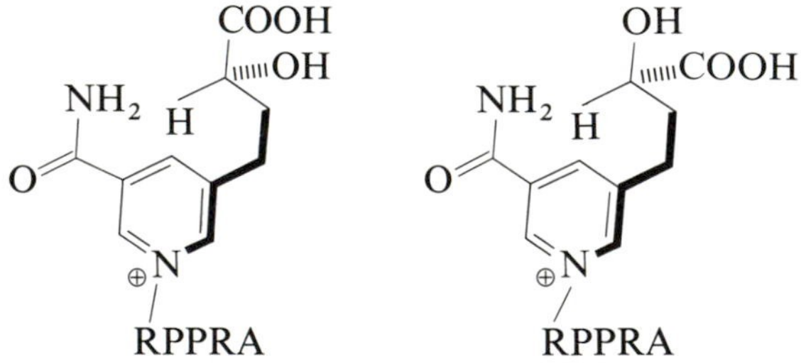

S-isomer   R-isomer

The two synthetic diastereomeric nicotinamide adenine dinucleotide derivatives are attached via a methylene spacer at position 5 of the nicotinamide ring. Only the S-isomer undergoes the intramolecular hydride transfer, forming the corresponding pyruvate-nicotinamide analogue and NADH. Two (R)-lactate specific dehydrogenases, however, do not catalyze a similar reaction with either one of the two diastereoisomers. Consequently a possible arrangement of the substrates (lactate and pyruvate) at the active centers of these enzymes can be proposed:

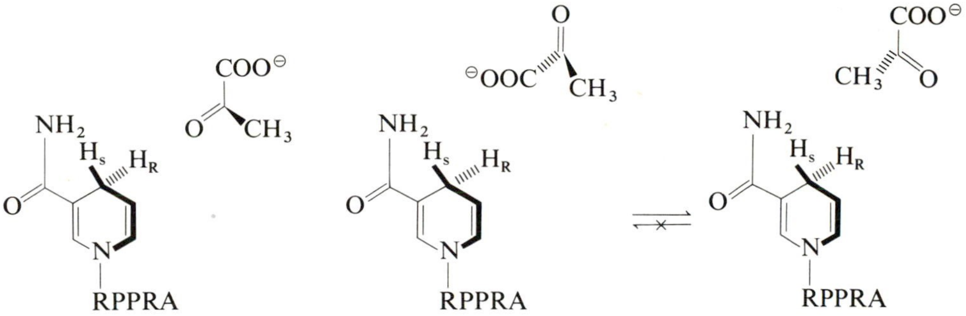

possible arrangement of pyruvate in the active center of (S)-lactate-specific dehydrogenase allowing for the transfer of the *pro-R* hydrogen from the dihydropyridine ring; this orientation agrees with X-ray data for ternary inhibitor complexes

this arrangement is not favored since no hydride transfer occurred with (R)-lactate-specific dehydrogenase

C_2–C_3 rotation gives this new orientation of pyruvate; this arrangement is postulated for (R)-lactate-specific dehydrogenase

Models for the reverse reaction (NADH → NAD$^+$) have also been prepared. An interesting example is the following crown ether NAD(P)H mimic (277).

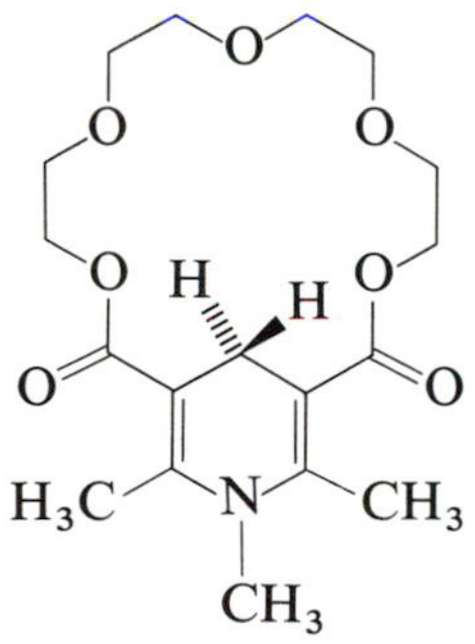

In the presence of sulfonium salts, transfer of hydride occurs 2700 times faster than with the corresponding analogue without the crown ether ring.

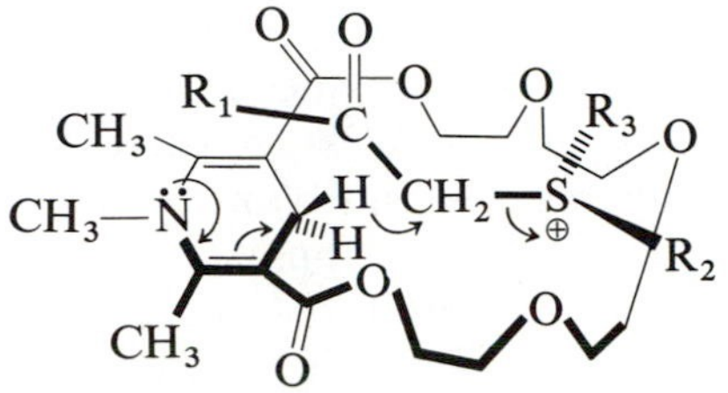

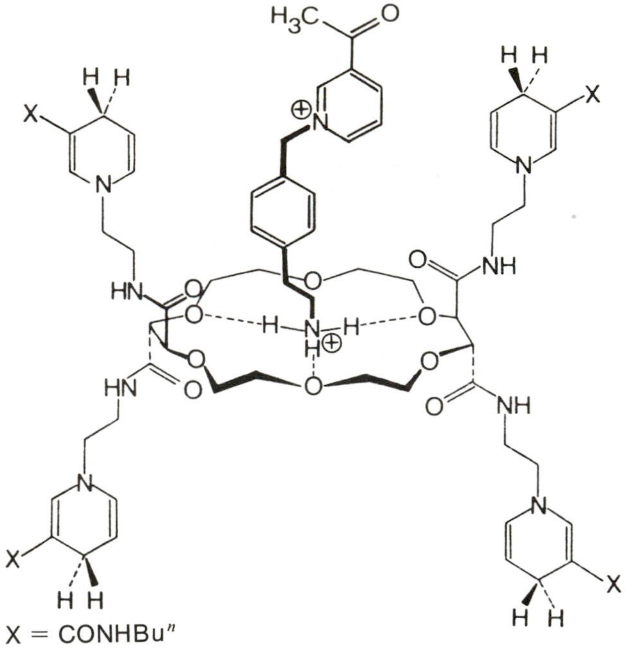

Figure 7.1. Lehn's crown ether model of NADH (278). Reprinted with permission. Copyright © 1978 by the American Chemical Society.

Obviously, the presence of the crown ether allows proper complexation of the cation.

Another model was developed, by the group of J. M. Lehn (278) of Strasbourg. They prepared a complex between a pyridinium substrate and a chiral crown ether receptor molecule having four dihydronicotinamide derivatives as the side chain (Fig. 7.1).

In this complex, enhanced rate of H-transfer to the bound pyridinium salt substrate is observed. This represents the first example of accelerated 1,4-dihydropyridine to pyridinium H-transfer (transreduction) in a synthetic molecular macrocyclic receptor–substrate complex. Therefore, such a synthetic catalyst displays some of the characteristic features which molecular catalysts should possess. It provides both a receptor site for substrate binding and a reactive site for transformation of the bound substrate. Consequently it is of interest as both as enzyme model, and as a new type of efficient and selective chemical reagent (278).

When designing models of coenzymes it is important to examine the structure of the corresponding enzymes. The three-dimensional structure of horse liver alcohol dehydrogenase has been resolved at 0.24 nm resolution by the group of C. I. Brändén from Uppsala (280) and has been correlated to a number of physical and chemical studies in solution. The active enzyme has a molecular weight of 80,000 and is a dimer of two identical subunits.

Each subunit is composed of a single polypeptide chain of 374 amino acids. Furthermore, each subunit is divided into two separate domains, each associated with a particular function. There is a coenzyme (NAD^+) binding domain where the adenine part of NAD^+ is oriented in a hydrophobic pocket and there is a catalytic domain which binds different substrate molecules. Three residues, Cys-46, His-67, and Cys-174, within this domain provide ligands to the catalytic zinc atom. Interestingly, a second zinc atom is present in this region and is liganded to four other cysteine residues. The function of this extra zinc atom remains unknown but it has been suggested that it might be essential for the structural stability of the enzyme. The similarities with ferredoxin (Section 6.4) also suggest that this region might be a catalytic center for a redox process possibly with one or two of the cysteine ligands as catalytic group (280). Finally, the two domains of the subunit are separated by a crevice that contains a wide and deep hydrophobic pocket with the catalytic zinc atom at the bottom of this pocket.

Obviously, the system is more complex than previously thought since coenzyme and substrate are bound in different domains on the enzyme. Future models of NAD^+-dependent enzymes will thus have to take these observations into consideration.

Last, let us consider the possibility of a mechanism other than a hydride transfer in NAD^+ chemistry. Indeed, G. A. Hamilton argued that if a direct hydride transfer process occurs in dehydrogenase reactions, it is unique in biology since proton transfer would be more favorable (279). However, it is not a simple task to distinguish between these two possibilities. Generally, it is simpler to say that the reduction reaction is analogous to a transfer of two electrons rather than postulating a hydride ion. More will be said on this subject in Section 7.1.3 on flavin coenzyme.

7.1.2 Nonenzymatic Recycling of Coenzymes and Some Applications in Organic Synthesis

In recent years, the requirements of synthetic chemists for reagents capable of effecting selective or asymmetric transformations have increased dramatically. Enzymes present unique opportunities in this regard, and the exploration of their properties as chiral catalysts is now receiving considerable attention. Of course, one of the great synthetic attractions of an enzyme is that the various facets of its specificity can endow it with the potential for effecting highly controlled and selective transformations in a single step (281).

The oxidoreductase of most current utility is horse liver alcohol dehydrogenase (HLADH), and its application in the stereoselective oxidoreduction of ketones and alcohols has been explored. It requires an NAD^+ coenzyme which becomes expensive if reactions have to be performed on a preparative (up to 5 g) scale. In fact, this cofactor sells for as much as $250,000 per mole!

To overcome this problem, methods of recycling this coenzyme were developed. One of these is a nonenzymatic method, studied by J. B. Jones

(282–284) of the University of Toronto, for regenerating catalytic amounts of the nicotinamide coenzyme continuously. (It should be recalled that Section 4.7, devoted to enzyme technology and its application to chemistry and medicine, has been presented.)

The method of Jones uses sodium dithionite to regenerate NADH from catalytic NAD^+. It has been shown to be preparatively viable for HLADH-catalyzed reductions of a broad range of aldehydes and ketones in high yields.

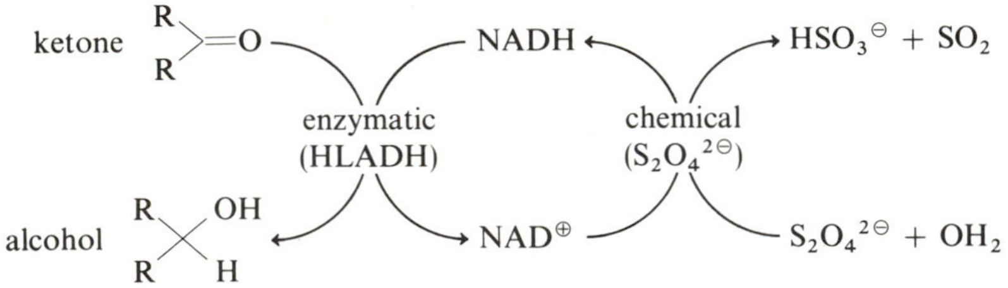

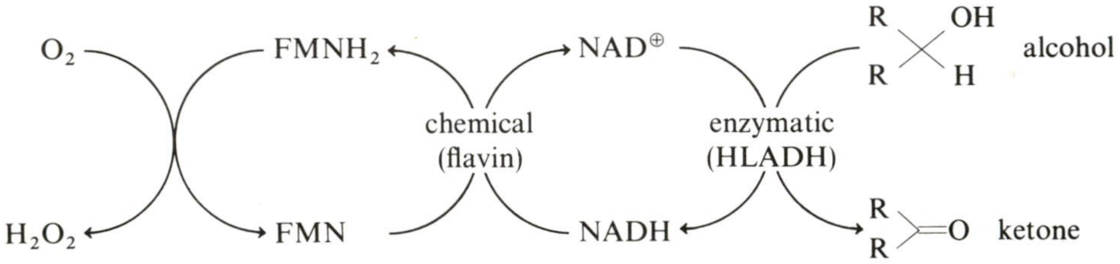

For the opposite mode (oxidation of alcohol to ketone) the system needs to be coupled to a flavin cofactor ($FMNH_2 \rightarrow FMN$) where oxygen is reduced to H_2O_2. The practicability of applying HLADH as a chiral oxidoreduction catalyst was demonstrated in many instances. Two examples are as follows:

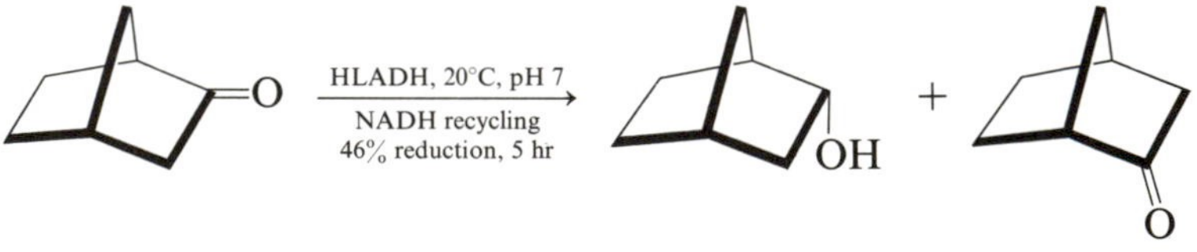

(±)-2-nonbornanone

(+)-(1S,2R,4R)-endo
exclusively
64% opt. pure
39% yield

(−)-(1R,4S)
47% opt. pure
31% yield

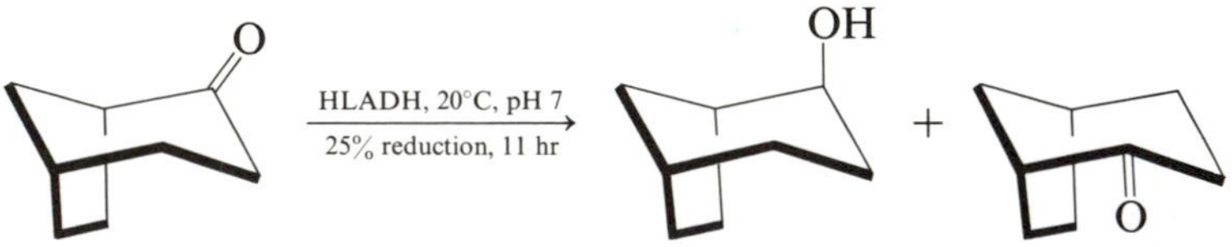

(±)-bicyclo[3.2.1]-
2-octanone

(−)-(1S,2S,5S)-exo
(difficult to get
otherwise)
83% opt. pure
36% yield

(−)-(1R,5R)
17% opt. pure
58% yield

Figure 7.2. Diamond lattice section for HLADH, as developed originally by Prelog (282). Reproduced with permission. Copyright © 1978 by the American Chemical Society.

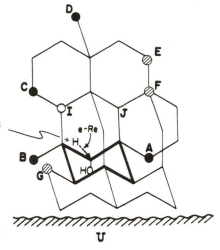

approach of $H^{\ominus}$ from the *re*-side of the ketone to deliver an equatorial hydrogen (e-*re* direction of HLADH)

Both reactions are performed with a high degree of stereospecificity. These stereochemical results are interpretable in terms of the diamond lattice section approach of V. Prelog (285), using the composite model shown in Fig. 7.2.

By fitting a cyclohexanone in the diamond lattice, Prelog has developed a step-by-step analysis of the HLADH catalyzed reduction. The positions marked ● (A to D) are "forbidden"; oxidoreduction will not take place if binding of a potential substrate places a group in one of these locations. Positions marked ⊘ (E to G) are "undesirable," although their occupation by part of a substrate does not necessarily preclude the oxidoreduction. The rate of reaction will be very slow. The positions under the lattice (U) are also in this category. The location ○ (I) is a newly identified "unsatisfactory" position. Placement of a group here is to be avoided if possible, but slow reaction will still take place if it is occupied.

Returning to the previous two examples, orientations of the enantiomers in their preferred "flat" positions within the diamond lattice section of HLADH allows an easy interpretation of the observed stereospecificity. This is illustrated in Figs. 7.3 and 7.4.

Jones has also examined the oxidation of a dihydroxycyclopentene and found that HLADH has the ability to retain its enantioselectivity while effecting regiospecific oxidation of only one of two unhindered hydroxyl groups within the same molecule.

OH ... OH (±)	HLADH, 20°C, pH 9 NAD⊕ recycling 40% oxidized, 4 hr →	O ... =O (+)-(1R,2S) 49% opt. pure 48% yield	+ ...OH OH (−)-(1S,2R) 23% opt. pure 35% yield

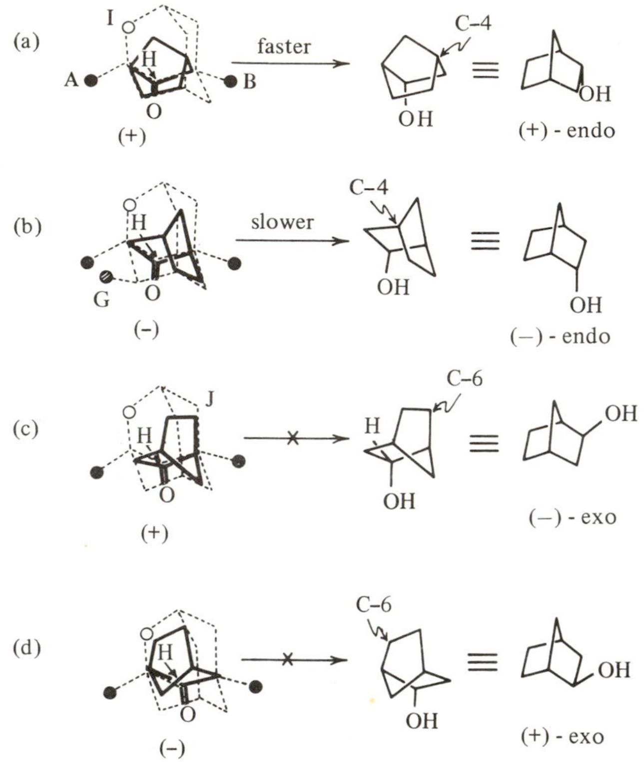

Figure 7.3. Diamond lattice section analysis of the stereochemical course of HLADH-catalyzed reduction of the 2-norbornanone enantiomers. The orientations shown are considered to be those of the "alcohol-like" transition states which would be involved, with H being delivered from the e-*re* direction of Fig. 7.2. There are no unfavorable lattice interactions when (+)- or (−)-isomers are positioned as shown in (a) and (b), respectively. Reductions to the endo-alcohols (+)- and (−)- are thus permitted processes. The predominant formation of the (+)-enantiomer is thought to be due to the preference of C-4 for its unhindered location in (a) over its position in (b) in which it approaches the forbidden A,I,C region of the lattice. The discrimination of HLADH against exo-alcohol formation is accounted for by the unfavorable interactions of C-6 with lattice positions J and I, as represented in (c) and (d), respectively (282). Reproduced with permission. Copyright © 1978 by the American Chemical Society.

This provides a synthetically useful combination of properties which cannot be duplicated in a single step by traditional chemical oxidation methods. Figure 7.5 shows that the regiospecificity observed is as predicted by the diamond lattice section of the active site.

The use of rigid molecular models is thus considered essential when analyzing or interpreting substrate behavior in diamond lattice terms. Hopefully, this approach and others with different enzymes and coenzymes will attract more organic chemists to use enzymes in difficult steps of a complicated organic synthesis.

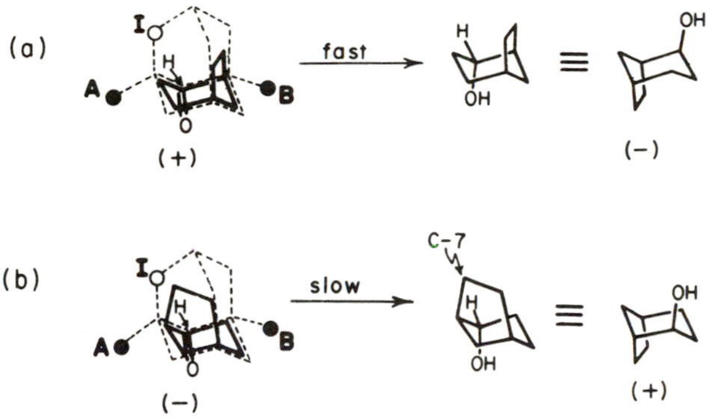

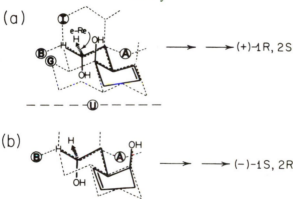

Figure 7.4. Schematic representation of the orientations of the enantiomers of bicy-clo[3.2.1]-2-octanone in their preferred "flat" positions within the diamond lattice section of HLADH. Delivery of H to the carbonyl group from the e-*re* direction ensures the formation of an exo-alcohol. In (a) orientation of (+)-isomer as shown does not place any substituents at undesirable positions. Reduction to the observed (−)-product is thus facile. The corresponding orientation of the enantiomeric (−)-ketone is shown in (b). Here, C-7 is required to locate close to the unsatisfactory position I and reduction in this mode is not favored. Formation of (+)-product is thus a relatively slow process (282). Reproduced with permission. Copyright © 1978 by the American Chemical Society.

Figure 7.5. Diamond lattice analysis of the regiospecificity of the cyclopentene diol. The relevant portions of the lattice, including the forbidden or undesirable positions A, B, G, I, and U are indicated by the dashed lines. When the hydroxyethyl groups of (+)-or (−)-compounds are oriented such that removal of the *pro-R* hydrogen from the e-*re* direction can occur as required by the model, unfavorable interactions with the lattice are completely avoidable. Substrate placements meeting these requirements and leading to the observed products are depicted in (a) for (+)-isomer and (b) for (−)-isomer. In contrast, positioning of the secondary alcohol function of (+)- or (−)-starting materials in a manner permitting hydride removal from the e-*re* direction would compel a methy-lene group of the hydroxyethyl function to violate position B or A, respectively, as shown in (c) for (+)-isomer and (d) for (−)-isomer. Oxidation of the C-1 hydroxyl groups is thus precluded for both enantiomers (283). Reproduced with permission. Copyright © 1977 by the American Chemical Society.

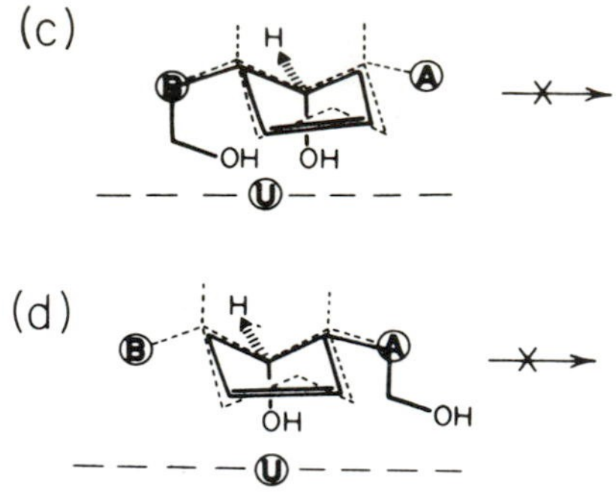

(c)

(d)

Figure 7.5. (continued)

7.1.3 Flavin Chemistry

Many hydrogenation–dehydrogenation processes are mediated by a flavin adeninedinucleotide (FAD) coenzyme where two electrons (equivalent to a hydride ion) from NADH are carried over to the respiratory chain (357). Flavins are the means of coupling one and two electron redox systems to cytochromes in the respiratory chain where oxygen is ultimately reduced to water. FAD is a coenzyme that like NAD^+ transfers electrons from a substrate, but in a much more versatile way. The structure of FAD was

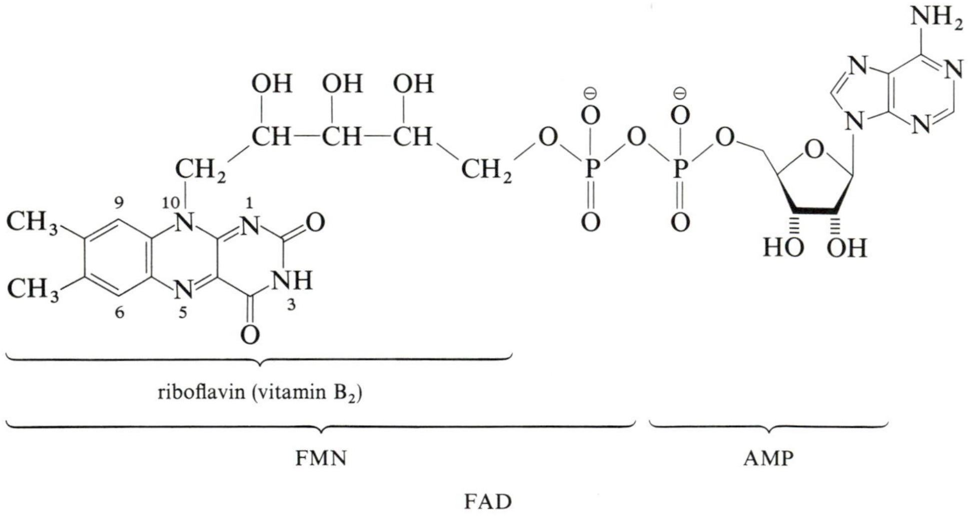

riboflavin (vitamin B$_2$)

FMN   AMP

FAD

determined in 1935 by R. Kuhn. The first mechanism of action was proposed in 1938 by O. Warburg and its total synthesis was accomplished in 1954 by Lord Todd's group. The hydrophilic nature of the substituent at N-10 is responsible for its solubility in an aqueous protein solution.

The site of fixation of an hydride ion from NADH (or two electrons) was suggested to be at the electrophilic nitrogen N-5 of the flavin nucleus.

$$NADH + FAD \xrightleftharpoons{H^\oplus} NAD^\oplus + FADH_2$$

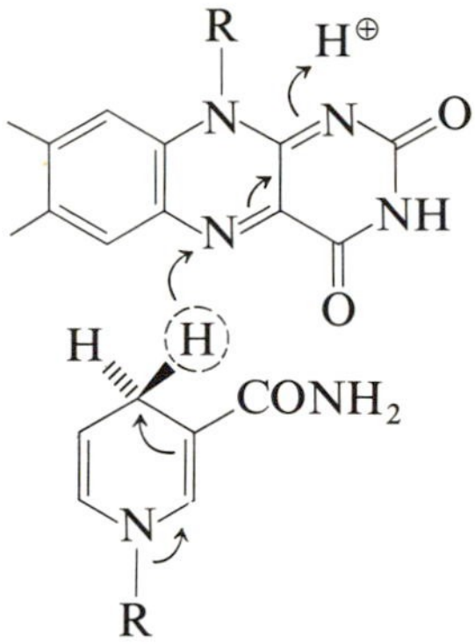

In 1973 Sigman showed that direct hydride transfer can occur from NADH (or d-4-NADH) by using an aromatic analogue; N-methyl acridinium salt (286).

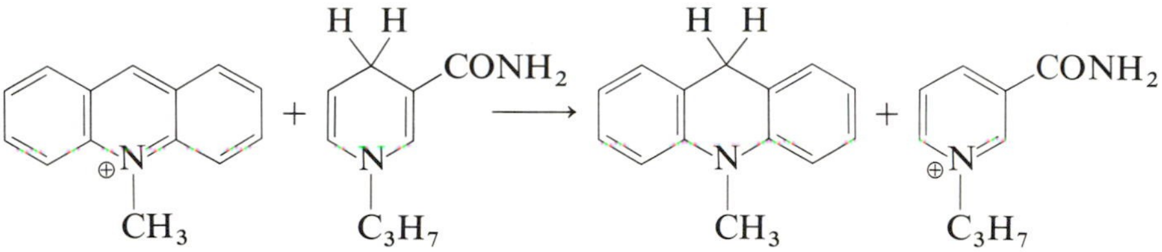

The reaction was also performed with a reduced nicotinamide molecule having a propyl side chain. The reverse reaction, however, does not take place. Free radical quenching agents had no effect on the rate of the reaction, suggesting an ionic rather than a radical mechanism. However, isotope effects with deuterated NADH show that under nonenzymatic conditions the reaction with FAD is not a simple bimolecular process.

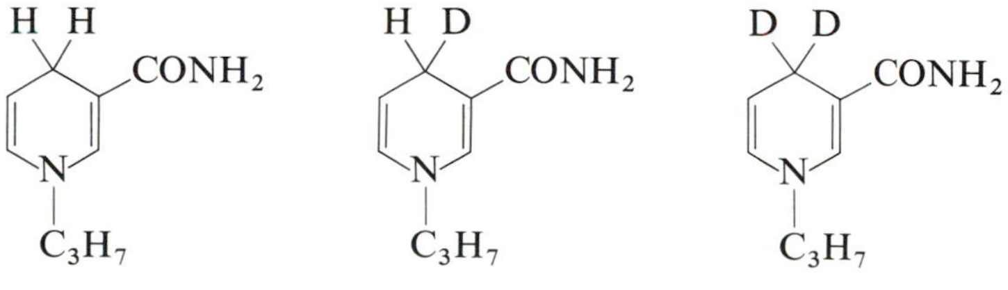

| $k(M^{-1}s^{-1})$ for loss of hydride at C-4: | 2040 | 1620 | 1398 |

From the above results a secondary isotope effect of 0.74 can be calculated. Since it is well known that conversion of an sp^3-carbon to sp^2 hybridization should give a secondary deuterium effect of greater than one, a simple bimolecular process is impossible. Consequently, a kinetically important intermediate must be postulated to account for the experimental facts.

Still no one general mechanism for flavin reactions exists. A charged-transfer complex cannot be excluded, but a free radical intermediate is also possible. It should be recalled that usually metal ions are required in flavin enzymes, and they may have an important role in the mechanism. In fact, the central position which flavoenzymes occupy in many biological pathways, after nicotinamide coenzymes (two electrons process) and before cytochromes (one electron process) in the respiratory chain,* may be caused by the versatility of the flavin structure, allowing both ionic and free radical mechanisms.

Free radical reactions are possible because the system can easily be oxidized or reduced by a one electron step via a stable intermediate called *semiquinone* (stabilized by ten resonance forms). In these planar structures,

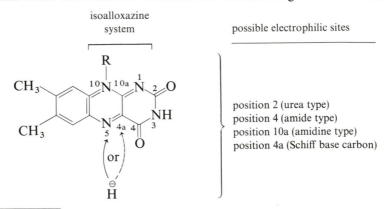

flavin radical   semiquinone radical

one ring always remains fully aromatic. On the other hand, if radicals are not intermediates, reactions must involve a hydride transfer or proceed by a PPC or PPM mechanisms. These two mechanistic classes were suggested by G. A. Hamilton, Pennsylvania State University (279): *PPC* (proton, proton, covalent bond compound) for systems that do not require a metal ion and *PPM* (proton, proton, metal) where a metal ion is involved. The flavin nucleus is thus structurally and electronically designed to stabilize both one- and two-electron transfers in redox reactions.

For the PPC mechanism (Hamilton mechanism) to be operative a covalent compound must be formed between the flavin coenzyme and the substrate, so that some mechanism is available for transmitting electrons from one

isoalloxazine
system   possible electrophilic sites

R
CH_3 10 N 10a N 1 2 O
                              NH
CH_3 N 4a 4 3
              5   O
           or
              H

position 2 (urea type)
position 4 (amide type)
position 10a (amidine type)
position 4a (Schiff base carbon)

---

* Cytochromes are electron carriers coupled to oxidative phosphorylation, a process in which ATP is formed as electrons are transferred from NADH or $FADH_2$ to molecular oxygen. The overall process consists of the oxidation of a substrate (glucose, for instance) where the flow of electrons cascade through a respiratory assembly (the cytochromes) to reach O_2 which is finally reduced to water.

molecule to the other. Since this reaction is ionic, there must be an electrophilic site on the flavin which can be attacked readily by the nucleophilic substrate. Examination of the flavin nucleus reveals four potential sites of attack, among which position 4a is believed to be the most reactive electrophilic site. The reactivity is in fact increased somewhat by the inductive effect of the adjacent amide and amidine functions.

Only two examples of well-known transformations are given to simply illustrate the point that these transformations are mechanistically feasible via the Hamilton mechanism.

Example 1: conversion of an alcohol to a ketone

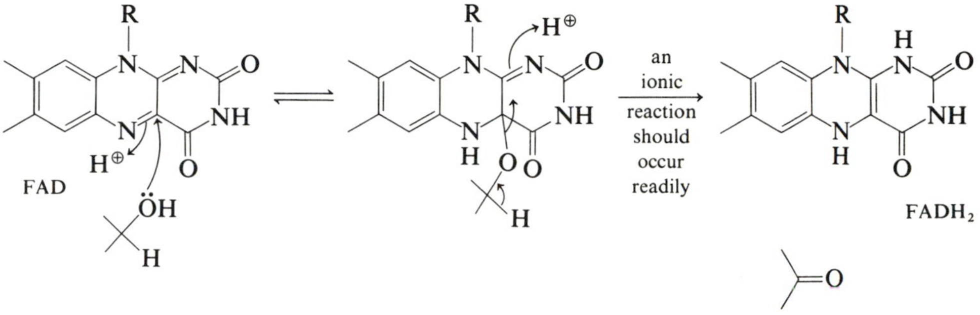

an ionic reaction should occur readily

FAD

FADH$_2$

Example 2: conversion of NADH to NAD$^+$

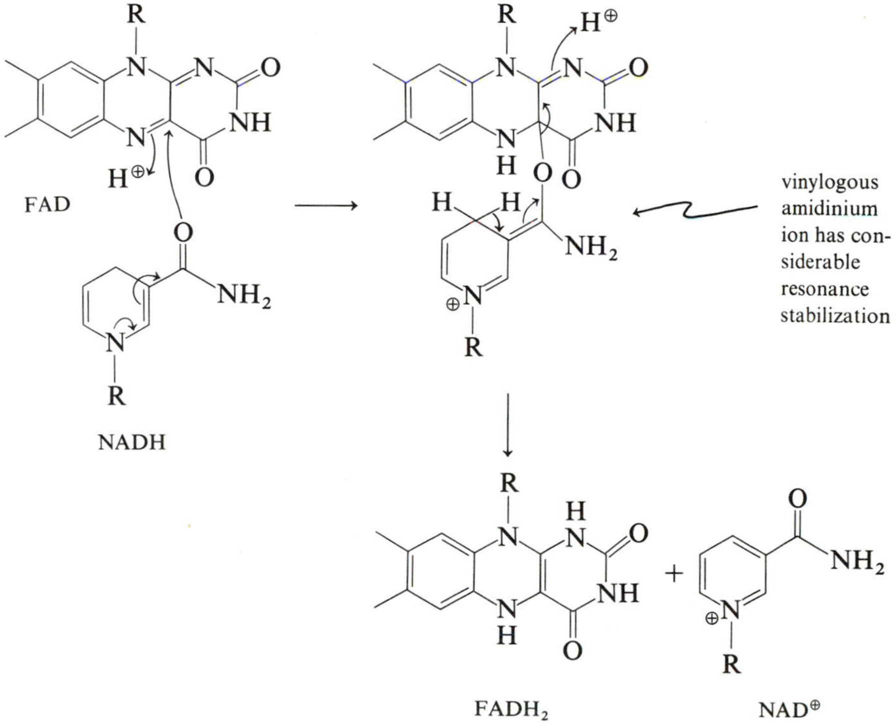

vinylogous amidinium ion has considerable resonance stabilization

FAD

NADH

FADH$_2$

NAD$^\oplus$

Clearly, no hydride transfer is involved, only proton shifts via the formation of a covalent intermediate at position 4a. Hamilton argues that biological oxidoreduction (dehydration) reactions rarely, if at all, involve hydride ions because protons are not shielded by electrons and thus travel much faster and more efficiently in biological media (279).

Here again model studies are necessary to clarify the situation.

Model 1

The first evidence of a 4a-adduct on a model compound appeared in 1967 (290).

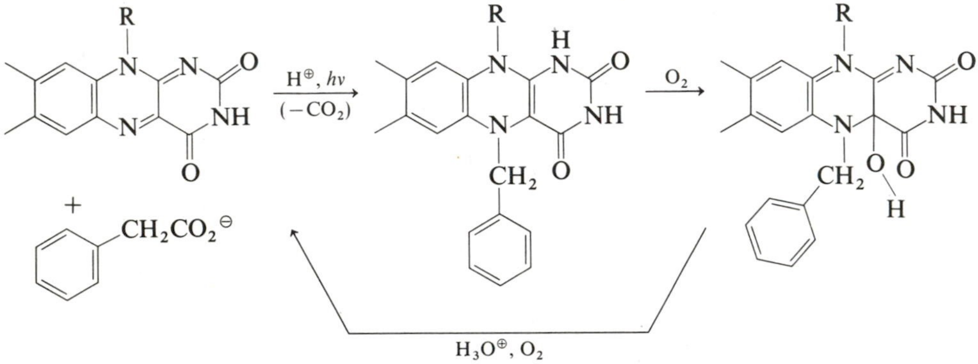

The compound was obtained by a photo-induced benzylation of flavin by phenylacetate. The resulting product reacts slowly with air and suggests that the reactivity of N-5 may be of importance in flavoprotein catalysis. It also indicates that a oxygen can add readily to position-4a if position N-5 is substituted, or protonated as it could be on the enzyme.

Although many models show reactivities at position-4a and N-5, the question as to what is the exact electrophilic site in enzymatic flavin chemistry has not been satisfactorily answered.

Model 2

The formation of a covalent bond between NADH and FAD has been suggested and previous studies indicated that a preequilibrium complex is probably formed with flavins. Second, theoretical calculations have shown that position N-5 should be a good electrophilic center. Therefore, Bruice (287) argued that changing that position for a carbon atom should also yield a strong electrophilic center with the advantage that the new C—H bond formed will be stable and not exchange with the solvent.

He synthesized the following flavin model compound and submitted it to the action of NADH in ^2H_2O for three days in the dark, under an argon atmosphere.

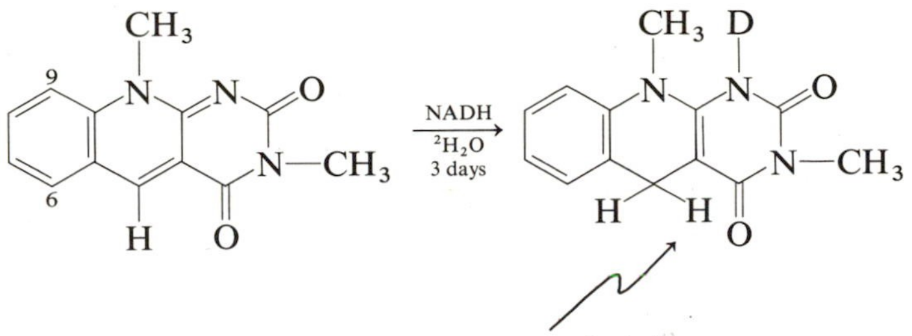

He obtained a reduced product with no deuterium incorporated at position-5. The results thus clearly shows that there is a direct hydrogen (or two-electron) transfer to position-5, and that in the preequilibrium complex, the NADH probably does not occupy the area adjacent to positions 1, 9, and 10. This result is a strong argument against the Hamilton proposal where this FAD analogue acts as an hydride acceptor.

Model 3

Similarly, a 5-deazariboflavin was synthesized and reacted in the presence of a NADH:FAD oxidoreductase (288). C. Walsh of M.I.T. showed that this 5-deaza-analogue functions coenzymatically, undergoing reduction by direct hydrogen transfer from NADH.

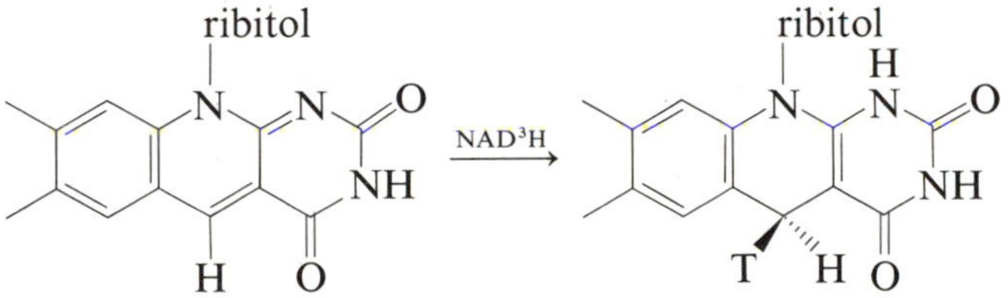

5-deazariboflavin

The analogue was reduced at 0.3% the rate of riboflavin. However, the reaction was not absolutely stereospecific but the oxidoreductase showed a preference for the R-isomer of $[4\text{-}^3H]NADH$ with attack on the *re*-face of 5-deazariboflavin. It turns out that a 5-deaza-FAM derivative (called factor F_{420}) has now been found as a natural cofactor in the anaerobic bacteria that produce CH_4 from CO_2 (289) and makes the study of analogous model systems even more significant.

Model 4

In 1979, Sayer et al. (291) used 1,3-dimethyl-5-(p-nitrophenylimino)-barbituric acid as a flavin model and were able to isolate a covalent intermediate in the reduction of this highly activated imine substrate by a thiol (methylthioglycolate or mercaptoethanol).

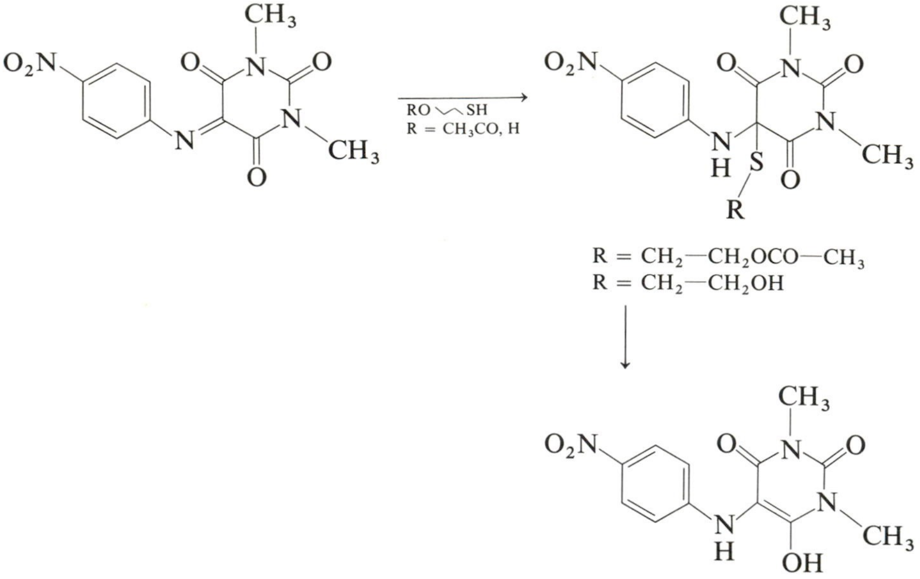

$$R = CH_2-CH_2OCO-CH_3$$
$$R = CH_2-CH_2OH$$

The thiol addition product was detectable spectrophotometrically and the intermediate with $R = CH_2CH_2OH$ was isolated and its structure proved by ^{1}H- and ^{13}C-NMR. The intermediate could then be converted to the dihydro compound showing that the addition must have occurred at position C-4a and not N-5. This work thus provides evidence for the existence of a thiol addition intermediate involving the C(4a)—N(5) bond in nonenzymatic reductions of flavins and analogues by thiols.

7.1.4 Oxene Reactions

Many oxidases (monooxygenases) are FAD-dependent and catalyze the oxidation of various substrates using molecular oxygen. FAD coenzymes are thus able to transfer oxygen to an organic substrate, another versatility in this cofactor not present in NAD^+.

The flavin-dependent monooxygenases bind and activate molecular oxygen, ultimately transferring one oxygen atom to substrate and releasing the second as water. Among the reactions catalyzed by the flavin monooxygenases are the hydroxylation of p-hydroxybenzoate, the oxidative decarboxylation of salicylate, and a variety of amine oxidations mediated by the mixed-function aminooxidase system (292). Apparently these reactions involve an "oxene" mechanism. This term is used because it is analogous to carbene and nitrene reactions, two other electrophilic species.

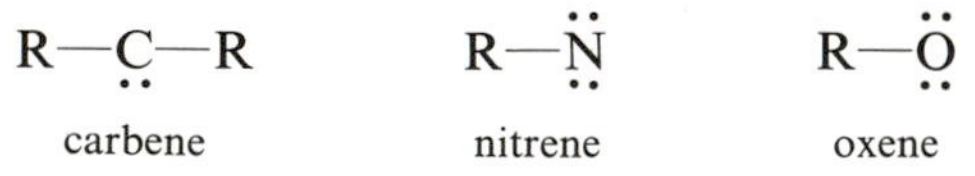

$$R-\overset{..}{\underset{..}{C}}-R \qquad R-\overset{..}{\underset{..}{N}} \qquad R-\overset{..}{\underset{..}{O}}$$

carbene   nitrene   oxene

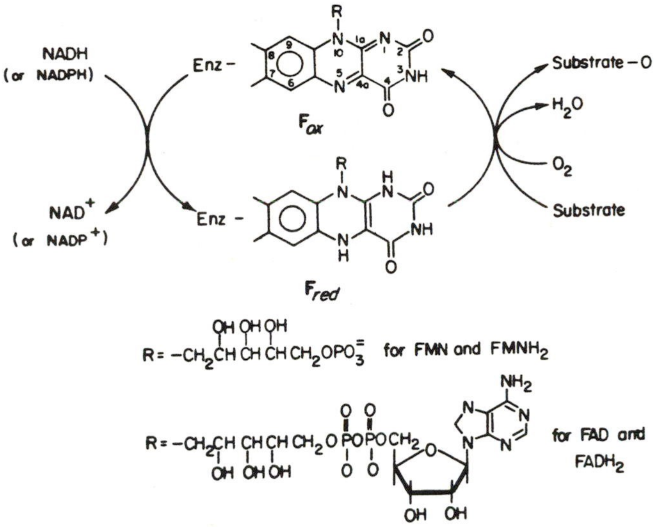

Figure 7.6. Oxygenation cycle for flavin monooxygenases (293). Reproduced with permission of D. Dolphin.

Some enzymes in these reactions require a transition metal ion while others do not. For simplicity, we will focus only on mechanisms not involving a metal ion.

One group of monooxygenases for which it is quite clear no metal ion is necessary requires a flavin as cofactor. Not requiring a metal ion also means that the reaction with oxygen could have a free radical character at some stage of the mechanism. Since substrate radicals are very unstable, it seems more likely that oxygen reacts with the reduced form of flavin to give an intermediate which then reacts by an ionic mechanism with the substrate. This way spin conservation is preserved. The oxygenation cycle for flavin monooxygenases is presented in Fig. 7.6.

In these aerobic flavin-linked monooxygenases the reduced flavin $FADH_2$ first reacts directly with oxygen to give a $FADH_2O_2$ adduct. This adduct is believed to rearrange to a flavin 4a-hydroperoxide, a good oxidizing agent, and a stronger oxidant than H_2O_2.

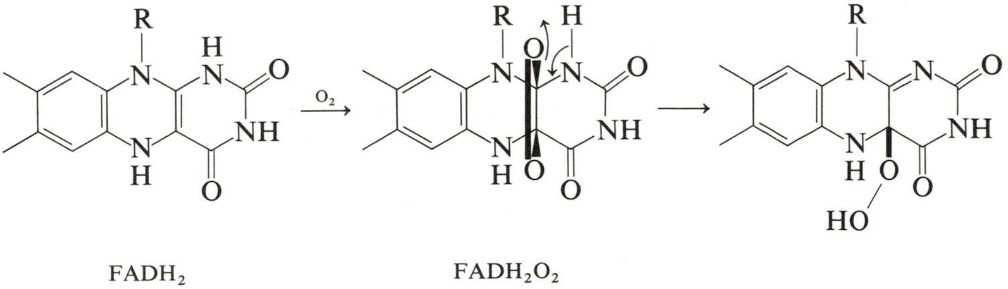

How does a flavin molecule activates molecular oxygen? It should be realized that in the transformation above, oxygen reacts as a ground state triplet molecule whereas organic molecules (flavin) are usually in the singlet state. However, the reaction of a singlet with a triplet to give a singlet product is a spin-forbidden process! Nevertheless, it is possible to obtain an ionic reaction of oxygen without requiring the formation of singlet oxygen if it is complexed to a transition metal ion which itself has unpaired electrons. Since, many oxidases do not require a metal ion to function, the situation is still not clear to say the least but a radical process with flavin is a possibility. In fact, the addition of oxygen to reduce flavin is similar to the reaction of oxygen with a substituted tetraaminoethylene, a strong electron-donating double bond.

a dioxetane   urea type

Furthermore, aromatic compounds (phenol) can be hydroxylated by oxygen in a nonenzymatic system when various reduced flavins are present. It is suggested that the hydroxylating agent is the flavin hydroperoxide which rearranges by a proton transfer to give the actual hydroxylating agent. It is a carbonyl oxide. Similar intermediates with the reaction of ozone on double bonds are known.

flavin hydroperoxide   carbonyl oxide

Since carbene and nitrene are electrophilic agents, the terminal oxygen will have to become more electropositive by stabilizing a negative charge on the rest of the molecule.

This mechanism was first proposed by Hamilton (279). It has, however, the disadvantage of involving an open form of the flavin ring.

Other ionic mechanisms can also be operative. The transfer of oxygen can occur via a radical type transfer rather than ionic mechanism (see page 413). At this point we could ask why nature selected flavin and not nicotinamide coenzyme to transfer oxygen to a substrate? The answer probably lies in the fact that NAD^+ can only transfer electrons via an hydride ion whereas

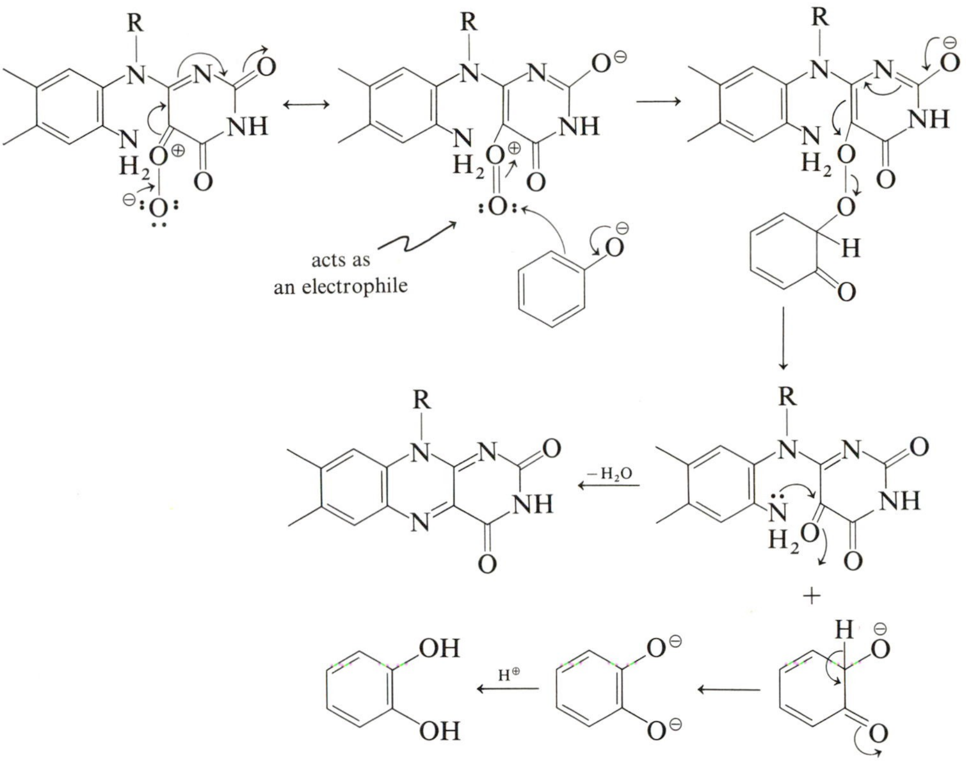

FAD has the chemical versatility to be converted to an oxygen-adduct intermediate which can then react with the substrate so that a direct transfer of oxygen to the substrate does not take place.

To overcome the difficulty encountered in Hamilton's mechanism, Dolphin in 1974 proposed an interesting alternative (293) where no open-ring product intermediate is formed. His mechanism involves a reactive oxaziridine-4a,5-flavin, obtainable from flavin hydroperoxide.*

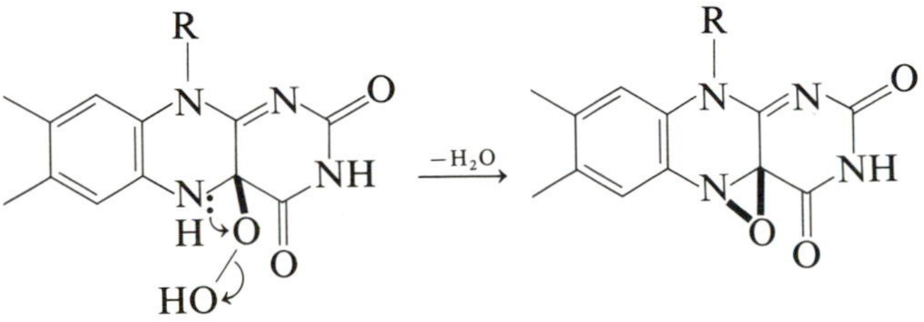

oxaziridine

---

* As mentioned earlier (page 405), 5-deazaflavin analogues are more susceptible to nucleophilic attack which typically occurs at position 5 to yield 1,5-dihydro adducts. In this respect, the formation of a 4a,5-epoxy-5-deazaflavin derivative by the reaction of 5-deaza-isoalloxazine with H_2O_2, *tert*-butyl hydroperoxide, or *m*-chloroperbenzoic acid has been reported recently (294).

This postulate is based on numerous chemical precedents,* among which is the following which shows the electrophilic character of the oxaziridine system:

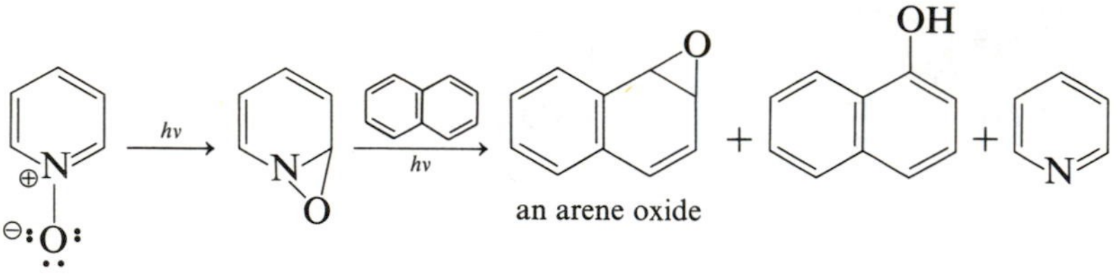

an arene oxide

So, the following rational mechanism can be written for the hydroxylation of phenol:

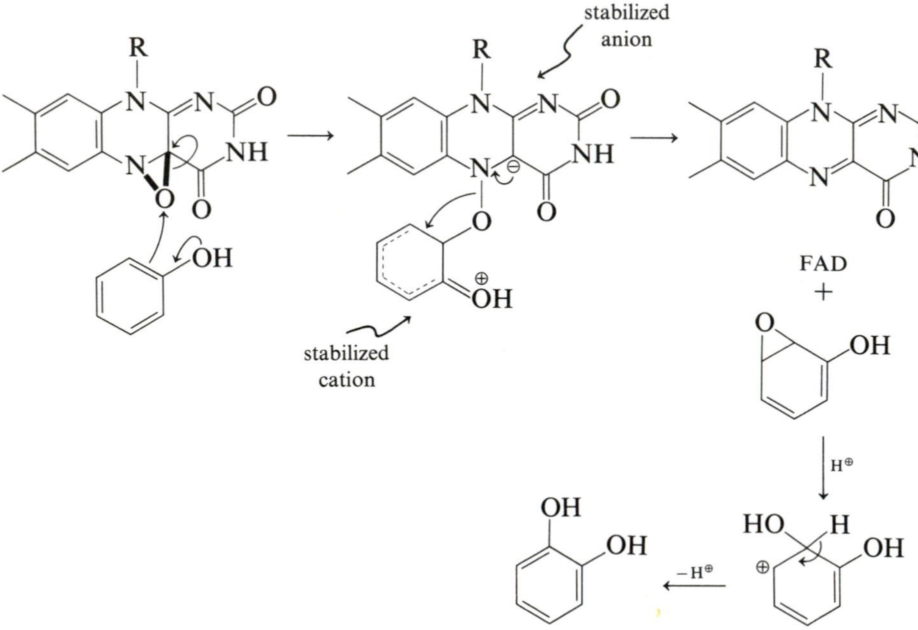

stabilized anion

stabilized cation

FAD +

Interestingly, transfer of oxygen from the oxaziridine reagent leads to a benzene epoxide intermediate.

B. Witkop and his group (295) from N.I.H. already suggested in 1969 that epoxides are intermediate in a number of monooxygenase catalyzed oxi-

---

* Recently, the synthesis of a new class of oxaziridines (arenesulfonyl-3-aryl-oxaziridines) was reported. These stable compounds were prepared by the oxidation of sulfonimines and were noted to have useful potential as oxidizing agents (359).

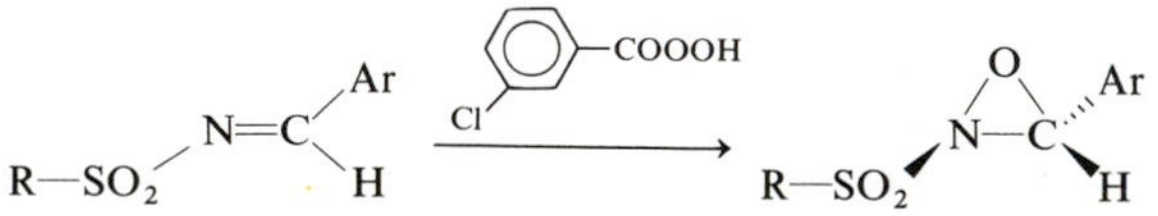

dations of aromatic compounds. For example:

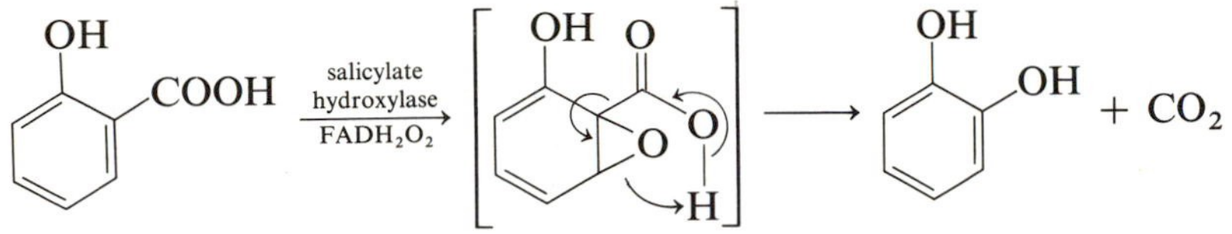

Once again, a flavin cofactor is being the agent of epoxidation.
Flavin hydroperoxide is also implicated in oxidative decarboxylation:

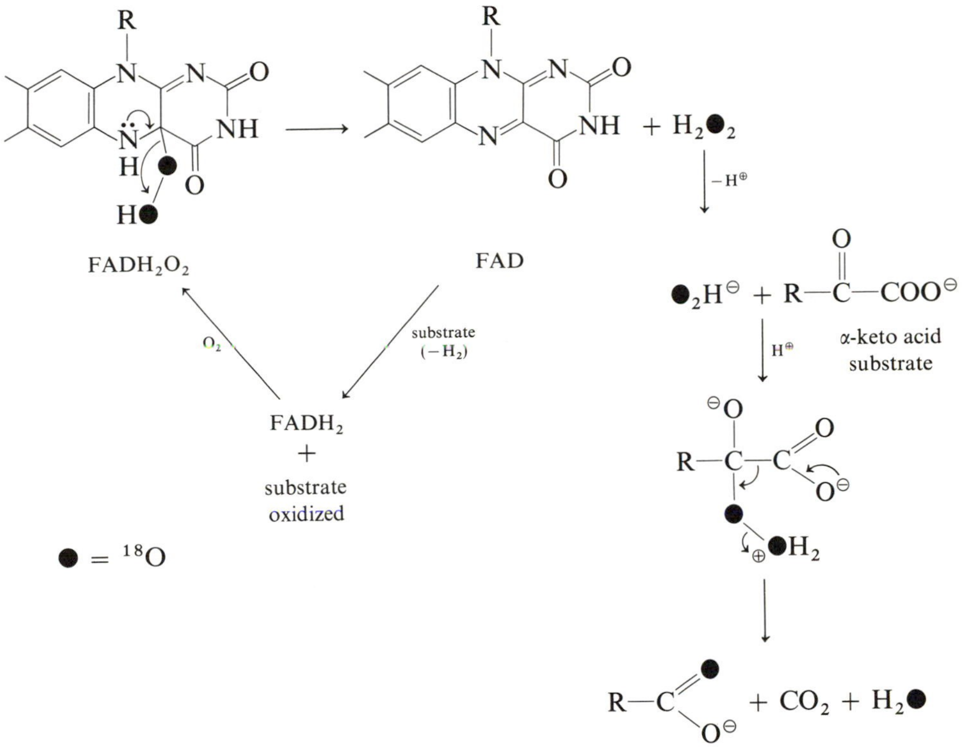

$$FADH_2O_2 \qquad\qquad FAD$$

$$\bullet_2 H^\ominus + R-\overset{\overset{\displaystyle O}{\|}}{C}-COO^\ominus$$

$$\text{substrate} \atop (-H_2)$$

$$\alpha\text{-keto acid substrate}$$

$$FADH_2 \\ + \\ \text{substrate} \\ \text{oxidized}$$

$$\bullet = {}^{18}O$$

$$R-\overset{\ominus O}{\underset{}{C}}-C\overset{\displaystyle O}{\underset{\displaystyle O^\ominus}{\diagup}}$$

$$\oplus \bullet H_2$$

$$R-C\overset{\displaystyle \bullet}{\underset{\displaystyle O^\ominus}{\diagup}} + CO_2 + H_2\bullet$$

It was shown by using $^{18}O_2$ that only one of the two oxygen atoms ends up in the carbonyl group. An interesting example of this operation is the conversion of p-hydroxyphenylpyruvic acid to homogentisic acid by a dioxygenase:

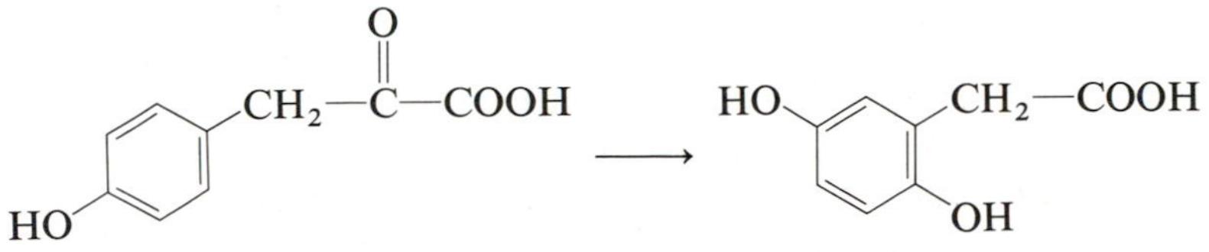

p-hydroxyphenylpyruvic acid   homogentisic acid

Witkop first proposed the following mechanism:

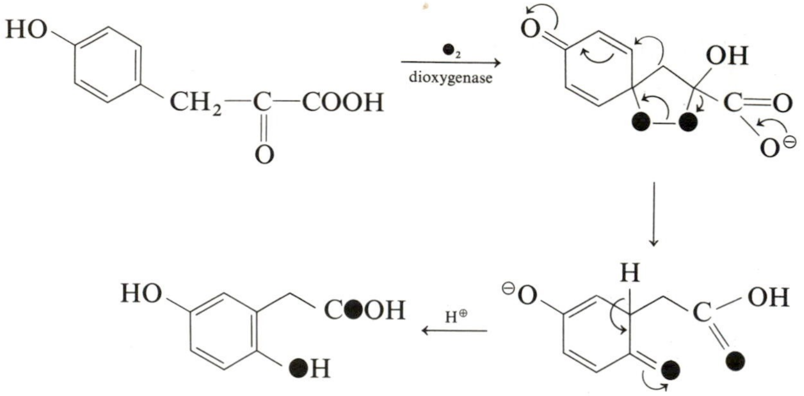

A side chain migration to the *ortho* position takes place in this rearrangement and this has been referred to as the *NIH shift* (296). However, this mechanism was questioned because the reaction still worked with phenylpyruvic acid (no *p*-OH). A more likely mechanism would be the involvement of a transition metal ion to form first a peracid which would then undergo an "oxene" mechanism via an epoxide intermediate:

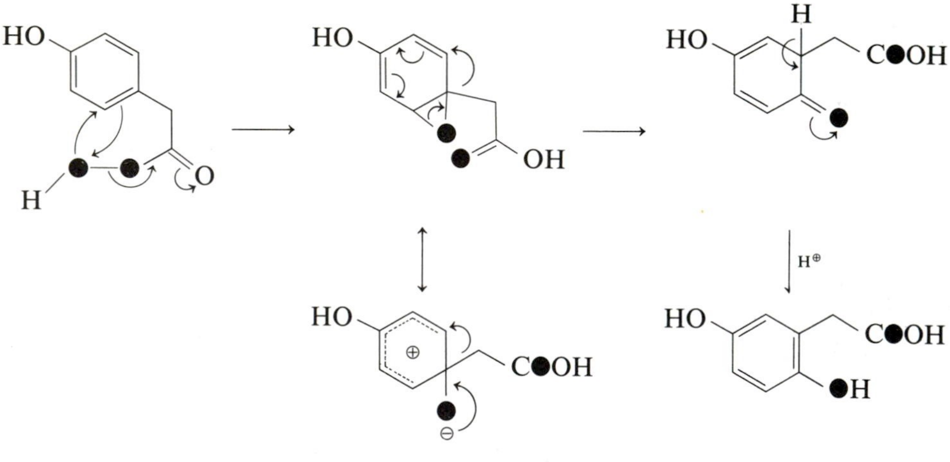

The proposal by Dolphin of an oxaziridine mechanism for flavin mediated oxidations has been criticized by Rastetter's group (292) since oxygen insertion in enzymatic oxidations is not induced photochemically. What they have shown however is that nonenzymatic oxidations can be effected by a flavin N(5)-oxide (nitrone). Upon irradiation in the presence of a phenol, a flavin nitroxyl radical and a phenoxy radical are generated, as detected by EPR spectroscopy. The coupling of these two species leads to flavin and hydroquinone.

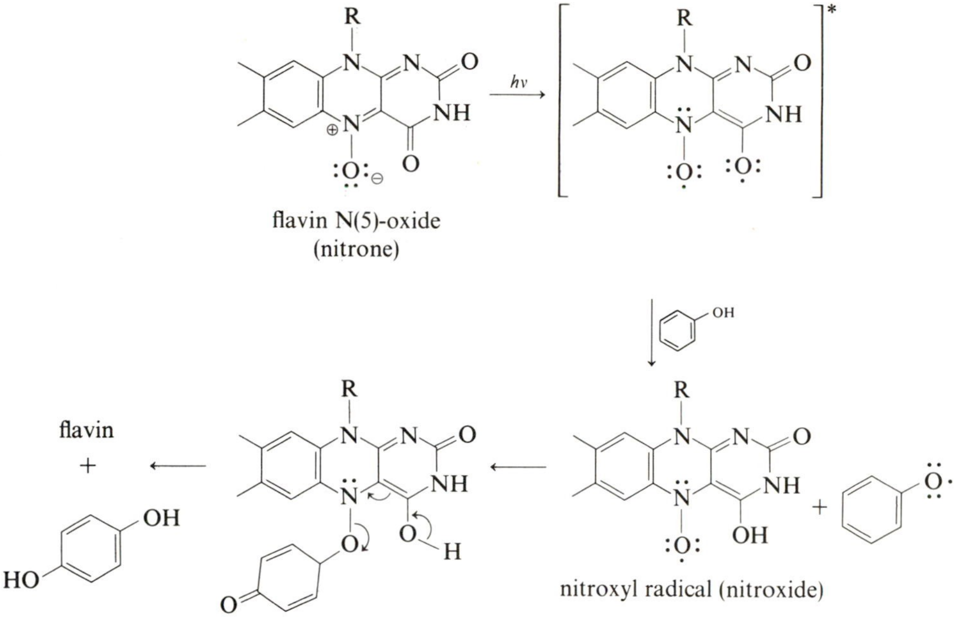

flavin N(5)-oxide
(nitrone)

nitroxyl radical (nitroxide)

The authors argue that flavin nitroxyl radical may play a role in flavo-enzyme chemistry and suggest that the nitroxyl radical intermediate may be derived from enzyme-bound flavin 4a-hydroperoxide.

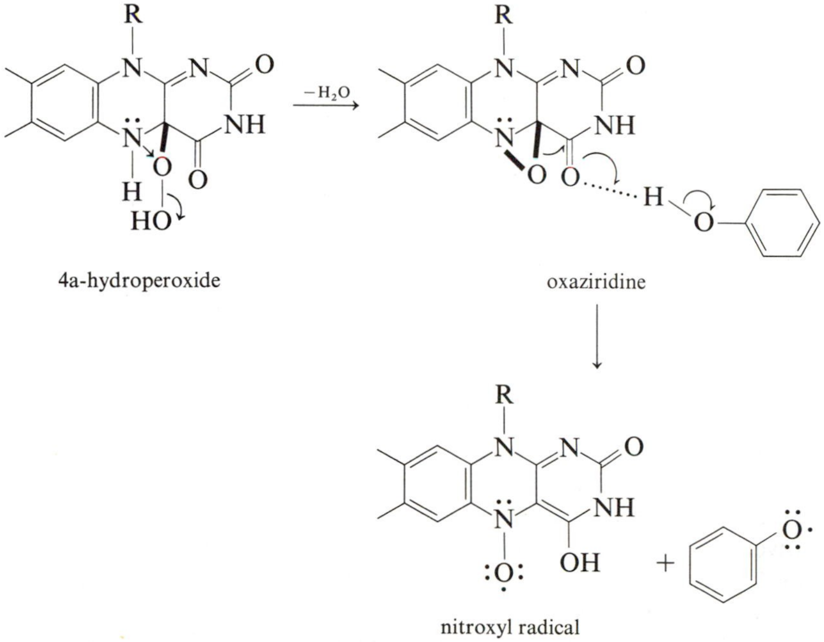

4a-hydroperoxide

oxaziridine

nitroxyl radical

Such a radical-pair mechanism avoids the need to invoke a "oxene" intermediate during the oxygen transfer by the flavoenzyme, and at the same time releases the ring strain of the oxaziridine species.

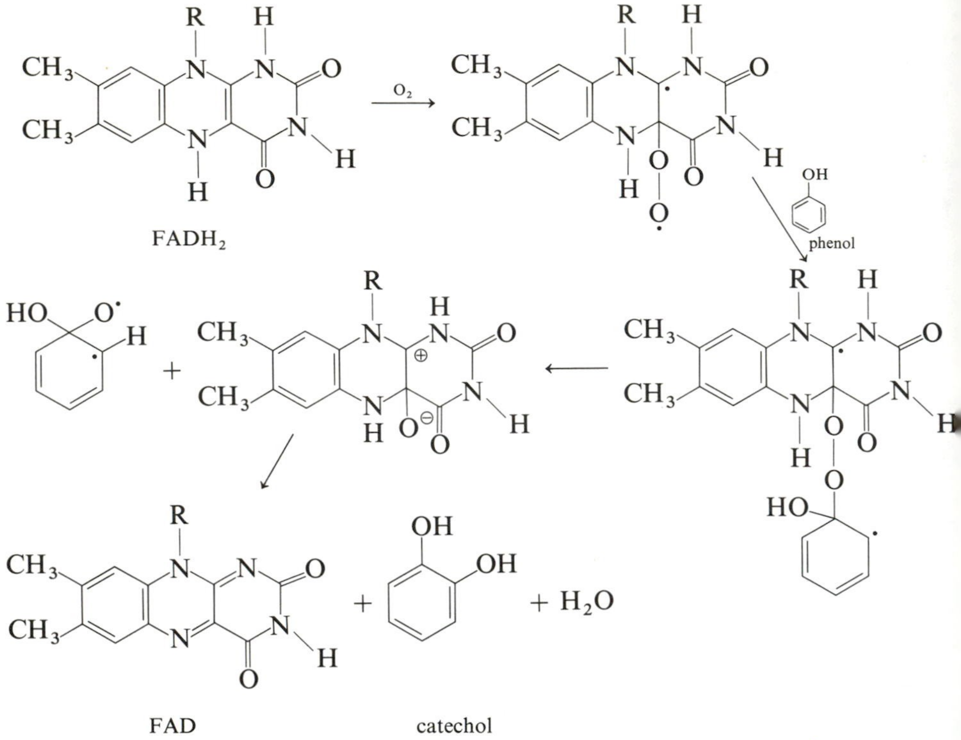

Figure 7.7. Goddard's proposal of how flavin coenzyme carries out the oxidation of phenol(297).

In 1978, Caltech's W. A. Goddard proposed a different mechanism for the hydroxylation of phenolic compounds and attempted to show how flavin coenzymes carry out such oxidations. It is a theoretical proposal based on wave functions and quantum mechanics using generalized valence bond theory, applied to biological problems (297). An example is shown in Fig. 7.7 for the oxidation of phenol to catechol.

In the first step of this oxidation process, triplet oxygen attacks flavin, forming a biradical. In a way flavin helps to activate oxygen and for this, one of the unpaired electrons of the biradical is stabilized by the flavin ring, and the other remains associated with the oxygen atom that reacts with the phenol in the second step. This leads to an intermediate in which the flavin and phenol are joined through the two oxygen atoms of molecular oxygen. Then, the flavin molecule helps drive the reaction to the next step where the lone-pair electrons on the nitrogen atoms at positions 5 and 1 stabilize the positive charge of the intermediate. According to Goddard, the nitrogen atoms in flavin assist in breaking the oxygen–oxygen bonds, an uphill (energetically) step in the reaction.

Interestingly, the mechanism postulated involves an attack on phenol by an oxygen atom at the carbon bearing the hydroxyl group. That radical intermediate is believed to be lower in energy than the one with oxygen on the adjacent ring carbon, according to bond energy calculations. The oxygen later migrates in a rearrangement reaction. At this point, the catechol product has been made, but the flavin still contains an extra oxygen. It is believed to be removed from the flavin by picking up protons and forming water. The process would certainly be enzyme mediated. The proof of this proposal awaits model studies.

Among the miscellaneous redox reactions in nature, the emission of visible light by living beings is one of the most fascinating of natural phenomena. Luminescent bacteria, worms, and the amazing firefly have all been the objects of the biochemists' curiosity. A natural question is: What kind of chemical reaction can lead to emission of energy in the visible region of the spectrum? Although an AMP-intermediate is found, the process requires far too much energy to be provided by the hydrolysis of ATP itself. A clue comes from the fact that chemiluminescence is very common when O_2 is used as an oxidant and extraction of luminous materials from organisms showed that one of the active principle in the firefly system is luciferin. This compound is decarboxylated by the enzyme luciferase (298).

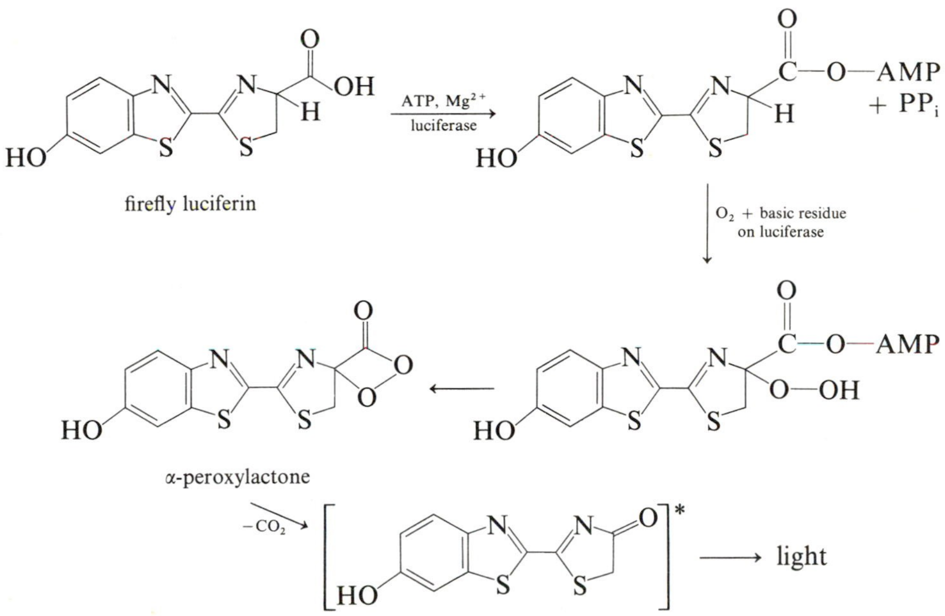

firefly luciferin

α-peroxylactone

electronically excited oxyluciferin

After activation by ATP, luciferin is believed to be converted to a α-peroxylactone by dehydrative cyclization. This intermediate serves as a chemienergizer for the formation of an electronically excited oxyluciferin by decarboxylation.

This enzymatic oxygeneation by carbanion attack on O_2 followed by decarboxylation occurs without conjugated-cofactor assistance and is thus different from the mode of action of bacterial luciferase, a flavin-dependent monooxygenase (299). In the bacterial system, a 4a-flavin hydroperoxide undergoes a chemiluminescent reaction in the presence of an aldehyde.

A simple model compound for the oxidation–decarboxylation sequences is now known (300). Dehydration of the following very reactive α-peroxyacids by DCC yields the corresponding α-peroxylactone (dioxetanone).

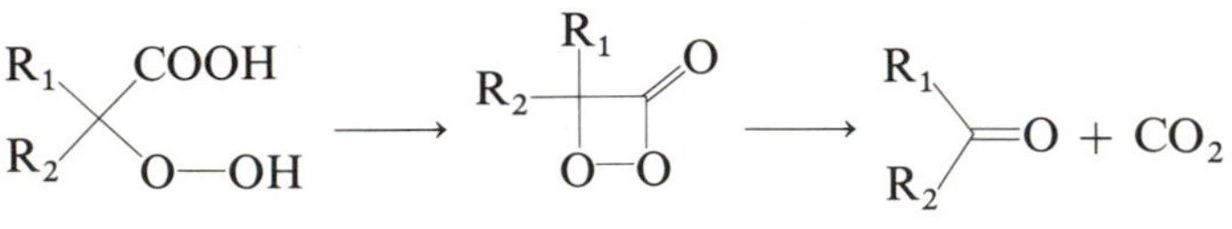

$R_1 = R_2 = CH_3$
$R_1 = R_2 = t$Bu
$R_1 = t$Bu, $R_2 = CH_3$

Thermolysis of the α-peroxylactone in CH_2Cl_2 at 25°C leads to the quantitative generation of the ketone and to light emission. The decarboxylation is believed to take place in a first step by a one-electron transfer from a catalytic amount of an aromatic activator present in the solution (300).

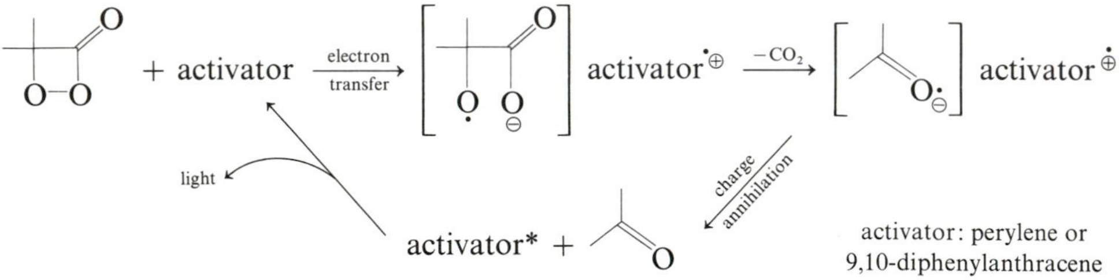

activator: perylene or
9,10-diphenylanthracene

The next step along the chemiluminescence path is the rapid loss of CO_2 from the reduced dioxetanone. The final step in the sequence is light emission from the excited activator, which is detected as chemiluminescence. It seems that the central feature of chemiluminescent reactions is radical anion formation.

7.1.5 Lipoic Acid

Lipoic acid (1,2-dithiolane-3-valeric acid) is widely distributed among microorganisms, plants, and animals. It belongs to the group of cofactors containing sulfur and in nature it is coupled to thiamine pyrophosphate (see Section 7.3). However, lipoic acid basically belongs to another class of electron transfer cofactors where the net oxidoreduction function is to produce ATP. The cofactor is needed in fatty acid synthesis and in the metabolism of carbohydrates.

Indeed, it has been recognized for a long time that lipoic acid is an interesting growth factor found in a number of microorganisms. Chemically, it can be reduced and the reduced form can be readily reoxidized to lipoic acid.

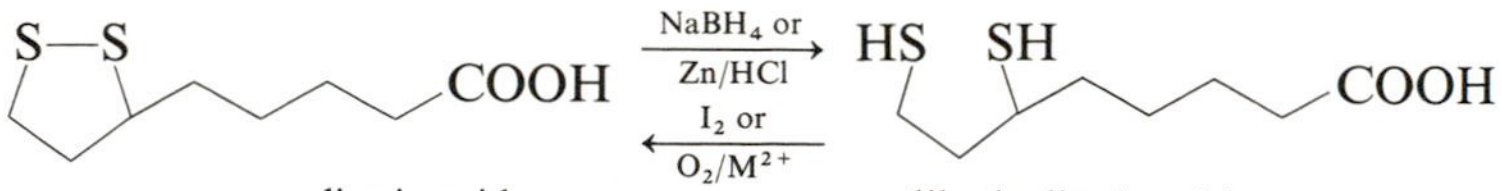

Dihydrolipoic acid is an efficient reducing agent of sulfate ion to sulfite. The sulfate is first activated as an adenyl sulfate, also called adenosine

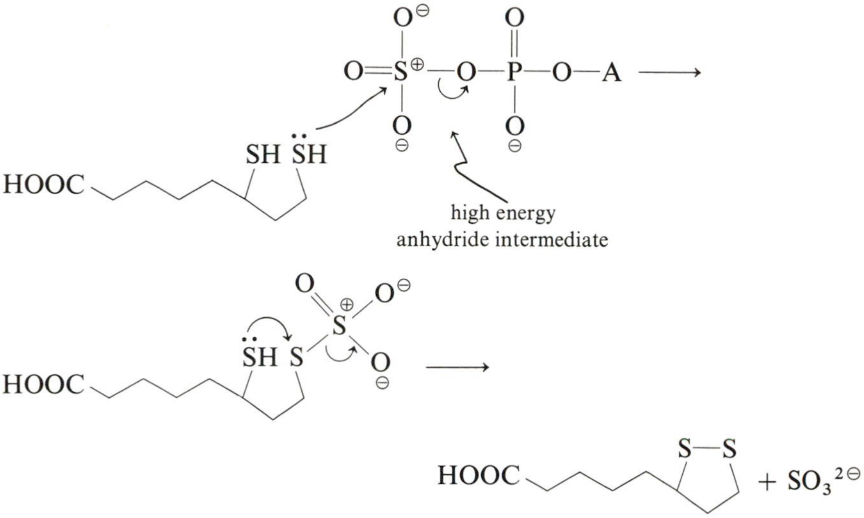

5'-phosphosulfate (APS). The formation of lipoic acid is an entropy favored process because both sulfur atoms are on the same molecule.

One should realize that the five-membered ring of lipoic acid is not planar but has a C—S—S—C dihedral angle of 26°. Normally, disulfides are colorless but lipoic acid is yellow. This property is attributed to the ring

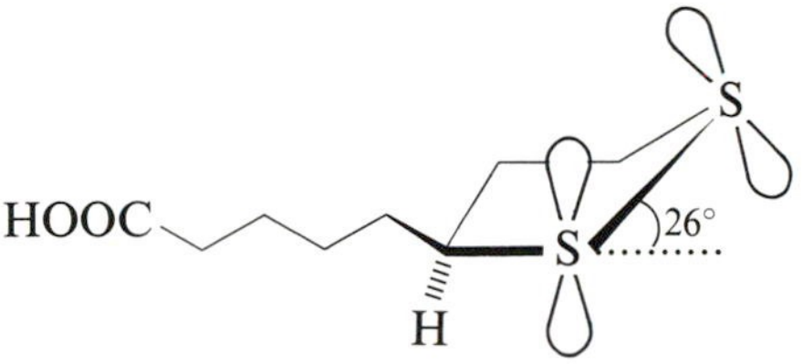

strain. This ring strain is caused in part by the repulsion of electron pairs on adjacent sulfurs and makes lipoic acid a better oxidizing agent than a less strained six-membered ring analogue. There is thus a structure–function relationship to its biological role (301).

Finally, on the multienzyme complex that performs the oxidative decarboxylation of an α-keto acid, lipoic acid is not found free but is covalently linked to a lysine residue, through an amide linkage (Fig. 7.8). The swinging arm of this prosthetic group coordinates other subsites of the complex so that the efficiency of the process is related to this unique chemical organization at the subcellular level.

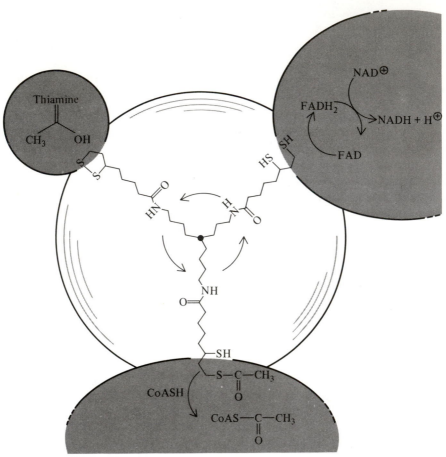

Figure 7.8. Diagram of the enzyme complex in the oxidative decarboxylation of pyruvic acid.

In the cell the thioester of dihydrolipoic acid is not stored but is coupled to another sulfur containing coenzyme: *coenzyme A*, CoA–SH, the universal acyl-transfer agent.

Acetyl-CoA is the real form of "activated acetate" where 36.9 kJ/mol (8.8 kcal/mol) is liberated upon hydrolysis (see Chapter 2). Reduced lipoic acid is reoxidized with an FAD coenzyme present on the complex.

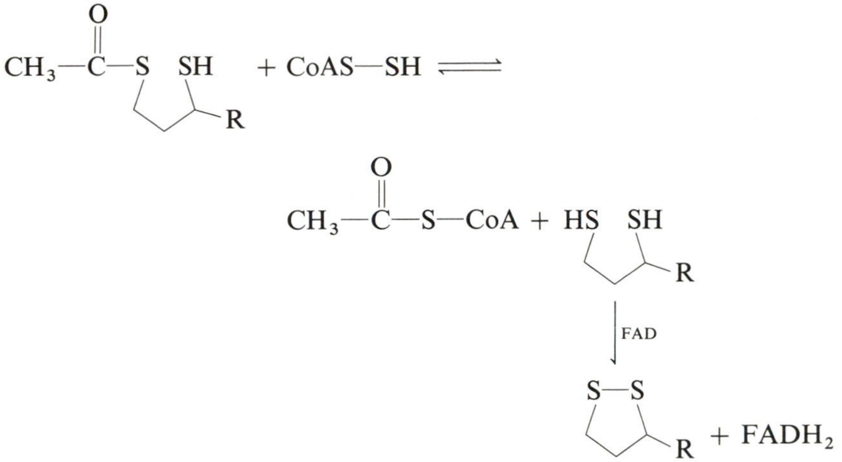

$$CH_3-\overset{O}{\overset{\|}{C}}-S \quad SH \quad + CoAS-SH \rightleftharpoons$$

with R

$$CH_3-\overset{O}{\overset{\|}{C}}-S-CoA + HS \quad SH$$

R

FAD

S—S

R + FADH$_2$

7.2 Pyridoxal Phosphate

Pyridoxine or vitamin B$_6$ is an important dietary requirement. The aldehyde form is called pyridoxal and its phosphate ester is implicated in many enzyme catalyzed reactions of amino acids and amines. The reactions are numerous and pyridoxal phosphate (pyridoxal-P) is surely one of nature's most versatile catalysts or coenzymes. The chemistry that will be emphasized here is one of proton transfer. In transamination (equation 7-1), a process of central importance in nitrogen metabolism, it is converted to pyridoxamine.

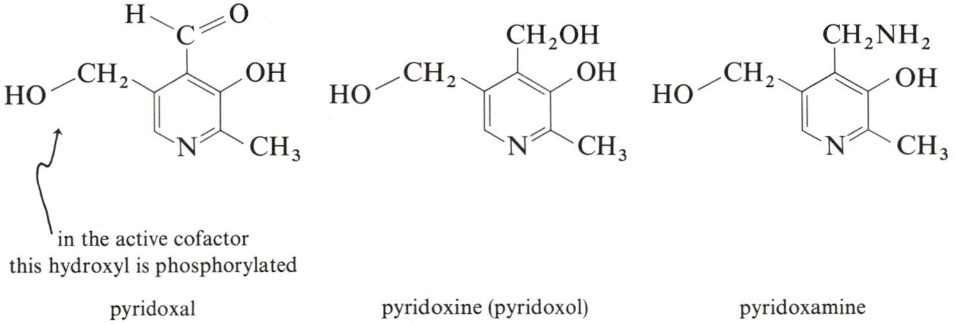

in the active cofactor
this hydroxyl is phosphorylated

pyridoxal   pyridoxine (pyridoxol)   pyridoxamine

Structure of
vitamin B$_6$
compounds

In fact, pyridoxal-P coenzyme catalyzes at least seven very different reactions where acid–base chemistry and tautomerism is fully exploited.

$$R-\underset{\underset{O}{\parallel}}{C}-CO_2^{\ominus} + R'-\underset{\underset{\overset{\oplus}{N}H_3}{|}}{CH}-CO_2^{\ominus} \rightleftharpoons R-\underset{\underset{\overset{\oplus}{N}H_3}{|}}{CH}-CO_2^{\ominus} + R'-\underset{\underset{O}{\parallel}}{C}-CO_2^{\ominus}$$

transamination

$$\text{(7-1)}$$

$$HOCH_2-\underset{\underset{\overset{\oplus}{N}H_3}{|}}{CH}-CO_2^{\ominus} \longrightarrow CH_3-\underset{\underset{O}{\parallel}}{C}-CO_2^{\ominus} \qquad \text{(7-2)}$$

serine   pyruvate

elimination–hydration

$$\overset{2\ominus}{HO_3PO}-CH_2CH_2-\underset{\underset{\overset{\oplus}{N}H_3}{|}}{CH}-CO_2^{\ominus} \longrightarrow CH_3-\underset{}{CH}-\underset{\underset{\overset{\oplus}{N}H_3}{|}}{\overset{\overset{OH}{|}}{CH}}-CO_2^{\ominus} \quad \text{(7-3)}$$

homoserine phosphate   threonine

elimination–hydration

$$HO_2C-CH_2CH_2-\underset{\underset{\overset{\oplus}{N}H_3}{|}}{CH}-CO_2^{\ominus} \longrightarrow HO_2C-CH_2CH_2-CH_2-NH_2 + CO$$

glutamic acid   γ-aminobutyric acid
(GABA)

$$\text{(7-4)}$$

decarboxylation

$$HOCH_2-\underset{\underset{\overset{\oplus}{N}H_3}{|}}{CH}-CO_2^{\ominus} \longrightarrow \underset{H}{\overset{H}{>}}C=O + CH_2-\underset{\underset{\overset{\oplus}{N}H_3}{|}}{}CO_2^{\ominus} \quad \text{(7-5)}$$

serine   glycine

reverse condensation

$$R-\underset{\underset{\overset{\oplus}{N}H_3}{|}}{\overset{\overset{H}{|}}{C}}\!\!\!-\!\!\!CO_2^{\ominus} \rightleftharpoons R-\underset{\underset{\overset{\oplus}{N}H_3}{|}}{\overset{\overset{H}{|}}{C}}\!\!\!-\!\!\!CO_2^{\ominus} + R-\underset{\underset{\overset{\oplus}{N}H_3}{|}}{C}\!\!\!-\!\!\!H \quad \text{(7-6)}$$

racemization

$$\text{indole} + \underset{\substack{| \quad | \\ \text{OH} \quad {}^{\oplus}\text{NH}_3}}{\text{CH}_2-\text{CH}-\text{CO}_2{}^{\ominus}} \longrightarrow \text{tryptophan} \tag{7-7}$$

indole   serine   tryptophan

tryptophan synthesis

While it may be surprising that the above diverse reactions require the same cofactor, this will be readily understood when it is realized that these reactions have certain common features. All require imine (Schiff base) formation between the aldehyde carbonyl of the cofactor and the amino group of the substrate. The pyridoxal phosphate becomes an electrophilic catalyst or "electron sink," as electrons may be delocalized from the amino acid into the ring structure. It is the direction of this delocalization that dictates the reaction type and in model systems more than one reaction pathway is often observed. Thus the enzyme both enhances the rate of reaction and gives direction to that reaction (see page 428).

process a: leads to decarboxylation
process b: leads to transamination
process c: leads to aldol-type condensation and retrocondensation

In the transamination process, the pyridoxal coenzyme is transferred from the enzyme–imine intermediate to the substrate–imine. The evidence of an imine function comes from reduction by borohydride which does not yield pyridoxine but shows that a covalent bond is formed with a lysine residue of the enzyme. The protonation of the pyridine ring is also essential for catalysis.

What advantage, if any, is served by formation of a Schiff base between the enzyme and coenzyme? Enzyme-bound imine must provide a more rapid pathway than the substrate-bound imine (301). There is thus a structural reason linked with the greater reactivity of imines as compared to the corresponding aldehydes. A more basic nitrogen forms a stronger hydrogen bond (i.e., with an appropriate hydrogen bond donor on the enzyme surface) and is protonated to a much greater extent than an oxygen. Furthermore, an imine carbon is more electrophilic than a carbonyl carbon; hence it is attacked more readily by nucleophiles.

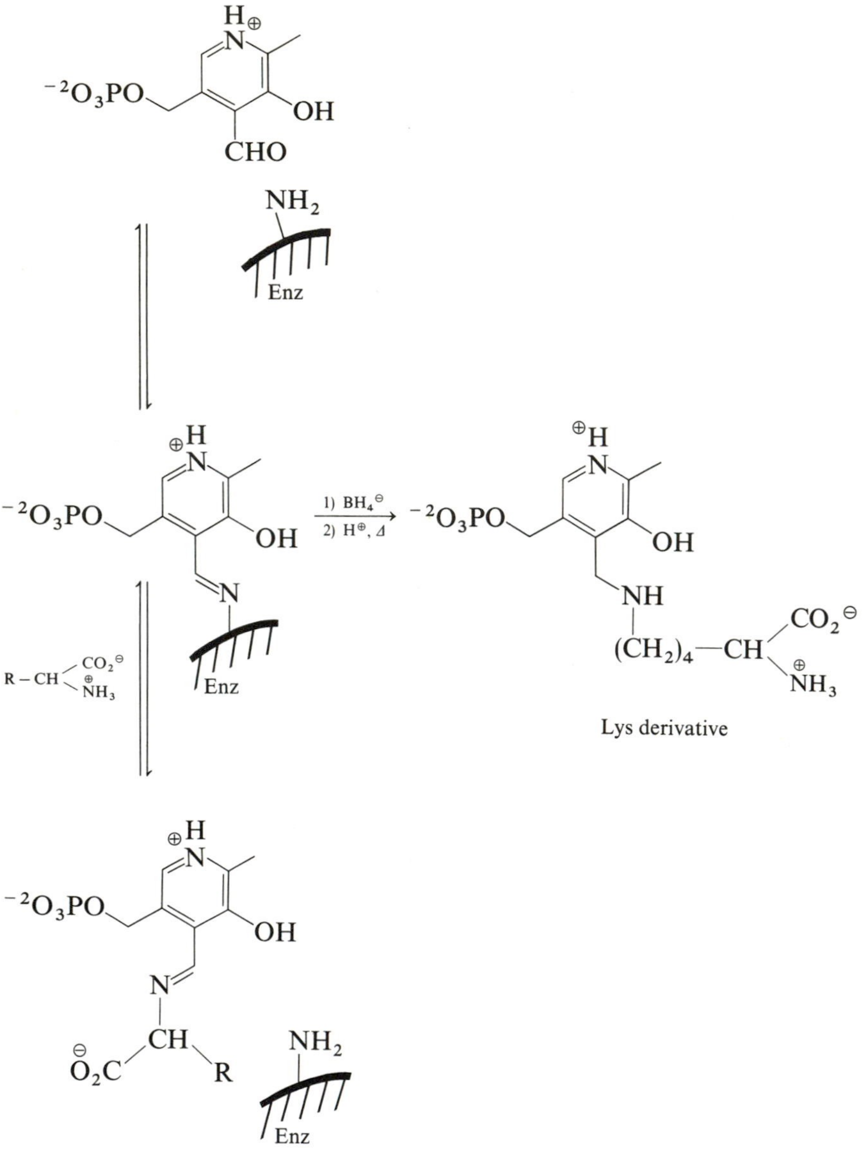

Lys derivative

Thus, an enzyme–imine intermediate facilitates a more rapid formation of a covalent intermediate between substrate and coenzyme.

The role of the phosphate is to bind the coenzyme to the corresponding apoenzyme. Lowe and Ingraham (301) and D. Arigoni of E.T.H., Zurich, proposed an attractive hypothesis (Fig. 7.9) where the phosphate and the methyl groups form an axis about which pyridoxal could pivot between enzyme–imine and substrate–imine covalent structures.

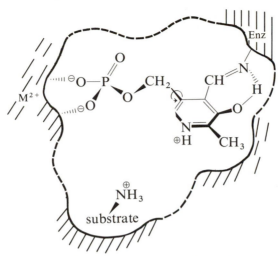

Figure 7.9. Hypothetical mode of action of pyridoxal phosphate. Adapted from Lowe and Ingraham (301) with permission.

As pointed out by the authors, if everything in nature exists for a reason, this may indeed be the case (301).

7.2.1 Biological Role

Pyridoxal phosphate has several features which make it an excellent catalyst for transamination reactions. In particular, the hydroxyl group is ideally located to provide general acid and base catalysis. Being intramolecular it is particularly effective. Second, the positively charged nitrogen of the pyridine ring acts as an electron sink to lower the free energy of C—H bond tautomerization. Finally, the groups are positioned in just the right geometry on the enzyme surface.

Addition of a metal ion such as Al(III) to the nonenzymatic system considerably enhances the catalytic effect (302). It complexes with the imine and acts as a general acid catalyst.

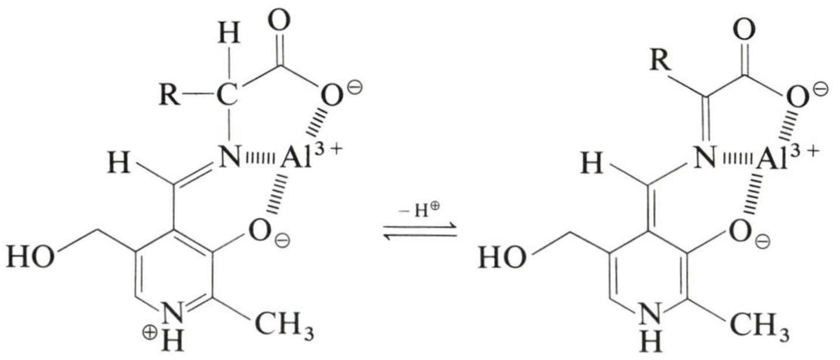

It has been demonstrated that Cu(II) and Fe(III) can also greatly accelerate the rate of reaction and actual chelate intermediates have been isolated. In model systems, all reactions catalyzed by enzymes except decarboxylation have been reproduced: transamination, oxidative deamination, elimination of β- and γ-substituents, etc. Many substitutes for pyridoxal have been synthesized and examined in the biological system. The following compounds were observed to be ineffective:

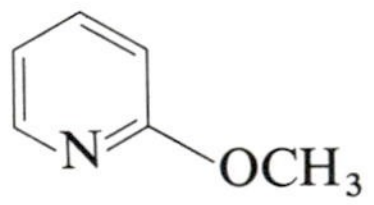

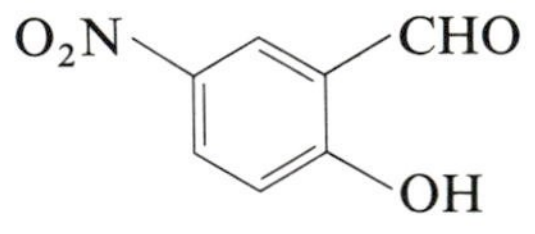

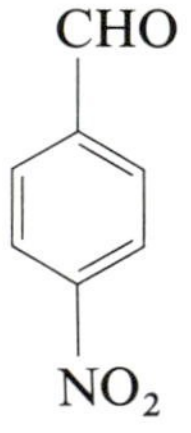

whereas the following were effective catalysts:

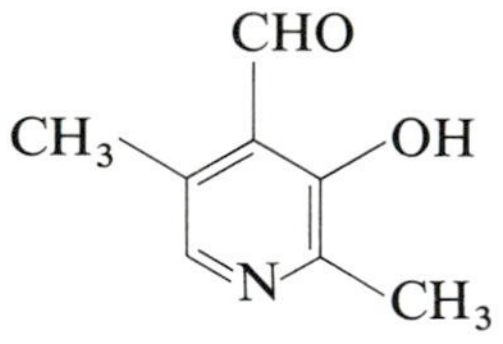

5′-deoxypyridoxal
analogue

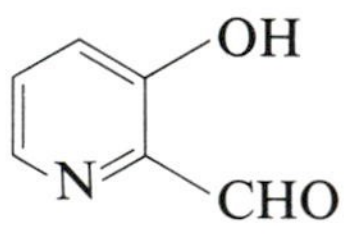

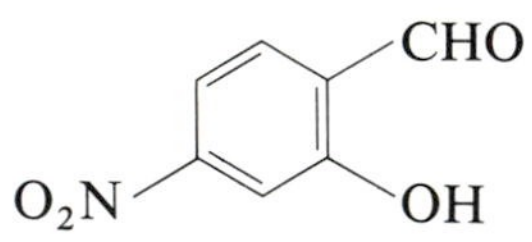

Proper electron delocalization is thus important. Furthermore, the presence of a hydroxyl group for chelation with metal ions appears essential. Surprisingly, with α-phenyl-α-aminomalonic acid in the presence of Cu(II) and appropriate model compounds (see below), an oxidation reaction takes place which has no biological precedent (303).

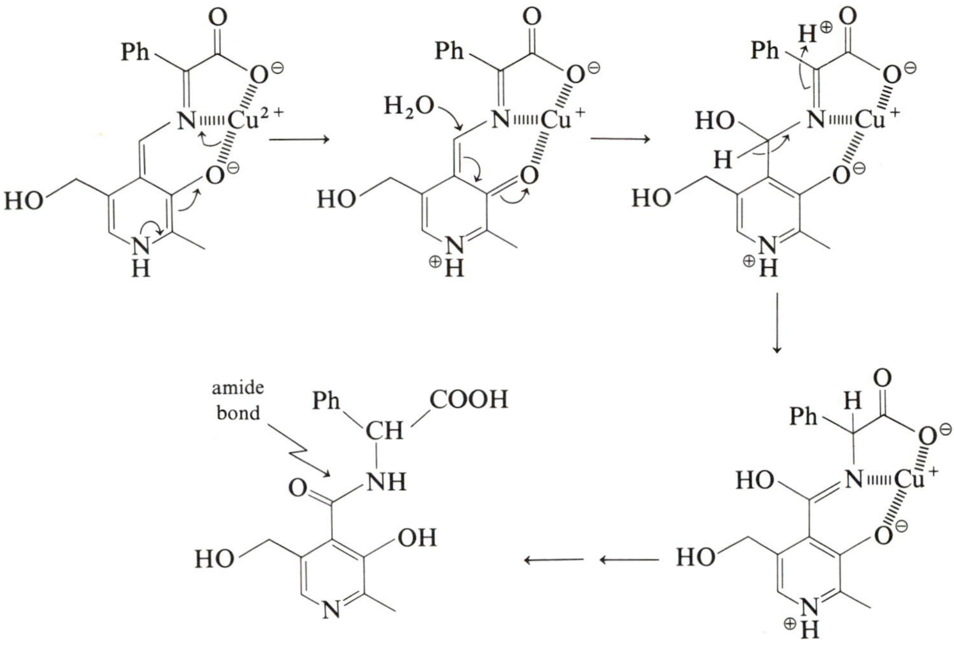

It should be noted that the role of the metal ion has often been over-estimated in pyridoxal-P-dependent enzymes because most of these enzymes are not metalloenzymes. Rather, it is the enzyme itself that plays a template role, as the metal ion.

To better understand the mode of action of pyridoxal-P, let us examine in detail equation 7-3; the conversion of homosereine phosphate to threonine. It is an elimination–hydration transformation. The first process (Fig. 7.10) involves an aldimine $\rightleftharpoons$ ketimime tautomerization which is subjected to general-acid catalysis intramolecularly by the hydroxyl group, followed by the slow breaking of a C—H bond. The latter is the rate-determining step.

The presence of a negative charge on the β-carbon can be shown by trapping the anion with N-methylmaleimide (NMM) via a Michael addition. Interestingly, pyridoxal phosphate helps stabilize a negative charge (anionic form) on α-, β-, and γ-carbon positions. Conjugation with the ring nitrogen stabilizes the negative charge on α and γ positions while the imine nitrogen stabilizes the negative charge on the β position. In the presence of ^2H_2O, positions α and γ can be deuterated. This particularly clear cut sequence of protonations shows that the role of pyridoxal phosphate is to stabilize car-banionic intermediates by acting as a delocalizing electron sink for the extra electron density (315).

It should be realized that since pyridoxal is converted to pyridoxamine or the corresponding enzyme–imine intermediate, it is not really a true catalyst! Rather, it is a reactant.

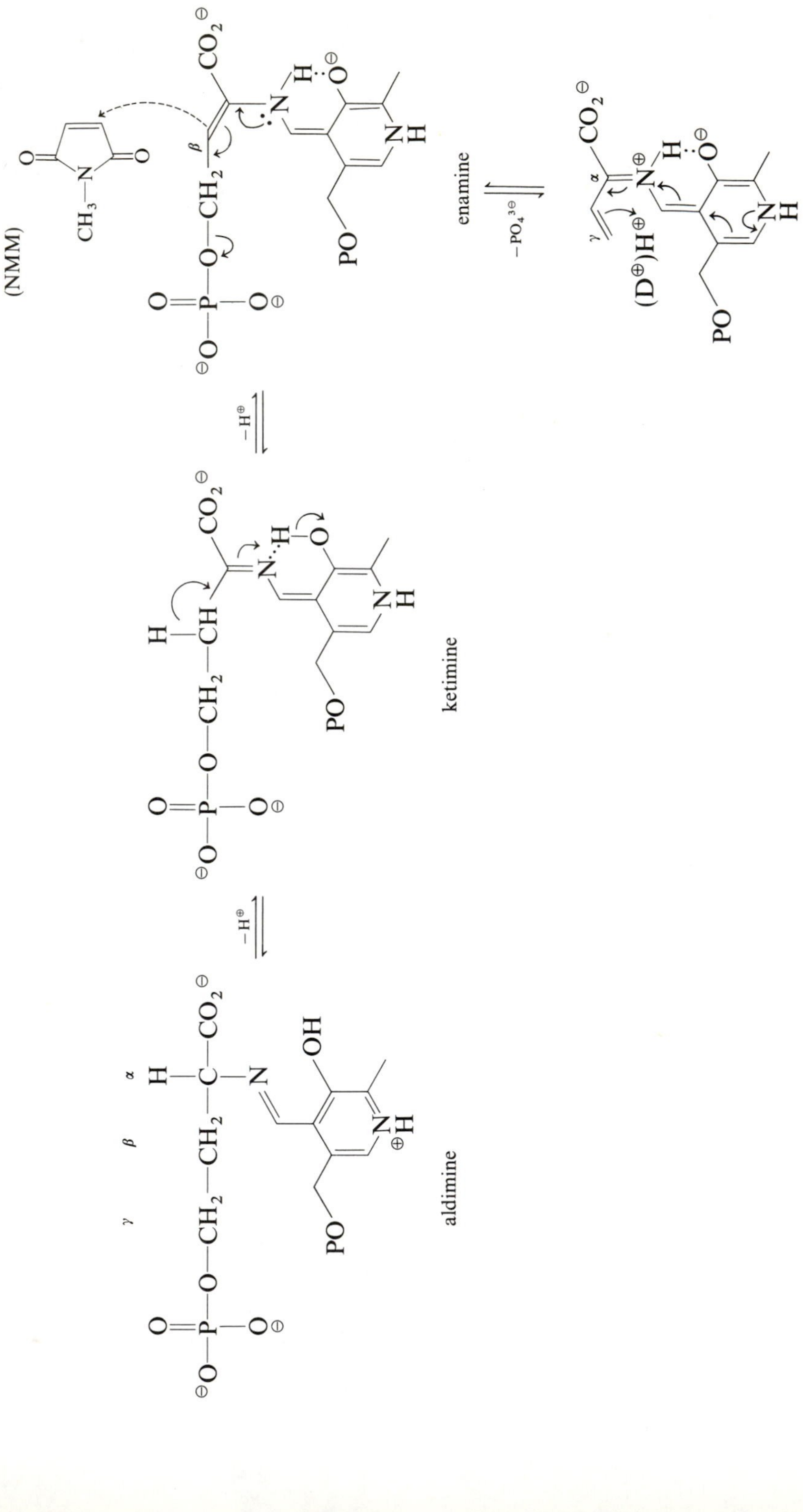

(NMM)

enamine

$-PO_4^{3\ominus}$

$-H^{\oplus}$

ketimine

$-H^{\oplus}$

aldimine

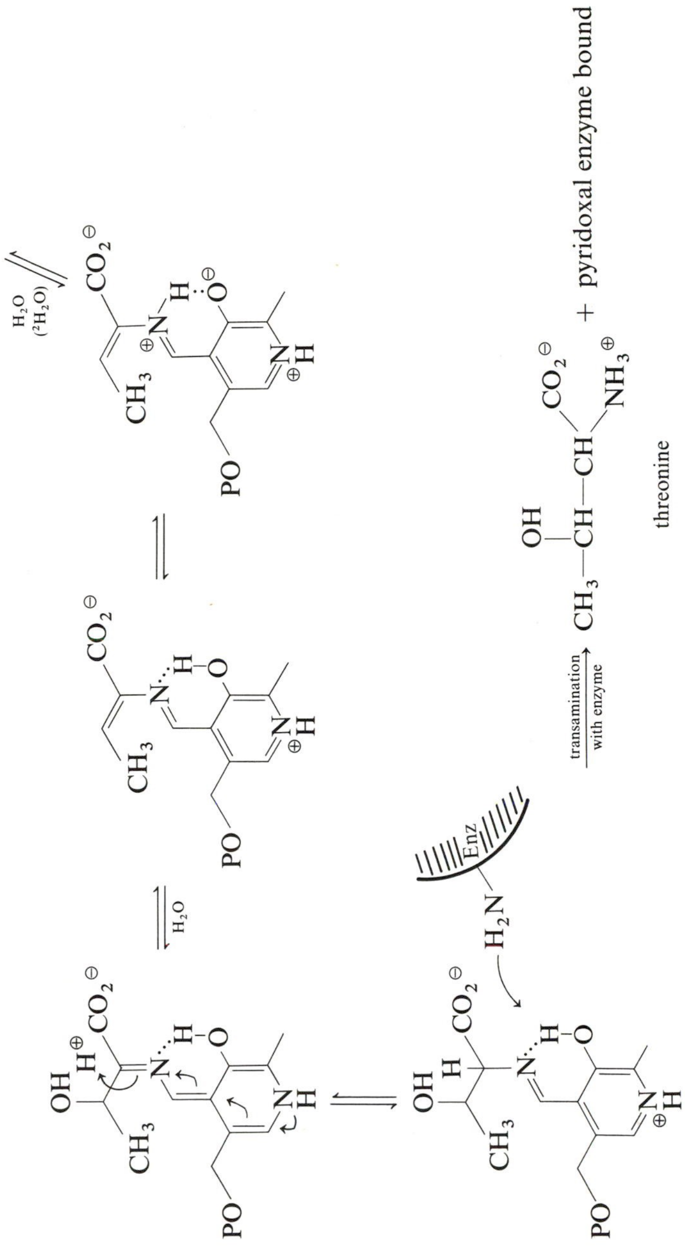

Figure 7.10. The conversion of homoserine phosphate to L-threonine (301).

The most important central problem in transamination processes is a consideration of stereochemistry. The enzyme–coenzyme complex, depending on the reaction and enzyme type, can remove from the substrate amino acid the R group, the carboxylate group, or the hydrogen of the α-carbon. What structural features determine which bonds to be broken? This depends on the enzyme, as does the reaction rate. The critical factor is the lowest energy pathway of the transition state of the covalent intermediate. In other words, the correct conformation of the coenzyme-bound substrate on the enzyme must have a dominant influence (301).

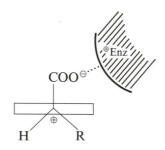

transaminase   aldolase   decarboxylase

Orientational dependency of pyridoxal phosphate containing enzyme catalyzed reactions

The labile bond is always the one perpendicular to the pyridine ring and combined ionic, polar, and hydrophobic interactions on the enzyme determine which conformer predominates. This is easily seen for example in the Newman projection of enzymatic decarboxylation. The conformation required for decarboxylation places the carboxyl group substantially out of the plane of the conjugated system. Consequently, the specificity of the reaction is manifested principally at this stage. For instance, enzymatic decarboxylation of amino acids occurs with retention of configuration and thus allows the preparation of optically pure α-deuterated amines if the reaction is carried out in heavy water (304).

Newman projection for the transition state of a pyridoxal phosphate enzyme catalyzed decarboxylation

Metal binding, if it occurs, can give different pathways. An interesting example of this effect is observed with L-serine hydroxymethyltransferase which can also catalyze the transamination of D-serine. The same binding site is implicated but different products are obtained:

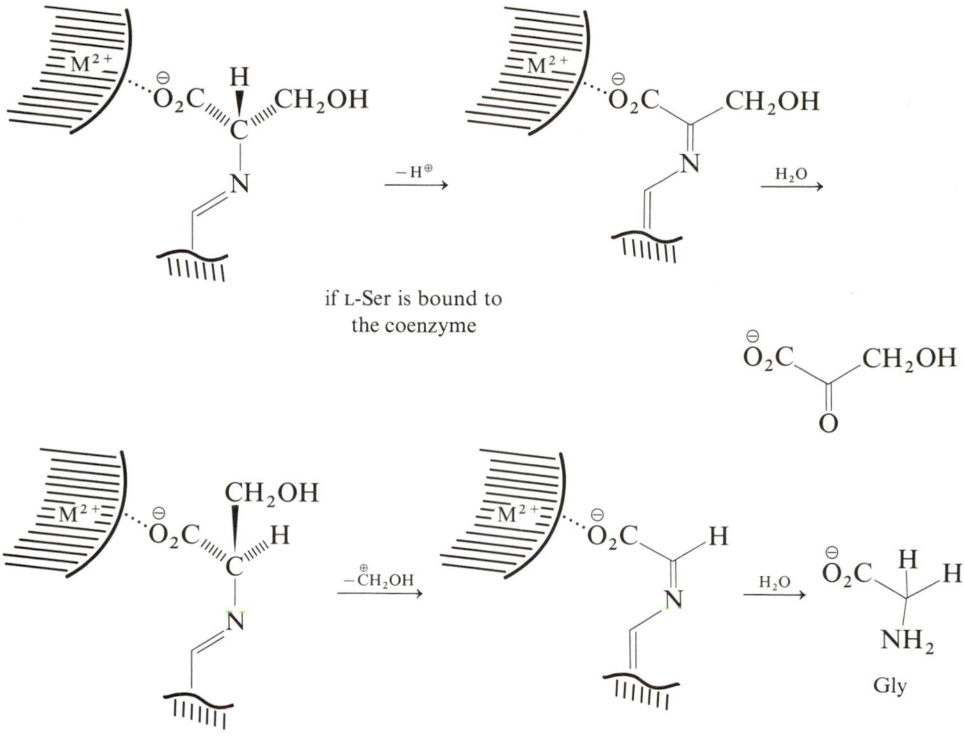

if L-Ser is bound to
the coenzyme

if D-Ser is bound to
the coenzyme

The orientation of the carboxyl group is thus determined here by the structure of the binding site (301). It shows also the important role of vitamins because excess glycine and serine in the system can be toxic but not if enough vitamin B_6 is administered.

Clearly, besides proton transfer, pyridoxal phosphate is also implicated in mechanisms where carbanion chemistry is involved. By leaving a negative charge on the α-carbon of an amino acid substrate, a new problem is introduced: one of stereochemistry. On the enzyme, does this carbanion (negative charge) finally get protonated by proton exchange with the medium or by a tautomeric form of the coenzyme? What kind of models can one select to best imitate such processes?

7.2.2 Model Systems

The following results on models are evidence of the existence of carbanion intermediates in pyridoxal enzymes. For instance, NMR studies by Gansow and Holm (302) in 1969 on pyridoxamine in the presence of Zn(II) or Al(III) ions in ^2H_2O, pH 1–13 range, showed the condensation of substrate (pyruvate) with a B_6 vitamin (pyridoxamine). The rate of exchange of hydrogen atoms increases with pH.

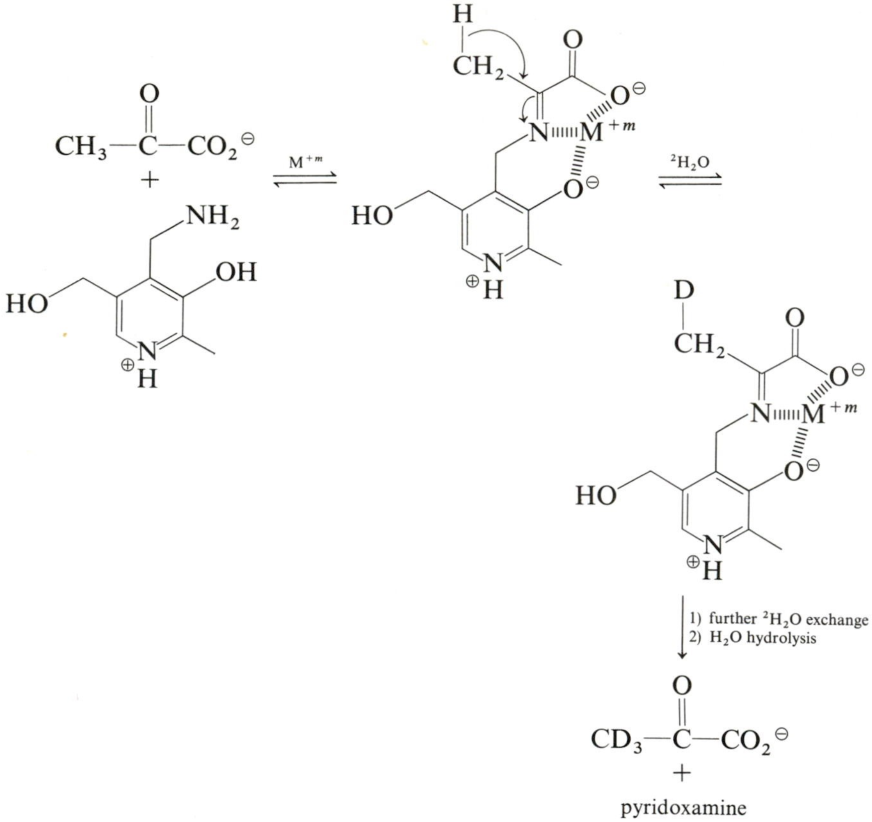

$$CH_3-\overset{\overset{\textstyle O}{\|}}{C}-CO_2^{\ominus}$$

pyridoxamine

Earlier it has been observed that the complex of pyridoxamine, ethyl pyruvate, and $Al(NO_3)_3$ in methanol gives a species absorbing at 488 nm.

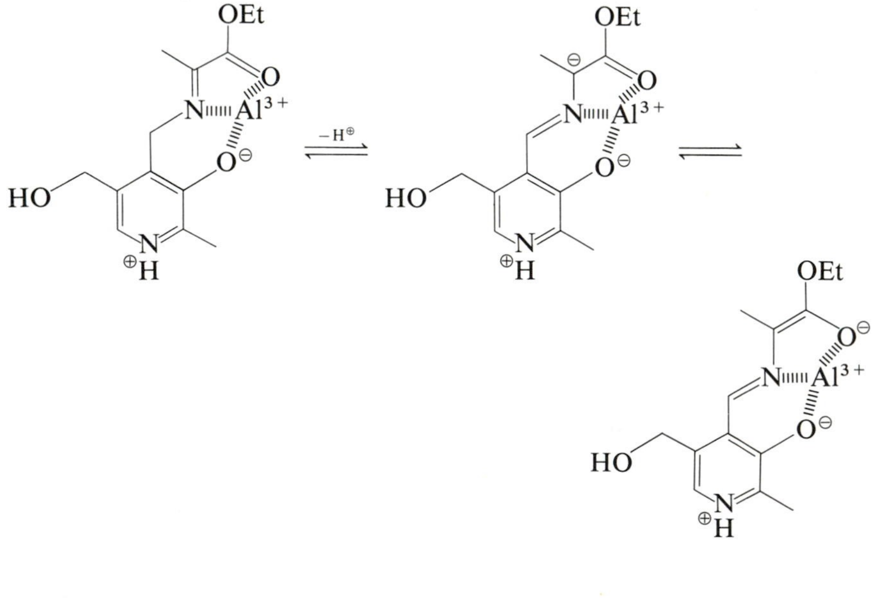

This visible band corresponds to the presence of a carbanion intermediate which disappears after a few hours to give transaminated products.

Other data should also be examined. For example, pyridoxamine in the presence of an α-keto acid and appropriate enzyme gives pyridoxal plus the corresponding L-amino acid. If the reaction is undertaken in ^2H_2O, half of the methylene hydrogens of pyridoxamine are exchanged and only one monodeuterated pyridoxamine isomer is formed.

one enantiomer (*R*-isomer)

the other enantiomer (*S*-isomer)

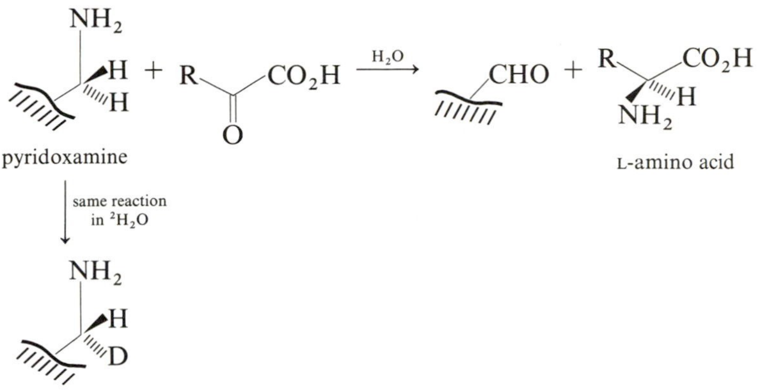

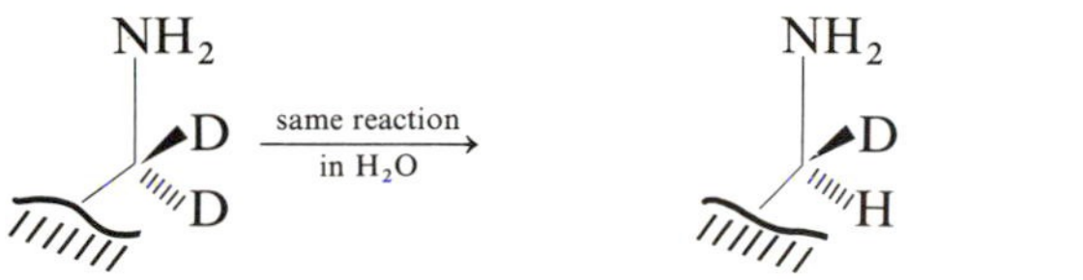

If double-deuterated pyridoxamine is used for the same reaction in water, half of the deuterium is lost and the other monodeuterated pyridoxamine enantiomer is obtained. Furthermore, if an α-deuterated L-amino acid is used, the deuterium is transferred stereospecifically to give only one of the two possible stereoisomers of monodeuterated pyridoxamine.

only the *R*-isomer

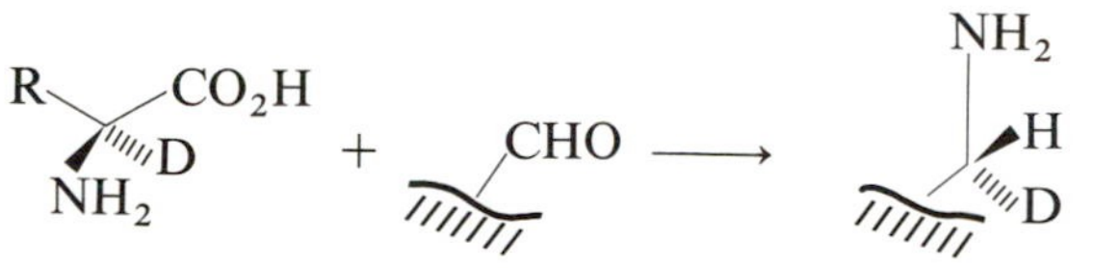

A. C. Dunathan and his co-workers (305, 306) at Haverford College in Pennsylvania contributed information about these processes by studying the mechanism and stereochemistry of transamination reactions. They defined the problem by three stereochemical variables:

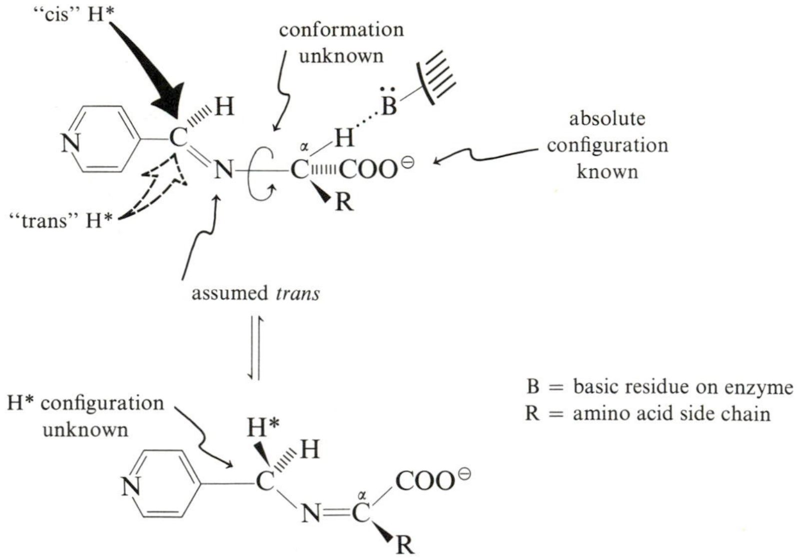

Therefore, the tautomerization process can be pictured as either an intramolecular (enzyme assisted) *cis* transfer or a *trans* transfer of hydrogen. This rearrangement is referred to as a tautomeric 1,3-prototropic shift and model studies have tried to solve this stereochemical ambiguity:

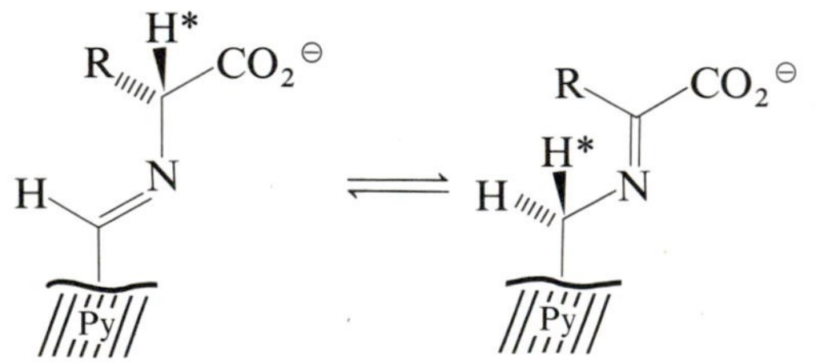

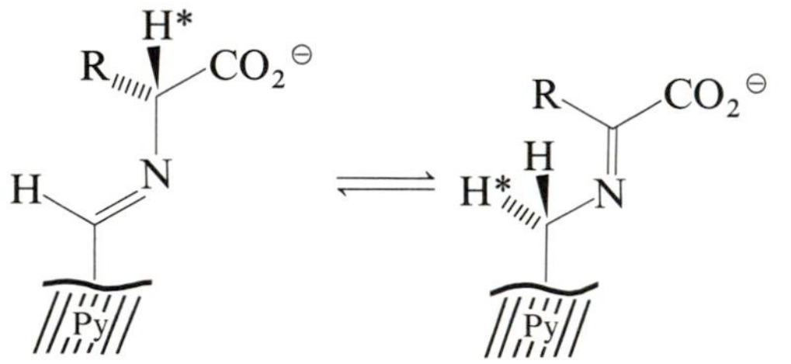

First, the biochemical observations of Dunathan (306) and others led to the hypothesis that the prototropic shift is stereospecific with a *cis* internal transfer. If so, a basic residue on the apoenzyme must be implicated in the process where an azaallylic anion intermediate is formed (Fig. 7.11).

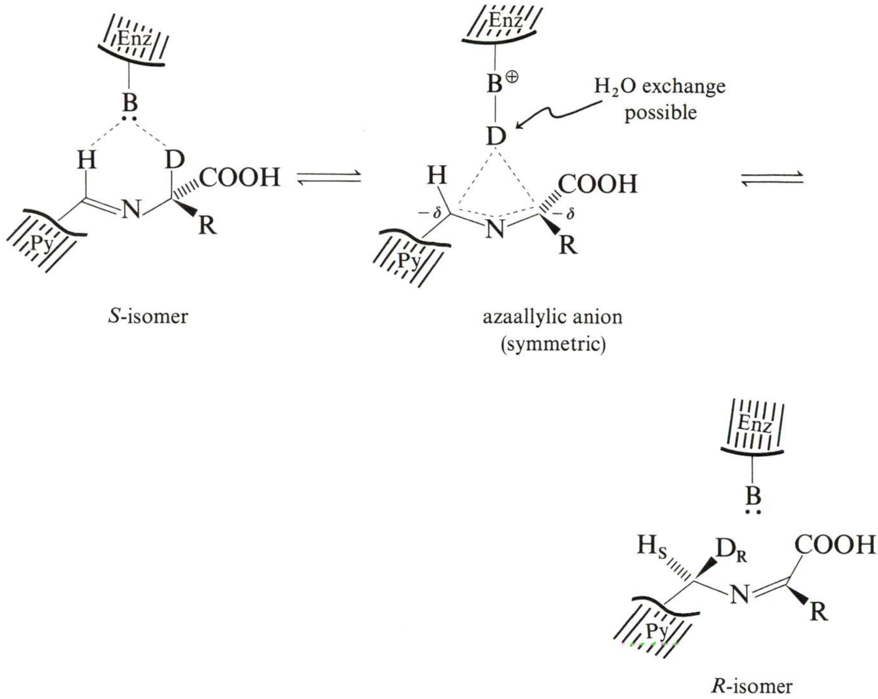

Figure 7.11. The *cis* internal 1,3-prototropic shift involved in enzyme-mediated pyridoxal phosphate catalyzed reaction. Py, the rest of the coenzyme molecule.

The group of D. J. Cram (307–310) studied in detail this base catalyzed methylene–azamethine rearrangement.

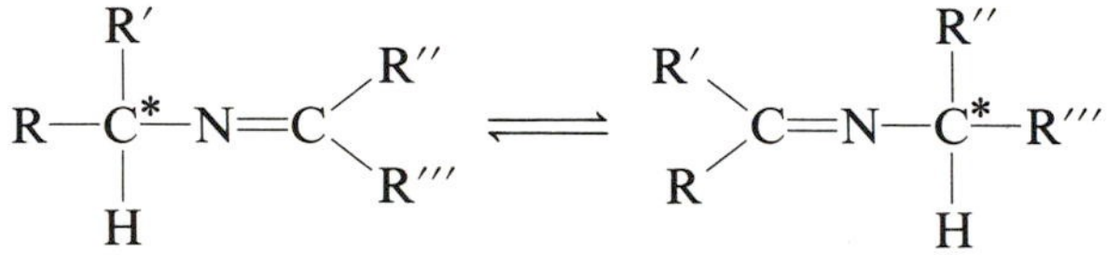

For such carbanionic rearrangement, two mechanisms in $EtO^- - EtOH$ can be proposed. One stage concerted:

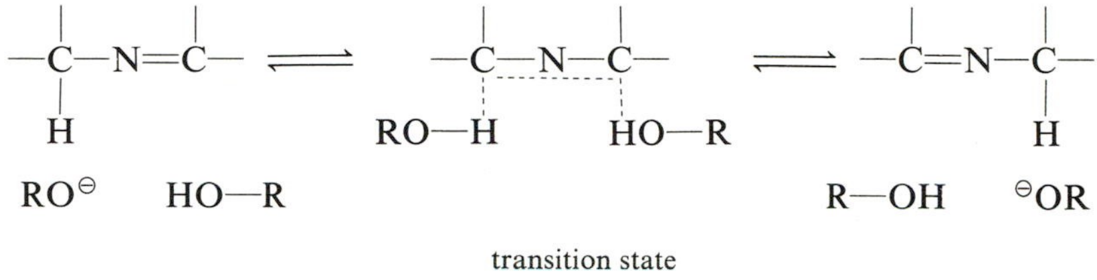

transition state

or two-stages or anionic mechanism:

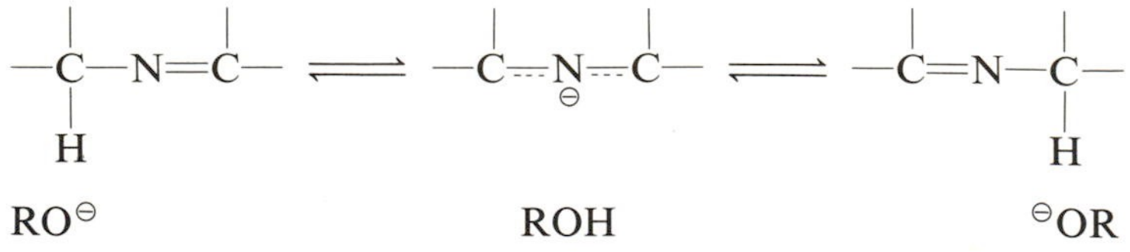

$$-\overset{|}{\underset{H}{C}}-N=\overset{|}{C}-\quad\rightleftharpoons\quad-\overset{|}{C}\overset{..}{\underset{\ominus}{N}}\overset{|}{C}-\quad\rightleftharpoons\quad-\overset{|}{C}=N-\overset{|}{\underset{H}{C}}-$$

$$RO^{\ominus}\qquad\qquad\qquad ROH\qquad\qquad\qquad {}^{\ominus}OR$$

azaallylic anion

Monitoring the isomerization process in deuterated solvent allows the determination of the mechanism operative in the model system. For instance, in the concerted mechanism, that proceeds through a symmetric transition state, for each deuteron incorporated from the solvent, an isomerization must occur. On the other hand, via the azaallylic intermediate, it is possible for a deuteron to be incorporated without a subsequent isomerization. It is this latter process which has been observed, thus strongly supporting the existence of the azaallylic intermediate (307, 308).

Since the kinetics of this reaction was studied in ^2H_2O and in H_2O, a few comments should be made concerning H—D exchange. Starting with a compound having an acidic hydrogen at a chiral center, three different products are possible via a carbanion intermediate (307):

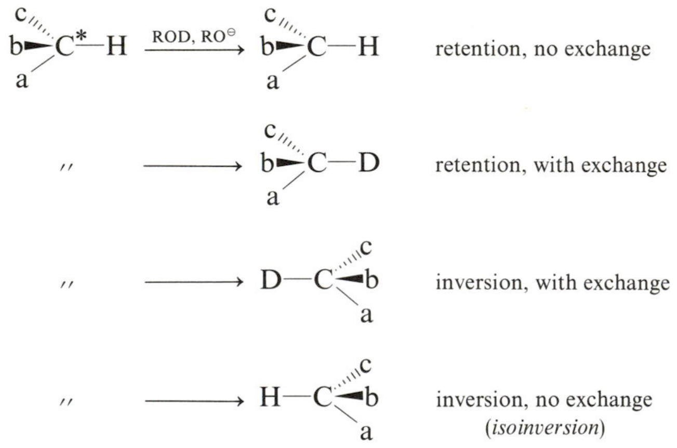

The process of inversion (racemization) without exchange is called *iso-inversion*. This involves an ion-pair mechanism. Indeed, in the presence of a crown ether, to separate the ions, the percentage of racemate increases (307). In fact, the stereochemical course of many metal-alkoxide catalyzed reactions in nonpolar solvents can be drastically modified by addition of catalytic amounts of crown ethers to the medium. For this reason, ion pairs, in low dielectric media, play a remarkable role as intermediates in reactions in which the negative ion is a carbanion. For example, Cram studied the rate of exchange (k_e) and racemization (k_α) as a function of the substituents present with deuterium-labeled 9-methylfluorene (309). The most interesting case was $k_e/k_\alpha < 0.5$, racemization without exchange. This happens for X = di-

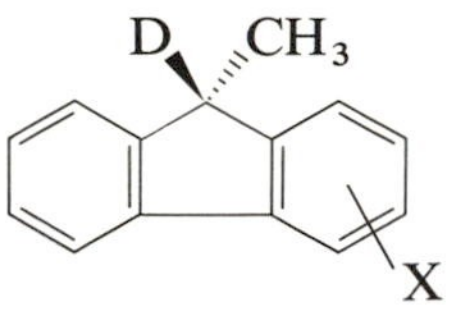

	k_e/k_α	Reaction solvent-base
	0.6 (exchange with inversion)	MeOH-N(Pr)$_3$
	1 (exchange with racemization)	tBuOH-tBuO$^-$K$^+$
	> 1 (exchange with retention)	C$_6$H$_6$-C$_6$H$_5$OH-C$_6$H$_5$O$^-$K$^+$
	< 0.5 (isoinversion)	tBuOH-THF-N(Pr)$_3$

methylamide in the presence of *n*-propylamine base and in *tert*-butanol-THF solvent replacing methanol.

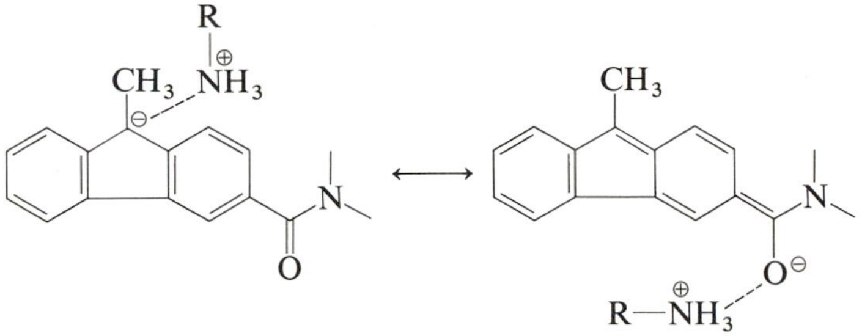

Cram reasoned that the deuterium is removed by the amine, and the ammonium ion formed remained paired with the carbanion via an ionic bond. The cation need not remain at the original site because the resonance of the ring system allows the negative charge to be delocalized all the way to the oxygen of the substituent. Then the ion pair which now lies in the plane of the ring may slide to the other side of the planar structure or return to the original position without exchanging its deuterium with the solvent protons. For this particular process, Cram coined the name *conducted-tour mechanism* (the base has taken a conducted tour), to explain the phenomenon of isoinversion. Notice that in methanol (which is a stronger acid than *tert*-butanol) the carbanion is more readily protonated and so does not have a long enough half-life to allow a conducted-tour process.

It should be noted that isoinversion can also occur without a conducted tour mechanism. This has been observed again with 9-methylfluorene and *tert*-butanol solvent, but with *n*-propylamine replaced by the base pentamethylguanidine. That an alternative mechanism is operative is indicated by the fact that a substituent such as dimethylamide no longer needs to be present. The base can provide a charge delocalization pathway for travel of the deuteron from one side of the planar ring system to the other without solvent exchange. The deuteron can thus add to either side of the plane, as is illustrated below:

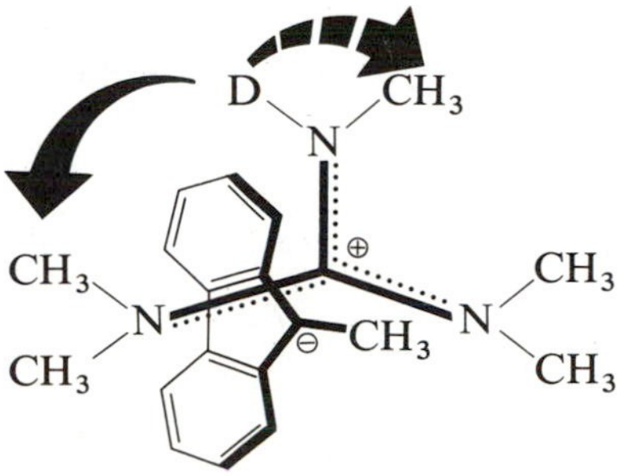

With this brief excursion into carbanion chemistry we can now examine the models that Cram and co-workers used to study the stereochemical outcome of the proton transfer stages of pyridoxal-mediated biological transaminations (310).

The following model systems were examined (311):

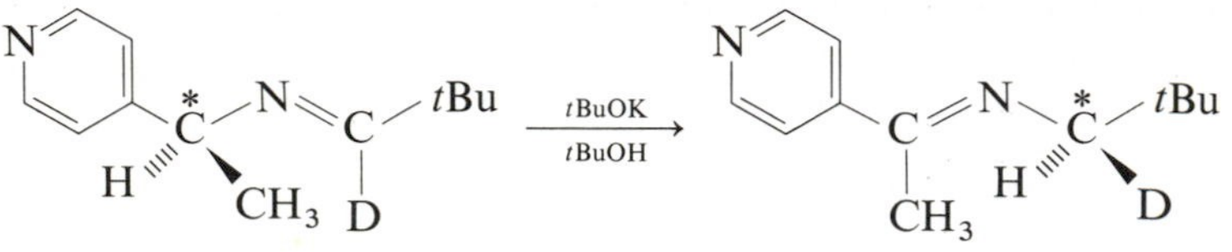

(S)-$(-)$-**7-1** $\xrightarrow[tBuOH]{tBuOK}$ (R)-$(+)$-**7-2**

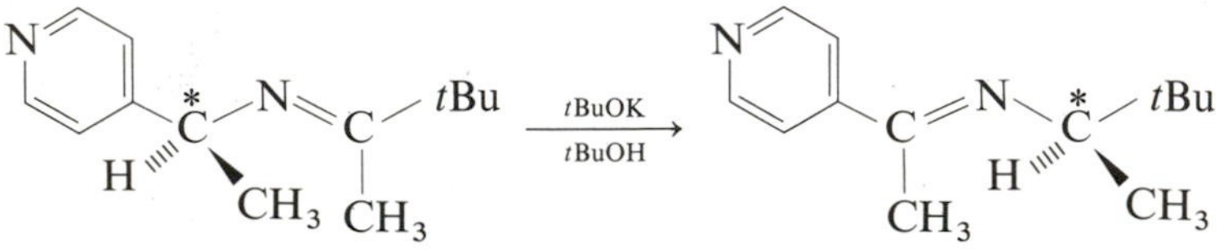

(S)-$(-)$-**7-3** $\xrightarrow[tBuOH]{tBuOK}$ (R)-$(-)$-**7-4**

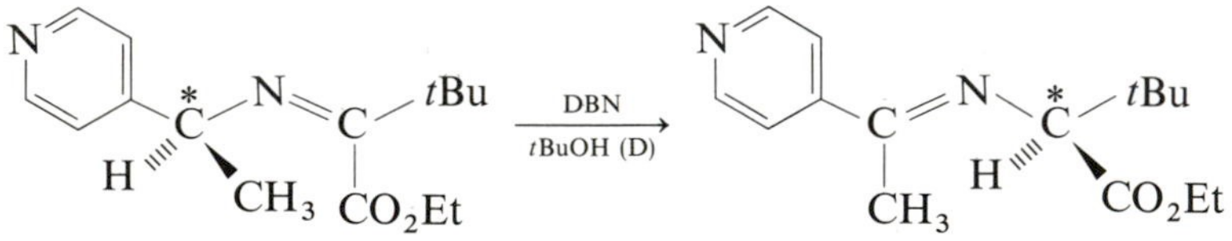

(S)-$(-)$-**7-5** $\xrightarrow[tBuOH (D)]{DBN}$ (S)-$(-)$-**7-6**

With all the processes, isomerizations were stereospecific as indicated. Reaction in *tert*-butyl alcohol-*O-d* allows isotopic exchange and helps to monitor the course of the reactions.

For instance, starting with pure $(-)$-(S)-**7-5**, in the presence of the base diazobicyclo [4.3.0] nonene (DBN), after 811 hr of equilibration, less than 0.5% of optically pure **7-5** remains and a 30–60% conversion to $(-)$-(S)-**7-6** is observed. This result indicates that the interconversion is sterospecific and that *suprafacial* (same side) dominates *antarafacial* (opposite side) isomerization of **7-5** → **7-6**. The large pyridine ring and *tert*-butyl group enforce conformational homogeneity of the azaallylic system. The chiral center of **7-5** racemizes without exchange with the solvent. This corresponds to the phenomenon of isoinversion via a conducted-tour mechanism. The inversion is believed to occur via a bridge between DBN and the pyridine nucleus:

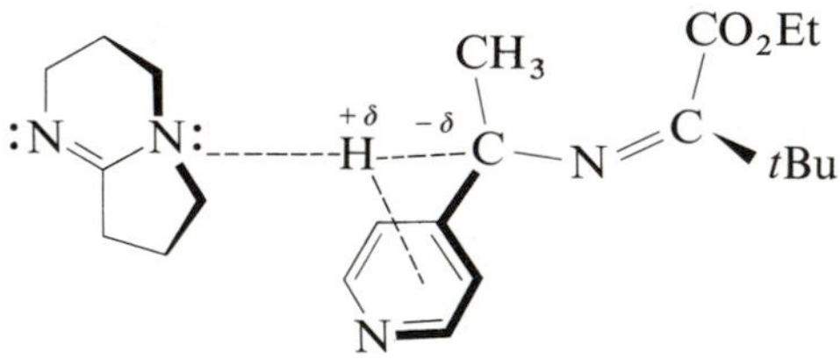

The chiral center of **7-6**, however, exchanges with retention of configuration; it cannot invert because it is too bulky.

In conclusion, the tautomerization is intramolecular and the suprafacial 1,3 proton shift occurs across a azaallylic anion. The model differs slightly, however, from the biological system by providing competing stereochemical and isotope-labeling reactions pathways. Therefore, coenzymes carry out stereospecific reactions due to their apoenzymes, while nonenzymatic model reactions are not as stereospecific (310).

This stereospecific transformation can be used advantageously in the case of a bulky amino acid such as L-threonine for the preferential asymmetric synthesis of a monodeuterated pyridoxamine:

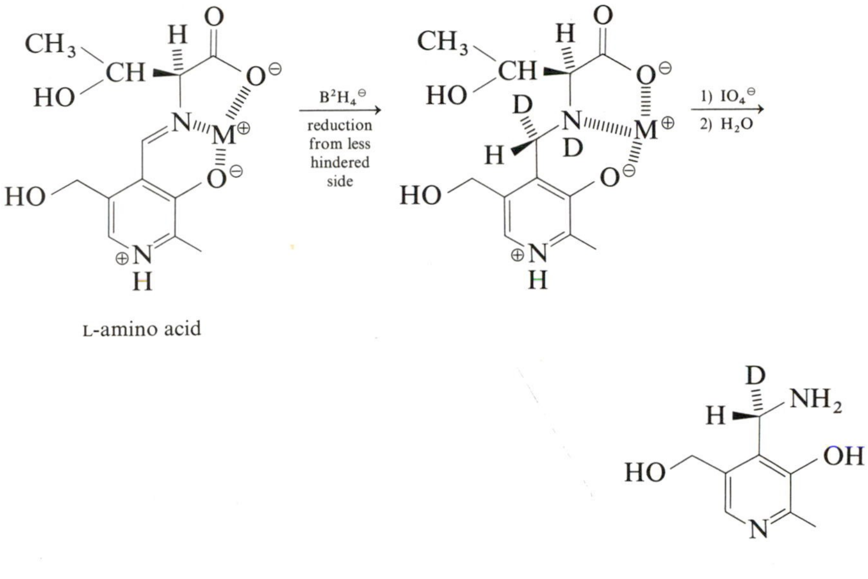

L-amino acid

S-isomer

From D- and L-threonine, the two pyridoxamine products give different isotopic primary effects with glutamic-oxaloacetic transaminase.

7.2.3 Suicide Enzyme Inactivators and Affinity Labels

In a broad sense an enzyme is specifically inhibited when its active site is blocked physically and/or chemically without significant alteration of the rest of the molecule. For this, many types of covalent inhibitors have been developed. The desired goal is chemical modification of an active site amino acid residue of the enzyme and subsequent loss of catalytic activity. The most common approach has been the synthesis of structurally and chemically reactive analogues of a substrate of the target enzyme. Such inhibitors have been referred to as *active-site-directed irreversible inhibitors* or *affinity labels* (312, 313). Generally the affinity label has a reactive electrophilic substituent that can generate a stable covalent bond with an active site nucleophilic group. Such experiments may serve to identify a nucleophile important in catalysis.

The principles of the method can be illustrated by one of the first affinity labeling experiments developed by E. Shaw, the reaction of *N*-tosyl-L-phenylalanine chloromethylketone (TPCK) with α-chymotrypsin (314).

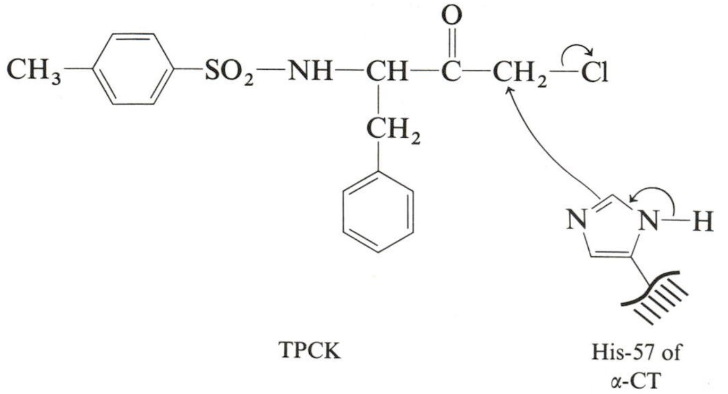

TPCK   His-57 of
                                          α-CT

The tosyl-phenylalanine moiety mimics a substrate like *N*-tosyl-phenyl-alanine methyl ester and the chloroketone function (see Section 2.3) acts as an electrophile where the chloride ion is displaced by His-57 at the active site of the enzyme. Similarly, a chloromethylketone analogue of lysine inhibits trypsin.

Diisopropylphosphorofluoridate (DFP, Section 4.4) is also an active-site irreversible inhibitor that blocks the active serine residue of serine proteases. It is easy to show that the inhibition is irreversible since exhautive dialysis still produces an inactive enzyme.

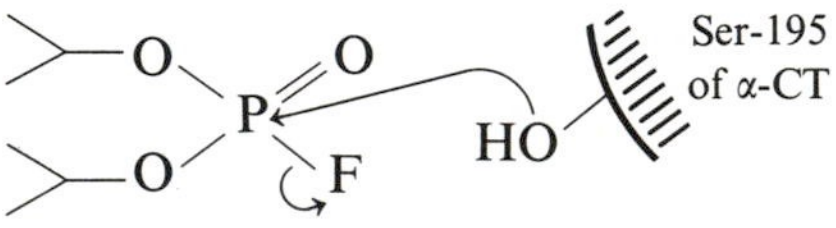

Alternatively, diazomethylketone substrate derivatives can be efficiently used as active-site-directed inhibitors of thiol proteases. For instance, the carbobenzoxyphenylalanine analogue reacts stoichiometrically at the active center cysteine residue of papain.

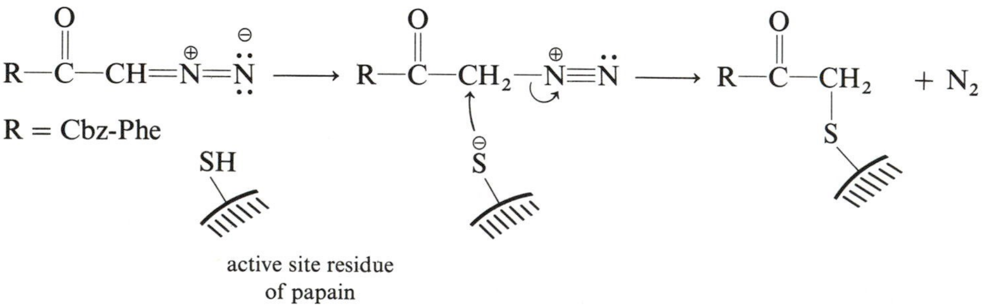

R = Cbz-Phe

active site residue
of papain

However, active-site-directed inhibitors have two disadvantages. First, they are intrinsically reactive molecules and a large portion is simply hydrolyzed by the aqueous medium. Second, they might react nonspecifically with other reactive moieties on the protein surface.

Many more examples of active-site-directed inhibitors can be found in recent books and review articles (312, 313, 315). Besides chemical modification of enzyme and affinity labeling, a few new techniques have been developed in the past ten years. Although not directly relevant to bioorganic models of enzymes, these techniques should be mentioned since in general they have given useful information when applied to biological problems. Among those are *photoaffinity labeling* (316) and *fluorescence energy transfer* as a spectroscopic ruler (317). These recent methods involve mainly biophysical techniques that are beyond the scope of the present book but are a useful aid for a better understanding of biological processes. The information obtained can be a valuable guide for the planning and design of new bioorganic models of macromolecules of biological interest.

It became apparent in the 1970s that new kinds of inhibitors with increased selectivity was needed. For instance, it has long been recognized that carbonyl reagents such as hydroxylamine and hydrazine are interesting inhibitors (irreversible Schiff base formation) of pyridoxal phosphate-dependent enzymes. A well-known example is isonicotinyl hydrazide (isoniazid), one of the most effective drugs against tuberculosis. Apparently, this drug competes with pyridoxal to form a hydrazone which blocks the corresponding kinase enzyme. The latter catalyzes the biosynthesis of pyridoxal phosphate from pyridoxal and ATP.

isoniazid

Many naturally occurring irreversible enzyme inhibitors also exist. These are referred to as *toxins*. A well-known and intriguing toxin is the β,γ-unsaturated amino acid rhizobitoxine, produced by *Rhizobium japonicum*. This natural metabolite is a highly specific irreversible inhibitor of pyridoxal-linked β-cystathionase from bacteria and plants (318).

$$\text{HOCH}_2\text{—CH—CH}_2\text{—O—CH}\overset{trans}{=}\text{CH—CH—CO}_2\text{H}$$
$$\overset{|}{\text{NH}_2}\qquad\qquad\qquad\overset{|}{\text{NH}_2}$$

rhizobitoxine

$$\text{HO}_2\text{C—CH—CH}_2\text{—S—CH}_2\text{—CH}_2\text{—CH—CO}_2\text{H}$$
$$\overset{|}{\text{NH}_2}\qquad\qquad\qquad\qquad\overset{|}{\text{NH}_2}$$

cystathionine

The role of β-cystathionase is to degrade cystathionine as follows:

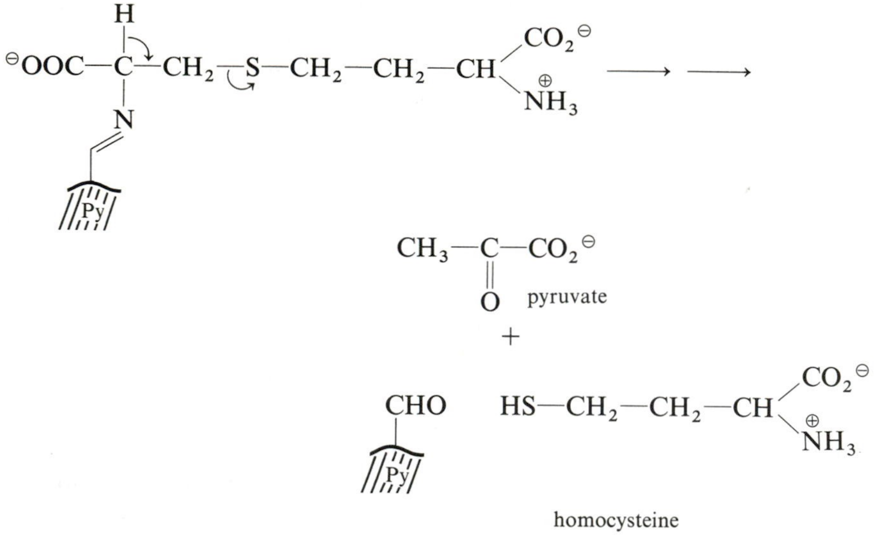

$$CH_3-\underset{\underset{O}{\|}}{C}-CO_2^{\ominus} \quad \text{pyruvate}$$

+

homocysteine

The inhibitor binds similarly but follows a different chemical pathway. The introduction of a conjugated system greatly facilitates attack by a neighboring nucleophile on the apoenzyme. This forms a new stable covalent bond resulting in an irreversible inhibition of the enzyme, preventing the system from normal metabolic functioning.

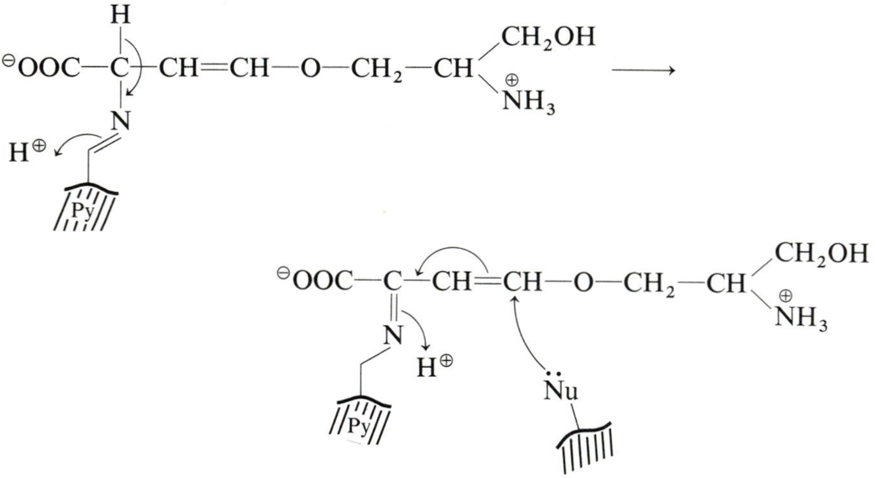

It is important to realize that the natural toxin is so constructed that it requires chemical activation by the target enzyme. Upon activation, a chemical reaction ensues between the inhibitor and enzyme resulting in the irreversible inhibition of the latter. Thus, the enzyme by its specific mode of action catalyzes its own inactivation or "suicide."

In fact, many well-known drugs have inhibitory properties on one or another enzyme, but only a few were conceived with that goal in mind. The

knowledge of the structure and mechanism of action of enzymes has progressed tremendously over the past two decades so that the search for specific
enzyme inhibitors as potential new drugs became more attractive. In order
to do so, it is necessary to know as much as possible about the specificity of
the enzyme and about the secondary binding sites, if any, near its active site.
This has led in recent years to the development of a fascinating new group of
irreversible enzyme inhibitors: the *enzyme activated irreversible inhibitors*,
also referred to as *suicide enzyme inactivators* (313, 318, 319). In a sense, the
expression suicide inactivator is rather ambiguous but nonetheless useful.

There is considerable promise in this type of inhibitor because the potentially reactive group may be rather innocuous in the presence of a number of
enzymes or *in vivo* until it reaches and is activated by the target enzyme. It is
really a question of exploiting the organic and physical chemistry of enzymes
(315).

This concept of *suicide substrates* or *enzyme inactivators* has many potential implications whereby organic chemists can synthesize well-designed
substrate analogues of specific enzymes. Of course, the design of new suicide
enzyme inactivators involves higher and higher levels of chemical sophistication. The majority of the work done with suicide inhibitors has been on non-
proteolytic enzymes, especially pyridoxal- and flavin-dependent enzymes.

Pargyline is a potent irreversible inhibitor of a flavin-linked monoamine
oxidase (MAO) and has found clinical application. The latter catalyzes the
inactivation of biologically important catecholamines. It forms a covalent
bond with the enzyme via the flavin cofactor and the mode of action is
believed to be as shown in Scheme 7.1.

Alternatively, β,γ-unsaturated amino acids, γ-acetylenic amino acids, or
amino acids having a leaving group at the β-position are potentially good
inhibitors of pyridoxal-dependent enzymes involved in amino acid metabolism. The first two become activated either by carbanion formation

adjacent to the unsaturated function or by two electron oxidation to the
bound conjugated ketimine. For the halogenated compounds, an enzyme-
mediated loss of HX produces an aminoacrylate Schiff base. In all cases the
activated electrophiles are attacked by enzyme nucleophiles at or near the
active site.

Needless to say, the design of such inhibitors is of therapeutic importance
and three factors are necessary for a successful design:

(a) The enzyme or the enzyme–coenzyme complex must convert a chemically
unreactive molecule to a reactive one.
(b) The reactive molecule must be generated within bonding distance of a
crucial active site residue.

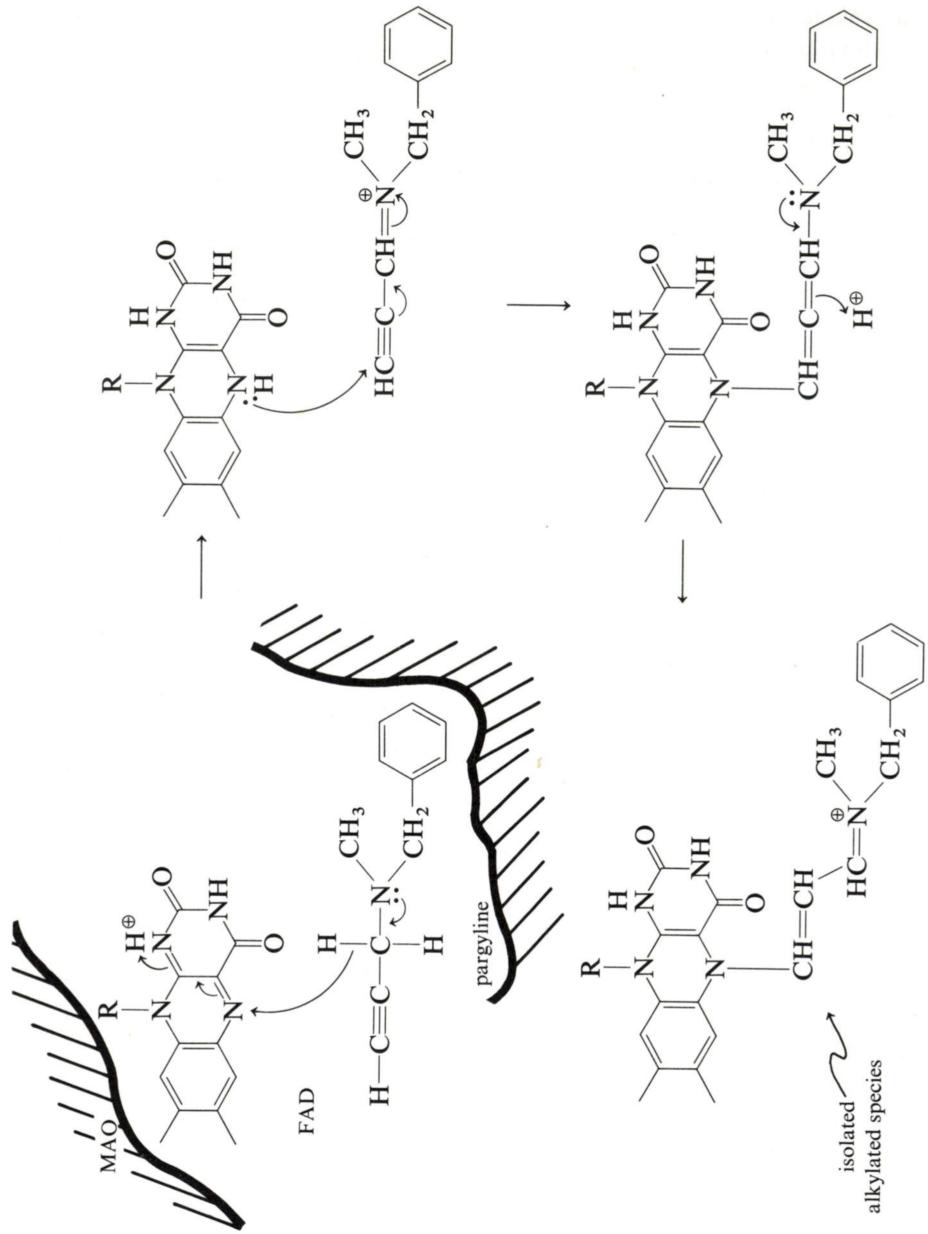

Scheme 7.1

(c) The activated species created must be designed in such a way that reaction with an active site residue can occur rather than with an external nucleophile.

In other words, a harmless compound must be converted into a potent inhibitor, the enzyme serving as the agent of its own destruction. A few examples will serve to illustrate this point.*

Example 1:
Serine *O*-sulfate and β-chloroalanine are both inhibitors of aspartate aminotransferase (320) and aspartate-β-decarboxylase (321).

Normal process:

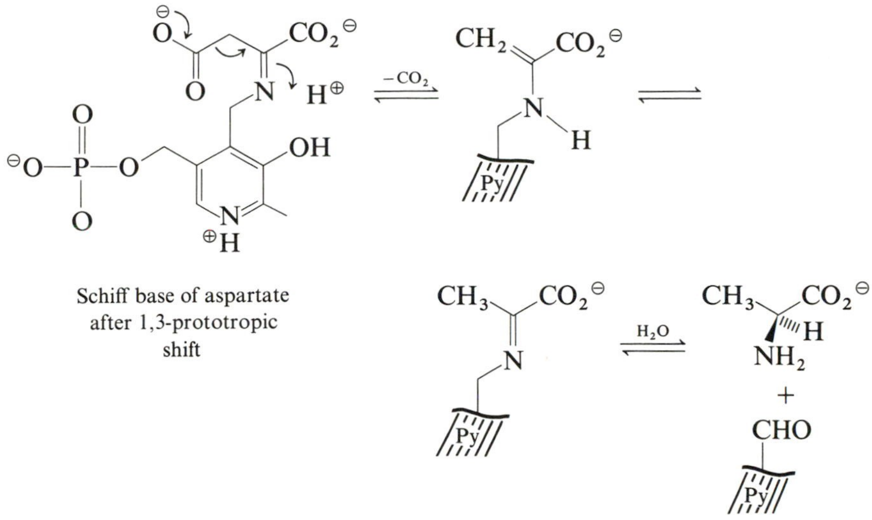

Schiff base of aspartate
after 1,3-prototropic
shift

---

* Most recent of these is the design of a halo-sulfoxide inhibitor which in the presence of some pyridoxal phosphate dependent enzymes converts to an active allyl sulfenate (360).

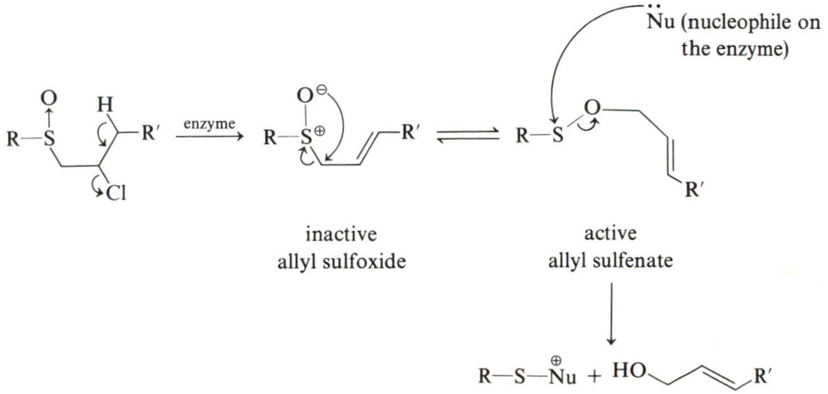

inactive
allyl sulfoxide

active
allyl sulfenate

$R—S—\overset{\oplus}{Nu}$ + HO $\diagup\diagdown$ R′

In the presence of the inhibitors:

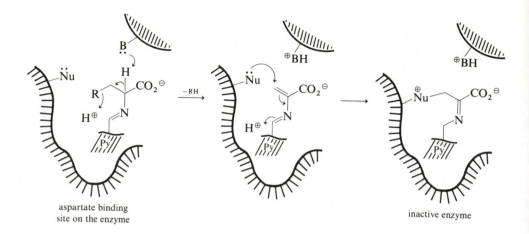

aspartate binding
site on the enzyme

inactive enzyme

$\overset{..}{B}$ = basic group on the enzyme
$\overset{..}{Nu}$ = nucleophile on the enzyme
R = $-Cl$, $-SO_4^{2\ominus}$
Py = pyridoxal nucleus

Example 2:

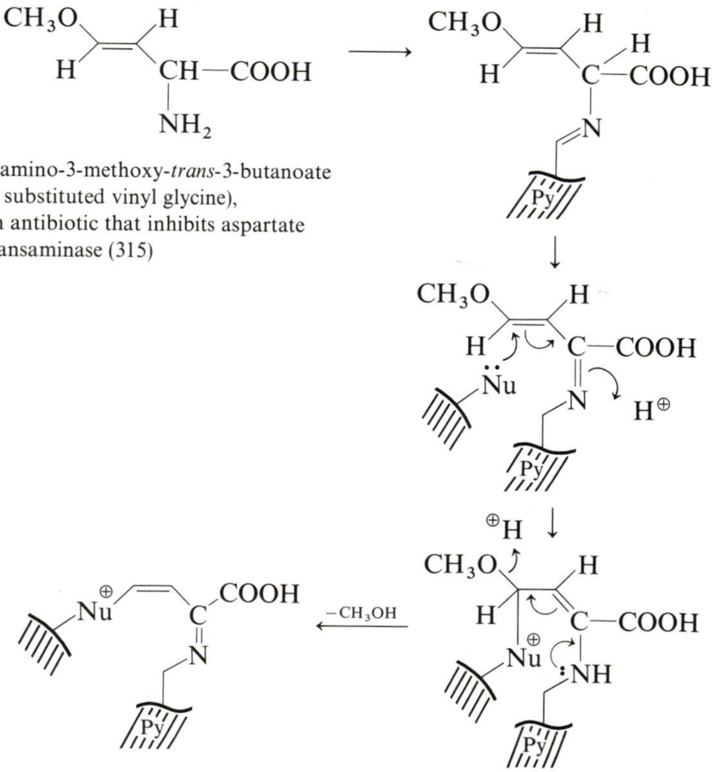

2-amino-3-methoxy-*trans*-3-butanoate
(a substituted vinyl glycine),
an antibiotic that inhibits aspartate
transaminase (315)

Example 3:

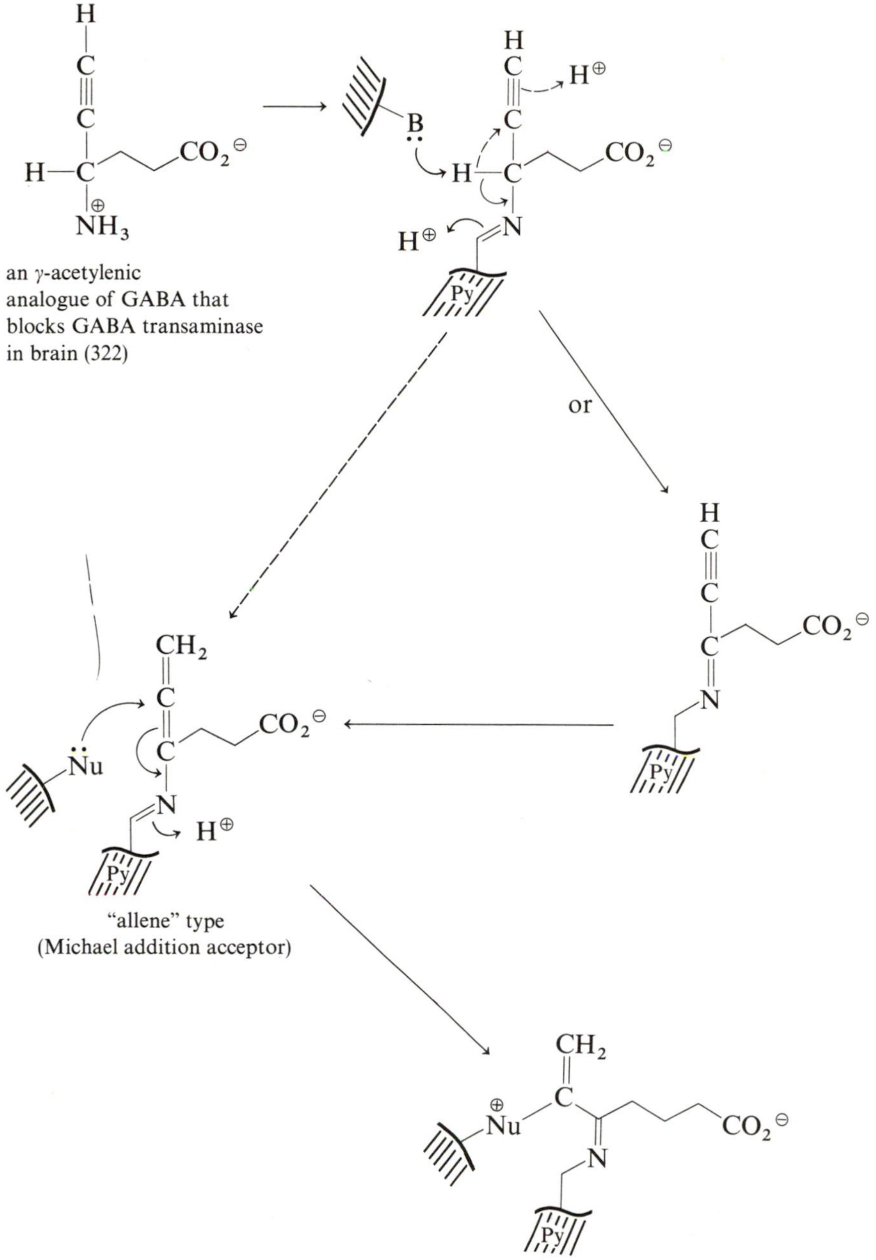

an γ-acetylenic
analogue of GABA that
blocks GABA transaminase
in brain (322)

or

"allene" type
(Michael addition acceptor)

This analogue of γ-aminobutyric acid GABA (322) appears to be a promising clinical candidate as an antiepileptic drug.

Example 4:

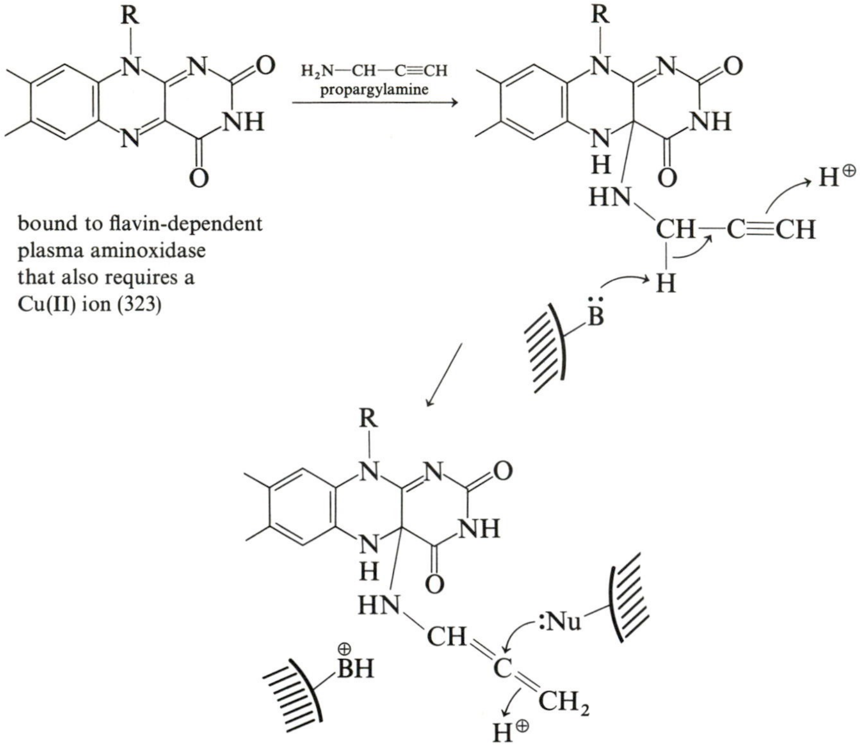

bound to flavin-dependent
plasma aminoxidase
that also requires a
Cu(II) ion (323)

Example 5 (318):

with
thiamine-
dependent
system

α-dichloro-
pyruvate derivative

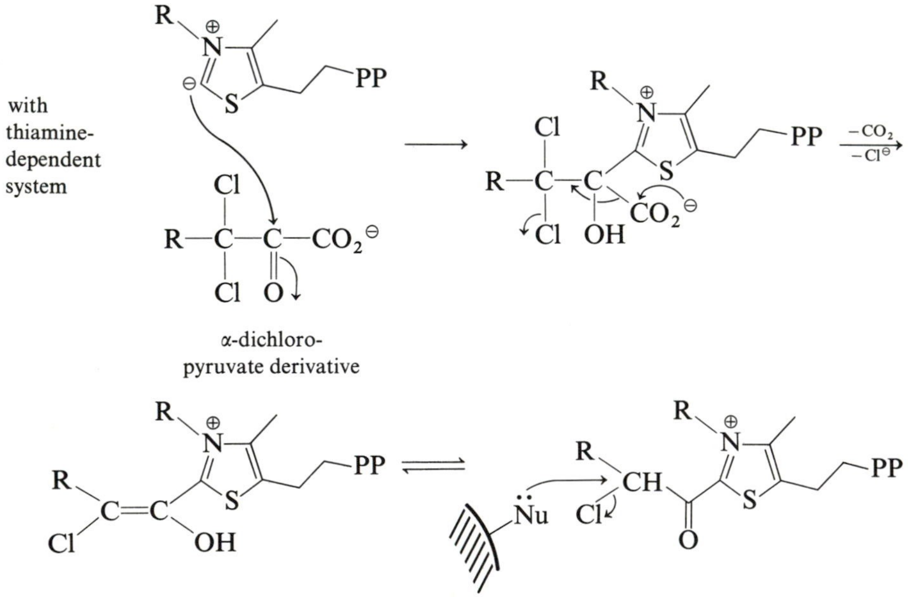

Most of these examples illustrate that proton abstraction is an important concept in suicide enzyme inactivation. For this, pyridoxal-dependent enzymes are obvious potential victims. In the future one can expect more developments on mechanism-based inactivators of pyridoxal-dependent enzymes based on functional group activation achieved by way of carbanion chemistry of intermediates (315). It is very likely the design of more selective active-site inhibitors that brought about the development of suicide enzyme inhibitors or inactivators. Compared to the active-site-directed irreversible inhibitors evocated earlier, they have the advantage of being relatively unreactive but becoming so only after interaction with the enzyme active site residues. The reactive form depends on the specific catalytic capacities of the active site. It is thus an enzyme-catalyzed conversion. However, both types of inhibition allow labeling and identification of active site residues and functional groups on enzymes.

7.3 Thiamine

As seen in the first section of this chapter a large percentage of the chemical reactions occurring in biological systems involve oxidation or reduction of one or more reactants. However, a particularly important type of reaction which apparently occurs in the majority of nonredox enzymatic reactions involves proton transfer which is aided by general-acid or -base catalysis. Of course, many of these enzymatic transformations are accomplished with the aid of a nonproteinic cofactor or coenzyme. In this category are some sulfur-containing coenzymes among which thiamine pyrophosphate (often referred to as vitamin B_1) is the most important. It will soon became apparent that thiamine pyrophosphate involves carbanion chemistry as an intermediate but with a degree of sophistication not yet encountered.

The coenzyme participates in reactions involving formation and breaking of carbon–carbon bonds immediately adjacent to a carbonyl group. Examples include nonoxidative and oxidative decarboxylations and aldol condensations. For instance it is involved in the nonoxidative decarboxylation of pyruvic acid to acetaldelyde:

$$CH_3-\underset{\underset{O}{\|}}{C}-COOH \xrightarrow[\text{decarboxylase}]{\text{pyruvate}} CH_3-C\overset{\nearrow O}{\underset{\searrow H}{}} + CO_2$$

pyruvic acid   acetaldehyde

It also participates in the formation of acetyl phosphate:

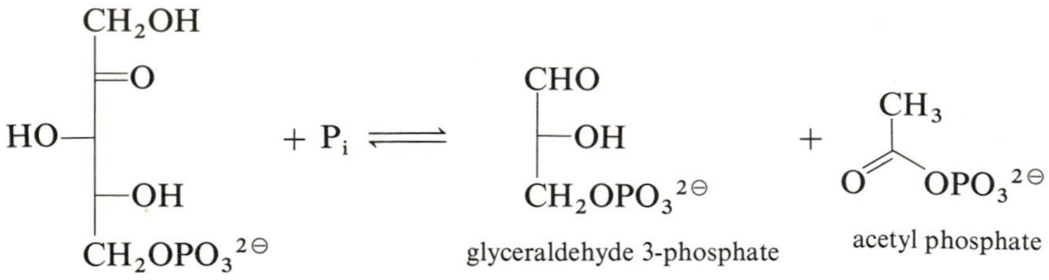

$$
\begin{array}{c}
CH_2OH \\
\parallel O \\
HO\!-\!\!\!\rule{0pt}{0pt} \\
-OH \\
CH_2OPO_3{}^{2\ominus}
\end{array}
\quad + P_i \rightleftharpoons \quad
\begin{array}{c}
CHO \\
-OH \\
CH_2OPO_3{}^{2\ominus}
\end{array}
\quad + \quad
\begin{array}{c}
CH_3 \\
O \diagdown\!\! OPO_3{}^{2\ominus}
\end{array}
$$

xylulose 5-phosphate   glyceraldehyde 3-phosphate   acetyl phosphate

or in the following condensation:

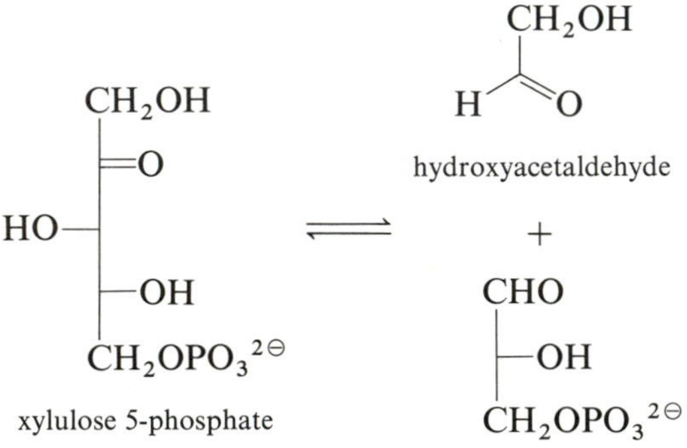

$$
\begin{array}{c}
CH_2OH \\
\parallel O \\
HO\!- \\
-OH \\
CH_2OPO_3{}^{2\ominus}
\end{array}
\quad \rightleftharpoons \quad
\begin{array}{c}
CH_2OH \\
H \diagdown\!\! O \\
\text{hydroxyacetaldehyde} \\
+ \\
CHO \\
-OH \\
CH_2OPO_3{}^{2\ominus}
\end{array}
$$

xylulose 5-phosphate   glyceraldehyde 3-phosphate

In attempting to understand the mechanism of these condensations, the simplest analogy to be made by the chemist is that thiamine behaves like a cyanide ion in the catalysis of a benzoin-type condensation (301):

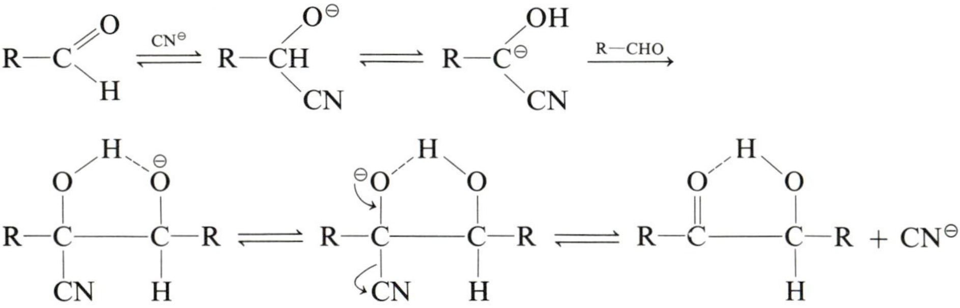

$$
R\!-\!\underset{H}{\overset{O}{C}} \xrightarrow{CN^{\ominus}} R\!-\!\underset{CN}{\overset{O^{\ominus}}{CH}} \rightleftharpoons R\!-\!\underset{CN}{\overset{OH}{C^{\ominus}}} \xrightarrow{R\!-\!CHO}
$$

$$
R\!-\!\underset{CN}{\overset{\overset{H}{O}\cdots O}{C}}\!\!-\!\underset{H}{C}\!-\!R \rightleftharpoons R\!-\!\underset{CN}{\overset{\overset{H}{{}^{\ominus}O}\cdots O}{C}}\!\!-\!\underset{H}{C}\!-\!R \rightleftharpoons R\!-\!\underset{}{\overset{\overset{H}{O}\cdots O}{C}}\!\!-\!\underset{H}{C}\!-\!R + CN^{\ominus}
$$

R = C_6H_5, benzoin

To explain this behavior, we have to examine its structure more carefully. The coenzyme thiamine pyrophosphate (thiamine-PP) contains a thiazolium ring system:

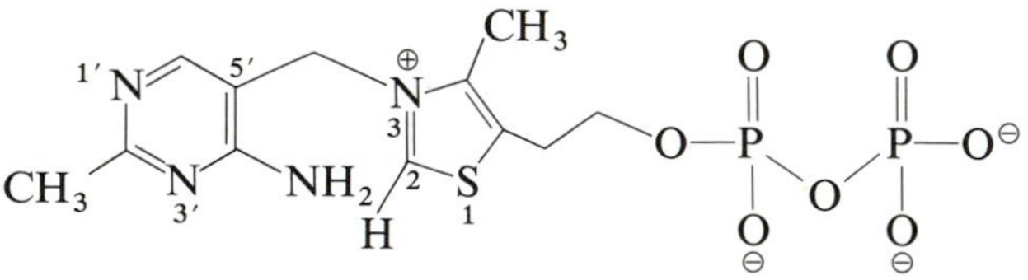

thiamine-PP (TPP)

R. Breslow (324) was the first to recognize, from NMR studies, that the thiazolium ion is acidic and the hydrogen at the C-2 position can be exchanged by deuterium in basic heavy water.

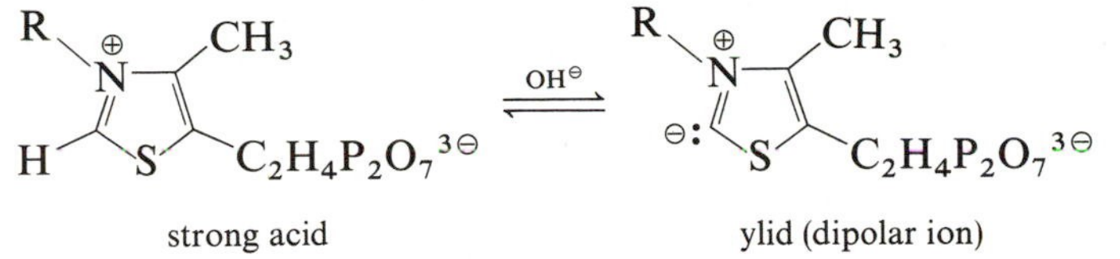

strong acid   ylid (dipolar ion)

What is the role of the pyrimidine portion of the coenzyme? One can only speculate that the C-4′ amino group can get close enough to the C-2 hydrogen and act as a weak base to facilate the generation of the thiazolium dipolar structure. Recent ^{13}C-NMR work on thiamine salts has indicated that this is the case and X-ray diffraction studies show that, in the crystalline form, the two rings of thiamine are oriented in a manner to favor this function (325). This process could be enzyme mediated and protonation at N-1′ would facilitate the process. In fact, introduction of a methyl group onto the N-1′ pyrimidine position, to mimic the presence of a single charge on this ring, confers to thiamine catalytic properties superior to thiamine itself in enzymatic reactions requiring the coenzyme (326). This positive charge on the pyrimidine very likely accelerates ylid formation, decarboxylation, and acetoin formation relative to the natural vitamin. As pointed out by

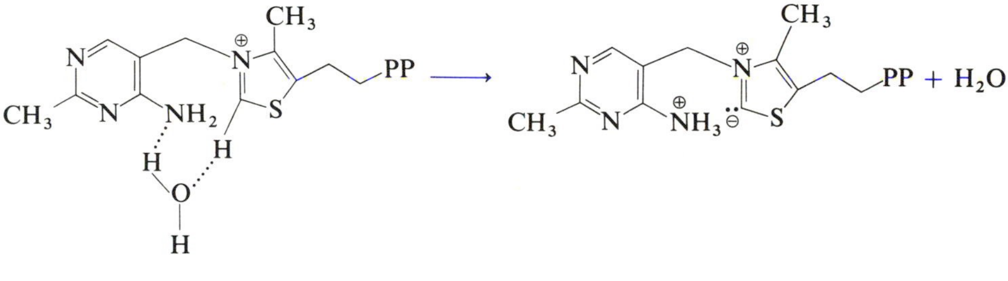

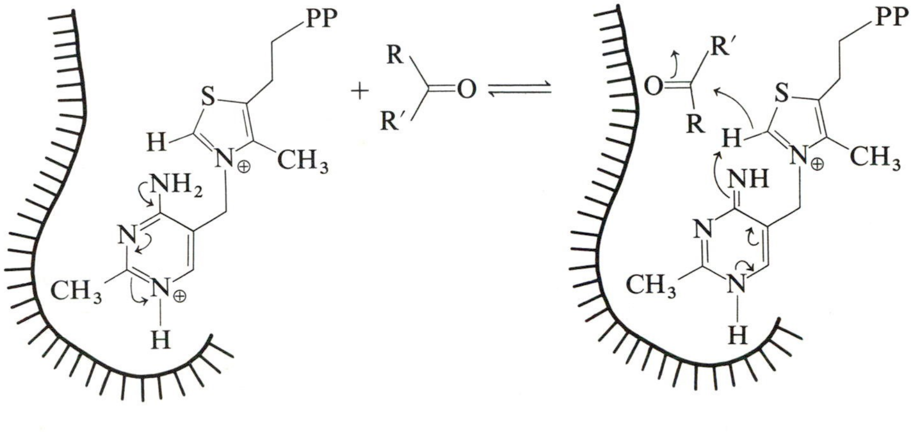

Metzler (273) the possibility that the enzyme might stabilize a minor tautomeric form of the aminopyrimidine ring is attractive. It would constitute a

charge-relay system in which the removal of the C-2 hydrogen of the thiazolium ring would be assisted by the amino group (in the imino form) of the pyrimidine ring.

One of the remarkable facets of thiamine chemistry is that the thiazolium ring forms a covalent intermediate with the substrate and can act either as an electrophile or a nucleophile. It is thus a catalyst with dual functionality. In other words, the electron flow can occur from the bond to be broken into the structure of the coenzyme (toward the $=\overset{+}{\underset{|}{N}}-$group) and then back toward

the new bond to be formed. For example, the ylid has a nucleophilic carbanion that can condense with pyruvic acid and other α-keto acids:

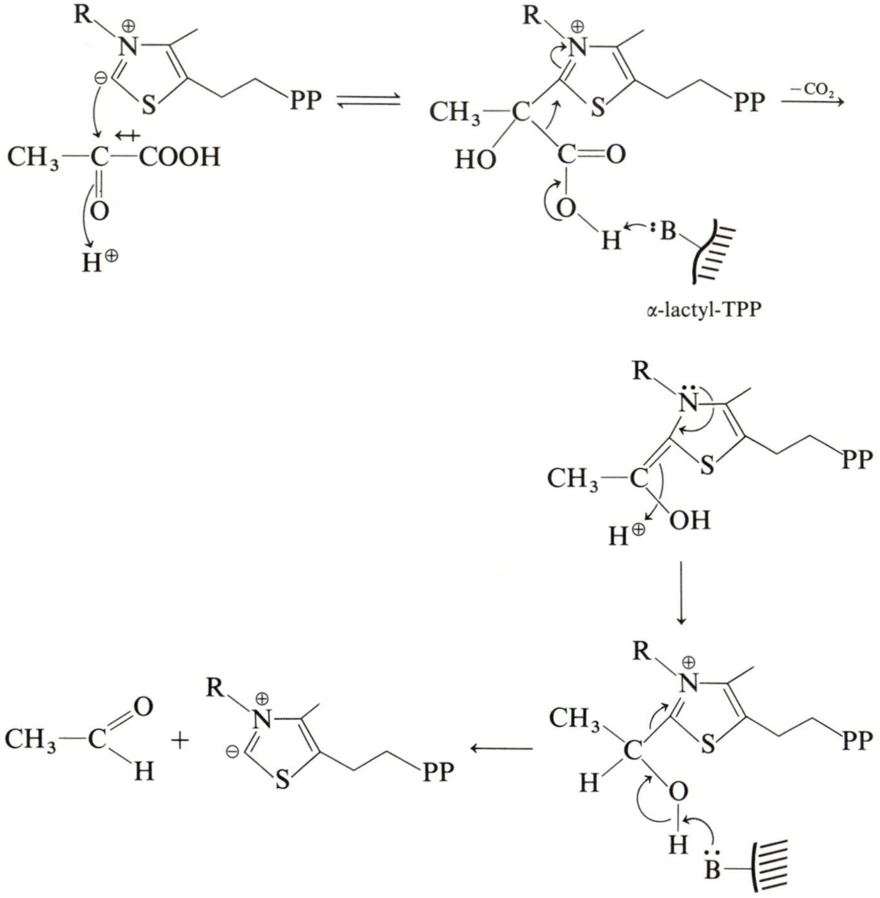

α-lactyl-TPP

$$R = $$

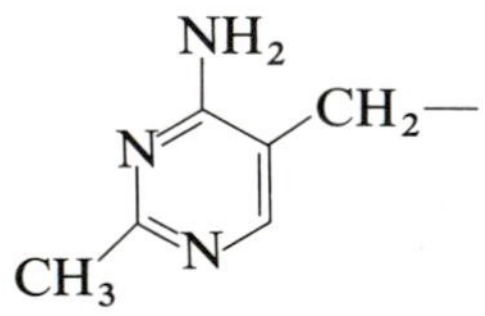

The thiazolium ion then behaves as an "electron sink" or electrophile and decarboxylation follows. The enolic intermediate, on the other hand, acts as a nucleophile which can be protonated. This intermediate has been isolated. Finally, acetaldelyde is formed and the coenzyme (ylid form) is regenerated at the same time. The liberation of acetaldelyde is the rate-limiting step in the pyruvate decarboxylase mechanism.

It should be realized that the nucleophilic enolic intermediate (hydroxy-ethyl-thiamine pyrophosphate) is a form in which much of the coenzyme is found *in vivo* and can react with other electrophiles such as a molecule of acetaldehyde (aldol condensation):

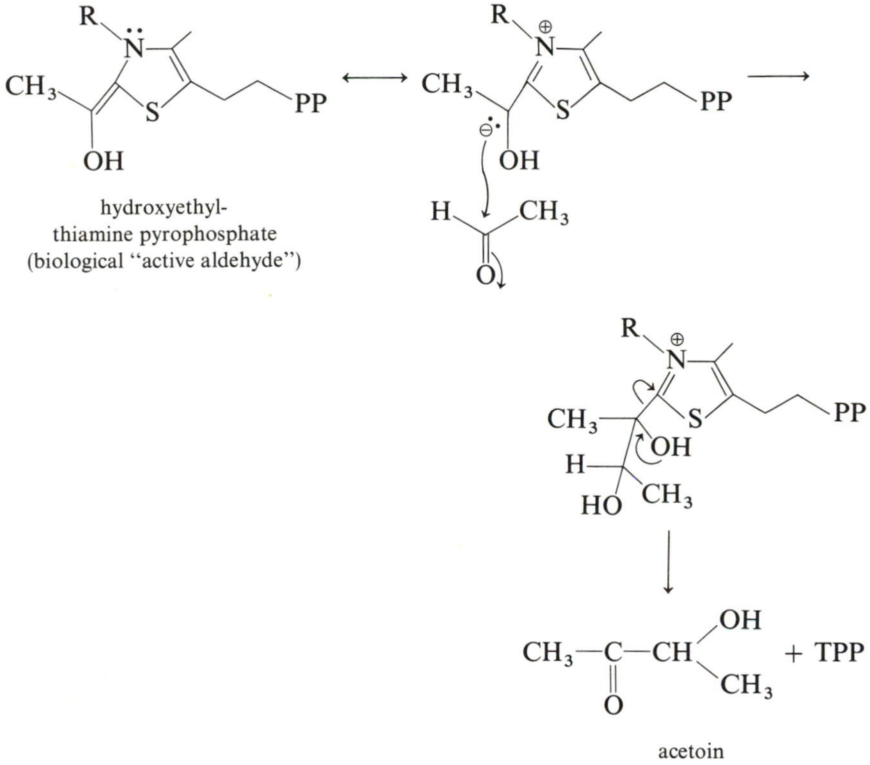

hydroxyethyl-
thiamine pyrophosphate
(biological "active aldehyde")

acetoin

The intermediate is thus a potent nucleophile and adds to carbonyl compounds to form carbon–carbon bonds. The medium has to be basic enough to ensure significant amount of the ylid, but not too basic or significant thiazolium ring opening will occur.

A second molecule of pyruvic acid can also add to the same intermediate to generate a molecule of acetoin after decarboxylation. With pyruvate decarboxylase only trace amounts of acetoin are reported in the enzymatic reaction.

Hydroxyethyl-thiamine pyrophosphate is also nucleophilic toward lipoic acid. In nature lipoic acid operates in tandem with thiamine pyrophosphate.

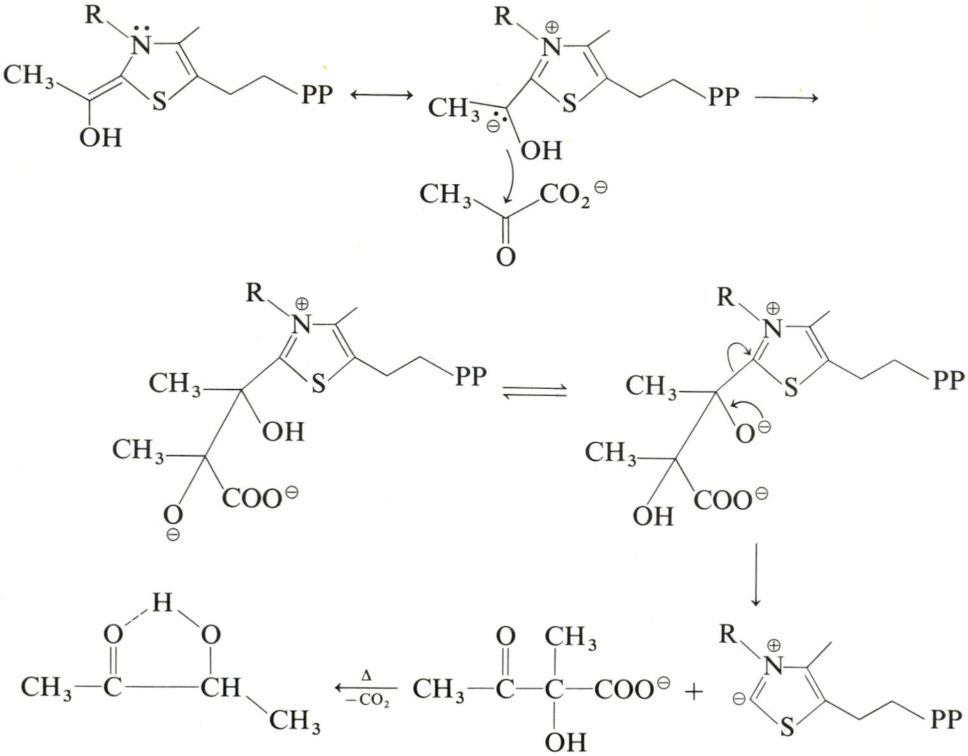

acetoin

This transformation corresponds to a reductive acylation of lipoic acid and was formulated first by L. L. Ingraham (327).

An interesting aspect of the organic chemistry of these reactions is the transformation of pyruvic acid with an $R\overset{\ominus}{-}C{=}O$ character to an acetyl thioester of lipoic acid with an $R\overset{\oplus}{-}C{=}O$ character. This behavior is not very common in organic chemistry. The explanation comes from the fact that the sulfur atom of the thioester can expand its valence shell by using orbitals.*

---

* The synthon dithiane is one of the closest organic analogues where a $>\overset{\oplus}{C}{=}O$ character is transformed to a $>\overset{\ominus}{C}{=}O$ character.

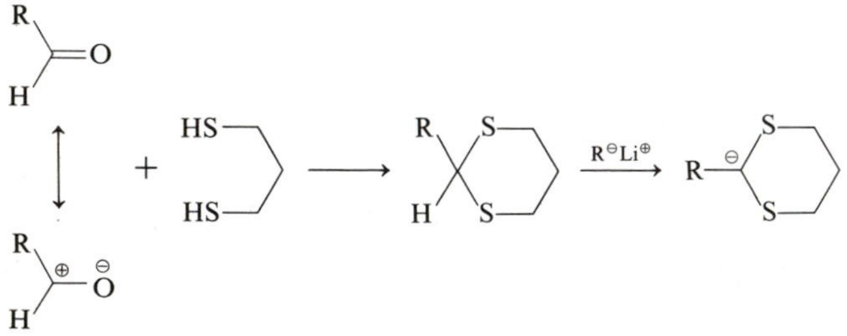

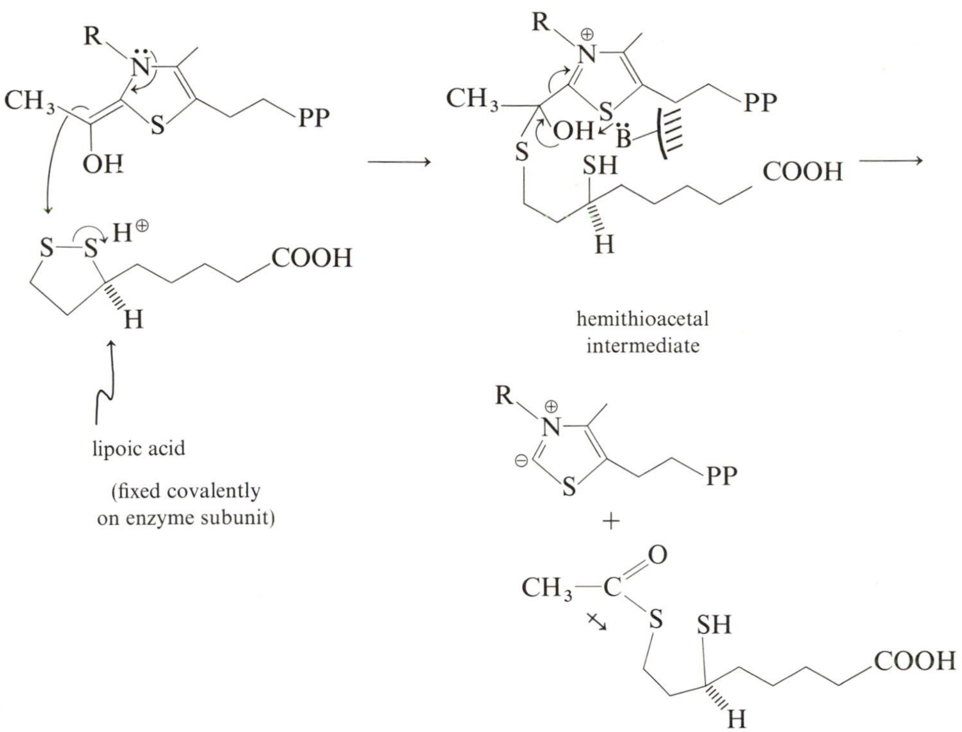

hemithioacetal
intermediate

lipoic acid

(fixed covalently
on enzyme subunit)

acetyl lipoamide
(thioester)

In connection with lipoic acid chemistry, mention should be made of arsenic compounds, some of the oldest and best-known poisons used throughout history. More recently, organic derivatives have been used as fungicides and insecticides. The more important arsenic compounds from a toxic stand point are trivalent compounds. Arsenite ($O{=}As{-}O^{\ominus}$) for instance is noted for its tendency to react rapidly with thiol groups, especially dithiols such as reduced lipoic acid. The result is that by blocking oxidative enzymes which require lipoic acid, arsenite causes the accumulation of pyruvate and other α-keto acids.

Other organic examples in the same trend are:

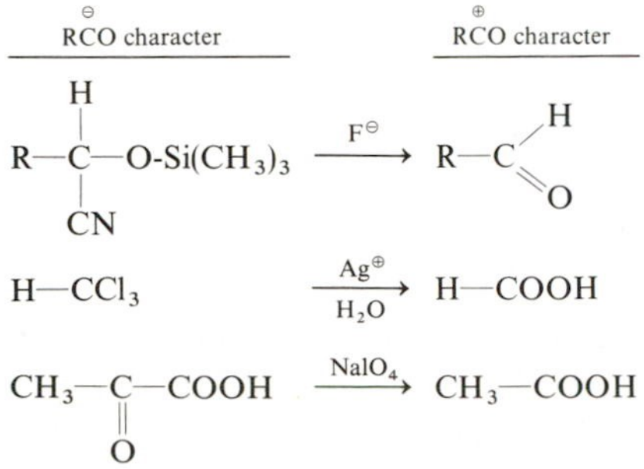

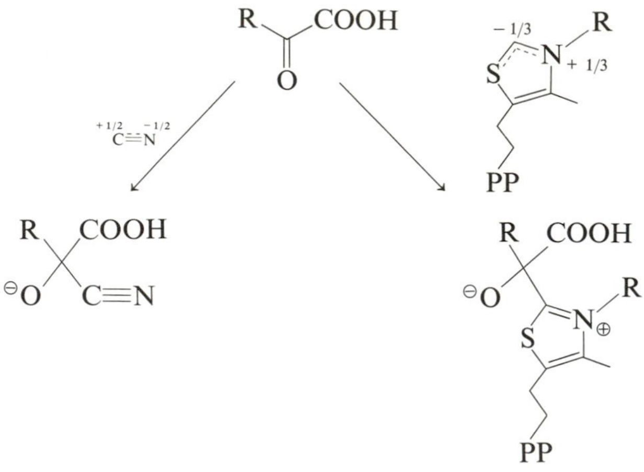

Figure 7.12. Analogy between cyanide ion and thiazolium ion of thiamine-PP.

The thiamine transformations described above show clearly the analogy of thiamine with a cyanide ion where the resulting carbanion is relatively stable because the negative charge is partly shared by the nitrogen atom (Fig. 7.12). Since the nitrogen atom in the thiazolium ring is already positively charged, it probably takes on an even greater share of the negative charge than in the cyanide ion. Because of this it can be called "biological cyanide" (301).

Furthermore, the adducts formed between thiamine-PP and substrates could be compared with a β-keto acid and a β-keto alcohol which readily undergoes β-cleavage:

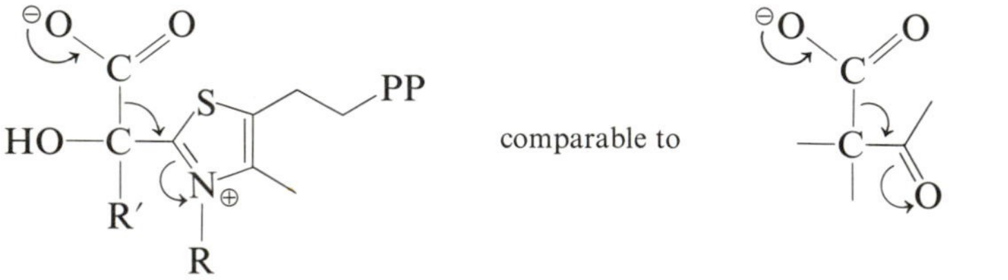

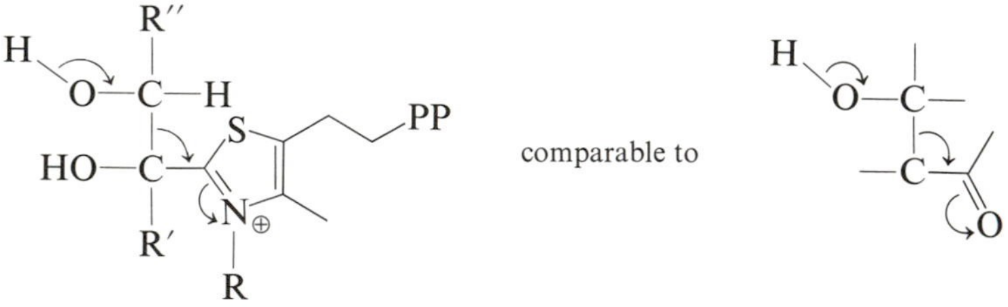

comparable to

comparable to

7.3.1 Model Design

Only a few biomodels of thiamine have been reported among which is N-benzyl thiazolium ion. At pH 8.0 and 25°C, it can readily decarboxylate pyruvic acid and modify other substrates:

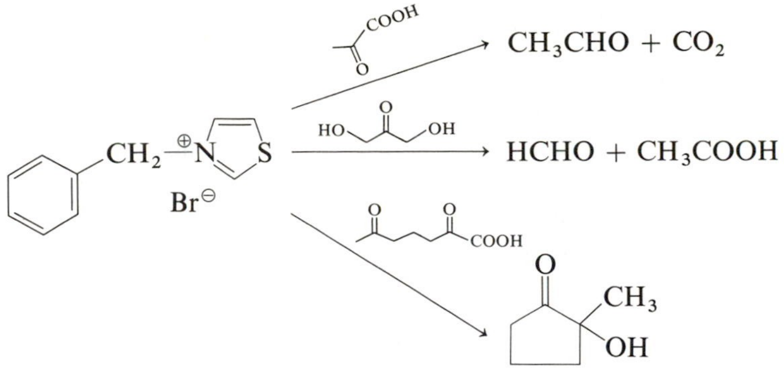

These reactions amplify the role of the active portion of thiamine as a condensing agent.

By studying other thiazolium compounds in the catalyzed acetoin condensation, it was found that blocking the 2-position either with a bulky neighboring substituent (isopropyl), or by substitution with a methyl group at this position destroys the catalytic abilities of the thiazolium salt (328).

active inactive

inactive active

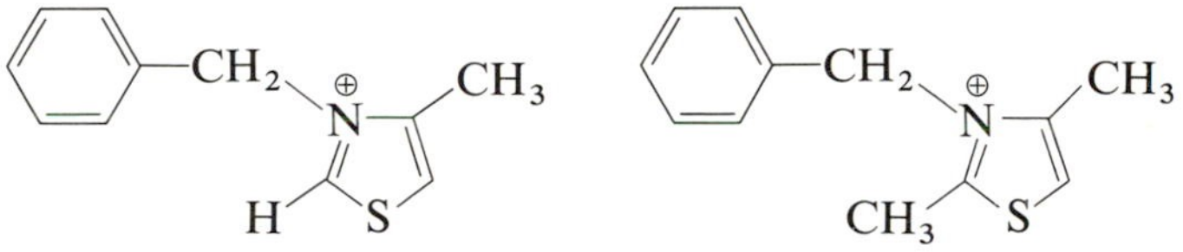

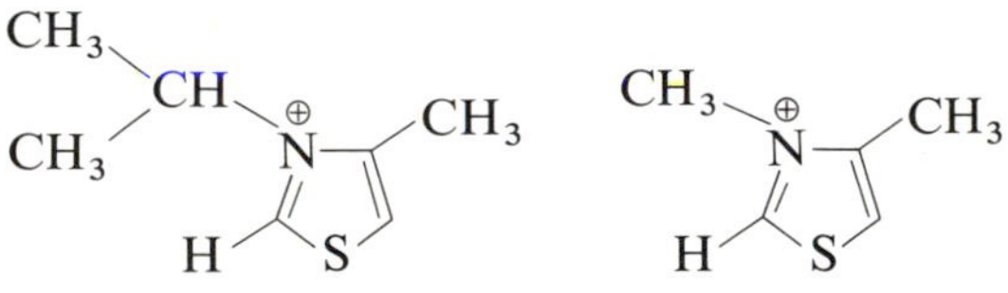

Furthermore, the greater efficiency of catalysis of the *N*-benzyl analogue to the corresponding *N*-methyl analogue led Breslow to suggest that the benzyl group, through an inductive effect, increases the reactivity. A similar role is postulated for the pyrimidine ring of thiamine.

Another interesting model is the following:

$\Delta G = -63$ kJ/mol (-15 kcal/mol)

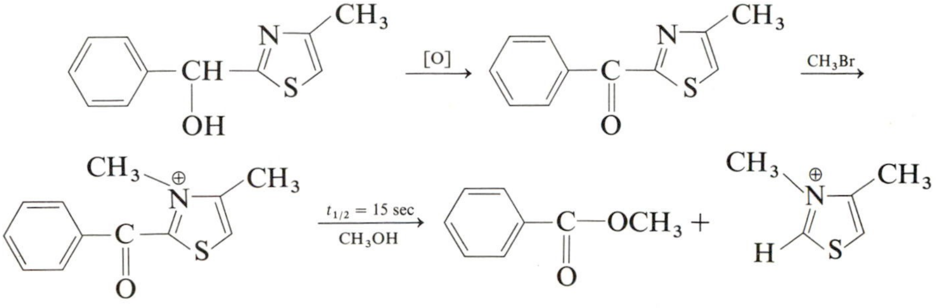

Here, the thiamine analogue, as a good leaving group, behaves as a "high energy" acylating agent (327).

In a different approach, the reactivity of thiazolium salts derived acyl anion equivalents (biological "active aldehyde") toward sulfur electrophiles has been examined recently (329) and provides a model for the thioester-forming step catalyzed by the lipoic acid containing enzymes. The results suggest that the biological generation of thioesters of coenzyme A from α-keto acids occurs via the direct reductive acylation of enzyme-bound lipoic acid by the "active aldehyde," as already shown on page 453.

In order to define the chemical basis of the binding function of thiamine-dependent enzymes, Kluger and Pike (337) studied the reaction of thiamine-PP and methyl acetylphosphonate in aqueous sodium carbonate. This substrate analogue of pyruvic acid binds to pyruvate dehydrogenase but cannot undergo all the enzyme multistep reaction processes.

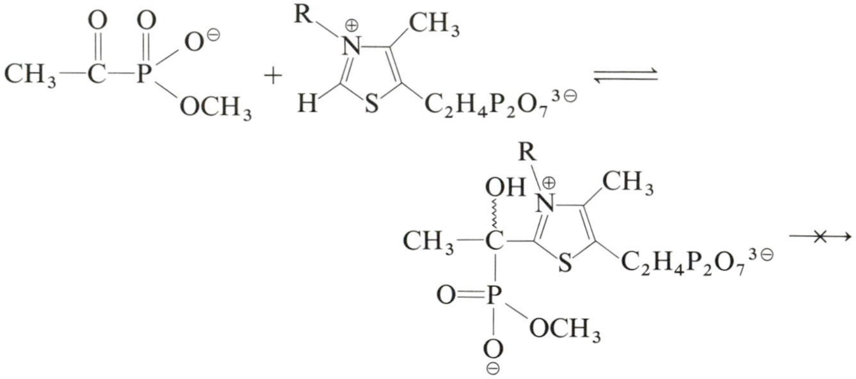

Indeed, the phosphonate adduct resembles the reactive intermediate α-lactyl-TPP (see normal process, page 450) but the reaction cannot proceed further because the necessary leaving group species would be methyl metaphosphate, and that certainly involves a large energy barrier (337). Furthermore, this reaction generates a chiral center where the thiamine-PP adduct is a racemic mixture. Kinetic data showed that both enantiomers bind to the apoenzyme but only one is being converted back to thiamine-PP. This provides information about the enzyme's active site. Of course, the absolute stereochemistry is unknown. This example illustrates a concept of growing importance in bioorganic chemistry; the formation of *enzyme-generated "reactive-intermediate analogues."*

According to Lowe and Ingraham (301), it is very likely that thiamine-PP must have optimized, through chemical adaptation, its chemical structure

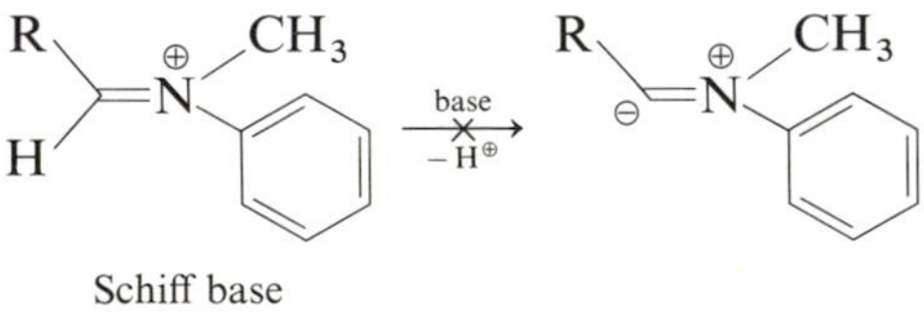

Schiff base

for its biological role as a result of chemical evolution. For instance, an alkylated Schiff base is not basic enough to be a possible catalyst in the benzoin condensation.

On the other hand, the cyclic form of a thioamide (enolic form) could be a possible candidate but was found to be extremely unstable in water, reacting like a Schiff base to give the open ring system.

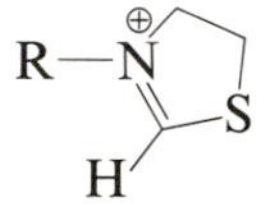

cyclic thioamide

Possibly if the sulfur is replaced by a nitrogen, then the positive charge will be stabilized by both nitrogens and the hydrogen will remain acidic enough for the compound to act as a catalyst.

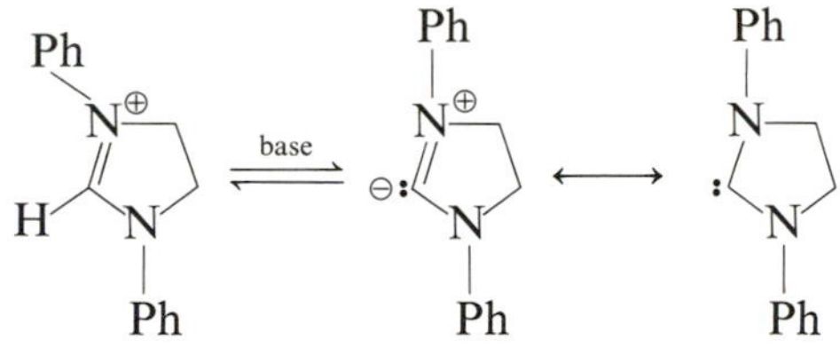

ylid   carbene

Such a compound is found to be stable in water and the hydrogen acidic enough to be exchanged with ^2H_2O. The acidity is ascribed to the carbene resonance structure which helps stabilize the carbanion intermediate. However, the compound has no catalytic activity toward benzaldehyde. Instead, an addition product is formed which can be explained by the following mechanism (301):

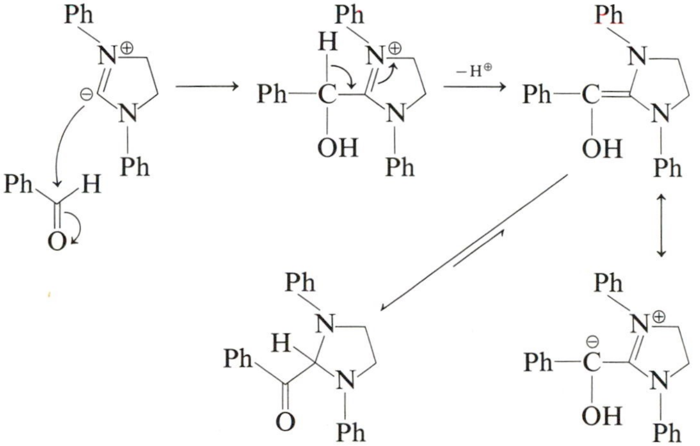

Therefore, the carbanionic intermediate is not nucleophilic enough to react with another molecule of benzaldehyde. It prefers to ketonize. What is needed is a compound that can do the same thing without so much carbanion stabilization. If too much resonance contribution is present, a good nucleophile does not form and internal ketonization takes place. Obviously, the solution of the problem is to have only partial aromatic character in the ring. A sulfur atom can help to do this because it does not have much π-character. So a thiazolium ion is the only remaining choice.

thiazolium ion

As mentioned earlier the compound with R = benzyl for instance has been synthesized and found to act as an excellent catalyst in the benzoin condensation. Interestingly, oxazolium salts are also well known by organic chemists and they exchange protons faster than thiazolium salts. π-Overlap to stabilize the cation and σ-electron withdrawal (inductive effect) to stabilize the anion are both superior in oxygen than in sulfur chemistry.

oxazolium ion

Consequently, one can wonder if nature made a mistake in choosing a thiazolium ring system rather than an oxazolium ion? Of course the answer is no because oxazolium salts simply *do not* catalyze the benzoin condensation. No products are formed because the system is too stable and thus unreactive toward benzaldehyde. In other words, the oxazolium system is not unstable enough to react with a carbonyl group. Consequently, thiamine-PP is a particularly well designed catalyst by nature and its conception corresponds to a compromise among many alternatives (301). This analysis on how thiamine-PP has adapted to its role should help to understand how to plan and design future bioorganic models of coenzymes in general.

7.4 Biotin

Biotin was first isolated and identified as a yeast growth factor in 1935. Contrary to the rapid progress made in the field of water soluble vitamins in the 1940s, the function of biotin remained a mystery until 1959 when F. Lynen and his colleagues from Munich observed that bacterial β-methylcrotonyl-CoA carboxylase carries out the carboxylation of free (+)-biotin in the absence of its natural CoA-thioester substrate.

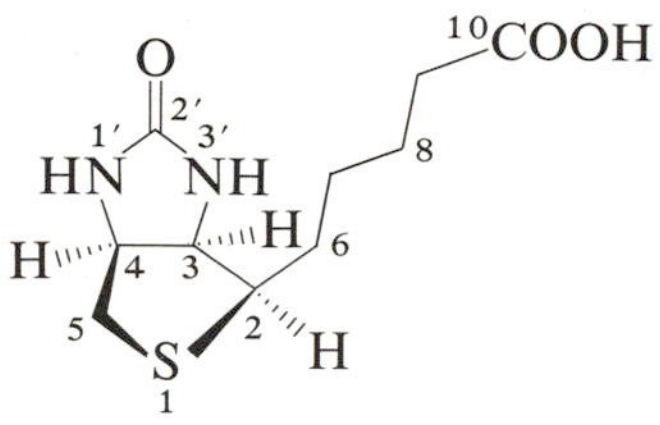

(+)-biotin

At the same period, X-ray work on the active ($+$)-form revealed that the two rings are *cis* fused, and the hydrogens of all three asymmetric carbons are in a *cis* relationship. The structure shows that the N-3′ of the ureido function is hindered from reaction by the five carbon side chain of valeric acid. The distance between N-3′ and C-6 is only 0.28 nm. Only then did the mechanism of action of biotin become clear; a CO_2-transfer reaction is achieved through the reversible formation of 1′-N-carboxybiotin.

The sulfur atom of biotin can be readily oxidized to the sulfoxide or the sulfone which both have biological activity. Desthiobiotin, a biosynthetic precursor of biotin, in which the sulfur is not yet present and is thus replaced by two hydrogen atoms, and oxybiotin, in which the sulfur has been replaced by an oxygen atom, are also both biologically active for many organisms. Consequently, the sulfur atom does not seem to be a stringent requirement for biological activity.

Biotin (referred to as vitamin H in humans) is an essential cofactor for a number of enzymes that have diverse metabolic functions. Almost a dozen different enzymes use biotin. Among the most well-known are acetyl-CoA carboxylase, pyruvate carboxylase, propionyl-CoA carboxylase, urea carboxylase, methylmalonyl-CoA decarboxylase, and oxaloacetate decarboxylase. Biotin serves as a covalent bound "CO_2 carrier" for reactions in which CO_2 is fixed into an acceptor by carboxylases. Then this carboxyl group in an independent reaction can be transferred from the acceptor substrate to a new acceptor substrate by transcarboxylases, or the carboxyl group can be removed as CO_2 by decarboxylases.

Later, in 1968, Lynen showed that ^{14}C-bicarbonate ion is a better substrate than free $^{14}CO_2$, being incorporated into the product much more rapidly. ATP and Mg(II) are also essential for the transformation to occur. Presently there is ample evidence that the overall reaction catalyzed by the biotin enzyme complex proceeds in two discrete steps (E refers to the enzyme):

$$\text{E-biotin} + \text{ATP} + \text{HCO}_3^{\ominus} \xrightleftharpoons{\text{Mg}^{2+}} \text{E-biotin}-\text{CO}_2^{\ominus} + \text{ADP} + \text{P}_i \quad (7\text{-}8)$$

$$\text{E-biotin}-\text{CO}_2^{\ominus} + \text{acceptor} \rightleftharpoons \text{E-biotin} + \text{acceptor}-\text{CO}_2^{\ominus} \quad (7\text{-}9)$$

$$\text{ATP} + \text{HCO}_3^{\ominus} + \text{acceptor} \xrightleftharpoons{\text{Mg}^{2+}} \text{ADP} + \text{P}_i + \text{acceptor}-\text{CO}_2^{\ominus} \quad (7\text{-}10)$$

The initial step involves the formation of carboxybiotinyl enzyme. In the second step, carboxyl transfer from carboxybiotinyl enzyme to an appropriate acceptor substrate takes place, the nature of this acceptor being dependent on the specific enzyme involved. In brief, the function of biotin is to mediate the coupling of ATP cleavage to carboxylation. This is accomplished in two stages in which a carboxybiotin intermediate is formed. In transcarboxylation ATP is not needed because "activated carbonate," and not HCO_3^-, is the substrate.

These biochemical transformations occur on a multienzyme complex composed of at least three dissimilar proteins: biotin carrier protein (MW = 22,000), biotin carboxylase (MW = 100,000) and biotin transferase (MW = 90,000). Each partial reaction is specifically catalyzed at a separate subsite and the biotin is covalently attached to the carrier protein through an amide linkage to a lysyl ε-amino group of the carrier protein (338, 339). In 1971, J. Moss and M. D. Lane, from Johns Hopkins University proposed a model for acetyl-CoA carboxylase of *E. coli* where the essential role of biotin in catalysis is to transfer the fixed CO_2, or carboxyl, back and forth between two subsites. Consequently, reactions catalyzed by a biotin-dependent carboxylase proceed though a carboxylated enzyme complex intermediate in which the covalently bound biotinyl prosthetic group acts as a mobile carboxyl carrier between remote catalytic sites (Fig. 7.13).

Basically, biotin behaves as a CO_2-carrier between two sites. Schematically, the biotin carboxylase subsite catalyzes the carboxylation of the biotinyl prosthetic group on the carrier protein. Following the translocation of the carboxylated functional group from the carboxylase subsite to the carboxyl transferase subsite, carboxyl transfer from CO_2-biotin to acetyl-CoA occurs. Presumably free (+)-biotin can gain access to the carboxylation

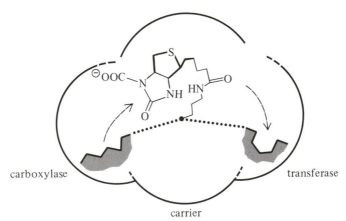

Figure 7.13. A pictorial model of acetyl-CoA carboxylase of *E. coli* (338).

site because the prosthetic group is attached to the carrier protein through a 1.4 nm side chain and can oscillate in and out of the site. However, intensive biophysical studies on the transcarboxylase that catalyzes the transfer of a carboxyl group from methylmalonyl-CoA to pyruvate show that the carboxy-biotin moves, at most, only 0.7 nm during the transfer of CO_2 (340). The role of the long arm (1.4 nm) thus appears to be to place the transferred carboxyl group at the end of a long probe, permitting it to traverse the gap that occurs at the interface of the three subunits and to be located between the coenzyme A and pyruvate sites.

What kind of chemical mechanism can explain the two partial reactions (carboxylation and transcarboxylation) mentioned previously? Biotin is a CO_2 activator where the CO_2 molecule is covalently bound to an enolizable position of the cofactor. Any mechanism must agree with the experimental observation that if oxygen-labeled bicarbonate is used ($HC^{18}O_3^-$), two of the labeled oxygens ($\frac{2}{3}$) are found in the end product and the third one ($\frac{1}{3}$) ends up in the inorganic phosphate that is split from ATP.

In fact three schools of thought exist, all consistent with the above observations. In 1962, S. Ochoa first proposed a two-step mechanism which involves the activation of bicarbonate by ATP to produce carbonyl phosphate.

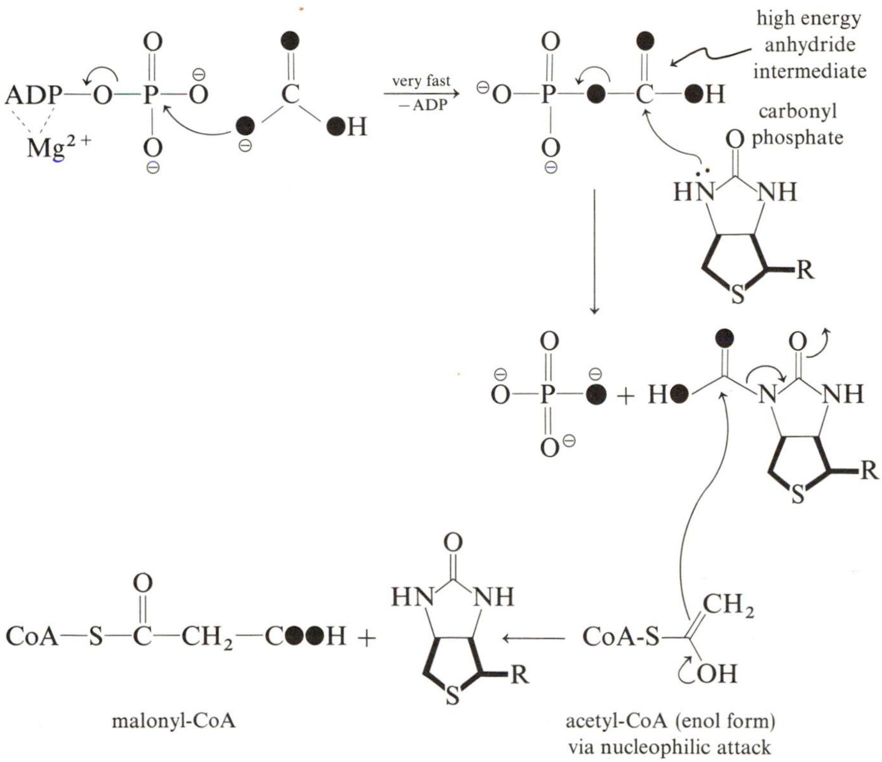

This activated bicarbonate then reacts with biotin-bound enzyme to give 1′-N-carboxybiotin, another activated form of carbonate. These two processes could occur in a concerted way and have been generally accepted for some time. The 1′-N-carboxybiotin intermediate is stable under basic conditions but decarboxylates readily in acid medium. In the above example, carboxybiotin reacts with the enol form of acetyl-CoA to give malonyl-CoA and regenerated biotin. As expected, one oxygen from bicarbonate appears in the form of phosphate ion, the other two are in the carboxyl group of malonyl-CoA.

In 1976, Kluger and Adawadkar (341), from the University of Toronto, proposed an interesting variation where the first step involves the formation of an O-phosphobiotin, a "masked" form of carbodiimide (Fig. 7.14). A "masked carbodiimide" was already described in Chapter 3 as a O-phosphoryl urea.

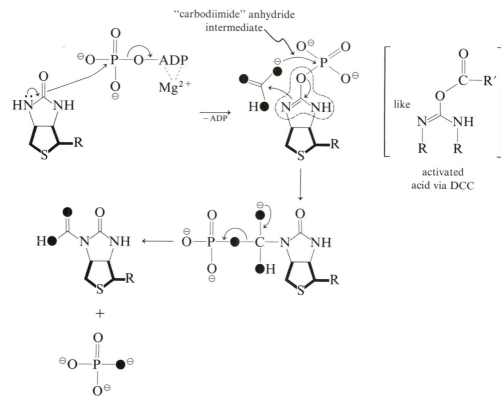

Figure 7.14. Possible mode of action of biotin according to Kluger and Adawadkar (341).

The isoimide intermediate (*O*-acyl urea) reacts with bicarbonate and then decomposes to give 1′-*N*-carboxybiotin and phosphate ion. Kluger synthesized an interesting model compound to prove the existence of his intermediate (see page 470).

Finally, in 1977, W. C. Stallings (342), from the Institute for Cancer Research in Philadelphia, proposed a third mechanism based on the analysis of the hydrogen bonding observed in the crystal structure of biotin (Fig. 7.15). He argues that biotin and bicarbonate make complementary hydrogen bonds to form a "stable" intermediate. Approach of ATP triggers carboxylation by bond polarization and a keto–enol tautomerism of the ureido moiety, which enhances the nucleophiticity of the N-1′ position. The involvement of an enol form of biotin is supported by the crystallographic studies on biotin. Notice that the previous mechanism also displays an enol intermediate.

Therefore, the ATP participation shifts the equilibrium toward the enol form. This last mechanism represents an example of activation by a substrate. The binding of bicarbonate, the substrate in question, increases the nucleophilicity of the biotin cofactor. However, the formation of an activated bicarbonate, as proposed in the first mechanism, cannot be excluded. In all three cases examined the role of Mg(II) ions could be one of binding to ATP and promoting the formation of ADP. It is also remarkable that in all three cases the action of biotin is mediated by activation by phosphorylation, reminiscent of many reactions presented in Chapter 3.

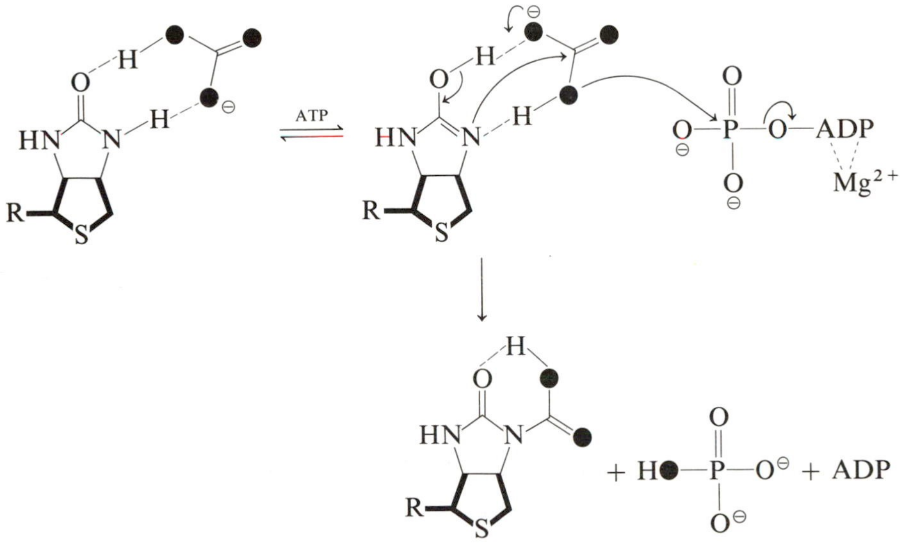

Figure 7.15. Stallings' hypothetical mode of action of biotin (342).

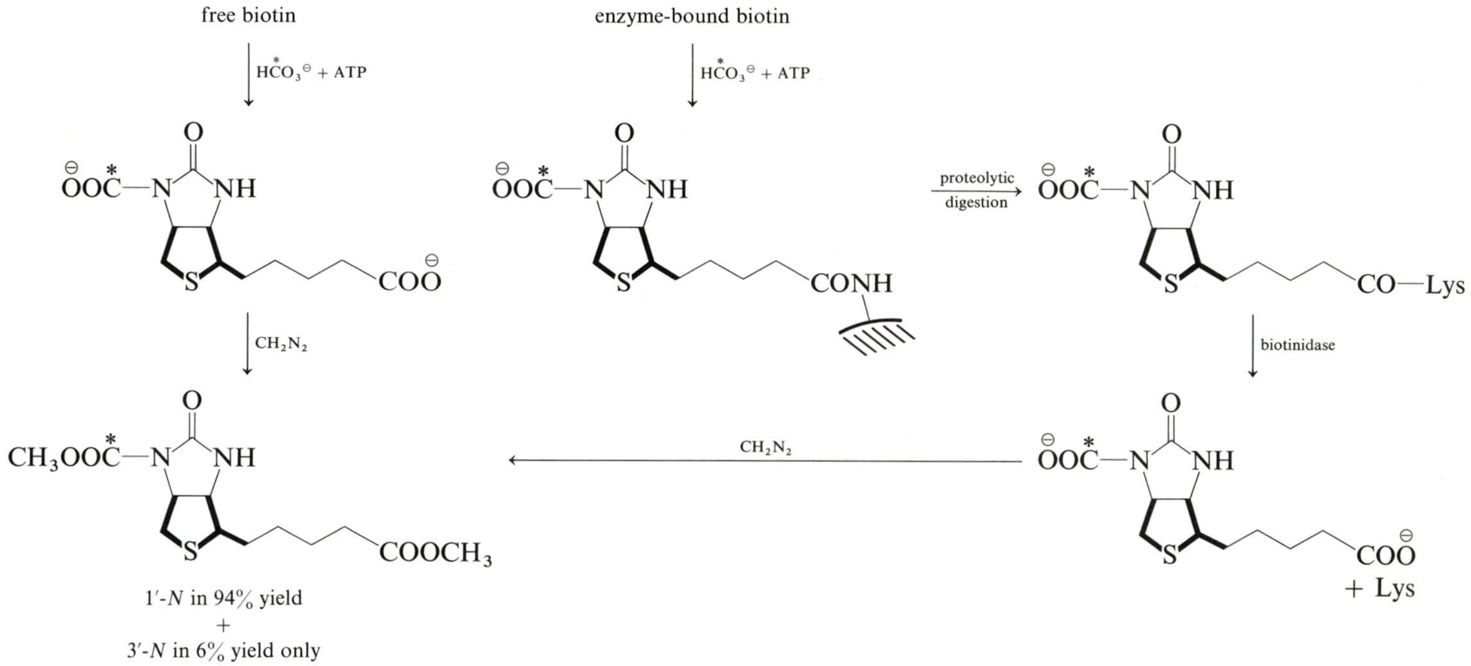

Figure 7.16. Identification of the site of carboxylation of free and enzyme-bound biotin (343).

Now that we have examined objectively three mechanisms that agree with the experimental results, we should examine the real problem in biotin chemistry. Indeed, in recent years a controversy has arisen regarding the exact site on biotin to which the carboxyl group is attached. During isolation and characterization by Lynen of the relatively unstable free carboxybiotin, particularly at acidic pH, the product was converted to the more stable dimethyl ester with diazomethane (343, 344). This derivative was subsequently identified as 1'-N-methoxycarbonyl-(+)-biotin methyl ester. The same product was also obtained by enzymatic degradation of enzyme-bound biotin. Figure 7.16 shows some of these transformations.

All these experiments with several different biotin enzymes concluded that the site of carboxylation on biotin is the 1'-ureido-N atom, in conformity with the original observations made by Lynen in 1959 (344). However, in 1970, T. C. Bruice and A. F. Hegarty (330) from the University of California, pointed out that this structural assignment of the 1'-N-carboxyl as the reactive group is equivocal. They argued that carboxylation occurred first at the 2'-ureido-O atom and that, during the methylation used to isolate carboxybiotin, the O-carboxylated product could have rearranged by transferring the carboxyl group to the more thermodynamically stable 1'-ureido-N-substituted biotin derivative.

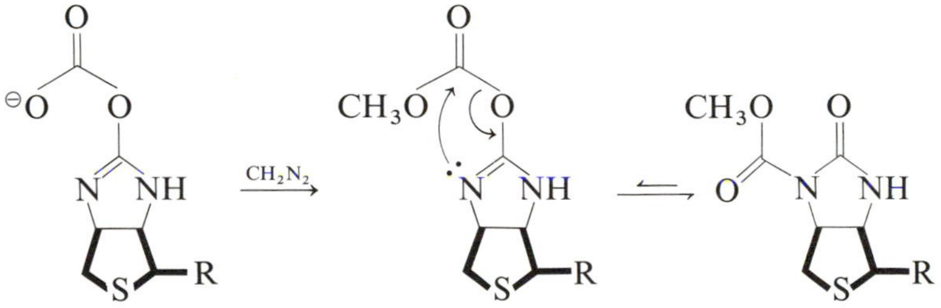

The question is: What is involved first, an N-carboxylation (Lynen) or an O-carboxylation (Bruice) on the most nucleophilic site, the urea oxygen, followed by migration to the nitrogen? Obviously, bioorganic models are needed to understand the pivotal role of the ureido function.

7.4.1 Model Studies

Since biotin appears to participate in carboxylation reactions as a nucleophilic catalyst, Caplow and Yager (331) used a 2-imidazolidone structure as a biotin analogue.

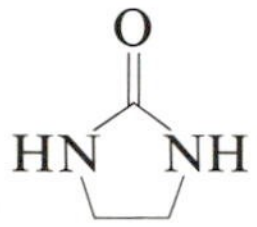

2-imidazolidone

Decarboxylation studies of *N*-carboxy-2-imidazolidone revealed the poor leaving ability of the imidazolidone anion. Metal ions such as Cu(II) or Mn(II) prevent decarboxylation. The sensitivity of carboxyimidazolidone to specific acid catalysis is probably the most remarkable feature of this compound and could account for the fact that carboxybiotin–enzyme intermediates are unstable. The neutral form however can decarboxylate in a unimolecular pathway via a six-membered transition state.

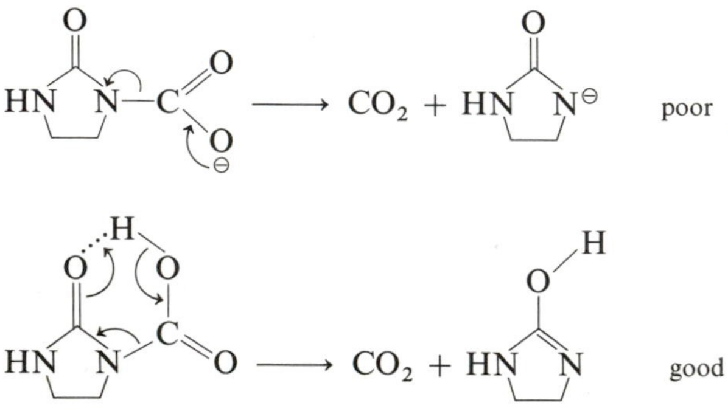

The failure for the model compound to be acylated by *p*-nitrophenyl acetate, acetylimidazole or acetyl-3-methylimidazolium chloride also indicated that the ureido-N atom of biotin is not highly nucleophilic. Nonetheless, an intramolecular model reaction in which the interacting functional groups are properly juxtaposed might more closely mimic the enzymatic process.

For instance, Bruice's group studied the intramolecular nucleophilic attack of a neutral or anionic ureido function on acyl substrates with leaving groups of varying basicity (330).

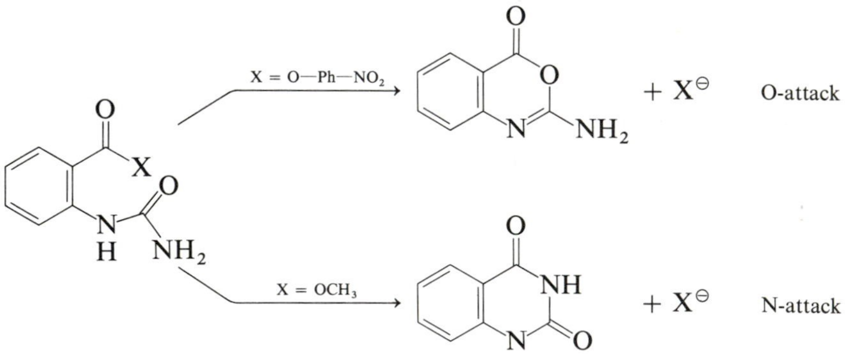

It was observed that O-attack was favored over N-attack if a good leaving group is present. These results support his proposal of O-carboxylation of biotin where a high energy carbodiimide-type compound (isoimide or isourea) is first formed and rapidly rearranges to the more stable *N*-acyl form.

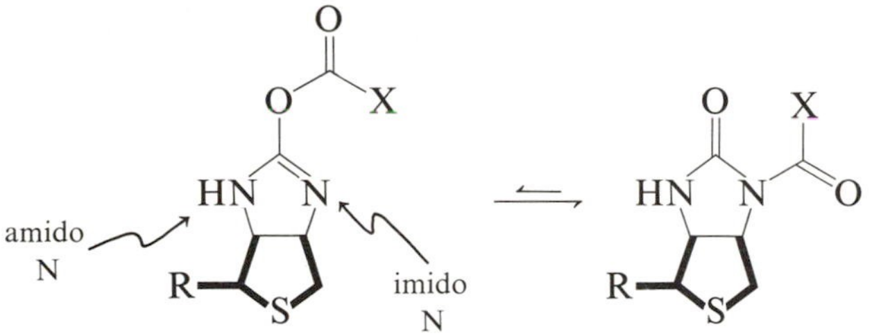

isoimide (*O*-acylurea or isourea)

X = metal–substrate or ATP-bound

However, since it is not known to what extent the enzyme contributes to the stabilization of the leaving group anion X in the biotin-dependent carboxylation reactions, a conclusion cannot be drawn *a priori* as to whether O- or N-attack occurs. Clearly, simple model compounds are not always reliable indicators of reactivity in the environment of an enzyme (343). Maybe enzyme-bound biotin reacts in its high energy isourea form to increase the nucleophilicity of the nitrogen atom. After all an imido nitrogen is known to be a better nucleophile than an amido nitrogen.

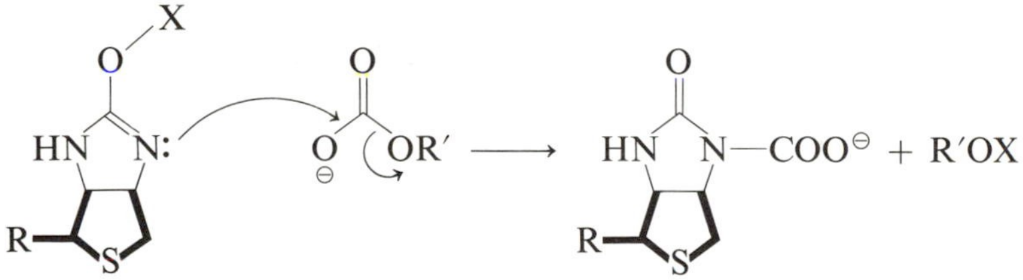

R′ = phosphate or ATP
X = electrophile
(H$^{\oplus}$ or metal ion)

Returning to the original question as to whether the postulate of *O*-carboxybiotin by Bruice is the true intermediate in the first half-reaction of biotin chemistry, the group of M. D. Lane from Johns Hopkins provided very convincing evidence in 1974 that in fact 1′-*N*-carboxybiotin must be a true biochemical intermediate (332). Using two subunits of acetyl-CoA carboxylase of *E. coli*, the biotin carboxylase and the carboxytransferase,* they showed quite conclusively, that l′-*N*-carboxybiotin is active as a carboxyl

---

* Each subunit of the multienzyme complex can be used separately in model studies.

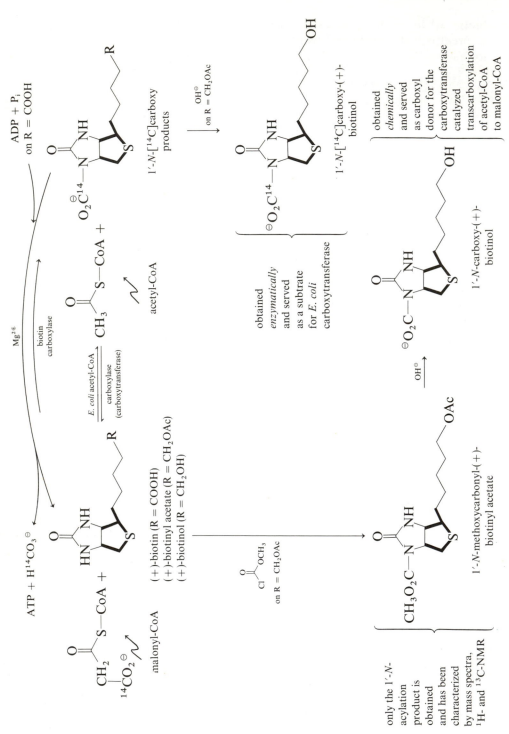

Figure 7.17. The methodology used by Lane (332) to prove that biotin is carboxylated at position N-1′.

donor (Fig. 7.17). They synthesized 1'-N-methoxycarbonyl-(+)-biotinyl acetate and characterized it chemically and spectroscopically by mass spectroscopy and ^{1}H- and ^{13}C-NMR. Alkaline hydrolysis afforded 1'-N-carboxy-(+)-biotinol. It was observed that biotinol, the reduced form (alcohol) of the valeric acid group of biotin, is much more effective than biotin. They also showed that the reaction mediated by the carboxylase occurred from right to left with synthesis of ATP in the presence of authentic 1'-N-carboxy-(+)-biotin. Furthermore, authentic 1'-N-carboxy-(+)-biotin had stability properties indistinguishable from those of the carboxybiotin formed enzymatically.

These convincing data showed that the 1'-N-ureido position of biotin serves as the site for carboxyl transfer with biotin enzymes. Lane also correctly pointed out that $N \rightarrow O$ carboxyl migration might have preceded the participation of carboxybiotin in the enzymatic process. However, the well-established thermodynamic and kinetic stabilities of N-acyl and N-carboxy-2-imidazolidone derivatives render this possibility unlikely. Moreover, the urea carboxylase component of ATP-amidolyase, also a biotin-dependent enzyme, reversibly carboxylates urea to form N-carboxyurea, a known example of carboxylation at the N-ureido position (333).

$$H_2N \overset{\displaystyle O}{\underset{}{\|}} NH_2 + HCO_3^{\ominus} \xrightarrow[\substack{\text{urea} \\ \text{carboxylase}}]{\text{ATP}} H_2N \overset{\displaystyle O}{\underset{}{\|}} NH-COO^{\ominus} \xrightarrow[\substack{\text{allophanate} \\ \text{amidolyase}}]{H_2O}$$

N-carboxyurea
(allophanate)

$$2 \, HCO_3^{\ominus} + 2 \, NH_4^{\oplus}$$

Lane and co-workers also speculated as to why the 1'-N, rather than the apparently more reactive 2'-O, carboxybiotinyl prosthetic group evolved as the carboxylating agent in these enzymatic processes. They suggest that possibly the consequent exposure to the mobile carboxyl carrier chain to the solvent would oblige the carboxylate group to possess low, rather than high, carboxyl transfer potential to prevent $O \rightarrow N$ migration during the transfer process.

We should now return to the unsolved original problem of finding the method by which biotin and carboxybiotin are activated. Earlier in this section we presented three possible mechanisms for carboxylation of a substrate by biotin. One of them, by Kluger and Adawadkar (341), suggests the presence of an O-phosphobiotin intermediate. Kluger has now developed an interesting model for such a mechanism, in which a phosphoryl group, via a methylene bridge, is attached to a ureido nitrogen. The starting material was synthesized from dimethylurea, and LiOH hydrolysis led to a model compound possessing the reactive portions of biotin and ATP bound in the same disposition:

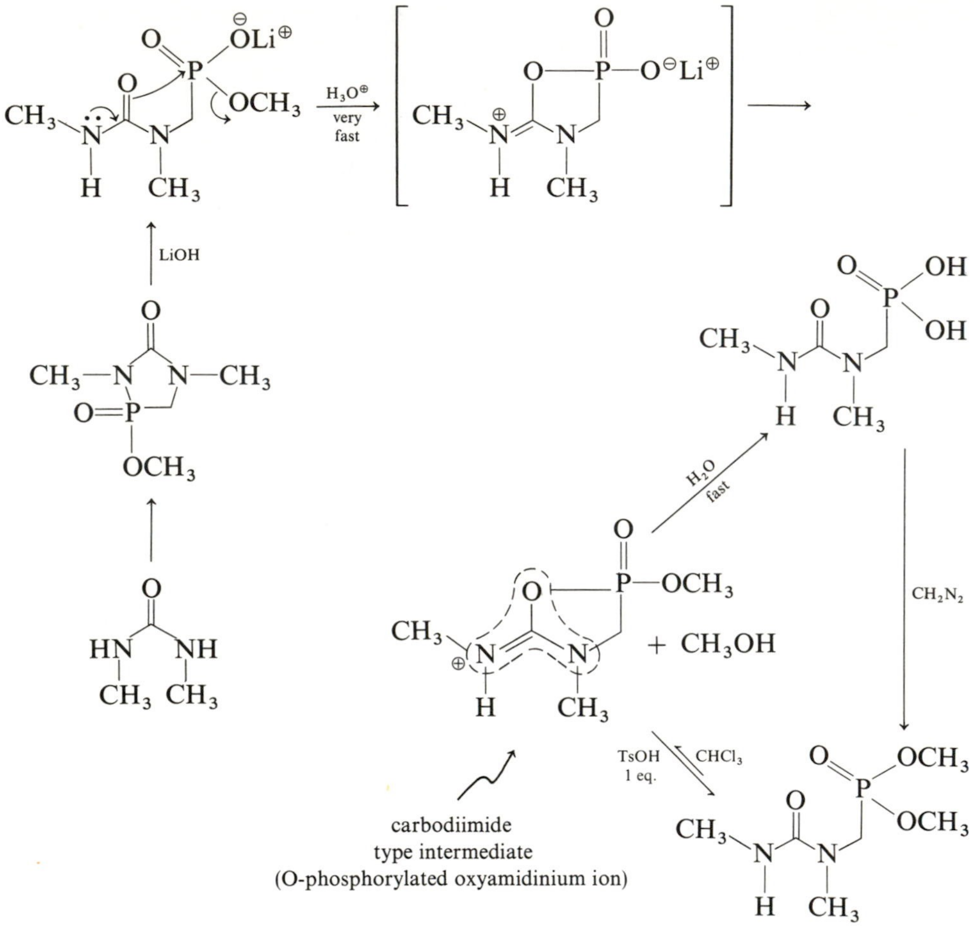

carbodiimide
type intermediate
(O-phosphorylated oxyamidinium ion)

Aqueous acid hydrolysis of the product is very fast and proceeds through the formation of a cyclic phosphourea intermediate. On the basis of its hydrolysis and other properties of related compounds, he proposed that the urea moiety of biotin is nucleophilic toward phosphate derivatives. Thus, ATP, HCO_3^-, and biotin can bind in close proximity at the carboxylation subunit to give an O-phosphobiotin, a "masked" carbodiimide dehydrating agent, that can condense with HCO_3^- and rearrange to $1'$-N-carboxybiotin (see Fig. 7.14). In fact, the possibility of a O-phosphobiotin mechanism was first proposed by Calvin in 1959. However, these new available data are appealing because they demostrate that the $2'$-O-ureido function of biotin is a nucleophile toward phosphorus.

In 1977, R. M. Kellogg (347), from the University of Groningen, synthesized an interesting model to mimic an essential aspect of the second half-reaction, namely, the transfer of the carboxyl group from biotin nitrogen to an acceptor molecule. He found that the lithio derivative of N-methylethyleneurea (or thiourea), a model for the isourea form of biotin, is capable of causing the rearrangement of 1-methyl-4-methylene-3, 1-benzoxazin-2-one to 4-hydroxy-1-methyl-2-quinolinone. The *N*-methyl derivative improves the solubility in THF and allows the abstraction of only one hydrogen. This isomerization does not occur with organic bases such as $R_4N^+OH^-$, pyridine, or LiOH. But if the isourea is activated by BuLi, the reaction proceeds in 49% yield in refluxing THF for 15 min and the following adduct was isolated in 10% yield:

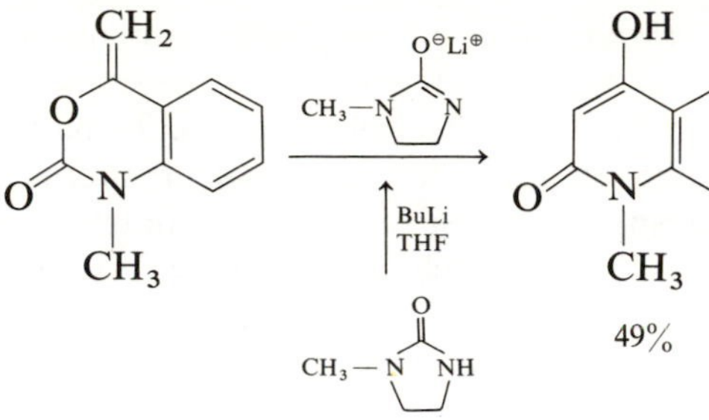

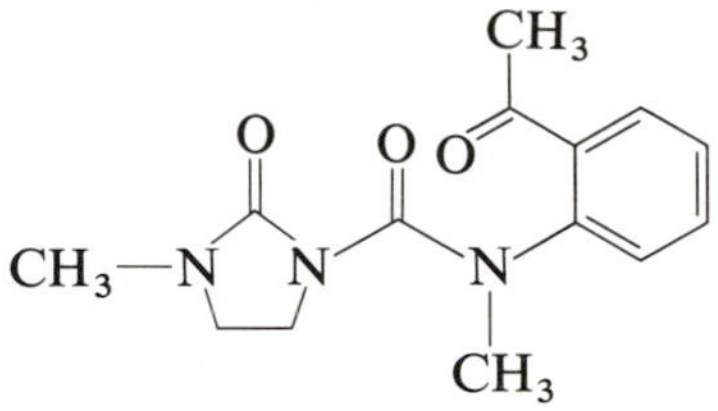

A reasonable mechanism for this cooperative process is:

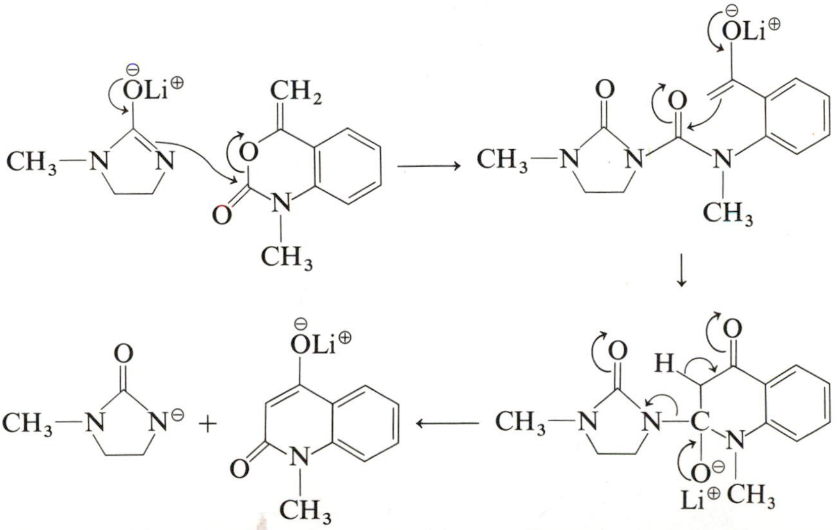

Therefore, in a number of respects this transformation mimics the carboxylation of biotin on nitrogen and the subsequent carboxyl transfer to a carbon atom bonded to a carbonyl group.

In 1979, Kluger's group examined in more detail his mechanism of urea participation in phosphate ester hydrolysis (334). In addition to contributing further evidence for the potential involvement of *O*-phosphobiotin in ATP-dependent carboxylations in enzymatic systems, they have also considered the stereochemical pathways of carboxylation of *O*-phosphobiotin in the enzymatic reactions, leading to the formation of *N*-carboxybiotin and phosphate ion. Since a pentacovalent intermediate must be involved, two mechanisms are possible (refer to Chapter 3 for notions of pseudo-rotation): an adjacent mechanism, not involving free carboxyphosphate, and an in-line mechanism, involving free carboxyphosphate.

Adjacent mechanism leads to retention of configuration at phosphorus:

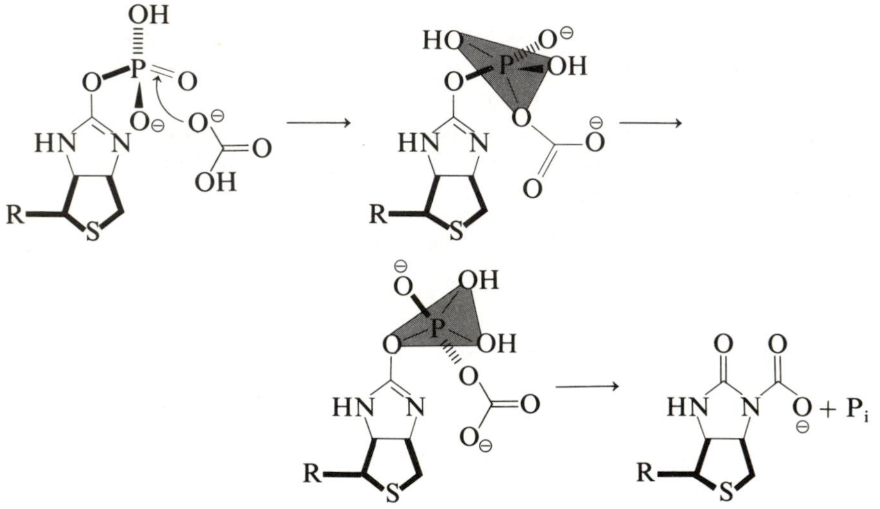

In-line mechanism causes an inversion of configuration at phosphorus:

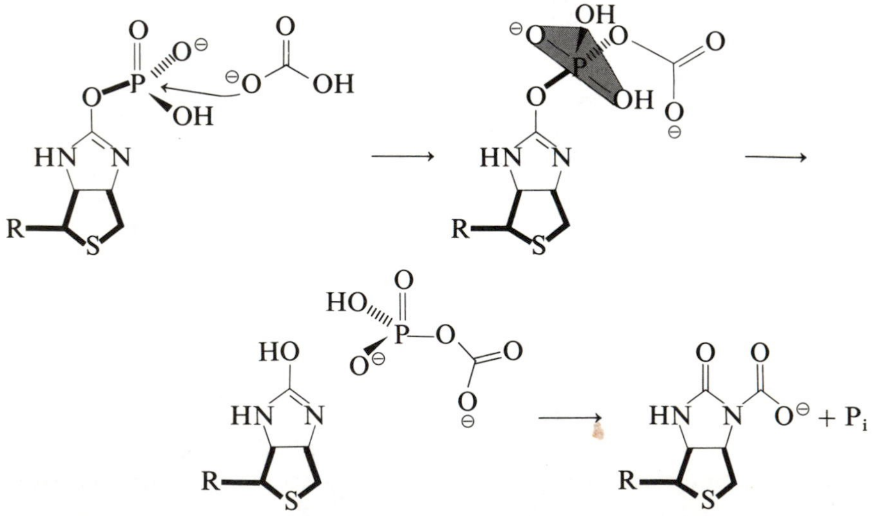

In the in-line mechanism, bicarbonate attacks *O*-phosphobiotin *trans* to the biotin ureido function. If this is the case, a concerted in-line attack followed by carboxylation is stereochemically impossible. Consequently, in the in-line mechanism carboxyphosphate must be formed. On the other hand, the adjacent mechanism allows the reaction to proceed without formation of carboxyphosphate because of a more favorable stereochemical disposition of bicarbonate. Although this mechanism corresponds to the one they had previously suggested (341), more work is needed to solve this remaining ambiguity related to the stereochemistry of carboxylation of O-phosphobiotin in enzymatic processes. Structural analysis of a biotin-dependent enzyme may resolve this problem.

In conclusion, the formation of O-phosphobiotin from ATP is likely to occur with inversion of configuration at the γ-phosphorus. Since the subsequent transfer to bicarbonate can occur with retention or inversion, the route via O-phosphobiotin can account for net inversion (inversion and retention) or net retention (inversion and inversion) at the phosphorus atom (334).

Finally, a word should be said about the stereochemistry of the trans-carboxylation reaction. In 1966, Arigoni and co-workers (335) first showed by tritium labeling that carboxylation of propionic acid operates by *retention* of configuration.

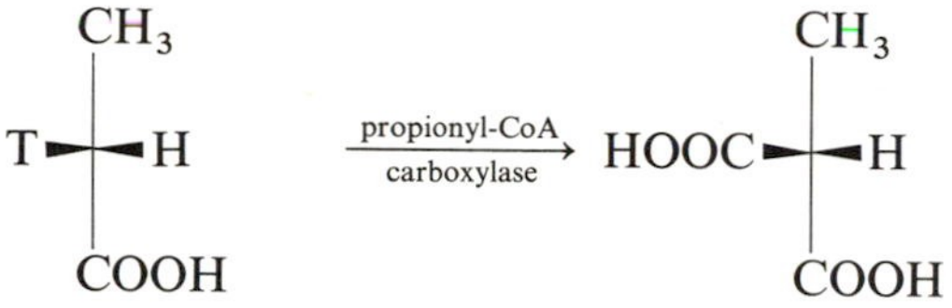

$2S$-[2-³H]propionic acid   methylmalonic acid

Like other biotin-dependent carboxylations, this reaction involves the abstraction of the α-proton (tritium here) and its replacement by CO_2. This aspect was also analyzed by Rétey and Lynen (336) and by Rose and co-workers (345). They proposed that proton abstraction and carboxylation occur in a concerted process (Fig. 7.18).

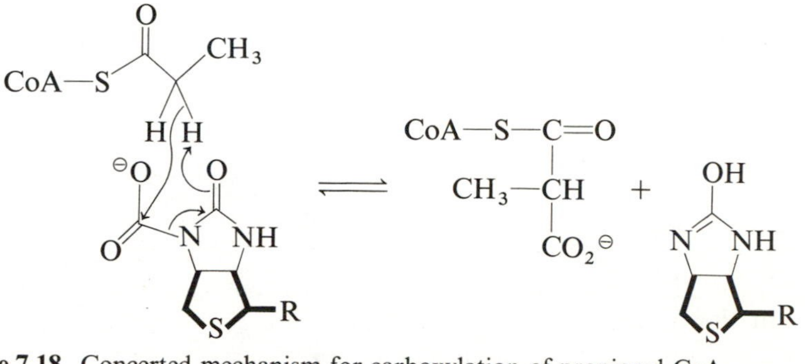

Figure 7.18. Concerted mechanism for carboxylation of propionyl-CoA.

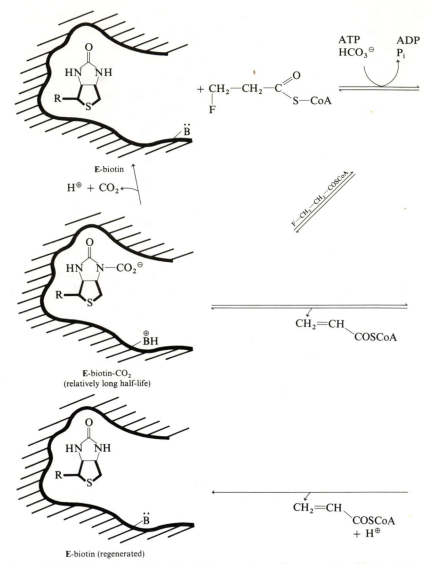

Figure 7.19. Mechanism of fluoride elimination from fluoropropionyl-CoA (346).

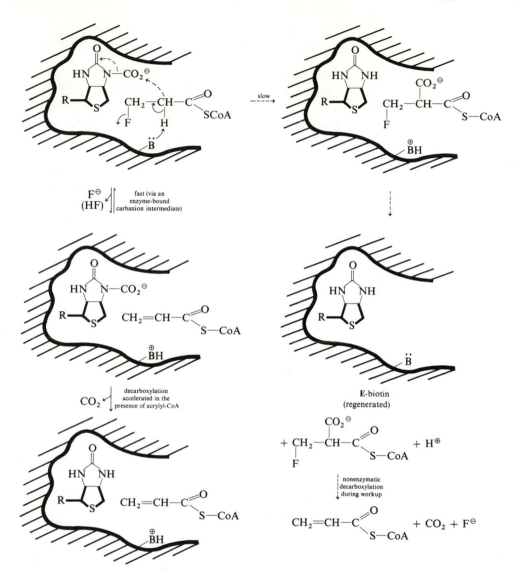

The concerted mechanism, although it could account for the retention of configuration in the carboxyl transfer during the process, is stereochemically unreasonable.

Such transformations imply that the carbonyl group of biotin serves as a proton acceptor in one direction and a proton donor in the other. A better mechanism is to have an external base. Thus, an alternative to the concerted mechanism is a stepwise process involving α-proton abstraction followed by carboxylation. To shed light on this possibility, Stubbe and Abeles in 1979 (346) examined the action of propionyl-CoA carboxylase on β-fluoropropionyl-CoA. This substrate is particularly suitable to determine whether proton abstraction occurs in the absence of carboxylation because (a) fluorine is a small atom and will not cause steric problem and (b) when a carbanion will be generated β to the carbon bearing the fluorine atom, fluoride will be eliminated. Fluoride elimination will thus be indicative of carbanion formation.

When β-fluoropropionyl-CoA is incubated in the presence of HCO_3^-, ATP, and Mg(II) and added to the enzyme, ADP is formed and F^- is released. However, the rate of F^- release is 6 times that of ADP formation and no evidence for the formation of fluoromethylmalonyl-CoA was obtained. The release of an F^- ion is an indication of an abstraction of an α-proton of the substrate, and the formation of ADP corresponds to the formation of biotin-CO_2. Therefore, these results indicate that hydrogen abstraction can occur without concomitant CO_2 transfer from biotin-CO_2 to the substrate. The concerted mechanism proposed earlier is thus not applicable when propionyl-CoA carboxylase acts on β-fluoropropionyl-CoA. The authors also suggest that the *nonconcerted process should occur* with "normal" substrates and that there must be a group at the active site of the enzyme which functions as proton acceptor. Furthermore, it was shown that this site does not exchange protons with the solvent.

In summary, the mechanism (Fig. 7.19) of F^- elimination from β-fluoropropionyl-CoA can be proposed (346) where a base (B) on the enzyme abstracts the α-proton of the substrate to form an enzyme-bound carbanion.

The elimination of F^- is more rapid than the transfer of CO_2 from biotin to the carbanion. After the elimination, the complex can be decomposed through several possible routes such as decarboxylation and release of acrylyl-CoA.

We would like to end this chapter by mentioning an interesting speculation made by Visser and Kellogg (347) about the origin of biotin and other coenzymes. First, model studies indicated that biotin-like molecules do not have catalytic power in carboxylation unless they have been activated first to a high energy tautomer, which is more nucleophilic. Consequently, biotin seems to act only as a CO_2 shuttle, linking two different active subunits together. Contrary to other transferring coenzymes, it does not contribute itself to the lowering of the activation energy of the reaction. Second, and more peculiar, biotin and the structurally related enzyme-bound lipoic acid bear no resemblance to a nucleotide. All other coenzymes contain at least a purine or a pyrimidine ring in their structures. In the speculation about the early stages of the origin of life (Section 3.7) it is believed that self-reproducing systems were made of nucleic acid-like molecules coding for nucleic acid enzymes, proteins being not yet important. According to Kellogg, present-day coenzymes may well be the vestiges from this poly-nucleotide enzyme period and thus still possess the catalytic capacity of group transfer reactions. However, biotin and lipoic acid have no tendency to act as a coenzyme unless they are covalently attached to a carrier protein and thus, they behave only as transferring agents in multienzyme complexes. Therefore, as prosthetic groups they may well have originated from essentially late evolutionary developments.

References

General Reading

R. Barker (1971), *Organic Chemistry of Biological Compounds*. Prentice-Hall, Englewood Cliffs, New Jersey.

M. L. Bender (1971), *Mechanisms of Homogeneous Catalysis from Protons to Proteins*. Wiley-Interscience, New York.

M. L. Bender and L. J. Brubacher (1973), *Catalysis and Enzyme Action*. McGraw-Hill, New York.

T. C. Bruice and S. Benkovic (1966), *Bioorganic Mechanisms*, Vols. 1 and 2. Benjamin, New York.

A. Fersht (1977), *Enzyme Structure and Mechanism*. Freeman, San Francisco.

R. P. Hanzlik (1976), *Inorganic Aspects of Biological and Organic Chemistry*. Academic Press, New York.

W. P. Jencks (1969), *Catalysis in Chemistry and Enzymology*. McGraw-Hill, New York.

J. B. Jones, C. J. Sih, and D. Perlman, Eds. (1976), *Application of Biochemical Systems in Organic Chemistry*, Vol. 10, Parts I and II. Techniques of Chemistry Series. Wiley-Interscience, New York.

K. D. Kopple (1971), *Peptides and Amino Acids*. Benjamin, New York.

J. N. Lowe and L. L. Ingraham (1974), *An Introduction to Biochemical Reaction Mechanisms*. Prentice-Hall, Englewood Cliffs, New Jersey.

D. E. Metzler (1977), *Biochemistry, The Chemical Reactions of Living Cells*. Academic Press, New York.

E. Ochai (1977), *Bioinorganic Chemistry, an Introduction*. Allyn and Bacon, Boston.

E. E. van Tamelen, Ed. (1977), *Bioorganic Chemistry*, Vols. I–IV. Academic Press, New York.

C. Walsh (1979), *Enzymatic Reaction Mechanisms*. Freeman, San Francisco.

J. D. Watson (1976), *Molecular Biology of the Gene*, 3rd ed. Benjamin, New York.

Selected References

1. V. B. Schatz (1960), Isosterism and bio-isosterism as guides to structural variations. In: *Medicinal Chemistry* (A. Burger, Ed.), 2nd ed., pp. 72–88. Interscience, New York.

2. A. Korolkovas (1970), *Essentials of Molecular Pharmacology*, pp. 55–59. Wiley-Interscience, New York.

3. R. Engel (1979), Phosphonates as analogues of natural phosphates. *Chem. Rev.* **77**, 349–367.

4. K. C. Tang, B. E. Tropp, and R. Engel (1978), The synthesis of phosphonic acid and phosphate analogues of glycerol-3-phosphate and related metabolites. *Tetrahedron* **34**, 2873–2878.

5. F. Lipmann (1973), Nonribosomal polypeptide synthesis on polyenzyme templates. *Acc. Chem. Res.* **6**, 361–367.

6. J. A. Walder, R. Y. Walder, M. J. Heller, S. M. Freier, R. L. Letsinger, and I. M. Klotz (1979), Complementary carrier peptide synthesis: General strategy and implications for prebiotic origin of peptide synthesis. *Proc. Nat. Acad. Sci. USA*, **76**, 51–55.

7. B. Belleau and G. Malek (1967), A new convenient reagent for peptide synthesis. *J. Amer. Chem. Soc.* **90**, 1651.

8. J. M. Stewart and J. D. Young (1969), *Solid Phase Peptide Synthesis*. Freeman, San Francisco.

9. J. Rebek and D. Feitler (1974), Mechanism of the carbodiimide reaction. II. Peptide synthesis on the solid phase. *J. Amer. Chem. Soc.* **96**, 1606–1607.

10. R. Kalir, M. Fridkin, and A. Patchornik (1974), (4-Hydroxy-3-nitro)benzylated polystyrene. An improved polymeric nitrophenyl derivative for peptide synthesis. *Eur. J. Biochem.* **42**, 151–156.

11. J. P. Vigneron, H. Kagan, and A. Horeau (1968), Synthèse asymétrique de l'acide aspartique optiquement pur. *Tetrahedron Lett.* 5681–5683.

12. E. J. Corey, R. J. McCaully, and H. S. Sachdev (1970), Studies on the asymmetric synthesis of α-amino acids. I. A new approach. *J. Amer. Chem. Soc.* **92**, 2476–2488.

13. M. D. Fryzuk and B. Bosnich (1977), Asymmetric synthesis. Production of optically active amino acids by catalytic hydrogenation. *J. Amer. Chem. Soc.* **99**, 6262–6267.

14. M. D. Fryzuk and B. Bosnich (1979), Asymmetric synthesis. Preparation of chiral methyl chiral lactic acid by catalytic asymmetric hydrogenation. *J. Amer. Chem. Soc.* **101**, 3043–3049.

15. N. Takaishi, H. Imai, C. A. Bertelo, and J. K. Stille (1978), Transition metal catalyzed asymmetric organic synthesis via polymer-attached optically active phosphine ligands. Synthesis of *R* amino acid and hydratopic acid by hydrogenation. *J. Amer. Chem. Soc.* **100**, 264–267.

16. M. E. Wilson and G. M. Whitesides (1978), Conversion of a protein to a homogenous asymmetric hydrogenation catalyst by site-specific modification with a diphosphinerhodium(I) moiety. *J. Amer. Chem. Soc.* **100**, 306–307.

17. M. E. Wilson, R. G. Nuzzo, and G. M. Whitesides (1978), Bis(2-diphenylphosphinoethyl)amine. A flexible synthesis of functionalized chelating diphosphines. *J. Amer. Chem. Soc.* **100**, 2269–2270.

18. M. Feughelman, R. Langridge, W. E. Seeds, A. R. Stokes, H. R. Wilson, C. W.

Hooper, M. H. F. Wilkins, R. K. Barclay, and L. D. Hamilton (1955), Molecular structure of deoxyribose nucleic acid and nucleoprotein. *Nature* **175**, 834–838.

19. J. D. Watson and F. H. C. Crick (1973), The structure of DNA. *Cold Spring Harbor Symp. Quant. Biol.* **18**, 123–131.

20. S. K. Dube and K. A. Marker (1969), The nucleotide sequence of N-formyl methionyl transfer RNA. Partial digestion with pancreatic and T_1 ribonuclease and derivation of the total primary structure. *Eur. J. Biochem.* **8**, 256–262.

21. A. Rich and S. H. Kim (1978), The three-dimensional structure of transfer RNA. *Sci. Amer.* **238**, 52–62.

22. R. Singleton, Jr. (1973), Bioorganic chemistry of phosphorus. *J. Chem. Educ.* **50**, 538–544.

23. F. H. Westheimer (1968), Pseudo-rotation in the hydrolysis of phosphate esters. *Acc. Chem. Res.* **1**, 70–78.

24. C. A. Deakyne and L. C. Allen (1979), Role of active-site residues in the catalytic mechanism of ribonuclease A. *J. Amer. Chem. Soc.* **101**, 3951–3959.

25. D. A. Usher, E. S. Erenrich, and F. Eckstein (1972), Geometry of the first step in the action of ribonuclease A. *Proc. Nat. Acad. Sci. USA.* **69**, 115–118.

26. D. A. Usher and D. I. Richardson, Jr. (1970), Absolute stereochemistry of the second step of ribonuclease action. *Nature* **228**, 663–665.

27. D. Yee, V. W. Armstrong, and F. Eckstein (1979), Mechanistic studies on deoxyribonucleic acid dependent ribonucleic acid polymerase from *E. coli* using phosphorothioate analogues. I. Initiation and pyrophosphate exchange reactions. *Biochemistry* **18**, 4116–4120.

28. P. M. J. Burgers, F. Eckstein, and D. H. Hunneman (1979), Stereochemistry of hydrolysis by snake venom phosphodiesterase. *J. Biol. Chem.* **254**, 7476–7478.

29. D. E. Hoard and D. G. Ott (1965), Conversion of mono- and oligodeoxyribonucleotides to 5'-triphosphates. *J. Amer. Chem. Soc.* **87**, 1785–1788.

30. R. L. Baughn, O. Adelsteinsson, and G. M. Whitesides (1978), Large-scale enzyme-catalyzed synthesis of ATP from adenosine and acetyl phosphate. Regeneration of ATP from AMP. *J. Amer. Chem. Soc.* **100**, 304–306.

31. W. A. Blättler and J. R. Knowles (1979), The stereochemical course of glycerol kinase-phosphoryl transfer from chiral $[\gamma\text{-}(S)\text{-}^{16}O, ^{17}O, ^{18}O]$ATP, *J. Amer. Chem. Soc.* **101**, 510–511.

32. W. A. Blättler and J. R. Knowles (1979), Stereochemical course of phosphokinases. The use of adenosine $[\gamma\text{-}(S)\text{-}^{16}O, ^{17}O, ^{18}O]$ triphosphate and the mechanistic consequences for the reactions catalyzed by glycerol kinase, hexokinase, pyruvate kinase and acetate kinase. *Biochemistry* **18**, 3927–3933.

33. P. Greengard and J. A. Nathanson (1977), "Second messengers" in the brain. *Sci. Amer.* **237**, 108–119.

34. T. M. Lincoln, D. A. Flockhart, and J. D. Corbin (1978), Studies on the structure and mechanism of activation of the guanosine 3',5'-monophosphate-dependent protein kinase. *J. Biol. Chem.* **253**, 6002–6009.

35. J. A. Gerlt, N. I. Gutterson, P. Dalta, B. Belleau, and C. L. Penney (1980). Thermochemical identification of the structural factors responsible for the thermodynamic instability of 3',5'-cyclic nucleotides. *J. Amer. Chem. Soc.* **102**, 1655–1660.

36. J. A. Fyfe, P. M. Keller, P. A. Furman, R. L. Miller, and G. B. Elion (1978), Thymidine kinase from herpes simplex virus phosphorylates the new antiviral compound, 9-(2-hydroxyethoxymethyl) guanine. *J. Biol. Chem.* **253**, 8721–8727.

37. V. Amarnath and A. D. Broom (1977), Chemical synthesis of oligonucleotides. *Chem. Rev.* **77**, 183–245.
38. L. A. Slotin (1977), Current methods of phosphorylation of biological molecules. *Synthesis* **11**, 737–752.
39. M. Ikehara, E. Ohtsuka, and A. F. Markham (1979), The synthesis of polynucleotides. *Advan. Carbohyd. Chem.* **36**, 135–213.
40. R. L. Letsinger and W. B. Lunsford (1976), Synthesis of thymidine oligonucleotides by phosphite triester intermediates. *J. Amer. Chem. Soc.* **98**, 3655–3661.
41. K. K. Ogilvie, A. L. Schifman, and C. L. Penney (1979), The synthesis of oligoribonucleotides. III. The use of silyl protecting groups in nucleoside and nucleotide chemistry. VIII. *Can. J. Chem.* **57**, 2230–2238.
42. A. L. Lehninger (1975), *Biochemistry*, Chapter 37. Worth, New York.
43. M. Calvin (1974), Biopolymers: Origin, chemistry and biology. *Angew. Chem. Int. Ed. Engl.* **13**, 121–131.
44. L. E. Orgel and R. Lohrmann (1974), Prebiotic chemistry and nucleic acid replication. *Acc. Chem. Res.* **7**, 368–377.
45. W. E. Elias (1972), The natural origin of optically active compounds. *J. Chem. Educ.* **49**, 448–454.
46. J. L. Abernethy (1972), The concept of dissymmetric worlds. *J. Chem. Educ.* **49**, 455–461.
47. H. Kuhn (1972), Self-organization of molecular systems and evolution of the genetic apparatus. *Angew. Chem., Int. Ed. Engl.* **11**, 798–820.
48. C. Hélène (1977), Specific recognition of guanine bases in protein–nucleic acid complexes. *FEBS Lett.* **74**, 10–13.
49. R. D. MacElroy, Y. Coekelenbergh, and R. Rein (1977), Macromolecular simulations as an approach to the study of the origins of self-replicating systems. *Biosystems* **9**, 111–119.
50. J. L. Fox (1978) Copolymer proposed as vital to evolution. *Chem Eng. News*, July 3, pp. 17, 18.
51. D. A. Usher (1977), Early chemical evolution of nucleic acids: A theoretical model. *Science* **296**, 311–313.
52. J. Hine (1978), Bifunctional catalysis of α-hydrogen exchange of aldehydes and ketones. *Acc. Chem. Res.* **11**, 1–7.
53. T. C. Bruice and S. Benkovic (1966) *Bioorganic Mechanisms*, Vol. 1, p. 134. Benjamin, New York.
54. P. Cossee (1962), Stereoregularity in heterogeneous Ziegler-Natta catalysis. *Trans. Faraday Soc.* **58**, 1226–1232.
55. W. W. Cleland (1975), What limits the rate of an enzyme-catalyzed reaction? *Acc. Chem. Res.* **8**, 145–151.
56. K. R. Hanson and I. A. Rose (1975), Interpretations of enzyme reaction stereospecificity. *Acc. Chem. Res.* **8**, 1–10.
57. G. E. Schulz and R. H. Schimer (1979), *Principles of protein structure*, Springer-Verlag, New York.
58. A. Ferscht (1977), *Enzyme Structure and Mechanism*, pp. 44–48. Freeman, San Francisco.
59. J. R. Knowles and W. J. Albery (1977), Perfection in enzyme catalysis: The energetics of triosephosphate isomerase. *Acc. Chem. Res.* **10**, 105–111.
60. W. L. Alworth (1972), *Stereochemistry and Its Application in Biochemistry*, Chap. 3. Wiley-Interscience, New York.

61. A. G. Ogston (1948), Interpretation of experiments on metabolic processes, using isotopic tracer elements. *Nature* **162**, 963.

62. F. A. Loewus, F. H. Westheimer, and B. Vennesland (1953), Enzymatic synthesis of the enantiomorphs of ethanol-1-*d*. *J. Amer. Chem. Soc.* **75**, 5018–5023.

63. T. C. Bruice and V. K. Pandit (1960), The effect of general substitution ring size and rotamer distribution on the intramolecular nucleophilic catalysis of the hydrolysis of monophenyl esters of dibasic acids and the solvolysis of the intermediate anhydrides. *J. Amer. Chem. Soc.* **85**, 5858–5865.

64. W. P. Jencks (1975), Binding energy, specificity, and enzymic catalysis: The circle effect. *Adv. Enzymol.* **43**, 219–410.

65. D. R. Storm and D. E. Koshland, Jr. (1970), A source for the special catalytic power of enzymes: Orbital steering. *Proc. Nat. Acad. Sci. USA* **66**, 445–452.

66. T. C. Bruice, A. Brown, and D. C. Harris (1971), On the concept of orbital steering in catalytic reactions. *Proc. Nat. Acad. Sci. USA* **68**, 658–661.

67. D. R. Storm and D. E. Koshland, Jr. (1972), An indication of the magnitude of orientation factors in esterification. *J. Amer. Chem. Soc.* **94**, 5805–5814.

68. D. R. Storm and D. E. Koshland, Jr. (1972), Effect of small changes in orientation on reaction rate. *J. Amer. Chem. Soc.* **94**, 5815–5825.

69. M. Philipp, I. H. Tsai, and M. L. Bender (1979), Comparison of the kinetic specificity of subtilisin and thiolsubtilisin toward *n*-alkyl *p*-nitrophenyl esters. *Biochemistry* **18**, 3769–3773.

70. M. I. Page and W. P. Jencks (1971), Entropic contributions to rate accelerations in enzymic and intramolecular reactions and the chelate effect. *Proc. Nat. Acad. Sci. USA* **68**, 1678–1683.

71. C. G. Swain and J. F. Brown (1952), Concerted displacement reactions. VII. The mechanism of acid base catalysis in non-aqueous solvents. *J. Amer. Chem. Soc.* **74**, 2534–2537.

72. T. Higuchi, H. Takechi, I. H. Pitman, and H. L. Fung (1971), Intramolecular bifunctional facilitation in complex molecules, combined nucleophilic and general acid participation in hydrolysis of hexachlorophene monosuccinate. *J. Amer. Chem. Soc.* **93**, 539–540.

73. B. A. Cunningham and G. L. Schmir (1966), Iminolactones. II. Catalytic Effects on the nature of the products of hydrolysis. *J. Amer. Chem. Soc.* **88**, 551–558.

74. Y. N. Lee and G. L. Schmir (1978), Concurrent general acid and general base catalysis in the hydrolysis of an imidate ester. 1. Monofunctional catalysis. *J. Amer. Chem. Soc.* **100**, 6700–6707.

75. Y. N. Lee and G. L. Schmir (1979), Concurrent general acid and general base catalysis in the hydrolysis of an imidate ester. 2. Bifunctional catalysis. *J. Amer. Chem. Soc.* **101**, 3026–3035.

76. W. P. Jencks (1972), Requirements for general acid-base catalysis of complex reactions. *J. Amer. Chem. Soc.* **94**, 4731–4732.

77. D. M. Blow (1976), Structure and mechanism of chymotrypsin. *Acc. Chem. Res.* **9**, 145–152.

78. D. M. Blow, J. J. Birktoft, and B. S. Hartley (1969), Role of a buried acid group in the mechanism of action of chymotrypsin. *Nature* **221**, 337–340.

79. P. B. Sigler, D. M. Blow, B. W. Matthews, and R. Henderson (1968), Structure of crystalline α-chymotrypsin II. A preliminary report including a hypothesis for the activation mechanism. *J. Mol. Biol.* **35**, 143–164.

80. R. M. Garavito, M. G. Rossmann, P. Argos, and W. Eventoff (1977), Convergence of active center geometries. *Biochemistry* **16**, 5065–5071.

81. H. T. Wright (1973), Activation of chymotrypsinogen-A. An hypothesis based upon comparison of the crystal structures of chymotrypsinogen-A and α-chymotrypsin. *J. Mol. Biol.* **79**, 13–23.

82. M. W. Hunkapiller, M. D. Forgac, and J. H. Richards (1976), Mechanism of action of serine proteases: Tetrahedral intermediate and concerted proton transfer. *Biochemistry* **15**, 5581–5588.

83. H. Kaplan, V. B. Symonds, H. Dugas, and D. R. Whitaker (1970), A comparison of properties of α-lytic protease of *Sorangium* sp. and porcine elastase. *Can. J. Chem.* **47**, 649–658.

84. M. W. Hunkapiller, S. H. Smallcombe, D. R. Whitaker, and J. H. Richards (1973), Carbon nuclear magnetic resonance studies of the histidine residue in α-lytic protease. Implication for the catalytic mechanism of serine proteases. *Biochemistry* **12**, 4732–4743.

85. J. L. Markley and I. B. Ibañez (1978), Zymogen activation in serine proteinases. Proton magnetic resonance pH titration studies of the two histidines of bovine chymotrypsinogen A and chymotrypsin Aα. *Biochemistry* **17**, 4627–4639.

86. G. Robillard and R. G. Shulman (1972), High resolution nuclear magnetic resonance study of the histidine-aspartate hydrogen bond in chymotrypsin and chymotrypsinogen. *J. Mol. Biol.* **71**, 507–511.

87. G. D. Brayer, L. T. J. Delbaere, and M. N. G. James (1979), Molecular structure of the α-lytic protease from *Myxobacter* 495 at 2.8 Å resolution. *J. Mol. Biol.* **131**, 743–775.

88. W. W. Bachovchin and J. D. Roberts (1978), Nitrogen-15 nuclear magnetic resonance spectroscopy. The state of histidine in the catalytic triad of α-lytic protease. Implications for the charge-relay mechanism of peptide bond cleavage by serine proteases. *J. Amer. Chem. Soc.* **100**, 8041–8047.

89. A. L. Fink and P. Meehan (1979), Detection and accumulation of tetrahedral intermediates in elastase catalysis. *Proc. Nat. Acad. Sci. USA* **76**, 1566–1569.

90. M. A. Porubcan, W. M. Westler, I. B. Ibañez, and J. L. Markley (1979), (Diisopropylphosphoryl) serine proteinases. Proton on phosphorus-31 nuclear magnetic resonance-pH titration studies. *Biochemistry* **18**, 4108–4115.

91. J. Kraut (1977), Serine proteases: Structure and mechanism of catalysis. *Annu. Rev. Biochem.* **46**, 331–358.

92. M. Komiyama and M. L. Bender (1979), Do cleavages of amides by serine proteases occur through a stepwise pathway involving tetrahedral intermediates? *Proc. Nat. Acad. Sci. USA* **76**, 557–560.

93. T. C. Bruice and J. M. Slurtevant (1959), Imidazole catalysis. V. The intramolecular participation of the imidazolyl group in the hydrolysis of some esters and the amide of γ-(4-imidazolyl)-butyric acid and 4-(2′-acetoxyethyl)-imidazole. *J. Amer. Chem. Soc.* **81**, 2860–2870.

94. G. A. Rogers and T. C. Bruice (1973), Isolation of a tetrahedral intermediate in an acetyl transfer reaction. *J. Amer. Chem. Soc.* **95**, 4452–4453.

95. G. A. Rogers and T. C. Bruice (1974), Control of modes of intramolecular imidazole catalysis of ester hydrolysis by steric and electronic effects. *J. Amer. Chem. Soc.* **96**, 2463–2472.

96. G. A. Rogers and T. C. Bruice (1974), Synthesis and evaluation of a model for the so-called "charge-relay" system of the serine esterases. *J. Amer. Chem. Soc.* **96**, 2473–2480.

97. G. A. Rogers and T. C. Bruice (1974), The mechanisms of acyl group transfer from a tetrahedral intermediate. *J. Amer. Chem. Soc.* **96**, 2481–2488.

98. M. Komiyama and M. L. Bender (1977), General base-catalyzed ester hydrolysis as a model of the "charge-relay" system. *Bioorg. Chem.* **6**, 13–20.

99. L. D. Byers and D. E. Koshland, Jr. (1978), On the mechanism of action of methyl chymotrypsin. *Bioorg. Chem.* **7**, 15–33.

100. C. J. Belke, S. C. K. Su, and J. A. Shafer (1971), Imidazole catalyzed displacement of an amine from an amide by a neighbouring hydroxyl group. A model for the acylation of chymotrypsin. *J. Amer. Chem. Soc.* **93**, 4552–4561.

101. G. E. Hein and C. Niemann (1962), Steric course and specificity of α-chymotrypsin-catalyzed reactions. I. *J. Amer. Chem. Soc.* **84**, 4487–4494.

102. G. E. Hein and C. Niemann (1962), Steric course and specificity of α-chymotrypsin-catalyzed reactions. II. *J. Amer. Chem. Soc.* **84**, 4495–4503.

103. H. Dugas (1969), The stereospecificity of subtilisin BPN′ towards 1-keto-3-carbomethoxy-1,2,3,4-tetrahydroisoquinoline. *Can. J. Biochem.* **47**, 985–987.

104. M. S. Silver and T. Sone (1968), Stereospecificity in the hydrolysis of conformationally homogenous substrates by α-chymotrypsin. *J. Amer. Chem. Soc.* **90**, 6193–6198.

105. S. G. Cohen and R. M. Schultz (1968), The active site of α-chymotrypsin. *J. Mol. Biol.* **243**, 2607–2617.

106. B. Belleau and R. Chevalier (1968), The absolute conformation of chymotrypsin-bound substrates. Specific recognition by the enzyme of biphenyl asymmetry in a constrained substrate. *J. Amer. Chem. Soc.* **90**, 6864–6866.

107. P. Elie (1977), Ph.D. Dissertation, McGill University, Montreal, Canada.

108. D. C. Phillips (1967), The hen egg-white lysozyme molecule. *Proc. Nat. Acad. Sci. USA* **57**, 484–495.

109. R. E. Dickerson and I. Geis (1969), *The Structure and Action of Proteins* p. 71. Harper and Row, New York.

110. M. R. Pincus and H. A. Scheraga (1979), Conformational energy calculation of enzyme-substrate and enzyme-inhibitor complexes of lysozyme. 2. Calculation of the structures of complexes with a flexible enzyme. *Macromolecules* **12**, 633–644.

111. B. Capon and M. C. Smith (1965), Intramolecular catalysis in acetal hydrolysis. *Chem. Comm.* 523–524.

112. R. Kluger and C. H. Lam (1978), Carboxylic acid participation in amide hydrolysis. External general base catalysis and general acid catalysis in reactions of norbornenylanilic acids. *J. Amer. Chem. Soc.* **100**, 2191–2197.

113. I. Hsu, T. J. Delbaere, M. M. G. James, and T. Hofmann (1977), Penicillopepsin from *Penicillium janthinellum*. Crystal structure at 2.8 Å and sequence. Homology with porcine pepsin. *Nature* **266**; 140–145.

114. P. Deslongchamps (1975), Stereoelectronic control in the cleavage of tetrahedral intermediates in the hydrolysis of esters and amides. *Tetrahedron* **31**, 2463–2490.

115. P. Deslongchamps (1977), Stereoelectronic control in hydrolytic reactions. *26th IUPAC*, Tokyo, Japan.

116. P. Deslongchamps (1977); Stereoelectronic Control in Hydrolytic Reactions. *Heterocycles* **7**, 1271–1317.

117. P. Deslongchamps, U. O. Cheriyan, J. -P. Pradère, P. Soucy, and R. J. Taillefer (1979), Hydrolysis and isomerization of syn unsymmetrical *N,N*-dialkylated immidate salts. Experimental evidence for conformational changes and for stereoelectronically controlled cleaves in hemi-orthoamide tetrahedral intermediates. *Nouv. J. Chim.* **3**, 343–350.

118. S. A. Bizzozero and B. O. Zweifel (1975), The importance of the conformation of the tetrahedral intermediate for the α-chymotrypsin-catalyzed hydrolysis of peptide substrates. *FEBS Lett.* **59**, 105–108.

119. D. Petkov, E. Christova, and I. Stoineva (1978), Catalysis and leaving group binding in anilide hydrolysis by chymotrypsin. *Biochim. Biophys. Acta* **527**, 131–141.

120. D. G. Gorenstein, J. B. Findlay, B. A. Luxon, and D. Kar (1977), Stereoelectronic control in carbon-oxygen and phosphorus-oxygen bond breaking processes. *Ab initio* calculation and speculations on the mechanism of ribonuclease A, staphylococcal nuclease and lysozyme. *J. Amer. Chem. Soc.* **99**, 3473–3479.

121. W. L. Mock (1976), Torsional-strain considerations in enzymology. Some applications to proteases and ensuing mechanistic consequences. *Bioorg. Chem.* **5**, 403–414.

122. B. J. F. Hudson (1975), Immobilized enzymes. *Chem. Ind.* pp. 1059–1060.

123. K. J. Skimer (1975), Enzymes technology. *Chem. Eng. News.* August 18, pp. 23–42.

124. K. Mosbach (1976), Applications of biochemical systems in organic chemistry, In: *Techniques of Chemistry Series* (J. B. Jones, C. J. Sih, and D. Perlman, Eds.), Vol. 10, Part II, Chap. 9. Wiley-Interscience, New York.

125. C. J. Sucking (1977), Immobilized enzymes. *Chem. Soc. Rev.* **6**, 215–233.

126. A. Pollak, R. L. Baughn, O. Adalsteinsson, and G. M. Whitesides (1978), Immobilization of synthetically useful enzymes by condensation polymerization. *J. Amer. Chem. Soc.* **100**, 302–304.

127. G. G. Guilbault and M. H. Sadar (1979), Preparation and analytical uses of immobilized enzymes. *Acc. Chem. Res.* **12**, 344–359.

128. K. Mosbach and P. O. Larsson (1970), Preparation and application of polymer-entrapped enzymes and microorganisms in microbial transformation processes with special reference to steroid 11-β-hydroxylation and Δ^1-dehydrogenation. *Biotechnol. Bioeng.* **13**, 19–27.

129. P. A. Suf, S. Kay, and M. D. Lilly (1969), The conversion of benzyl penicillin to 6-aminopenicillanic acid using an insoluble derivative of penicillin amidase. *Biotechnol. Bioeng.* **12**, 337–348.

130. T. H. Fife (1977), Intramolecular nucleophilic attack on esters and amides. In: *Bioorganic Chemistry* (E.E. van Tamelen, Ed.), Vol. 1, Chap. 5, pp. 93–116. Academic Press, New York.

131. M. L. Bender and M. Komiyama (1977), In: *"Bioorganic Chemistry"* (E. E. van Tamelen, Ed.), Vol. I, Chap. 2, pp. 19–57. Academic Press, New York.

132. C. J. Pedersen (1967), Cyclic polyethers and their complexes with metal salts. *J. Amer. Chem. Soc.* **89**, 2495–2496.

133. C. J. Pedersen (1967), Cyclic polyethers and their complexes with metal salts, *J. Amer. Chem. Soc.* **89**, 7017–7036.

134. D. J. Cram and J. M. Cram (1974), Host-guest chemistry. *Science* **183**, 803–809.

135. D. J. Cram and J. M. Cram (1978), Design of complexes between synthetic hosts and organic guests. *Acc. Chem. Res.* **11**, 8–14.

136. D. J. Cram (1976), Applications of Biochemical Systems in Organic Chemistry, In: *Techniques of Chemistry Series* (J. B. Jones, C. S. Sih, and D. Perlman, Eds.), Vol. 10, Part II, Chap. 5, pp. 815–874. Wiley-Interscience, New York.

137. D. J. Cram *et al.* (1977), Host-guest complexation. 1. Concept and illustration. *J. Amer. Chem. Soc.* **99**, 2564–2571.

138. D. J. Cram *et al.* (1977), Host-guest complexation. 3. Organization of pyridyl binding sites. *J. Amer. Chem. Soc.* **99**, 6392–6398.

139. A. Warshawsky, R. Kalir, A. Deshe, H. Berkovitz, and A. Patchornik (1979), Polymeric pseudocrown ethers. 1. Synthesis and complexation with transition metal anions. *J. Amer. Chem. Soc.* **101**, 4249–4258.

140. W. D. Curtis, D. A. Laidler, and J. F. Stoddart (1977), To enzyme analogues by lock and key chemistry with crown compounds. I. Enantiomeric differentation by configurationally chiral cryptands synthesized from L-tartaric acid and D-mannitol. *J. Chem. Soc. Perkin Trans. 1*, 1756–1769.

141. J. F. Stoddart (1979), From carbohydrates to enzyme analogues. *Chem. Soc. Rev.* **18**, 85–142.

142. S. Hanessian (1979), Approaches to the total synthesis of natural products using "chiral templates" derived from carbohydrates. *Acc. Chem. Res.* **12**, 159–165.

143. Y. Chao and D. J. Cram (1976), Catalysis and chiral recognition through designed complexation of transition states in transacylations of amino ester salts, *J. Amer. Chem. Soc.* **98**, 1015–1017.

144. R. G. Pearson (1968), Hard and soft acids and bases, HSAB, Part I. *J. Chem. Educ.* **48**, 581–587.

145. R. G. Pearson (1968), Hard and soft acids and bases, HSAB, Part II. *J. Chem. Educ.* **48**, 643–648.

146. J. Sunamoto, H. Kondo, H. Okamoto, and Y. Murakami (1977), Catalysis of the deacylation of *p*-nitrophenyl hexadecanoate by 11-amino [20] paracyclophane-10-ol in neutral and alkaline media. *Tetrahedron Lett.* 1329–1332.

147. Y. Murakami, Y. Aoyama, M. Kada, and J. I. Kikuchi (1978), Macrocyclic enzyme model system: Catalysis of ester degradation by a [20] paracyclophane bearing nucleophilic and metal-binding sites. *Chem. Commun.* 494–496.

148. J. M. Lehn and C. Sirlin (1978), Molecular catalysis: Enhanced rate of thiolysis with high structural and chiral recognition in complexes of a reactive receptor molecule. *Chem. Commun.* 949–951.

149. J. M. Lehn (1978), Cryptates: The chemistry of macropolycyclic inclusion complexes. *Acc. Chem. Res.* **11**, 49–57.

150. K. Kirch and J. M. Lehn (1975), Selective transport of alkali metal cations through a liquid membrane by macrobicyclic carriers. *Angew. Chem., Int. Ed. Engl.* **14**, 555–556.

151. F. Vogtle and E. Weber (1979), Multidentate acyclic neutral ligands and their complexation. *Angew. Chem., Int. Ed. Engl.* **18**, 753–776.

152. C. M. Deber and E. R. Blout (1974), Amino acid-cyclic peptide complexes. *J. Amer. Chem. Soc.* **96**, 7566–7568.

153. F. M. Menger (1979), On the structure of micelles. *Acc. Chem. Res.* **12**, 111–117.

154. C. A. Bunton (1976), Applications of Biochemical Systems in Organic Chemistry, In: *Techniques of Chemistry Series.* (J. B. Jones, C. J. Sih, and D. Perlman, Eds.), Vol. 10, Part II, Chap. 4, pp. 731–814. Wiley-Interscience, New York.

155. E. H. Cordes and R. B. Dunlap (1969), Kinetics of organic reactions in micellar systems. *Acc. Chem. Res.* **2**, 329–337.

156. L. R. Fisher and D. G. Oakenfull (1977), Micelles in aqueous solution. *Chem. Soc. Rev.* **6**, 25–42.

157. F. M. Menger (1977), In: *Bioorganic Chemistry*, (E. E. von Tamelen, Ed.) Vol. III, Chap. 7, pp. 137–152. Academic Press, New York.

158. J. H. Fendler and E. J. Fendler (1975), Catalysis in micelles and macromolecular systems. Academic Press, New York.

159. J. Baumucker, M. Calzadilla, M. Centeno, G. Lehrmann, M. Urdanela, P. Lindquist, D. Dunham, M. Price, B. Sears, and E. H. Cordes (1972), Secondary valence force catalysis. XII. Enhanced reactivity and affinity of cyanide ion toward N-substituted 3-carbamoylpyridinium ions elicted by ionic surfactants and biological lipids. *J. Amer. Chem. Soc.* **94**, 8164–8172.

160. F. M. Menger, J. A. Donohue, and R. F. Williams (1973), Catalysis in water pools. *J. Amer. Chem. Soc.* **95**, 286–288.

161. J. R. Escabi-Pérez and J. H. Fendler (1978), Ultrafast excited state proton transfer in reversed micelles. *J. Amer. Chem. Soc.* **100**, 2234–2236.

162. R. G. Shorenstein, L. S. Pratt, C. J. Hsu, and T. E. Wagner (1968), A model system for the study of equilibrium hydrophobic bond formation. *J. Amer. Chem. Soc.* **90**, 6199–6207.

163. R. A. Moss, R. G. Mahas, and T. L. Lukas (1978), A cysteine-functionalized micellar catalyst. *Tetrahedron Lett.* 507–510.

164. W. Tagaki and H. Hara (1973), Acyloin condensation of aldehydes catalyzed by *N*-laurylthiazolium bromide. *Chem. Comm.* 891–892.

165. C. A. Bunton, L. Robinson, and M. F. Stam (1971), Stereospecific micellar catalyzed ester hydrolysis. *Tetrahedron Lett.* 121–124.

166. J. M. Brown and C. A. Bunton (1974), Stereoselective micelle-promoted ester hydrolysis. *Chem. Comm.* 969–971.

167. R. A. Moss, Y. S. Lee, and T. J. Lukas (1979), Micellar stereoselectivity cleavage of diastereomeric substrates by functional surfactant micelles. *J. Amer. Chem. Soc.* **101**, 2499–2501.

168. Y. Imanishi (1979) Intramolecular reactions on polymer chains. *J. Polym. Sci. Macromo. Rev.* **14**, 1–205.

169. G. Manecke and W. Storck (1978), Polymeric catalysis. *Angew Chem., Int. Ed. Engl.* **17**, 657–670.

170. C. G. Overberger and J. C. Salamone (1969), Esterolytic action of synthetic macromolecules. *Acc. Chem. Res.* **2**, 217–224.

171. H. C. Kiefer, W. I. Congdon, I. S. Scarpa, and I. M. Klotz (1972), Catalytic accelerations of 10^{12}-fold by an enzyme-like synthetic polymer. *Proc. Nat. Acad. Sci. USA* **69**, 2155–2159.

172. H. Frank, G. C. Nicholson, and E. Bayer (1978), Chiral polysiloxanes for resolution of optical antipodes. *Angew. Chem., Int. Ed. Engl.* **17**, 363–365.

173. M. L. Bender and M. Komiyama (1978), *Cyclodextrin Chemistry*, Springer-Verlag, Berlin and New York.

174. R. Breslow and P. Campbell (1969), Selective aromatic substitution within a cyclodextrin mixed complex. *J. Amer. Chem. Soc.* **91**, 3085.

175. M. Komiyama, E. J. Breaux, and M. L. Bender (1977), The use of cycloamylose to probe the "charge-relay" system. *Bioorg. Chem.* **6**, 127–136.

176. J. Emert and R. Breslow (1975), Modification of the cavity of β-cyclodextrin by flexible capping. *J. Amer. Chem. Soc.* **97**, 670–672.

177. K. Fujita, A. Shinoda, and T. Imoto (1980), Hydrolysis of phenyl acetates with capped β-cyclodextrins: Reversion from *meta* to *para* selectivity. *J. Amer. Chem. Soc.* **102**, 1161–1163.

178. M. Komiyama and M. L. Bender (1978), Importance of apolar binding in complex formation of cyclodextrins with adamantanecarboxylate. *J. Amer. Chem. Soc.* **100**, 2259–2260.

179. R. Breslow, J. B. Doherty, G. Guillot, and C. Lipsey (1978), β-Cyclodextrinylbisimidazole, a model for ribonuclease. *J. Amer. Chem. Soc.* **100**, 3227–3229.

180. I. Tabushi, K. Shimokawa, N. Shimizu, H. Shirakata, and K. Fujita (1976), Capped cyclodextrin. *J. Amer. Chem. Soc.* **98**, 7855–5856.

181. I. Tabushi, K. Shimokawa, and K. Fujita (1977), Specific bifunctionalization on cyclodextrin. *Tetrahedron Lett.* 1527–1530.

182. I. Tabushi, Y. Kuroda, and A. Mochizuki (1980), The first successful carbonic anhydrase model prepared through a new route to regiospecifically bifunction-alized cyclodextrin. *J. Amer. Chem. Soc.* **102**, 1152–1153.

183. R. Breslow and L. E. Overman (1970), An "artificial enzyme" combining a metal catalytic group and a hydrophobic binding cavity. *J. Amer. Chem. Soc.* **92**, 1075–1077.

184. J. Boger, D. G. Brenner, and J. R. Knowles (1979), Symmetrical triamino-per-*O*-methyl-α-cyclodextrin: Preparation and characterization of primary trisubstituted α-cyclodextrins. *J. Amer. Chem. Soc.* **101**, 7630–7631.

185. J. Boger and J. R. Knowles (1979), Symmetrical triamino-per-*O*-methyl-α-cyclodextrin: A host for phosphate esters exploiting both hydrophobic and electrostatic interactions in aqueous solution. *J. Amer. Chem. Soc.* **101**, 7631–7633.

186. R. Breslow, M. Hammond, and M. Lauer (1980), Selective transamination and optical induction by a β-cyclodextrin-pyridoxamine artificial enzyme. *J. Amer. Chem. Soc.* **102**, 421–422.

187. N. Saenger, M. Noltemeyer, P. C. Manor, B. Hingerty, and E. B. Klar (1972), "Induced-fit"-type complex formation of the model enzyme α-cyclodextrin. *Bioorg. Chem.* **5**, 187–195.

188. I. Tabushi, Y. Kuroda, K. Fujita, and H. Kawakubo (1978), Cyclodextrin as a ligase-oxidase model. Specific allylation-oxidation of hydroquinone derivatives included by β-cyclodextrin. *Tetrahedron Lett.* 2083–2086.

189. I. Tabushi, Y. Kuroda, and K. Shimokawa (1979), Cyclodextrin having an amino group as a rhodopsin model. *J. Amer. Chem. Soc.* **101**, 4759–4760.

190. I. Tabushi, K. Yamamura, K. Fujita, and H. Kawakubo (1979), Specific inclusion catalysis by β-cyclodextrin in the one-step preparation of vitamin K_1 or K_2 analogues. *J. Amer. Chem. Soc.* **101**, 1019–1026.

191. J. P. Guthrie (1976), Application of Biochemical Systems in Organic Chemistry, In: *Techniques of Chemistry Series* (J. B. Jones, C. J. Sih, and D. Perlman, Eds.), Vol. 10, Part II, Chap. 3, pp. 627–730. Wiley-Interscience, New York.

192. J. P. Guthrie and S. O'Leary (1975), General base catalysis of β-elimination by a steroidal enzyme model. *Can. J. Chem.* **53**, 2150–2156.

193. J. P. Guthrie and Y. Ueda (1974), Electrostatic catalysis and inhibition in aqueous solution. Rate effects on the reactions of charged esters with a cationic steroid bearing an imidazolyl substituent. *Chem. Comm.* 111–112.

194. R. Breslow and M. A. Winnick (1969), Remote oxidation of unactivated methylene groups. *J. Amer. Chem. Soc.* **91**, 3083–3084.

195. R. Breslow, J. Rothbard, F. Herman, and M. L. Rodriguez (1978), Remote functionalization reactions as conformational probes for flexible alkyl chains. *J. Amer. Chem. Soc.* **100**, 1213–1218.

196. M. F. Czarniecki and R. Breslow (1979), Photochemical probes for model membrane structures. *J. Amer. Chem. Soc.* **101**, 3675–3676.

197. R. Breslow (1972), Biomimetic chemistry. *Chem. Soc. Rev.* **1**, 553–580.

198. R. Breslow, R. J. Corcoran, and B. B. Snider (1974), Remote functionalization of steroids by a radical relay mechanism. *J. Amer. Chem. Soc.* **96**, 6791–6792.

199. R. Breslow, B. B. Snider, and R. J. Corcoran (1974), A cortisone synthesis using remote oxidation. *J. Amer. Chem. Soc.* **96**, 6792–6794.

200. M. A. Schwartz and I. S. Mami (1975), A biologically patterned synthesis of the morphine alkaloids. *J. Amer. Chem. Soc.* **97**, 1239–1240.
201. B. Samuelsson and R. Paoletti (1976), *Advances in Prostaglandin and Thromboxane Research*, Vols. 1 and 2. Raven Press, New York.
202. K. C. Nicolaou, G. P. Gasic, and W. E. Barrette (1978), Synthesis and biological properties of prostaglandin endoperoxides, thromboxanes and prostacyclins. *Angew. Chem., Int. Ed. Engl* **17**, 293–312.
203. J. M. Bailey (1979), Prostaglandins, thromboxanes and cardiovascular disease. *Trends Biochem. Sci.* **4**, 68–71.
204. M. O. Funk, R. Isaac, and N. A. Porter (1975), Free radical cyclization of unsaturated hydroperoxides. *J. Amer. Chem. Soc.* **97**, 1281–1282.
205. N. Adam, A. Birke, C. Cádiz, Sv. Diaz, and A. Rodriguez (1978), Prostonoid endoperoxide model compounds: Preparation of 1,2-dioxolanes from cyclopropanes. *J. Org. Chem.* **43**, 1154–1158.
206. D. J. Coughlin, R. S. Brown, and R. G. Salomon (1979), The prostaglandin endoperoxide nucleus and related bicyclic peroxides. Synthetic and spectroscopic studies. *J. Amer. Chem. Soc.* **101**, 1533–1539.
207. E. E. van Tamelen (1975), Bioorganic chemistry; Total synthesis of tetra- and pentacyclic triterpenoids. *Acc. Chem. Res.* **8**, 152–158.
208. W. S. Johnson (1976), Biomimetic polyene cyclizations. *Angew. Chem., Int. Ed. Engl.* **15**, 9–17.
209. W. S. Johnson (1976), Biomimetic Polyene Cyclizations. *Bioorg. Chem.* **5**, 51–98.
210. E. E. van Tamelen, J. D. Willett, R. B. Clayton, and R. E. Lord (1966), Enzymic conversion of squalene 2,3-oxide to lanosterol and cholesterol. *J. Amer. Chem. Soc.* **88**, 4752–4754.
211. E. E. van Tamelen and J. H. Freed (1970), Biochemical conversion of partially cyclized squalene 2,3-oxide types to the lanosterol system. Views on the normal enzymic cyclization process. *J. Amer. Chem. Soc.* **92**, 7206–7207.
212. E. J. Corey, W. E. Russey, and P. R. Ortiz de Montellano (1966), 2,3-Oxido-squalene, an intermediate in the biological synthesis of sterols from squalene. *J. Amer. Chem. Soc.* **88**, 4750–4751.
213. R. B. Woodward and K. Bloch (1953), The cyclization of squalene in cholesterol synthesis. *J. Amer. Chem. Soc.* **75**, 2023–2024.
214. B. L. Vallee and R. J. P. Williams (1968), Enzyme action: Views derived from metalloenzyme studies. *Chem. Br.* **4**, 397–402.
215. M. M. Jones and T. H. Pratt (1976), Therapeutic chelating agents. *J. Chem. Educ.* **53**, 342–347.
216. E. W. Ainscough and A. M. Brodie (1976), The role of metal ions in proteins and other biological molecules. *J. Chem. Educ.* **53**, 156–158.
217. Y. Pocker and D. W. Bjorkquist (1977), Comparitive studies of bovine carbonic anhydrase in H_2O and D_2O. Stopped-flow studies of the kinetics of interconversion of CO_2 and HCO_3^-. *Biochemistry* **16**, 5698–5707.
218. C. C. Tang, D. Davalian, P. Huand, and R. Breslow (1978), Models of metal binding sites in zinc enzymes. Synthesis of tris [4(5)-imidazolyl] carbinol (4-TIC), tris (2-imidazolyl)-carbinol(2-TIC), and related ligands and studies on metal complex binding constants and spectra. *J. Amer. Chem. Soc.* **100**, 3918–3922.
219. F. A. Quiocho and W. N. Lipscomb (1971), Carboxypeptidase A: A protein and an enzyme. *Adv. Prot. Chem.* **25**, 1–78.

220. R. Breslow, D. E. McClure, R. S. Brown, and J. Elsenach (1975), Very fast zinc catalyzed hydrolysis of an anhydride. A model for the rate and mechanism of carboxypeptidase A catalysis. *J. Amer. Chem. Soc.* **97**, 194–195.

221. R. Breslow and D. E. McClure (1976), Cooperative catalysis of the cleavage of an amide by carboxylate and phenolic groups in a carboxypeptidase A model. *J. Amer. Chem. Soc.* **98**, 258–259.

222. R. Breslow and D. Wernick (1976), On the mechanism of catalysis by carboxypeptidase A. *J. Amer. Chem. Soc.* **98**, 259–261.

223. T. H. Fife and V. L. Squillacote (1978), Metal ion effects on intramolecular nucleophilic carboxyl group participation in amide and ester hydrolysis. Hydrolysis of *N*-(8-quinolyl)phthalamic acid and 8-quinolyl hydrogen glutarate. *J. Amer. Chem. Soc.* **100**, 4787–4793.

224. M. W. Makinen, L. C. Kuo, J. J. Dymouski, and S. Jaffer (1979), Catalytic role of the metal ion of carboxypeptidase A in ester hydrolysis. *J. Biol. Chem.* **254**, 356–366.

225. R. Breslow and C. McAllister (1971), Intramolecular bifunctional catalysis of ester hydrolysis by metal ion and carboxylate in a carboxypeptidase model. *J. Amer. Chem. Soc.* **93**, 7096–7097.

226. G. J. Lloyd and B. S. Cooperman (1971), Nucleophilic attack by zinc(II)-pyridine-2-carbaldoxime anion on phosphorylimidazole. A model for enzymatic phosphate transfer. *J. Amer. Chem. Soc.* **93**, 4883–4889.

227. D. S. Sigman and C. T. Jorgenren (1972), Models for metalloenzymes. The zinc (II)-catalyzed transesterification of *N*-(β-hydroxyethyl)ethylene-diamine by *p*-nitrophenyl picolinate. *J. Amer. Chem. Soc.* **94**, 1724–1730.

228. D. A. Buckingham and J. P. Collman (1967), Hydrolysis of N-terminal peptide bonds and amino acid derivatives by the β-hydroxoaquotriethylenetetramine cobalt(III)ion. *J. Amer. Chem. Soc.* **89**, 1082–1087.

229. D. A. Buckingham, F. R. Keen, and A. M. Sargeson (1974); Facile intramolecular hydrolysis of dipeptides and glycinamide. *J. Amer. Chem. Soc.* **96**, 4981–4983.

230. E. Kimura (1974), Sequential hydrolysis of peptides with β-hydroxoaquotriethylenetetraminecobalt(III) Ion. *Inorg. Chem.* **13**, 951–954.

231. F. Frieden (1975), Non-covalent interactions. *J. Chem. Educ.* **52**, 754–761.

232. W. A. Hendrickson (1977), The molecular architecture of oxygen. Carrying proteins. *Trends Biochem. Sci.* **2**, 108–111.

233. J. C. Kendrew (1961), The three-dimensional structure of a protein molecule. *Sci. Amer.* **205**, December, 96–110.

234. E. F. Epstein, I. Bernal, and A. L. Balch (1970), Activation of molecular oxygen by a metal complex. The formation and structure of the anion $[Ph_3POFe-(S_2C_2\{CF_3\}_2)_2]^{\ominus}$. *Chem. Comm.* 136–138.

235. J. Almog, J. E. Baldwin, R. I. Dyer, and M. Peters (1975), Condensation of tetraaldehydes with pyrrole. Direct synthesis of "capped" porphyrins. *J. Amer. Chem. Soc.* **97**, 226–227.

236. J. Almog, J. E. Baldwin, and J. Huff (1975), Reversible oxygenation and autooxidation of a "capped" porphyrin iron(II) complex. *J. Amer. Chem. Soc.* **97**, 227–228.

237. T. G. Traylor (1978), *Bioorganic Chemistry* (E. E. van Tamelen, Ed.), Vol. IV, pp. 437–468. Academic Press, New York.

238. C. K. Chang and T. G. Traylor (1973), Synthesis of the myoglobin active site. *Proc. Nat. Acad. Sci. USA* **78**, 2647–2650.

239. (a) C. K. Chang and T. G. Traylor (1973), Solution behavior of a synthetic myoglobin active site. *J. Amer. Chem. Soc.* **95**, 5810–5811. (b) C. K. Chang and T. G. Traylor (1973), Neighboring group effect in heme-carbon monoxide binding. *J. Amer. Chem. Soc.* **95**, 8475–8477. (c) C. K. Chang and T. G. Traylor (1973), Proximal base influence on the binding of oxygen and carbon monoxide to heme. *J. Amer. Chem. Soc.* **95**, 8477–8479.

240. T. G. Traylor, D. Campbell, S. Tsuchiya (1979), Cyclophane porphyrin. 2. Models for steric hindrance to CO ligation in hemoproteins. *J. Amer. Chem. Soc.* **101**, 4748–4749.

241. E. Bayer and G. Holzbach (1977), Synthetic homopolymers for reversible binding of molecular oxygen. *Angew. Chem., Int. Ed. Engl.* **16**, 117–118.

242. J. P. Collman (1977), Synthetic models for the oxygen-binding hemoproteins. *Acc. Chem. Res.* **10**, 265–272.

243. J. P. Collman, J. L. Brauman, E. Rose, and K. S. Suslick (1978), Cooperativity in O_2 binding to iron porphyrins. *Proc. Nat. Acad. Sci. USA* **75**, 1052–1055.

244. N. Farrell, D. H. Dolphin, and B. R. James (1978), Reversible binding of dioxygen to ruthenium(II) porphyrins. *J. Amer. Chem. Soc.* **100**, 324–326.

245. F. S. Molinaro, R. G. Little, and J. A. Ibers (1977), Oxygen binding to a model for the active site in cobalt-substituted hemoglobin. *J. Amer. Chem. Soc.* **99**, 5628–5632.

246. J. P. Collman, J. I. Brauman, K. M. Doxsee, T. R. Halbert, S. E. Hayes, and K. S. Suslick (1978), Oxygen binding to cobalt porphyrins. *J. Amer. Chem. Soc.* **100**, 2761–2766.

247. M. R. Wasielewski, W. A. Svec, and B. T. Cope (1978), Bis(chlorophyll) cyclophanes, new models of special pair chlorophyll. *J. Amer. Chem. Soc.* **100**, 1961–1962.

248. R. Malkin (1973), *Iron–Sulfur Proteins*, Vol. II, Academic Press, New York.

249. R. H. Holm (1977), Synthetic approaches to the active sites of iron–sulfur proteins. *Acc. Chem. Res.* **10**, 427–434.

250. G. L. Eichhorn (1973), *Inorganic Biochemistry*, Vols. 1 and 2, Elsevier, New York.

251. A. R. Amundsen, J. Whelan, and B. Bosnich (1977), Biological analogues. On the nature of the binding sites of copper-containing proteins. *J. Amer. Chem. Soc.* **99**, 6730–6739.

252. R. R. Gagné, J. L. Allison, R. S. Gall, and C. A. Koval (1977), Models for copper-containing proteins: Structure and properties of novel five-coordinate copper(I) complexes. *J. Amer. Chem. Soc.* **99**, 7170–7178.

253. D. A. Buckingham, M. J. Gunter, and L. N. Mander (1978), Synthetic models for bis-metallo active sites, A prophyrin capped by a tetrakis (pyridine) ligand system. *J. Amer. Chem. Soc.* **100**, 2899–2901.

254. Y. Agnus, R. Louis, and R. Weiss (1979), Bimetallic copper(I) and (II) macrocyclic complexes as mimics for type 3 copper pairs in copper enzymes. *J. Amer. Chem. Soc.* **101**, 3381–3384.

255. P. K. Coughlin, J. C. Dewan, S. J. Lippard, E. Watanabe, and J.-M. Lehn (1979), Synthesis and structure of the imidazolate bridged dicopper(II) ion incorporated into a circular cryptate macrocycle. *J. Amer. Chem. Soc.* **101**, 265–266.

256. R. H. Abeles and D. Dolphin (1974), The vitamin B_{12} coenzyme. *Acc. Chem. Res.* **9**, 114–120.

257. J. Rétey, A. Ulmani-Ronchi, J. Seibl, and D. Arigoni (1966), Zum mechanismus der Propandioldehydrase-Reaktion. *Experentia* **22**, 502–503.

258. G. N. Schrauzer (1968), Organocobalt chemistry of vitamin B_{12} model compounds (cobaloximes). *Acc. Chem. Res.* **1**, 97–103.

259. G. N. Schrauzer (1976); New developments in the field of vitamin B_{12}: Reactions of the cobalt atom in corrins and in vitamin B_{12} model compounds. *Angew Chem., Int. Ed. Engl* **15**, 417–426.

260. G. N. Schrauzer (1977), New developments in the field of vitamin B_{12}: Enzymatic reactions dependent upon corrins and coenzyme B_{12}. *Angew. Chem., Int. Ed. Engl.* **16**, 233–244.

261. R. B. Silverman and D. Dolphin (1973), A direct method for cobalt-carbon bond formation in cobalt(III)-containing cobalamins and cobaloximes. Further support for cobalt(III) π-complexes in coenzyme B_{12} dependent rearrangements. *J. Amer. Chem. Soc.* **95**, 1686–1688.

262. R. B. Silverman and D. Dolphin (1974), Reaction of vinyl ethers with cobalamins and cobaloximes. *J. Amer. Chem. Soc.* **96**, 7094–7096.

263. R. B. Silverman, D. Dolphin, T. J. Carty, E. K. Krodel, and R. H. Abeles (1974), Formylmethylcobalamin. *J. Amer. Chem. Soc.* **96**, 7096–7097.

264. P. Dowd, B. K. Trivedi, M. Shapiro, and L. K. Marwaha (1976), Vitamin B_{12} model studies. Migration of the acrylate fragment in the carbon-skeleton rearrangement leading to α-methyleneglutaric acid. *J. Amer. Chem. Soc.* **98**, 7875–7877.

265. L. Salem, O. Eisentein, N. T. Anh, H. B. Burgi, A. Devaquet, G. Segal, and A. Veillard (1977), Enzymatic catalysis. A theoretically derived transition state for coenzyme B_{12}-catalyzed reaction. *Nouv. J. Chim.* **1**, 335–347.

266. H. Flohr, N. Paunhorst, and J. Rétey (1976), Synthesis, structure determination, and rearrangement of a model for the active site of methylmalonyl-CoA mutase with incorporated substrate. *Angew. Chem. Int. Ed. Engl.* **15**, 561–562.

267. R. Breslow and P. L. Khanna (1976), An intramolecular model for the enzymatic insertion of coenzyme B_{12} into unactivated carbon-hydrogen bonds. *J. Amer. Chem. Soc.* **98**, 1297–1299.

268. E. J. Corey, N. J. Cooper, and M. L. H. Green (1977), Biochemical catalysis involving coenzyme B_{12}: A rational stepwise mechanistic interpretation of vicinal interchange rearrangements. *Proc. Nat. Acad. Sci. USA* **74**, 811–815.

269. T. Toraya, E. Krodel, A. S. Mildran, and R. H. Abeles (1979), Role of peripheral side chains of vitamin B_{12} coenzymes in the reaction catalyzed by dioldehydrase. *Biochemistry* **18**, 417–426.

270. K. Sato, E. Hiei, S. Shimizu, and R. Abeles (1978), Affinity chromatography of N^5-methyltetrahydrofolate-homocysteine methyltransferase on a cobalamin-Sepharose. *FEBS Lett.* **85**, 73–76.

271. J. M. Wood (1974), Biological cycles for toxic elements in the environment. *Science* **183**, 1049–1052.

272. J. S. Thayer (1977), Teaching bio-organometal chemistry. II. The metals. *J. Chem. Educ.* **54**, 662–665.

273. A. E. Metzler (1977), *Biochemistry, The Chemical Reactions of Living Cells*, Chap. 8. Academic Press, New York.

274. G. DiSabato (1970), Adducts of diphosphopyridine nucleotide and carbonyl compounds. *Biochemistry* **9**, 4594–4600.

275. D. J. Creighton and D. S. Sigman (1971), A model for alcohol dehydrogenase. The zinc ion catalyzed reduction of 1,10-phenanthroline-2-carboxaldehyde by N-propyl-1,4-dihydronicotinamide. *J. Amer. Chem. Soc.* **93**, 6314–6316.

276. U. Grau, H. Kapmeyer, and W. E. Trommer (1978), Combined coenzyme-substrate analogues of various dehydrogenases. Synthesis of (3S)- and (3R)-5-(3-carboxy-3-hydroxypropyl) nicotinamide adenine dinucleotide and their interaction with (S)- and (R)-lactate-specific dehydrogenases. *Biochemistry* **17**, 4621–4626.

277. J. T. van Bergen and R. M. Kellogg (1977), A crown ether NAD(P)H mimic. Complexation with cations and enhanced hydride donating ability toward sulfonium salts. *J. Amer. Chem. Soc.* **99**, 3882–3884.

278. J. P. Behr and J. M. Lehn (1978), Enhanced rates of dihydropyridine to pyridinium hydrogen transfer in complexes of an active macrocyclic receptor molecule. *Chem. Comm.* 143–146.

279. G. A. Hamilton (1971), *Progress in Bioorganic Chemistry* (E. T. Kaiser and T. J. Kézdy, Eds.), Vol. 1, pp. 83–137. Wiley-Interscience. New York.

280. H. Eklund, B. Nordström, E. Zeppezauer, G. Söderlund, I. Ohlsson, T. Boiwe, B. O. Söderberg, O. Tapia, and C. I. Brändén (1976), Three-dimensional structure of horse liver alcohol dehydrogenase at 2.4 Å resolution. *J. Mol. Biol.* **102**, 27–59.

281. C. J. Suckling and H. C. S. Wood (1979), Should organic chemistry meddle in biochemistry? *Chem Br.* **5**, 243–248.

282. A. J. Irwin and J. B. Jones (1976), Stereoselective horse liver alcohol dehydrogenase catalyzed oxidoreductions of some bicyclic[2.2.1] and [3.2.1] ketones and alcohols. *J. Amer. Chem. Soc.* **98**, 8476–8482.

283. A. J. Irwin and J. B. Jones (1977), Regiospecific and enantioselective horse liver alcohol dehydrogenase catalyzed oxidations of some hydroxycyclopentanes. *J. Amer. Chem. Soc.* **99**, 1625–1630.

284. J. B. Jones and J. F. Beck (1976), Application of Biochemical Systems in Organic Chemistry, In: *Techniques of Chemistry Series* (J. B. Jones, C. J. Sih, and D. Perlman, Eds.), Part I, pp. 107–401. Wiley, New York.

285. V. Prelog (1964), Specification of the stereochemistry of some oxido-reductases by diamond lattice sections. *Pure Appl. Chem.* **9**, 119–130.

286. D. J. Creighton, J. Hajdu, G. Mooser, and D. S. Sigman (1973), Model dehydrogenase reactions. Reduction of N-methylacridinium ion by reduced nicotinamide adenine dinucleotide and its derivatives. *J. Amer. Chem. Soc.* **95**, 6855–6867.

287. M. Brüstlein and T. C. Bruice (1972), Demonstration of a direct hydrogen transfer between NADH and a deazaflavin. *J. Amer. Chem. Soc.* **94**, 6548–6549.

288. J. Fisher and C. Walsh (1974), Enzymatic reduction of 5-deazariboflavine from reduced nicotinamide adenine dinucleotide by direct hydrogen transfer. *J. Amer. Chem. Soc.* **96**, 4345–4346.

289. D. Eirich, G. Vogels, and R. Wolfe (1978), Proposed structure of coenzyme F_{420} from Methanobacterium. *Biochemistry* **17**, 4583–4593.

290. P. Hemmerich, V. Massey, and G. Weber (1967), Photo-induced benzyl substitution of flavins by phenylacetate: A possible model for flavoprotein catalysis. *Nature* **213**, 728–730.

291. J. M. Sayer, P. Conlon, J. Hupp, J. Fancher, R. Bélanger, and E. J. White (1979), Reduction of 1,3-dimethyl-5-(p-nitrophenylimino) barbituric acid by thiols. A high-velocity flavin model reaction with an isolable intermediate. *J. Amer. Chem. Soc.* **101**, 1890–1893.

292. W. H. Rastetter, T. R. Gadek, J. P. Tane, and J. W. Frost (1979), Oxidations and oxygen transfer effected by a flavin N(5)-oxide. A model for flavin-dependent monooxygenases. *J. Amer. Chem. Soc.* **101**, 2228–2231.

293. H. W. Orf and D. Dolphin (1974), Oxaziridines as possible intermediates in flavin monooxygenases. *Proc. Nat. Acad. Sci. USA* **71**, 2646–2650.

294. D. Vargo and M. S. Jorns (1979), Synthesis of a 4a,5-epoxy-5-deazaflavin derivative. *J. Amer. Chem. Soc.* **101**, 7623–7626.

295. D. M. Jerina, J. W. Daly, B. Witkop, S. Zaltzman-Nirenberg, and S. Udenfriend (1969), 1,2-Naphthalene oxide as an intermediate in the microsomal hydroxylation of naphthalene. *Biochemistry* **9**, 147–156.

296. G. Guroff, J. W. Daly, D. M. Jerina, J. Rensen, B. Witkop, and S. Udenfriend (1967), Hydroxylation-induced migration: The NIH shift. *Science* **157**, 1524–1530.

297. J. L. Fox (1978), Chemists attack complex organic mechanisms. *Chem. Eng. News* May 22, pp. 28–30.

298. W. Adam, A. Alzérreca, J. E. Liu, and F. Yany (1977), α-Peroxylactones via dehydrative cyclization of α-hydroperoxy acids. *J. Amer. Chem. Soc.* **99**, 5768–5773.

299. C. Kemal and T. C. Bruice (1976), Simple synthesis of a 4a-hydroperoxy adduct of a 1,5-dihydroflavine: Preliminary studies of a model for bacterial luciferase. *Proc. Nat. Acad. Sci. USA* **73**, 995–999.

300. S. P. Schmidt and G. B. Schuster (1978), Dioxetanone chemiluminescence by the chemically initiated electron exchange pathway. Efficient generation of excited singlet states. *J. Amer. Chem. Soc.* **100**, 1966–1968.

301. J. N. Lowe and L. L. Ingraham (1974), *An Introduction to Biochemical Reactions Mechanisms*, Chap. 3. Foundation of Molecular Biology Series. Prentice-Hall, Englewood Cliffs, New Jersey.

302. O. A. Gansow and R. H. Holm (1969), A proton resonance investigation of equilibra, solute structures, and transamination in the aqueous systems pyridoxamine-pyruvate-zinc(II) and aluminium(III). *J. Amer. Chem. Soc.* **91**, 5984–5993.

303. M. Blum and J. W. Thanassi (1977), Metal ion induced reaction specific in vitamin B_6 model systems. *Bioorg. Chem.* **6**, 31–41.

304. B. Belleau and J. Burba (1960), The stereochemistry of the enzymic decarboxylation of amino acids. *J. Amer. Chem. Soc.* **82**, 5751–5752.

305. H. C. Dunathan, L. Davis, P. G. Kury, and M. Kaplan (1968), The stereochemistry of enzymatic transamination. *Biochemistry* **7**, 4532–4536.

306. H. C. Dunathan and J. G. Voet (1974), Stereochemical evidence for the evolution of pyridoxal-phosphate enzymes of various function from a common ancestor. *Proc. Nat. Acad. Sci. USA* **71**, 3888–3891.

307. J. N. Roitenan and D. J. Cram (1971), Electrophilic substitution at saturated carbon. XLV. Dissection of mechanisms of base-catalyzed hydrogen-deuterium exchange of carbon acids into inverson, isoinversion, and racemization pathways. *J. Amer. Chem. Soc.* **90**, 2225–2230.

308. J. N. Roitenan and D. J. Cram (1971), Electrophilic substitution at saturated carbon. XLVI. Crown ethers' ability to alter role of metal cations in control of stereochemical fate of carbanions. *J. Amer. Chem. Soc.* **90**, 2231–2241.

309. D. J. Cram, W. T. Ford, and L. Gosser (1968), Electrophilic substitution and saturated carbon. XXXVIII. Survey of substituent effects on stereochemical fate of fluorenyl carbanions. *J. Amer. Chem. Soc.* **90**, 2598–2606.

310. M. D. Broadhurst and D. J. Cram (1974), A model for the proton transfer stages of the biological transaminations and isotopic exchange reactions of amino acids. *J. Amer. Chem. Soc.* **96**, 581–583.

311. D. A. Jaeger, M. D. Broadhurst, and D. J. Cram (1979), Electrophilic substitution

at saturated carbon. 52. A model for the proton transfer steps of biological transamination and the effect of a 4-pyridyl group on the base-catalyzed racemization of a carbon acid. *J. Amer. Chem. Soc.* **101**, 717–732.

312. B. R. Baker (1967), *Design of Active-Site-Directed Irreversible Enzyme Inhibitors.* Wiley, New York.

313. R. H. Abeles and A. L. Maycock (1976), Suicide enzyme inactivators. *Acc. Chem. Res.* **9**, 313–319.

314. G. Schoellmann and E. Shaw (1963), Direct evidence for the presence of histidine in the active center of chymotrypsin. *Biochemistry* **2**, 252–255.

315 C. Walsh (1978), Chemical approaches to the study of enzymes catalyzing redox transformations. *Annu. Rev. Biochem.* **47**, 881–931.

316. V. Chowdhry and F. H. Westheimer (1979), Photoaffinity labeling of biological systems. *Annu. Rev. Biochem.* **48**, 293–325.

317. L. Stryer (1978), Fluorescence energy transfer as a spectroscopic ruler. *Annu. Rev. Biochem.* **47**, 819–846.

318. R. R. Rando (1975), Mechanisms of action of naturally occurring irreversible enzyme inhibitors. *Acc. Chem. Res.* **8**, 281–288.

319. R. R. Rando (1974), Chemistry and enzymology of k_{cat} inhibitors. *Science* **185**, 320–324.

320. P. Fasella and R. John (1969), Substrate analogues as specific inhibitors of pyridoxal-dependent enzymes. *Proc. 4th Int. Congr. Pharmacol.* **5**, 184–186.

321. Y. Morino and M. Okamoto (1973), Labeling of the active site of cytoplasmic aspartate amino transferase by β-chloro-L-alanine. *Biochem. Biophys. Res. Commun.* **50**, 1061–1067.

322. M. J. Jung and B. W. Metcalf (1975), Catalytic inhibition of γ-aminobutyric acid α-ketoglutarate transaminase of bacterial origin by 4-aminohex-5-ynoic acid, a substrate analog. *Biochem. Biophys. Res. Commun.* **67**, 301–306.

323. R. R. Rando and J. de Mairena (1974), Propargyl amine-induced irreversible inhibition of non-flavin-linked amine oxidases. *Biochem. Pharmacol* **23**, 463–466.

324. R. Breslow (1958), On the mechanism of thiamine action. IV. Evidence from studies on model systems. *J. Amer. Chem. Soc.* **80**, 3719–3726.

325. A. A. Gallo and H. Z. Sable (1974), Coenzyme interactions. VIII. ^{13}C-NMR studies of thiamine and related compounds. *J. Biol. Chem.* **249**, 1382–1389.

326. F. Jordan and Y. H. Mariam (1978), $N^{1'}$-Methylthiaminium diiodide. Model study on the effect of a coenzyme bound positive charge on reaction mechanism requiring thiamine pyrophosphate. *J. Amer. Chem. Soc.* **100**, 2534–2541.

327. F. White and L. L. Ingraham (1962), Mechanism of thiamine action: A model of 2-acylthiamine. *J. Amer. Chem. Soc.* **84**, 3109–3111.

328. T. C. Bruice and S. Benkovic (1966), *Bioorganic Mechanisms.* Vol. 2, Chap. 8, p. 217. Benjamin, New York.

329. W. H. Rastetter, J. Adams, J. W. Frost, L. J. Nummy, J. E. Frommer, and K. B. Roberts (1979), On the involvement of lipoic acid in α-keto acid dehydrogenase complexes. *J. Amer. Chem. Soc.* **101**, 2752–2753.

330. T. C. Bruice and A. F. Hegarty (1970), Biotin-bound CO_2 and the mechanism of enzymatic carboxylation reactions. *Proc. Nat. Acad. Sci. USA* **65**, 805–809.

331. M. Caplow and M. Yager (1976), Studies on the mechanism of biotin catalysis. II. *J. Amer. Chem. Soc.* **89**, 4513–4521.

332. R. B. Guchhait, S. E. Polakis, D. Hollis, C. Fenselau, and M. D. Lane (1974), Acetyl coenzyme A carboxylase system of *E. coli*. Site of carboxylation of biotin

and enzymatic reactivity of 1'-N-(ureido)-carboxybiotin derivatives. *J. Biol. Chem.* **249**, 6646–6656.

333. P. A. Whitney and T. G. Cooper (1972), Urea carboxylase and allophanate hydroxylase. Two components of ATP: Urea-lyase in *S. cerevisiae. J. Biol. Chem.* **247**, 1349–1353.
334. R. Kluger, P. Davis, and P. D. Adawadkar (1979), Mechanism of urea participation in phosphonate ester hydrolysis. Mechanistic and stereochemical criteria for enzymic formation and reaction of phosphorylated biotin. *J. Amer. Chem. Soc.* **101**, 5995–6000.
335. D. Arigoni, F. Lynen, and J. Rétey (1966), Stereochemie der enzymatischen Carboxylierung von (2R)-2-³H-Propionyl-CoA. *Helv. Chim. Acta.* **49**, 311–316.
336. J. Rétey and F. Lynen (1965), Zur biochemischen Funktion des Biotins. IX. Der sterische Verlauf der Carboxylierung von Propionyl-CoA. *Biochem. Z.* **342**, 256–271.
337. R. Kluger and D. C. Pike (1979), Chemical synthesis of a proposed enzyme-generated "reactive intermediate analogue" derived from thiamine diphosphate. Self-activation of pyruvate dehydrogenase by conversion of the analogue in its components. *J. Amer. Chem. Soc.* **101**, 6425–6428.
338. J. Moss and M. D. Lane (1971), The biotin-dependent enzymes. *Adv. Enzymol.* **35**, 321–442.
339. H. G. Wood and R. E. Barden (1977), Biotin enzymes. *Annu. Rev. Biochem.* **46**, 385–413.
340. A. S. Mildvan (1977), Magnetic resonance studies of the conformations of enzyme-bound substrates. *Acc. Chem. Res.* **10**, 246–252.
341. R. Kluger and P. D. Adawadkar (1976), A reaction proceeding through intramolecular phosphorylation of a urea. A chemical mechanism for enzymic carboxylation of biotin involving cleavage of ATP. *J. Amer. Chem. Soc.* **98**, 3741–3742.
342. W. C. Stallings (1977), The carboxylation of biotin. Substrate recognition and activation by complementary hydrogen bonding. *Arch. Biochem. Biophys.* **183**, 179–199.
343. H. G. Wood (1976), The reactive group of biotin catalysis by biotin enzymes. *Trends. Biochem. Sci.* **1**, 4–6.
344. F. Lynen, J. Knappe, E. Lorch, G. Jütting, and E. Ringelmann (1959), Die biochemische Funktion des Biotins. *Angew. Chem.* **71**, 481–486.
345. I. A. Rose, E. L. O'Connell, and F. Solomon (1976), Intermolecular tritium transfer in the transcarboxylase reaction. *J. Biol. Chem.* **251**, 902–904.
346. J. Stubbe and R. H. Abeles (1979), Biotin carboxylations concerted or not concerted? That is the question! *J. Biol. Chem.* **252**, 8338–8340.
347. C. M. Visser and R. M. Kellogg (1977), Mimesis of the biotin mediated carboxyl transfer reactions. *Bioorg. Chem.* **6**, 79–88.
348. C. W. Wharton (1979), Synthetic polymers as models for enzyme catalysis—a review. *Int. J. Biol. Macromolecules* **1**, 3–16.
349. A. R. Fersht and A. J. Kirby (1980), Intramolecular catalysis and the mechanism of enzyme action. *Chem Br.* **16**, 136–142.
350. R. Breslow (1980), Biomimetic control of chemical selectivity. *Acc. Chem. Res.* **13**, 170–177.
351. R. Breslow (1980), Biomimetic chemistry in oriented systems. *Isr. J. Chem.* **18**, 187–191.

352. G. W. Gokel and H. D. Durst (1976), Principles and synthetic applications in crown ether chemistry. *Synthesis*, 168–184.
353. G. R. Newkome, J. D. Sauer, J. M. Roper, and D. C. Hager (1977), Construction of synthetic macrocyclic compounds possessing subheterocyclic rings, specifically pyridine, furan, and thiophene. *Chem Rev.* **77**, 513–597.
354. J. S. Bradshaw and P. E. Stott (1980), Preparation of derivatives and analogs of the macrocyclic oligomers of ethylene oxide (crown compounds). *Tetrahedron* **36**, 461–510.
355. J. A. Ibers and R. H. Holm (1980), Modeling coordination sites in metallobiomolecules, *Science* **209**, 223–235.
356. A. W. Sleight (1980), Heterogeneous catalysts. *Science* **208**, 895–900.
357. C. Walsh (1980), Flavin coenzymes; At the crossroads of biological redox chemistry. *Acc. Chem. Res.* **13**, 148–155.
358. M. L. Schiling, H. D. Roth, and W. C. Herndon (1980), Zwitterionic adducts between a strongly electrophilic ketone and tertiary amines. *J. Amer. Chem. Soc.* **102**, 4271–4272.
359. F. A. Davis, J. Lamendola, Jr., U. Nadir, E. W. Kluger, T. C. Sedergran, T. W. Panunto, R. Billmers, R. Jenkins, Jr., I. J. Turchi, and W. H. Watson (1980), Chemistry of oxaziridines. 1. Synthesis and structure of 2-arenesulfonyl-3-aryloxaziridines. A new class of oxaziridines. *J. Amer. Chem. Soc.* **102**, 2000–2005.
360. M. Johnston, R. Raines, C. Walsh, and R. A. Firestone (1980), Mechanism-based enzyme inactivation using an allyl sulfoxide-allyl sulfenate ester rearrangement, *J. Amer. Chem. Soc.* **102**, 4241–4250.
361. H. Ogura, S. Nagai, and K. Takeda (1980), A novel reagent (*N*-succinimidyl diphenylphosphate) for the synthesis of activated ester and peptide. *Tetrahedron Lett.* 1467–1468.
362. V. N. R. Pillai (1980), Photoremovable protecting groups in organic synthesis. *Synthesis* 1–26.
363. C. B. Reese and A. Ubasawa (1980), Reaction between 1-arenesulfonyl-3-nitro-1, 2,4-triazoles and nucleoside base residues. Elucidation of the nature of side-reactions during oligonucleotide synthesis. *Tetrahedron Lett.* pp. 2265–2268.
364. Y. Minematsu, M. Waki, K. Suwa, T. Kato, and N. Izumiya (1980), Facile synthesis of gramicidin S via cyclization of a linear pentapeptide. *Tetrahedron Lett.* 2179–2180.
365. D. J. H. Smith, K. K. Ogilvie, and M. F. Gillen (1980), The methyl group as phosphate protecting group in nucleotide synthesis. *Tetrahedron Lett.* 861–864.
366. W. Niewiarowski, W. J. Stec, and W. Zielinski (1980), Synthesis of 4-nitrophenyl esters of thymidine 3'-phosphate and 3'-phosphorothioate using a new phosphorylating agent. *Chem. Comm.* 524–525.
367. M. D. Matteucci and M. H. Caruthers (1980), The synthesis of oligodeoxypyrimidines on a polymer support, *Tetrahedron Lett.* 719–722.
368. G. W. Parshall (1980), Organometallic chemistry in homogeneous catalysis. *Science* **208**, 1221–1224.
369. J. P. Ferris, P. C. Joshi, and J. G. Lawless (1977), Chemical evolution XXIX. Pyrimidines from hydrogen cyanide. *BioSystems* **9**, 81–86.

Index